Climate Change: Observed Impacts on Planet Earth

Climate Change: Observed Impacts on Planet Earth

Edited by

Trevor M. Letcher
Emeritus Professor
University of KwaZulu-Natal
Durban, South Africa

ELSEVIER

Amsterdam • Boston • Heidelberg • London • New York • Oxford
Paris • San Diego • San Francisco • Singapore • Sydney • Tokyo

Elsevier
Radarweg 29, PO Box 211, 1000 AE Amsterdam, The Netherlands
Linacre House, Jordan Hill, Oxford OX2 8DP, UK

First edition 2009

British Library Cataloguing in Publication Data
A catalogue record for this book is available from the British Library

Library of Congress Cataloging-in-Publication Data
Climate change : observed impacts on planet Earth / edited by Trevor M. Letcher. – 1st ed.
p. cm.
Includes bibliographical references and index.
ISBN 978-0-444-53301-2
1. Climatic changes. 2. Climatic changes–Environmental aspects. 3. Global environmental change. I. Letcher, T. M. (Trevor M.)
QC903.C56 2009
551.6–dc22

2009006502

ISBN: 978-0-444-53301-2

For information on all Elsevier publications visit our website at elsevierdirect.com

Printed and bound by CPI Group (UK) Ltd, Croydon, CR0 4YY

Transferred to Digital Print 2011

Contents

Foreword xiii
Preface xv
Contributors xix
Introduction xxiii

Part I
Possible Causes of Climate Change

1. The Role of Atmospheric Gases in Global Warming

R. P. Tuckett

1. Introduction 3
2. Origin of the Greenhouse Effect: 'Primary' and 'Secondary' Effects 4
3. The Physical Chemistry Properties of Greenhouse Gases 9
4. The Lifetime of a Greenhouse Gas in the Earth's Atmosphere 15
5. General Comments on Long-Lived Greenhouse Gases 17
6. Conclusion 18
References 19

2. The Role of Widespread Surface Solar Radiation Trends in Climate Change: Dimming and Brightening

S. Cohen

1. Introduction 22
2. Solar Radiation and its Measurement 22
3. Trends in Surface Solar Radiation or Global Dimming and Brightening 26
4. The Causes of Dimming and Brightening 32
5. The Influence of Solar Radiation Changes (Dimming and Brightening) on Climate 33
6. Conclusions 38
References 38

3. The Role of Space Weather and Cosmic Ray Effects in Climate Change

L. I. Dorman

1. Introduction 44
2. Solar Activity, Cosmic Rays and Climate Change 45

3. The Influence on the Earth's Climate of the Solar System Moving Around the Galactic Centre and Crossing Galaxy Arms 65
4. The Influence of Molecular-Dust Galactic Clouds on the Earth's Climate 65
5. The Influence of Interplanetary Dust Sources on the Earth's Climate 67
6. Space Factors and Global Warming 68
7. The Influence of Asteroids on the Earth's Climate 70
8. The Influence of Nearby Supernova on the Earth's Climate 70
9. Discussion and Conclusions 71
References 74

4. The Role of Volcanic Activity in Climate and Global Change

G. Stenchikov

1. Introduction 77
2. Aerosol Loading, Spatial Distribution and Radiative Effect 79
3. Volcanoes and Climate 82
4. Summary 98
References 99

5. The Role of Variations of the Earth's Orbital Characteristics in Climate Change

L. J. Lourens and E. Tuenter

1. Introduction 103
2. Astronomical Parameters 104
3. Orbital-Induced Climate Change 112
4. Conclusion 120
References 121

Part II
A Geological History of Climate Change

6. A Geological History of Climate Change

J. Zalasiewicz and M. Williams

1. Introduction 127
2. Climate Models 128
3. Long-Term Climate Trends 129
4. Early Climate History 131
5. Phanerozoic Glaciations 132
6. The Mesozoic–Early Cenozoic Greenhouse 133
7. Development of the Quaternary Icehouse 134
8. Astronomical Modulation of Climate 135
9. Milankovitch Cyclicity in Quaternary (Pleistocene) Climate History 136

10. Quaternary Sub-Milankovitch Cyclicity 137
11. The Holocene 138
12. Climate of the Anthropocene 138
13. Conclusions 139
References 139

Part III
Indicators of Climate and Global Change

7. Changes in the Atmospheric Circulation as Indicator of Climate Change

T. Reichler

1. Introduction 145
2. The General Circulation of the Atmosphere 147
3. The Poleward Expansion of the Tropical Circulation 149
4. The Decreasing Intensity of the Tropical Circulation 155
5. Emerging Mechanisms 155
6. Connection to Extratropical Circulation Change 159
7. Outstanding Problems and Conclusions 160
References 162

8. Weather Pattern Changes in the Tropics and Mid-Latitudes as an Indicator of Global Changes

R. M. Trigo and L. Gimeno

1. Introduction 165
2. Observed Changes in Extra-Tropical Patterns 166
3. Changes in Tropical Patterns 170
4. Conclusion 178
References 179

9. Bird Ecology as an Indicator of Climate and Global Change

W. Fiedler

1. Introduction 181
2. Indicators of Change 182
3. Conclusion 193
References 193

10. Mammal Ecology as an Indicator of Climate Change

M. M. Humphries

1. Introduction: A Primer on Mammal Thermoregulation and Climate Impacts 197

2. Demonstrated Impacts of Climate Change on Mammals 199
3. Linking Time and Space in Mammal Climate Responses 210
References 211

11. Climate Change and Temporal and Spatial Mismatches in Insect Communities

S. L. Pelini, K. M. Prior, D. J. Parker, J. D. K. Dzurisin, R. L. Lindroth and J. J. Hellmann

1. Introduction 215
2. Direct Effects of Climate Change on Insects 217
3. Host Plant-Mediated Effects on Insects 219
4. Predator-Mediated Effects on Insect Populations 222
5. Climate Change and Insect Pests 225
6. Conclusion 226
References 227

12. Sea Life (Pelagic and Planktonic Ecosystems) as an Indicator of Climate and Global Change

M. Edwards

1. Pelagic and Planktonic Ecosystems 233
2. Observed Impacts on Pelagic and Planktonic Ecosystems 237
3. Conclusion and Summary of Key Indicators 246
References 248

13. Changes in Coral Reef Ecosystems as an Indicator of Climate and Global Change

M. J. Attrill

1. Introduction 253
2. Tropical Coral Reef Ecosystems 254
3. The Associated Fauna of Coral Reefs 258
4. Conclusion 260
References 260

14. Changes in Marine Biodiversity as an Indicator of Climate Change

B. Worm and H. K. Lotze

1. Introduction 263
2. Climate Change and the Oceans 264
3. Effects of Climate Change on Biodiversity 265
4. Cumulative Impacts and Indirect Effects of Climate Change 272
5. Biodiversity as Insurance against Climate Change Impacts 274

6. Conclusions 275
References 276

15. Intertidal Indicators of Climate and Global Change

N. Mieszkowska

1. Introduction 281
2. Climate Change and Biogeography 283
3. Mechanisms 289
4. Additional Impacts of Global Change 291
5. Conclusions 292
References 292

16. Plant Ecology as an Indicator of Climate and Global Change

M. D. Morecroft and S. A. Keith

1. Introduction 297
2. Changes in Phenology 299
3. Changes in Distribution 300
4. Community Composition 301
5. Plant Growth 302
6. Conclusions 303
References 304

17. The Impact of Climate and Global Change on Crop Production

G. R. Dixon

1. Introduction 307
2. Impact on Plant Growth and Reproduction 308
3. Scale of the Problems 313
4. Climate Change Models 314
5. Winners and Losers 315
6. Adaptation 320
References 322

18. Rising Sea Levels as an Indicator of Global Change

R. Gehrels

1. Introduction 325
2. Is Sea Level Rising? 325
3. Why is Sea Level Rising? 328
4. Are Contemporary Rates of Sea-level Rise Unusual? 333
5. Conclusion 334
References 334

19. Sea Temperature Change as an Indicator of Global Change

M. J. Attrill

1. Introduction: Role of Ocean, Mechanisms and Correction of Bias 337
2. Long-Term Trends in Sea Temperature: The Historical Context 341
3. Global and Regional Patterns of Sea Temperature over the Last 100–150 Years 343
4. Conclusion: Anthropogenic Influence 345
References 346

20. Ocean Current Changes as an Indicator of Global Change

T. Kanzow and M. Visbeck

1. Introduction 349
2. The Variable Ocean 350
3. Oceanographers' Tools 351
4. The Atlantic Meridional Overturning Circulation 353
5. The AMOC's Role in Heat Transport, Oceanic Uptake of Carbon and Ventilation of the Deep Ocean 357
6. Can we Detect Changes in the AMOC? Is the AMOC Changing Already? 361
7. Conclusion 362
References 364

21. Ocean Acidification as an Indicator for Climate Change

C. Turley and H. S. Findlay

1. Introduction 367
2. Evidence from Observations 370
3. Model Predictions of Future Change 374
4. Impacts 374
5. Biogeochemical Cycling and Feedback to Climate 381
6. Adaptation, Recovery and Mitigation 383
7. Conclusion 385
References 387

22. Ice Sheets: Indicators and Instruments of Climate Change

D. G. Vaughan

1. Introduction 391
2. Sea-level and Ice 391
3. How Ice Sheets Work 394
4. Summary 398
References 399

23. Lichens as an Indicator of Climate and Global Change

A. Aptroot

1. Introduction 401
2. Predicted Effects 402
3. Observed Effects 402
4. Uncertain Effects 403
5. Habitats with Vulnerable Lichens 405
6. Conclusion 407
References 408

24. Coastline Degradation as an Indicator of Global Change

R. J. Nicholls, C. Woodroffe and V. Burkett

1. Introduction 409
2. Sea-Level Rise and Coastal Systems 411
3. Climate Change and Global/Relative Sea-Level Rise 412
4. Increasing Human Utilisation of the Coastal Zone 413
5. Climate Change, Sea-Level Rise and Resulting Impacts 415
6. Recent Impacts of Sea-Level Rise and Climate Change 416
7. Global Warming and Coasts at Latitudinal Extremes 418
8. The Challenge to Understand Contemporary Impacts 420
9. Concluding Remarks 421
References 422

25. Plant Pathogens as Indicators of Climate Change

K. A. Garrett, M. Nita, E. D. De Wolf, L. Gomez and A. H. Sparks

1. Introduction 425
2. Climate Variables and Plant Disease 426
3. Evidence that Simulated Climate Change Affects Plant Disease in Experiments 430
4. Evidence that Plant Disease Patterns have Changed due to Climate Change 431
References 436

Index 439

Foreword

The 2007 Assessment of the Intergovernmental Panel on Climate Change drew two substantially new conclusions which have had a marked effect on policymakers. The first was that current climate change is 'unequivocal' and is due largely to emissions of greenhouse gases resulting from human activity. The second was that the effects of this observed global warming can now be detected on every continent in the form of altered hydrology and biology. The positive response by policymakers was due not only to the higher levels of certainty surrounding the issue, but that empirically observed evidence now supported the simulation modelling of the future that had generally characterised the previous three IPCC assessments. Now, the policymakers could say, we are beginning to see come true just what the scientists had been predicting.

The concreteness of that case is examined in great detail in this book. Its chapters on the various processes that may affect the world's climate and on the detected changes in atmospheric, ocean and terrestrial (especially biological) systems serve to unfold this scientific narrative for the reader. But the book's strength lies in this not just being a summary of the IPCC because many of its authors were not involved extensively in the 2007 assessment, and thus they are able both to evaluate afresh the nature of the evidence and to bring new perspectives to bear on the issue.

As the editor says in his Introduction, if there ever was doubt about climate change then this should be dispelled in this book. I would add that if there ever was a case to be made for action, then this case has been made here in this volume and others like it which has followed the IPCC 2007 assessment. It is clear that stringent and immediate action is needed to curb greenhouse emissions and that we also need to start, now, on building our capacity to adapt to climate change impacts. It will take both massive mitigation and adaptation to meet the challenge of climate change.

Martin Parry
Visiting Professor,
Centre for Environmental Policy,
Imperial College London
Co-Chair Working Group II Assessment on Impacts,
Adaptation and Vulnerability,
Intergovernmental Panel on Climate Change 2007

Preface

Despite the many signs of global warming, global dimming, and changes to the climate, there are still many people who will not accept that something very ominous is taking place. This book is a very positive contribution to the problem. Professor Letcher is to be congratulated for inspiring so many world class experts into compiling such a wide reaching volume aimed at assessing and accounting for our changing climate. The book on climate change answers the basic questions: what can possibly cause global warming and climate change; and what evidence do we have that such changes are taking place?

The first five chapters focus on the possible causes of climate change with the first salvo being fired by Richard Tuckett who has put forward an eloquent chapter on the possible effects of anthropogenic greenhouse gases on the climate. Although we cannot prove conclusively that there is a cause-and-effect correlation between rising global temperature and atmospheric carbon dioxide concentration, the correlation over the past 100 years is very convincing.

To put the whole idea of climate change in perspective, there is a chapter on a Geological History of Climate Change. In this chapter Jan Zalasiewicz and Mark Williams traces the climate of the Earth over the past billion years. One really needs to understand the past in order to assess the present and indeed to attempt to predict the future consequences of climate change.

The remaining 19 chapters focus on a variety of global changes brought about by climate change. These include detailed scientific observations on weather patterns, plants and plant pathogens, lichens, bird, insect and animal ecology, sea temperature and ocean currents, rising sea levels and coastal erosion, and ice sheets. The chapter by Geoff Dixon on the impact of climate change on crop production is highly relevant, particularly in the developing world.

The evidence from the book that global warming and all the resultant changes, is due to human activity makes one appreciate just how fragile our environment is as we spin round the sun and move through space with our galaxy and all the other galaxies. Life was created on earth in an environment with more or less fixed physical properties which include: the level of radiation from the sun; the degree of shielding of the sun's radiation by the atmospheric gases; the level of the Earth's internal heat; and above all, the properties of the atmosphere such as pressure, temperature and gaseous composition. The Earth and its atmosphere form a very fragile system, which is in equilibrium with the life forms it supports. These conditions have developed over billions of years and any disturbance of this equilibrium could spell disaster to life on earth. The rapid rise of the world's human population, together with the need for more energy, protein rich food, and greater wealth, has led us to a situation in which this equilibrium is now being seriously threatened.

This is not the first time that life on Earth has been responsible for altering the composition of the atmosphere. About 2.45 billion years ago enormous numbers of cyanobacteria began changing the composition of the atmosphere by producing oxygen and hence ozone, and together with later plant life, photosynthetically produced most of the oxygen we now have in our atmosphere. This change led to the demise of many life forms which were unable to live or adapt to an atmosphere rich in oxygen.

Climate Change can be considered as a sequel to Professor Letcher's recent book *Future Energy* (published in 2008, also by Elsevier). Our quest for more energy is strongly linked to the problems of climate change and the bottom line is that we must reduce our dependency on fossil fuels and move to more sustainable and cleaner forms of energy which do not produce carbon dioxide. Both books pose huge questions which this and future generations must focus on.

As the main causes of global warming and climate change are largely chemical in nature it is right that the International Union of Pure and Applied Chemistry should take a lead in highlighting the problem with the hope of directing and influencing governments and world leaders to take a stand in reducing the burning of fossil fuels and the manufacture of other greenhouse gases. IUPAC has supported Professor Letcher's work and the production of the book is indeed a Project done through its Chemistry and the Environment Division (V1). *Climate Change* confirms IUPAC's commitment, as a leading scientific union, to pure and applied science.

This book supports the work done by the Intergovernmental Panel on Climate Change and presents experimental evidence for both the cause of the problem and the problem itself, with little attempt at computer modelling and predicting possible future scenarios. The evidence in this book should alert an anxious population of what is happening and the next step is to close ranks and change our ways. *Climate Change* with its 25 chapters is an excellent source book and will be an important guide for all who wish to know the truth of global warming, global dimming and climate change.

The conclusions to be derived from this book make it abundantly clear that we are challenged to significantly reduce the greenhouse gas emissions from human activity. This is not an easy task especially with our rapidly increasing population and the need for more energy to fuel our growing economies and associated wealth generating projects. Our future and that of our grandchildren will be severely compromised unless we take heed and act now. I believe this book is a shining light in the drive to educate the public in what is really taking place in the world of climate change and will be a beacon for many years to come.

I warmly thank Professor Letcher and his team for making such a phenomenal contribution to one of our greatest challenges. Climate change is a crisis which affects all living species. We have only one globe; let us all care for it.

Professor Piet Steyn
Past President of the International Union of Pure and Applied Chemistry
Department of Chemistry and Polymer Science
Stellenbosch University
Stellenbosch

Contributors

Numbers in parentheses indicate the pages on which the authors' contributions begin.

Richard P. Tuckett (3), School of Chemistry, University of Birmingham, Edgbaston, Birmingham B15 2TT, United Kingdom

Shabtai Cohen (21), Department of Environmental Physics and Irrigation. Institute of Soil, Water and Environmental Sciences, Agricultural Research Organization, The Volcani Centre, P.O. Box 6, Bet Dagan 50250, Israel

Lev I. Dorman (43), Head of Cosmic Ray and Space Weather Center with Emilio Sègre Observatory, Tel Aviv University, TECHNION and Israel Space Agency, P.O. Box 2217, Qazrin 12900, Israel and Chief Scientist of Cosmic Ray Department of IZMIRAN Russian Academy of Science, Troitsk 142090, Moscow region, Russia

Georgiy Stenchikov (77), Department of Environmental Sciences, Rutgers University, New Brunswick, New Jersey 08901-855, USA

Lucas J. Lourens (103), Department of Earth Sciences, Faculty of Geosciences, Utrecht University, Budapestlaan 4, 3584 CD Utrecht, The Netherlands

Erik Tuenter (103), Institute for Marine and Atmospheric Research Utrecht (IMAU), P.O. Box 80,000, 3508 TA Utrecht, The Netherlands

Jan Zalasiewicz (127), Department of Geology, University of Leicester, University Road, Leicester LE1 7RH, United Kingdom

Mark Williams (127), Department of Geology, University of Leicester, University Road, Leicester LE1 7RH, United Kingdom

Thomas Reichler (145), Department of Meteorology, University of Utah, Salt Lake City, Utah 84112-0110, USA

Ricardo M. Trigo (165), Centro de Geofisica da Universidade de Lisboa, IDL, Faculty of Sciences, University of Lisbon, Campo Grande, Ed C8, Piso 3, 1749-016 Lisbon, Portugal

Luis Gimeno (165), Departamento de Física Aplicada, Faculty of Sciences, University of Vigo, 32004 Ourense, Spain

Wolfgang Fiedler (181), Max Planck Institute for Ornithology, Vogelwarte Radolfzell, Schlossallee 2, D-78315, Radolfzell, Germany

Murray M. Humphries (197), Department of Natural Resource Sciences, McGill University, Ste-Anne-de-Bellevue, Quebec, Canada H9X 3V9

Shannon L. Pelini (215), Department of Biological Sciences, University of Notre Dame, Notre Dame, Indiana 46556, USA

Kirsten M. Prior (215), Department of Biological Sciences, University of Notre Dame, Notre Dame, Indiana 46556, USA

Derrick J. Parker (215), Department of Biological Sciences, University of Notre Dame, Notre Dame, Indiana 46556, USA

Jason D.K. Dzurisin (215), Department of Biological Sciences, University of Notre Dame, Notre Dame, Indiana 46556, USA

Jessica J. Hellmann (215), Department of Biological Sciences, University of Notre Dame, Notre Dame, Indiana 46556, USA

Richard L. Lindroth (215), Department of Entomology, University of Wisconsin, Madison, Wisconsin, USA

Martin Edwards (233), Sir Alister Hardy Foundation for Ocean Science, Citadel Hill, The Hoe, Plymouth PL1 2PB, United Kingdom and Marine Institute, University of Plymouth, Drake Circus, Plymouth PL4 8AA, United Kingdom

Martin J. Attrill (253), Marine Biology and Ecology Research Centre, Marine Institute, University of Plymouth, Drake Circus, Plymouth PL4 8AA, United Kingdom

Boris Worm (263), Biology Department, Dalhousie University, Halifax, Nova Scotia, Canada B3H 4J1

Heike K. Lotze (263), Biology Department, Dalhousie University, Halifax, Nova Scotia, Canada B3H 4J1

Nova Mieszkowska (281), Marine Biological Association of the UK, The Laboratory, Citadel Hill, Plymouth PL1 2PB, United Kingdom

Michael D. Morecroft (297), NERC Centre for Ecology and Hydrology, Crowmarsh Gifford, Wallingford OX10 8 BB, United Kingdom

Sally A. Keith (297), Centre for Conservation Ecology and Environmental Change, Bournemouth University, Fern Barrow, Poole BH12 5BB, United Kingdom

Geoffrey R. Dixon (307), Centre for Horticulture and Landscape, School of Biological Sciences, Whiteknights, The University of Reading, Reading, Berkshire RG6 6AS, United Kingdom

Roland Gehrels (325), School of Geography, University of Plymouth, Drake Circus, Plymouth PL4 8AA, United Kingdom

Martin J. Attrill (337), Marine Biology and Ecology Research Centre, Marine Institute, University of Plymouth, Drake Circus, Plymouth PL4 8AA, United Kingdom

T. Kanzow (349), National Oceanographic Centre, Empress Dock, Southampton SO14 3ZH, United Kingdom; Leibniz Institute of Marine Sciences, Kiel, Germany

M. Visbeck (349), Leibniz Institute of Marine Sciences, Kiel, Germany

Carol Turley (367), Plymouth Marine Laboratory, Prospect Place, The Hoe, Plymouth PL1 3DH, United Kingdom

Helen S. Findlay (367), Plymouth Marine Laboratory, Prospect Place, The Hoe, Plymouth PL1 3DH, United Kingdom

David G. Vaughan (391), British Antarctic Survey, Natural Environment Research Council, Madingley Road, Cambridge CB3 0ET, United Kingdom

Andre Aptroot (401), ABL Herbarium, Gerrit van der Veenstraat 107, NL-3762 XK Soest, The Netherlands

Robert J. Nicholls (409), School of Civil Engineering and the Environment and the Tyndall Centre for Climate Change Research, University of Southampton, Southampton SO17 1BJ, United Kingdom

Colin Woodroffe (409), School of Earth and Environmental Sciences, University of Wollongong, NSW 2522, Australia

Virginia Burkett (409), U.S. Geological Survey, 540 North Courthouse Street, Many, LA 71449 USA

Karen A. Garrett (425), Department of Plant Pathology, Kansas State University, Manhattan, Kansas 66506, USA

M. Nita (425), Department of Plant Pathology, Kansas State University, Manhattan, Kansas 66506, USA

E.D. De Wolf (425), Department of Plant Pathology, Kansas State University, Manhattan, Kansas 66506, USA

L. Gomez (425), Department of Plant Pathology, Kansas State University, Manhattan, Kansas 66506, USA

A.H. Sparks (425), Department of Plant Pathology, Kansas State University, Manhattan, Kansas 66506, USA

Introduction

The phrases CLIMATE CHANGE and GLOBAL WARMING and more recently GLOBAL COOLING are now part of our lives and rarely does a day go by without a mention in the press or on the radio of the possible causes of climate change and its consequences. Climate change has come upon us in a relatively short space of time and is accelerating with alarming speed. It is perhaps the most serious problem that the civilized world has had to face. It is the subject of major international co-operation through the Intergovernmental Panel on Climate Change (IPCC) which was set up in 1988 by the World Meteorological Organization and the United Nations Environment Programme. The IPCC has reported its findings in 1990, 1996, 2001 and 2007. The intention of this book is not to compete with the IPCC reports but to offer support through a different approach. This book does not focus on predicting the outcomes of climate change but presents both the facts relating to the possible causes of climate change and the evidence that climate and global changes are taking place.

In spite of all the publicity and coverage and indeed in the face of real evidence, there are many dissenting voices who either do not accept that climate change is taking place or that anthropogenic gases and compounds, such as carbon dioxide, are responsible for the major effect. One of the aims of this book is to counteract these comments and to present the evidence for climate change, in an unemotional, non-political, readable and scientific manner.

The book is divided into 25 chapters, each one written by an expert in the field. Five chapters have been devoted to answering the questions surrounding possible causes of climate change and the role being played by anthropogenic gases, compounds and particles. The five include solar effects, space weather, volcanic activity, variations in the earth's orbit, the role of cosmic radiation and the effect of changes in atmospheric carbon dioxide, nitrogen oxides, water vapour and man made gases such as freons.

To put climate change into perspective, there is a chapter on the geological history of the earth's climate. There is evidence of slow changes in climate, taking place over millions of years, and also of abrupt reorientations of the Earth's climate, the latter perhaps foreshadowing the way climate is responding to the present human activity.

If there ever was doubt about whether global and climate changes are taking place or not, then the last section of nineteen chapters should put pay to such thinking. These chapters give expert interpretations of the changes taking place in diverse areas such as weather patterns; bird, mammal and insect ecology; sea life and marine biodiversity; the inter-tidal zone; impacts on food supply; sea level rising; sea temperature rising; ocean current and ocean acidification; glacial and polar cap melting; plants, lichen, and plant pathogens and coastline degradation.

Little or no attempt has been made to present climate models or to predict climate changes in the future. This book focuses more on the experimental observational and presents the reader with the likelihood, through statistically significant evidence, of a climate changing future.

An aim of the book is to have all the scientific details of possible causes and scientific evidence for climate change written by experts in a language accessible by all, brought together in one volume. In this way comparisons can be made and issues put into perspective. The book will benefit both the non-specialist and the serious student. Each chapter begins with an Introduction and finishes with a Conclusion, written in lay-person's language and each chapter contains references to all the relevant and latest scientific publications. In this way the book will be of great benefit to students and researchers in each of the topics as well as making an excellent source and textbook for University and College courses in 'Climate Change'.

The International Union of Pure and Applied Chemistry supports the book, through its 'Chemistry and the Environmental' Division, and the IUPAC logo appears on the front cover. The IUPAC's adherence to the International System of Quantities is reflected in the book with the use of SI units where ever possible. One will, for example, notice that the symbols for 'hour', 'day' and 'year' are 'h', 'd' and 'a' respectively.

In spite of its title, the book does include indicators of global change such as 'ocean acidification' which, like climate change, is a result of excess carbon dioxide in the atmosphere.

The book is a scientific presentation of the facts surrounding climate change and no attempt has been made to offer solutions to climate change although the basic nature of the problem is obvious: the burning of oil, coal and gas is causing a significant rise in atmospheric carbon dioxide, water vapour, nitrogen oxides and also particulate matter. In this respect, climate change and our future energy are closely intertwined and this book will, I am sure, have a strong influence on deciding our future energy options.

"CLIMATE CHANGE: observed impacts on Planet Earth" is written not only for students and researchers and their professors, but for decision makers in government and in industry, journalists and editors, corporate leaders and all interested people who wish for a balanced, scientific and honest look at this major problem facing us in the 21st century.

I wish to thank all the authors for contributing chapters and for their many suggestions and discussions – all of which have helped to improve the book and its format. Special thanks must go to Professor Martin Attrill and to Dr Carol Turley for their suggestions, confidence and advice and to my wife, Dr. Valerie Letcher, for her help and support.

Trevor M Letcher
Stratton on the Fosse
1 November 2008

Part I

Possible Causes of Climate Change

Chapter 1

The Role of Atmospheric Gases in Global Warming

Richard P. Tuckett

School of Chemistry, University of Birmingham, Edgbaston, Birmingham B15 2TT, United Kingdom

1. Introduction
2. Origin of the Greenhouse Effect: 'Primary' and 'Secondary' Effects
3. The Physical Chemistry Properties of Greenhouse Gases
4. The Lifetime of a Greenhouse Gas in the Earth's Atmosphere
5. General Comments on Long-Lived Greenhouse Gases
6. Conclusion

Acknowledgements

References

1. INTRODUCTION

If the general public in the developed world is confused about what the greenhouse effect is, what the important greenhouse gases are, and whether greenhouse gases really are the predominant cause of the recent rise in temperature of the earth's atmosphere, it is hardly surprising. Nowadays, statements by one scientist are often immediately refuted by another, and both tend to state their claims with almost religious fervour. Furthermore, politicians and the media have not helped. It is only 14 a (years) ago that the newly appointed Secretary of State for the Environment in the United Kingdom made the cardinal sin of confusing the greenhouse effect with ozone depletion by saying they had the same scientific causes. (In retrospect, John Gummer was closer to the truth than he realised, in that one class of chemicals, the chlorofluorocarbons (CFCs), are both the principal cause of ozone depletion and are major greenhouse gases, but these two facts are scientifically unrelated.) Furthermore, to many, even in the respectable parts of the media, 'greenhouse gases' are two dirty words. In fact, nothing could be further from the truth, in that there has always been a greenhouse effect operative in the earth's atmosphere.

Without it we would inhabit a very cold planet, and not exist in the hospitable temperature of 290–300 K.

The purpose of this opening chapter of this book is to explain in simple terms what the greenhouse effect is, what its origins are and what the properties of greenhouse gases are. I will restrict this chapter to an explanation of the physical chemistry of greenhouse gases and the greenhouse effect, and not delve too much into the politics of 'what should or should not be done'. However, one simple message to convey at the onset is that the greenhouse effect is not just about concentration levels of carbon dioxide (CO_2), and it is too simplistic to believe that all our problems will be solved, if we can reduce CO_2 concentrations by *x*% in *y* years. Shine [1] has also commented many times that there is much more to the greenhouse effect than carbon dioxide levels.

2. ORIGIN OF THE GREENHOUSE EFFECT: 'PRIMARY' AND 'SECONDARY' EFFECTS

The earth is a planet in dynamic equilibrium, in that it continually absorbs and emits electromagnetic radiation. It receives ultra-violet and visible radiation from the sun, it emits infra-red radiation and energy balance says that 'energy in' must equal 'energy out' for the temperature of the planet to be constant. This equality can be used to determine what the average temperature of the planet should be. Both the sun and the earth are black-body emitters of electromagnetic radiation. That is, they are masses capable of emitting and absorbing all frequencies (or wavelengths) of electromagnetic radiation uniformly. The distribution curve of emitted energy per unit time per unit area versus wavelength for a black body was worked out by Planck in the first part of the twentieth century, and is shown pictorially in Fig. 1. Without mathematical detail, two points are relevant. First, the total energy emitted per unit time integrated over all wavelengths is proportional to $(T/K)^4$. Second, the wavelength of the maximum in the emission distribution curve varies inversely with (*T*/K), that is, $\lambda_{max} \propto (T/K)^{-1}$. These are Stefan's and Wien's

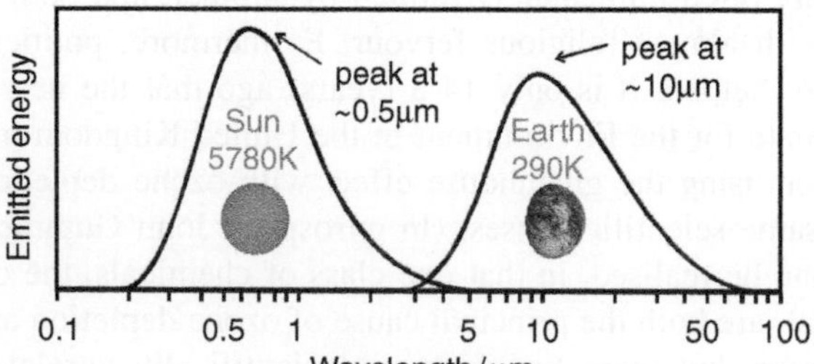

FIGURE 1 Black-body emission curves from the sun ($T \sim 5780$ K) and the earth ($T \sim 290$ K), showing the operation of Wien's Law that $\lambda_{max} \propto (1/T)$. The two graphs are not to scale. (See Color Plate 1).

Laws, respectively. Comparing the black-body curves of the sun and the earth, the sun emits UV/visible radiation with a peak at ca. 500 nm characteristic of $T_{sun} = 5780$ K. The temperature of the earth is a factor of 20 lower, so the earth's black-body emission curve peaks at a wavelength which is 20 times longer or ca. 10 μm. Thus the earth emits infra-red radiation with a range of wavelengths spanning ca. 4–50 μm, with the majority of the emission being in the range 5–25 μm (or 400–2000 cm^{-1}).

The solar flux energy intercepted per second by the earth's surface from the sun's emission can be written as $F_s(1-A)\pi R_e^2$, where F_s is the solar flux constant outside the Earth's atmosphere (1368 $J{\cdot}s^{-1}{\cdot}m^{-2}$), R_e is the radius of the Earth (6.38×10^6 m), and A is the earth's albedo, corresponding to the reduction of incoming solar flux by absorption and scattering of radiation by aerosol particles (average value 0.28). The infra-red energy emitted per second from the earth's surface is $4\pi R_e^2 s T_e^4$, where s is Stefan's constant (5.67×10^{-8} $J{\cdot}s^{-1}{\cdot}m^{-2}{\cdot}K^{-4}$) and $4\pi R_e^2$ is the surface area of the earth. At equilibrium, the temperature of the earth, T_e, can be written as:

$$T_e = \left[\frac{F_s(1-A)}{4s}\right]^{1/4} \tag{1}$$

Using the data above yields a value for T_e of ca. 256 K. Mercifully, the average temperature of the earth is not a Siberian −17 °C, otherwise life would be a very unpleasant experience for the majority of humans on this planet. The reason why our planet has a hospitable higher average value of ca. 290 K is the greenhouse effect. For thousands of years, absorption of some of the emitted infra-red radiation by molecules in the earth's atmosphere (mostly CO_2, O_3 and H_2O) has trapped this radiation from escaping out of the earth's atmosphere (just as a garden greenhouse operates), some is re-radiated back towards the earth's surface, thereby causing an elevation in the temperature of the surface of the earth. Thus, it is the greenhouse effect that has maintained our planet at this average temperature, and for this fact we should all be very grateful! This phenomenon is often called the 'primary' greenhouse effect. It is, therefore, a myth to portray all aspects of the greenhouse effect as bad news; it is the reverse that is true.

Evidence for the presence of greenhouse gases absorbing infra-red radiation in the atmosphere comes from satellite data. Figure 2 shows data collected by the Nimbus 4 satellite circum-navigating the earth at an altitude outside the earth's troposphere (0 < altitude, $h < 10$ km) and stratosphere ($10 < h < 50$ km). The infra-red emission spectrum in the range 6–25 μm escaping from earth represents a black-body emitter with a temperature of ca. 290 K, with absorptions (i.e., dips) between 12 and 17 μm, around 9.6 μm, and $\lambda < 8$ μm. These wavelengths correspond to infra-red absorption bands of CO_2, O_3 and H_2O, respectively, three atmospheric gases that have contributed to the primary greenhouse effect.

Of course, the argument that the primary greenhouse gases have maintained our planet at a constant temperature of ca. 290 K pre-supposes that their

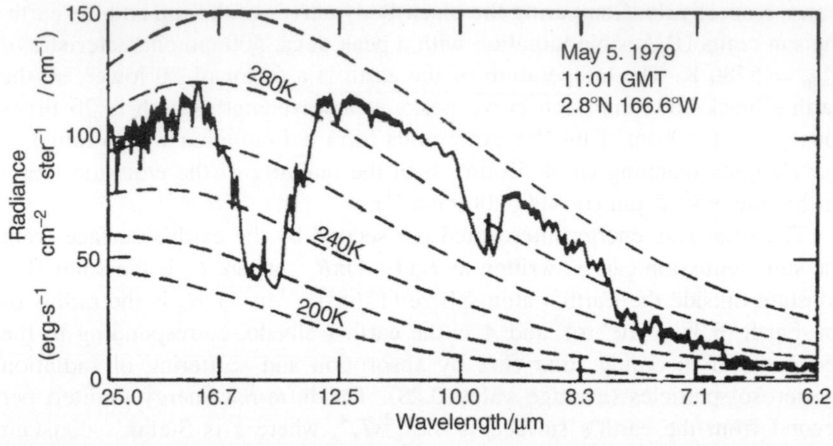

FIGURE 2 Infra-red emission spectrum escaping to space as observed by the Nimbus 4 satellite outside the earth's atmosphere. Absorptions due to CO_2 between 12 and 7 μm, O_3 (around 9.6 μm) and H_2O ($\lambda < 8$ μm) are shown. (With permission from Dickinson and Clark (eds.), Carbon dioxide Review, OUP, 1982.)

concentrations have remained approximately constant over very long periods of time. This has not happened with CO_2 and, to a lesser extent, with O_3 over the 260 a (years) since the start of the Industrial Revolution, ca. 1750, and it is changes in the concentrations of these and newer greenhouse gases that have caused a 'secondary' greenhouse effect to occur over this time window, leading to the temperature rises that we are all experiencing today. That, at least, is the main argument of the proponents of the 'greenhouse gases, mostly CO_2, equals global warming' school of thought. There is no doubt that the concentration of CO_2 in our atmosphere has risen from ca. 280 parts per million by volume (ppmv) to current levels of ca. 380 ppmv over the last 260 a. (1 ppmv is equivalent to a number density of 2.46×10^{13} molecules·cm^{-3} for a pressure of 1 bar and a temperature of 298 K.) It is also not in doubt that the average temperature of our planet has risen by ca. 0.5–0.8 K over this same time window (Fig. 3). What has not been proven is that there is a cause-and-effect correlation between these two facts, the main problem being that there is not sufficient structure or resolution with time in either the CO_2 concentration or the temperature data. Even more recent data of the last 100 a (Fig. 4), where the correlation seems to be better established will not convince the sceptic. That said, as demonstrated most clearly by the recent IPCC2007 report [2], the consensus of world scientists, and certainly physical scientists, is that a strong correlation does exist.

By contrast, an excellent example in atmospheric science of sufficient resolution being present to confirm a correlation between two sets of data occurred in 1989; the concentrations of O_3 and the ClO free radical in the stratosphere were shown to have a strong anti-correlation effect when data were collected by an aircraft as a function of latitude in the Antarctic (Fig. 5) [3].

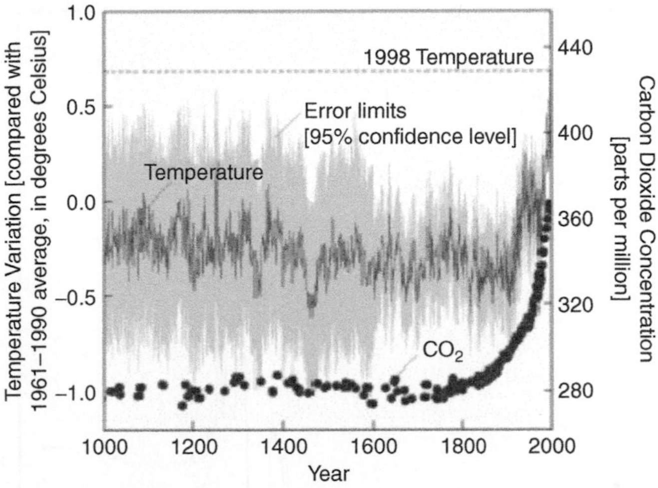

FIGURE 3 The average temperature of the earth and the concentration level of CO_2 in the earth's atmosphere during the last 1000 a. (With permission from www.env.gov.bc.ca/air/climate/indicat/images/appendnhtemp.gif and www.env.gov.bc.ca/air/climate/indicat/images/appendCO2.gif) (See Color Plate 2).

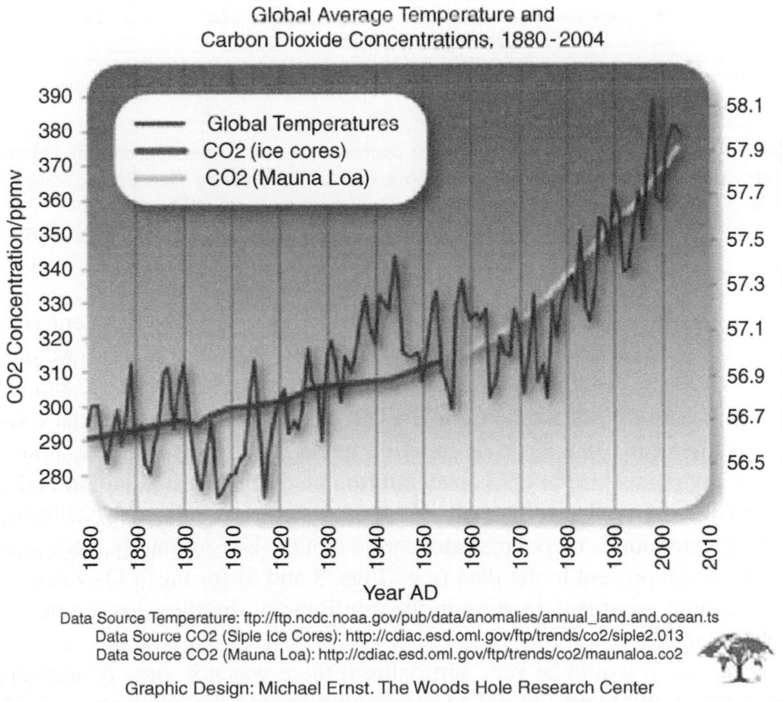

FIGURE 4 The average temperature of the earth and the concentration level of CO_2 in the earth's atmosphere during the 'recent' history of the last 100 a. (With permission from the web sites shown in the figure.) (See Color Plate 3).

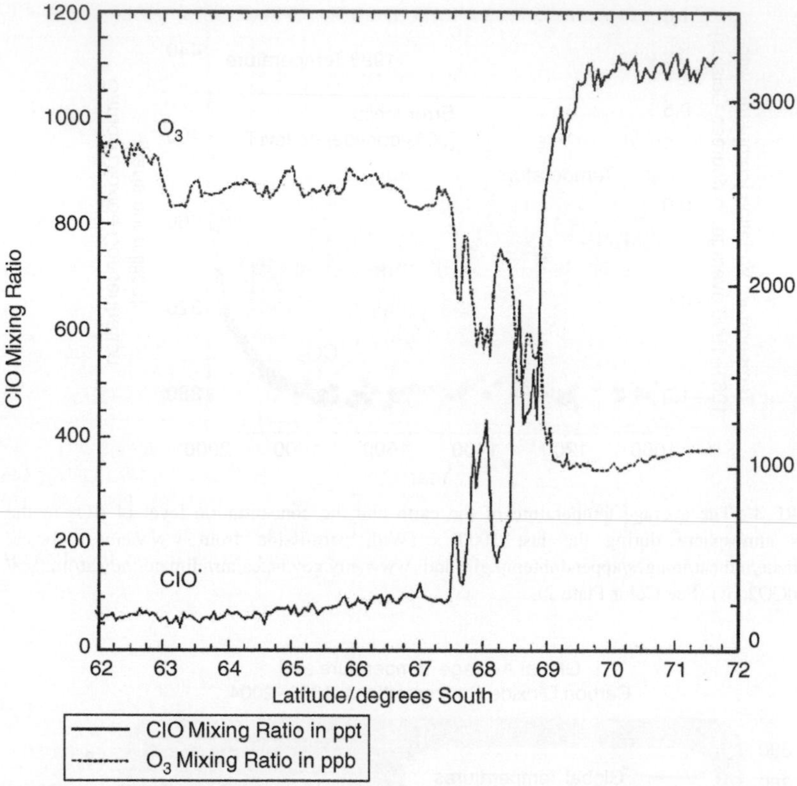

FIGURE 5 Clear anti-correlation between the concentrations of ozone, O_3, and the chlorine monoxide radical, $ClO^{\bullet}$, in the stratosphere above the Antarctic during their Spring season of 1987. (With permission from Anderson et al., J. Geophys. Res. D. 94 (1989) 11465.)

There was not only the general observation that a decrease of O_3 concentration correlated with an increase in ClO concentration, but also the resolution was sufficient to show that at certain latitudes dips in O_3 concentration corresponded exactly with rises in ClO concentration. Even the most doubting scientist could accept that the decrease in O_3 concentration in the Antarctic Spring was related somehow to the increase in ClO concentration, and this result led to an understanding over the next 10–15 a of the heterogeneous chemistry of chlorine-containing compounds on polar stratospheric clouds. Unfortunately, such good resolution is not present in the data (e.g., Figs. 3 and 4) for the 'CO_2 versus *T*' global warming argument, leading to the multitude of theories that are now in the public domain.

I accept that it would be very surprising if there was not some relationship between such rapid increases in CO_2 concentration and the temperature of the planet, nevertheless there are two aspects of Fig. 3 that remain unanswered by proponents of such a simple theory. First, the data suggest that the temperature of the earth actually decreased between 1750 and ca. 1920 whilst the CO_2 concentration increased from 280 to ca. 310 ppm over this time window. Second, the drop in temperature around 1480 AD in the 'little ice age' is not mirrored by a similar drop in CO_2 concentration. All that said, however, the apparent 'agreement' between rises of both CO_2 levels and T_e over the last 50 a is very striking. The most likely explanation surely is that there are a multitude of effects, one of which is the concentrations of greenhouse gases in the atmosphere, contributing to the temperature of the planet. At certain times of history, these effects are 'in phase' (as now), at other times they may have been in 'anti-phase' and working against each other.

3. THE PHYSICAL CHEMISTRY PROPERTIES OF GREENHOUSE GASES

The fundamental physical property of a greenhouse gas is that it must absorb infra-red radiation via one or more of its vibrational modes in the infra-red range of 5–25 μm. Furthermore, since the primary greenhouse gases of CO_2, O_3 and H_2O absorb in the range 12–17 μm (or 590–830 cm^{-1}), 9.6 μm (1040 cm^{-1}) and $\lambda < 8$ μm (>1250 cm^{-1}), an effective secondary greenhouse gas is one which absorbs infra-red radiation strongly outside these ranges of wavelengths (or wavenumbers). A molecular vibrational mode is only infra-red active if the motion of the atoms generates a dipole moment. That is, $d\mu/dQ \neq 0$, where μ is an instantaneous dipole moment and Q a displacement coordinate representing the vibration of interest. It is worth stating the obvious straightaway, that N_2 and O_2 which constitute 99% of the earth's atmosphere do not absorb infra-red radiation, their sole vibrational mode is infra-red inactive, so they play no part in the greenhouse effect and global warming. It is only trace gases in the atmosphere (Table 1) such as CO_2 (0.038%), CH_4 (0.0002%), O_3 (3×10^{-6}%) and CFCs such as CF_2Cl_2 (5×10^{-8}%) which contribute to the greenhouse effect. Put another way, the earth's atmosphere is particularly fragile if only 1% of the molecules present can have such a major effect on humans living on the planet. Furthermore, the most important molecular trace gas, CO_2, absorbs via its ν_2 bending vibrational mode at 667 cm^{-1} or 15.0 μm, which coincidentally is very close to the peak of the earth's black-body curve; the spectroscopic properties of CO_2 have not been particularly kind to the environment! Thus, infra-red spectroscopy of gas-phase molecules, in particular at what wavelengths and how strongly a molecule absorbs such radiation, will clearly be important properties to determine how effective a trace pollutant will be to the greenhouse effect.

The second property of interest is the lifetime of the pollutant in the earth's atmosphere: the longer the lifetime, the greater contribution a greenhouse gas will make to global warming. The main removal processes in the troposphere and stratosphere are reactions with OH free radicals and electronically excited

TABLE 1 Main constituents of ground-level clean air in the earth's atmosphere

Molecule	Mole fraction	ppmv[a] (2008)	ppmv (1748)
N_2	0.78 or 78%	780 900	780 900
O_2	0.21 or 21%	209 400	209 400
H_2O	0.03 (100% humidity, 298 K)	31 000	31 000
H_2O	0.01 (50% humidity, 298 K)	16 000	16 000
Ar	0.01 or 1%	9300	9300
CO_2	3.8×10^{-4} or 0.038%	379	280
Ne	1.8×10^{-5} or 0.002%	18	18
CH_4	1.77×10^{-6} or 0.0002%	1.77	0.72
N_2O	3.2×10^{-7} or 0.00003%	0.32	0.27
O_3[b]	3.4×10^{-8} or 0.000003%	0.034	0.025
All CFCs[c]	8.7×10^{-10} or 8.7×10^{-8}%	0.0009	0
All HCFCs[d]	1.9×10^{-10} or 1.9×10^{-8}%	0.0002	0
All PFCs[e]	8.3×10^{-11} or 8.3×10^{-9}%	0.00008	0
All HFCs[f]	6.1×10^{-11} or 6.1×10^{-9}%	0.00006	0

[a]*parts per million by volume. 1 ppmv is equivalent to a number density of 2.46×10^{13} molecules·cm^{-3} for a pressure of 1 bar and a temperature of 298 K.*
[b]*the concentration level of O_3 is very difficult to determine because it is poorly mixed in the troposphere. It shows large variation with both region and altitude.*
[c]*chlorofluorocarbons (e.g., CF_2Cl_2).*
[d]*hydrochlorofluorocarbons (e.g., $CHClF_2$).*
[e]*perfluorocarbons (e.g., CF_4, C_2F_6, SF_5CF_3, SF_6).*
[f]*hydrofluorocarbons (e.g., CH_3CF_3).*

oxygen atoms, O* (^{1}D), and photodissociation in the range 200–300 nm (in the stratosphere) or 300–500 nm (in the troposphere). Thus, the reaction kinetics of pollutant gases with OH and O* (^{1}D) and their photochemical properties in the UV/visible will yield important parameters to determine their effectiveness as greenhouse gases. All these data are incorportated into a dimensionless number, the global warming potential (GWP) or greenhouse potential (GHP) of a greenhouse gas. All values are calibrated with respect to CO_2 whose GWP value is 1. A molecule with a large GWP is one with strong infra-red absorption in the windows where the primary greenhouse gases such as CO_2, etc., do not absorb, long lifetimes, and concentrations rising rapidly due to human presence on the planet. GWP values of some of the most important secondary greenhouse gases are given in the bottom row of Table 2. Note that CO_2 has the lowest GWP value of the seven greenhouse gases shown.

TABLE 2 Examples of greenhouse gases and their contribution to global warming [2,20]

Greenhouse gas	CO_2	O_3	CH_4	N_2O	CF_2Cl_2 [all CFCs]	SF_6	SF_5CF_3
Concentration (2008)/ppmv	379	0.034[a]	1.77	0.32	0.0005 [0.0009]	5.6×10^{-6}	1.2×10^{-7}
ΔConcentration (1748–2008)/ppmv	99	0.009[a]	1.05	0.05	0.0005 [0.0009]	5.6×10^{-6}	1.2×10^{-7}
Radiative efficiency, a_o/W·m^{-2}·ppbv^{-1}	1.68×10^{-5}	3.33×10^{-2}	4.59×10^{-4}	3.41×10^{-3}	0.32 [0.18–0.32]	0.52	0.60
Total radiative forcing [b]/W·m^{-2}	1.66	ca. 0.30[c]	0.48	0.16	0.17 [0.27]	2.9×10^{-3}	7.2×10^{-5}
Contribution from long-lived greenhouse gases excluding ozone to overall greenhouse effect/%[d]	63 (57)	(10)	18 (16)	6 (5)	6 [10] (6 [9])	0.1 (0.1)	0.003 (0.003)
Lifetime, τ^e/a	ca. 50–200[f]	ca. days–weeks[g]	12	120	100 [45–1700]	3200	800
Global warming potential (100 a projection)	1	–[h]	25	298	10 900 [6130–14 400]	22 800	17 700

[a]*reference [20].*
[b]*due to change in concentration of long-lived greenhouse gas from the pre-Industrial era to the present time.*
[c]*an estimated positive radiative forcing of 0.35 W·m^{-2} in the troposphere is partially cancelled by a negative forcing of 0.05 W·m^{-2} in the stratosphere [2].*
[d]*assumes the latest value for the total radiative forcing of 2.63 ± 0.26 W·m^{-2} [2].*
The values in brackets show the percentage contributions when the estimated radiative forcing for ozone is included in the value for the total radiative forcing.
[e]*assumes a single-exponential decay for removal of greenhouse gas from the atmosphere.*
[f]*CO_2 does not show a single-exponential decay [4].*
[g]*O_3 is poorly mixed in the troposphere, so a single value for the lifetime is difficult to estimate. It is removed by the reaction, $OH + O_3 \rightarrow HO_2 + O_2$. Its concentration shows large variations both with region and altitude.*
[h]*GWP values are generally not applied to short-lived pollutants in the atmosphere, due to serious inhomogeneous changes in their concentration.*

Information in the previous two paragraphs is described in qualitative and descriptive terms. However, all the data can be quantified, and a mathematical description is now presented. The term that characterises the infra-red absorption properties of a greenhouse gas is the radiative efficiency, a_o. It measures the strength of the absorption bands of the greenhouse gas, x, integrated over the infra-red black-body region of ca. 400–2000 cm^{-1}. It is a (per molecule) microscopic property and is usually expressed in units of W·m^{-2}·ppbv^{-1}. If this value is multiplied by the change in concentration of pollutant over a defined time window, usually the 260 a from the start of the Industrial Revolution to the current day, the macroscopic radiative forcing in units of W·m^{-2} is obtained. (Clearly, a pollutant whose concentration has not changed over this long time window will have a macroscopic radiative forcing of zero.) One may then compare the radiative forcing of different pollutant molecules over this time window, showing the current contribution of different greenhouse gases to the total greenhouse effect. Thus, the IPCC 2007 report [2] quotes the radiative forcing for CO_2 and CH_4 in 2005 as 1.66 and 0.48 W·m^{-2}, respectively, out of a total for long-lived greenhouse gases of 2.63 W·m^{-2}. These two molecules, therefore, contribute 81% in total (63% and 18%, individually) to the global warming effect. Effectively, the radiative forcing value gives a current-day estimate of how serious a greenhouse gas is to the environment, using concentration data from the past.

The overall effect in the future of one molecule of pollutant on the earth's climate is described by its GWP (or GHP) value. It measures the radiative forcing, A_x, of a pulse emission of the greenhouse gas over a defined time period, t, usually 100 a, relative to the time-integrated radiative forcing of a pulse emission of an equal mass of CO_2:

$$\mathrm{GWP}_x(t) = \frac{\int_0^t A_x(t)\mathrm{d}t}{\int_0^t A_{\mathrm{CO}_2}(t)\mathrm{d}t} \tag{2}$$

The GWP value therefore informs how important one molecule of pollutant x is to global warming via the greenhouse effect compared to one molecule of CO_2, which is defined to have a GWP value of unity. It is an attempt to project into the future how serious the presence of a long-lived greenhouse gas will be in the atmosphere. (Thus, when the media state that CH_4 is 25 times as serious as CO_2 for global warming, what they are saying is that the GWP value of CH_4, looking 100 a into the future, is 25; one molecule of CH_4 is expected to cause 25 times as much 'damage' as one molecule of CO_2.) For most greenhouse gases, the radiative forcing following an emission at $t = 0$, takes a simple exponential form:

$$A_x(t) = A_{o,x} \exp\left(\frac{-t}{\tau_x}\right) \tag{3}$$

where τ_x is the lifetime for removal of species x from the atmosphere. For CO_2, a single-exponential decay is not appropriate since the lifetime ranges from 50 to 200 a, and we can write:

$$A_{CO_2}(t) = A_{o,CO_2}\left[b_o + \sum_i b_i \exp\left(\frac{-t}{\tau_i}\right)\right] \tag{4}$$

where the response function, the bracket in the right-hand side of Eq. (4), is derived from more complete carbon cycles. Values for b_i (i = 0–4) and τ_i (i = 1–4) have been given by Shine *et al.* [4]. It is important to note that the radiative forcing, A_o, in Eqs. (2)–(4) has units of $W{\cdot}m^{-2}{\cdot}kg^{-1}$. For this reason, it is given a different symbol to the microscopic radiative efficiency, a_o, with units of $W{\cdot}m^{-2}{\cdot}ppbv^{-1}$. Conversion between the two units is simple [4]. The time integral of the large bracket on the right-hand side of Eq. (4), defined K_{CO_2}, has dimensions of time, and takes values of 13.4 and 45.7 a for a time period of 20 and 100 a, respectively, the values of t for which GWP values are most often quoted. Within the approximation that the greenhouse gas, x, follows a single-exponential time decay in the atmosphere, it is then possible to parameterise Eq. (2) to give an exact analytical expression for the GWP of x over a time period t:

$$\frac{GWP_x(t)}{GWP_{CO_2}(t)} = \frac{MW_{CO_2}}{MW_x} \cdot \frac{a_{o,x}}{a_{o,CO_2}} \cdot \frac{\tau_x}{K_{CO_2}} \cdot \left[1 - \exp\left(\frac{-t}{\tau_x}\right)\right] \tag{5}$$

In this simple form, the GWP only incorporates values for the radiative efficiency of greenhouse gases x and CO_2, $a_{o,x}$ and a_{o,CO_2}; the molecular weights of x and CO_2; the lifetime of x in the atmosphere, τ_x; the time period into the future over which the effect of the pollutant is determined; and the constant K_{CO_2} which can easily be determined for any value of t. Thus the GWP value scales with both the lifetime and the microscopic radiative forcing of the greenhouse gas, but it remains a microscopic property of one molecule of the pollutant. The recent rate of increase in concentration of a pollutant (e.g., the rise in concentration per annum over the last decade), one of the factors of most concern to policymakers, does not contribute directly to the GWP value. This and other factors [4] have caused criticism of the use of GWPs in policy formulation.

Data for seven greenhouse gases are shown in Table 2. CO_2 and O_3 constitute naturally occurring greenhouse gases whose concentration levels ideally would have remained constant at pre-industrial revolution levels. Although H_2O vapour is the most abundant greenhouse gas in the atmosphere, it is neither long-lived nor well mixed: concentrations range 0–3% (i.e., 0–30 000 ppmv) over the planet, and the average lifetime is only a few days. Its average global concentration has not changed significantly in the last 260 a, and it therefore has zero radiative forcing. CH_4 and N_2O constitute naturally occurring greenhouse gases with larger a_o values than that of CO_2. The CH_4 concentration, although small, has increased by ca. 150% since pre-industrial times. After CO_2, it is the second most important greenhouse gas, and its current total radiative forcing is ca. 29% that of CO_2. N_2O concentration has increased only by ca. 16% over this same time period. It has the fourth highest total radiative forcing of all the naturally occurring greenhouse gases, following CO_2, CH_4 and O_3. Dichlorofluoromethane, CF_2Cl_2, is one of the most common of CFCs. These are man-made chemicals that have grown in concentration from zero in pre-industrial times to a current total concentration of 0.9 ppbv (1 ppbv is equivalent to 1 part per 10^9 (billion) by volume, or a number density of 2.46×10^{10} molecules·cm^{-3} at 1 bar pressure and a temperature of 298 K). Their concentration is now decreasing due to the 1987 Montreal and later International Protocols, introduced to halt stratospheric ozone destruction and (ironically) nothing to do with global warming! SF_6 and SF_5CF_3 are two long-lived halocarbons with currently very low concentration levels, but with high annual percentage increases and exceptionally long lifetimes in the atmosphere. They have very high a_o and GWP values, essentially because of their large number of strong infra-red-active vibrational modes and their long lifetimes.

It is noted that CO_2 and CH_4 have the lowest GWP values of all greenhouse gases. Why, then, is there such concern about levels of CO_2 in the atmosphere, and with the possible exception of CH_4 no other greenhouse gas is hardly ever mentioned in the media? The answer is that the overall contribution of a pollutant to the greenhouse effect, present and future, involves a convolution of its concentration with the GWP value. Thus CO_2 and CH_4 currently contribute most to the greenhouse effect (third bottom row of Table 2) simply due to their high change in atmospheric concentration since the Industrial Revolution; note, however, that the a_o and GWP values of both gases are relatively low. Indeed, the ν_2 bending mode of CO_2 at 15.0 μm, which is the vibrational mode most responsible for greenhouse activity in CO_2, is close to saturation. By contrast, SF_5CF_3 is a perfluorocarbon molecule with the highest microscopic radiative forcing of any known greenhouse gas (earning it the title 'super' greenhouse gas [5,6]), even higher than that of SF_6. SF_6 is an anthropogenic chemical used extensively as a dielectric insulator in high-voltage industrial applications, and the variations of concentration levels of SF_6 and SF_5CF_3 with time in the last 50 a have tracked each other very closely [7]. The GWP of these two molecules is very high, SF_6 being slightly higher because of its atmospheric lifetime, ca. 3200 a [8], is about four times greater than that of SF_5CF_3. However, the contribution of these two molecules to the overall greenhouse effect is still very small because their atmospheric concentrations, despite rising rapidly at the rate of ca. 6–7% per annum, are still very low, at the level of parts per 10^{12} (trillion) by volume; 1 pptv is equivalent to a number density of 2.46×10^7 molecules·cm^{-3} at 1 bar and 298 K).

In conclusion, the macroscopic properties of greenhouse gases, such as their method of production, their concentration and their annual rate of increase or decrease, are mainly controlled by environmental and sociological factors, such as industrial and agricultural methods, and ultimately population levels on the planet. The microscopic properties of these compounds, however, are controlled by factors that undergraduates world-wide learn about in science degree courses: infra-red spectroscopy, reaction kinetics and photochemistry. Data from such lab-based studies determine values for two of the most important parameters for determining the effectiveness of a greenhouse gas: the microscopic radiative efficiency, a_o, and the atmospheric lifetime, τ.

4. THE LIFETIME OF A GREENHOUSE GAS IN THE EARTH'S ATMOSPHERE

The microscopic radiative efficiency of a greenhouse gas is determined by measuring absolute absorption coefficients for infra-red-active vibrations in the range ca. 400–2000 cm^{-1} and integrating over this region of the electromagnetic spectrum. Its meaning is unambiguous. The lifetime, however, is a term that can mean different things to different scientists, according to their discipline. It is, therefore, pertinent to describe exactly what is meant by the lifetime of a greenhouse gas (penultimate row of Table 2), and how these values are determined.

To a physical chemist, the lifetime generally means the inverse of the pseudo-first-order rate constant of the dominant chemical or photolytic process that removes the pollutant from the atmosphere. Using CH_4 as an example, it is removed in the troposphere via oxidation by the OH free radical, $OH + CH_4 \rightarrow H_2O + CH_3$. The rate coefficient for this reaction at 298 K is 6.4×10^{-15} $cm^3 \cdot molecules^{-1} \cdot s^{-1}$ [9], so the lifetime is approximately equal to $(k_{298}[OH])^{-1}$. Assuming the tropospheric OH concentration to be 0.05 pptv or 1.2×10^6 $molecules \cdot cm^{-3}$ [2], the lifetime of CH_4 is calculated to be ca. 4 a. This is within a factor of three of the accepted value of 12 a (Table 2). The difference arises because CH_4 is not emitted uniformly from the earth's surface, a finite time is needed to transport CH_4 via convection and diffusion into the troposphere, and oxidation occurs at different altitudes in the troposphere where the OH concentration varies from its average value of 1.2×10^6 $molecules \cdot cm^{-3}$. We can regard this as an example of a two-step kinetic process,

$$A \rightarrow B \rightarrow C \tag{6}$$

with first-order rate constants k_1 and k_2. The first step, $A \rightarrow B$, represents the transport of the pollutant into the atmosphere, whilst the second step, $B \rightarrow C$, represents the chemical or photolytic process (e.g., reaction with an OH radical in the troposphere) that removes the pollutant from the atmosphere. In general, the overall rate of the process (whose inverse is called the lifetime) will be a function of both k_1 and k_2, but its value will be dominated by the slower of the two steps. Thus, in calculating the lifetime of CH_4 simply by determining $(k_{298}[OH])^{-1}$, we are assuming that the first step, transport into the region of the atmosphere where chemical reactions occurs, is infinitely fast compared to the removal process.

The exceptionally long-lived greenhouse gases in Tables 1 and 2 (e.g., SF_6, CF_4, SF_5CF_3) behave in the opposite sense. Now, the slow, rate-determining process is the first step, that is, transport of the greenhouse gas from the surface of the earth into the region of the atmosphere where chemical removal occurs. The chemical or photolytic processes that ultimately remove SF_6, etc., will have very little influence on the lifetime, that is, $k_1 \ll k_2$ in Eq. (6). These molecules do not react with OH or O* (^{1}D) to any significant extent, and are not photolysed by visible or UV radiation in the troposphere or stratosphere. They therefore rise higher into the mesosphere ($h > 60$ km) where the dominant processes that can remove pollutants are electron attachment and vacuum-UV photodissociation at the Lyman-α wavelength of 121.6 nm [6]. We can define a chemical lifetime, τ_{chemical}, for such species as:

$$\tau_{\text{chemical}} = [k_e[e^-] + \sigma_{121.6} J_{121.6} \Phi_{121.6}]^{-1} \tag{7}$$

k_e is the electron attachment rate coefficient, $\sigma_{121.6}$ is the absorption cross-section at this wavelength, $[e^-]$ is the average number density of electrons in the mesosphere, $J_{121.6}$ is the mesospheric solar flux and $\Phi_{121.6}$ the quantum yield for dissociation at 121.6 nm. Often, the photolysis term is much smaller than the electron-attachment term, and the second term of the squared bracket in Eq. (7) is ignored. It is important to appreciate that the value of τ_{chemical} is a function of position, particularly altitude, in the atmosphere. In the troposphere, τ_{chemical} will be infinite because both the concentration of electrons and $J_{121.6}$ are effectively zero, but in the mesosphere τ_{chemical} will be much less. However, multiplication of k_e for SF_6, etc., by a typical electron density in the mesosphere, ca. 10^4 cm^{-3} [10], yields a chemical lifetime which is far too small and bears no relation to the true atmospheric lifetime, simply because most of the SF_6, etc., does not reside in the mesosphere.

One may, therefore, ask where the quoted lifetimes for SF_6, CF_4 and SF_5CF_3 of 3200, 50 000 and 800 a, respectively, come from [8,11]. The lifetimes of such long-lived greenhouse gas can only be obtained from globally averaged loss frequencies. The psuedo-first-order destruction rate coefficient for each region of the atmosphere is weighted according to the number of molecules of compound in that region,

$$\langle k \rangle_{\text{global}} = \frac{\sum_i k_i V_i n_i}{\sum_i V_i n_i} \tag{8}$$

where i is a region, k_i is the pseudo-first-order removal rate coefficient for region i, V_i is the volume of region i, and n_i is the number density of the greenhouse gas under study in region i. The lifetime is then the inverse of $\langle k \rangle_{global}$. The averaging process thus needs input from a 2- or 3-dimensional model of the atmosphere in order to supply values for n_i. This is essentially a meteorological, and not a chemical problem. It may explain why meteorologists and physical chemists sometimes have different interpretations of what the lifetime of a greenhouse gas actually means.

Many such studies have been made for SF_6 [8,12,13], and differences in the kinetic model (k_i) and the atmospheric distributions (n_i) from different climate or transport models account for the variety of atmospheric lifetimes that have been reported. The importance of both these factors has also been explored by Hall and Waugh [14]. Their results show that because the fraction of the total number of SF_6 molecules in the mesosphere is very small, the global atmospheric lifetime given by Eq. (8) is very much longer than the mesospheric, chemical lifetime given by Eq. (7). Thus, they quote that if the mesospheric loss frequency is $9 \times 10^{-8}\ s^{-1}$, corresponding to a local lifetime of 129 d (days), then the global lifetime ranges between 1425 and 1975 a, according to which climate or transport model is used.

5. GENERAL COMMENTS ON LONG-LIVED GREENHOUSE GASES

In 1994, Ravishankara and Lovejoy wrote that the release of any long-lived species into the atmosphere should be viewed with great concern [15]. They noted that the CFCs, with relatively 'short' lifetimes of ca. 100 a, have had a disastrous effect over a relatively short period of time, ca. 30–50 a, on the ozone layer in the stratosphere that protects humans from harmful UV radiation. However, following implementation of international treaties (e.g., Montreal, 1987 [16]) it is now expected that the ozone layer will recover within 50–100 a [17]. At present, there are no known undesired chemical effects of low concentrations of perfluorocarbons such as CF_4 and SF_6 in the atmosphere. However, their rapidly increasing concentrations (ca. 7% per annum for SF_6) and their exceptionally long lifetimes (thousands, not hundreds of years) means that life on earth may not be able to adapt to any changes these gases may cause in the future. They suggested that all such long-lived molecules should be considered guilty, unless proven otherwise. If SF_6 is perceived potentially to be the major problem of this family of molecules, inert, dielectric gases with lower GWP values could be used as substitutes for SF_6 in industrial applications; ring-based perfluorocarbons, such as cyclic-C_4F_8 and cyclic-C_5F_8 are possibilities [18]. However, the simplest, possibly naïve, suggestion is that humans should not put up into the atmosphere any more pollutants than are absolutely necessary. The worldwide debate just starting, probably 50 a too late, is what constitutes 'absolutely necessary'.

6. CONCLUSION

In this chapter, I have only sought to explain the physical properties of greenhouse gases, and what are the factors that determine their effectiveness as pollutant gases that can cause global warming. I have not attempted to describe the natural or anthropogenic sources of these greenhouse gases, and why their concentrations have increased since the pre-Industrial era; this will be covered by other chapters in this book.

CO_2 and CH_4 currently contribute ca. 81% of the total radiative forcing of long-lived greenhouse gases (Table 2), but it is too simplistic to say that control of CO_2 levels will be the complete solution, as is often implied by politicians and the media. It is certainly true that concentration levels of CO_2 in the earth's atmosphere are a very serious cause for concern, and many countries are now putting in place targets and policies to reduce them. It is my personal belief that CO_2 levels in the atmosphere correlate strongly with lifestyle of many of the population, and with serious effort, especially in the developed world, huge reductions are possible. The challenge will be to effect policies to reduce significantly the concentration of CO_2 without seriously decreasing the standard of living of the population and negating all the benefits that technology has brought us in the last 50–100 a. I give two examples for possible policy change. First, I query whether the huge expansion in air travel within any one country at the expense of slower methods of transports (e.g., trains) is really worth all the social and economic benefits that are claimed. The price to be paid, of course, is hugely enhanced CO_2 emissions. Second, I query whether the benefits of 24 h shopping 7 days a week are really worth the extra CO_2 emissions that result from keeping shops open continuously. Would our standard of living drop significantly if shops opened for much fewer hours? Most of Switzerland closes at 4.00 p.m. on a Saturday for the rest of the weekend, yet this country is very close to the top of all international league tables for wealth creation, standard of living and levels of well-being/happiness.

CH_4 levels, however, in my opinion pose just as serious a threat to our planet as CO_2 simply because they will be much harder to reduce. Whilst it is surprising and remains unclear why the total radiative forcing of methane, 0.48 $W{\cdot}m^{-2}$, has remained unchanged over the last decade [2], a major component of methane emissions correlates strongly with the number of animal livestock which itself is dependent on the population of the planet. Controlling, let alone reducing world-wide population levels over the short period of time that is apparently available to 'save the planet' (ca. 20–40 a) [19] is a major task. Surely, this could and should be the major policy directive of the United Nations over the next few decades.

ACKNOWLEDGEMENTS

I thank members of my research group who participated in laboratory-based experiments on the long-lived 'super' greenhouse gas, SF_5CF_3, that are alluded to in this chapter. I also thank Professor Keith Shine (University of Reading, United Kingdom) for many useful discussions on radiative efficiency and global warming potentials.

REFERENCES

1. K.P. Shine, W.T. Sturges, Science, 315 (2007) 1804–1805.
2. Intergovernmental Panel on Climate Change (IPCC), 4th Assessment Report (2007), Working Group I, Chapters 1 and 2, Cambridge University Press, Cambridge, 2007.
3. J.G. Anderson, W.H. Brune, M.H. Proffitt, J. Geophys. Res. D, 94 (1989) 11465–11479.
4. K.P. Shine, J.S. Fuglestvedt, K. Hailemariam, N. Stuber, Clim. Change, 68 (2005) 281–302.
5. R.P. Tuckett, Educ. Chem., 45 (2008) 17–21.
6. R.P. Tuckett, Adv. Fluorine Sci., 1 (2006) 89–129 (Elsevier).
7. W.T. Sturges, T.J. Wallington, M.D. Hurley, K.P. Shine, K. Sihra, A. Engel, D.E. Oram, S.A. Penkett, R. Mulvaney, C.A.M. Brenninkmeijer, Science, 289 (2000) 611–613.
8. A.R. Ravishankara, S. Solomon, A.A. Turnipseed, R.F. Warren, Science, 259 (1993) 194–199.
9. T. Gierczak, R.K. Talukdar, S.C. Herndon, G.L. Vaghjiani, A.R. Ravishankara, J. Phys. Chem. A, 101 (1997) 3125–3134.
10. N.G. Adams, D. Smith, Contemp. Phys., 29 (1988) 559–578.
11. R.Y.L. Chim, R.A. Kennedy, R.P. Tuckett, Chem. Phys. Lett., 367 (2003) 697–703.
12. R.A. Morris, T.M. Miller, A.A. Viggiano, J.F. Paulson, S. Solomon, G. Reid, J. Geophys. Res. D, 100 (1995) 1287–1294.
13. T. Reddmann, R. Ruhnke, W. Kouker, J. Geophys. Res. D, 106 (2001) 14525–14537.
14. T.M. Hall, D.W. Waugh, J. Geophys. Res. D, 103 (1998) 13327–13336.
15. A.R. Ravishankara, E.R. Lovejoy, J. Chem. Soc. Farad. Trans., 90 (1994) 2159–2169.
16. United Nations Environment Program, Ozone Secretariat, 7th edition (2006), http://ozone.unep.org/publications/MP_Handbook/index.shtml
17. E.C. Weatherhead, S.B. Andersen, Nature, 441 (2006) 39–45.
18. M.A. Parkes, S. Ali, R.P. Tuckett, V.A. Mikhailov, C.A. Mayhew, Phys. Chem. Chem. Phys., 9 (2007) 5222–5231.
19. D.A. King, G. Walker, The Hot Topic: How to Tackle Global Warming and Still Keep the Lights on. Harvest Books, Washington, USA, 2008.
20. T.J. Blasing, Carbon Dioxide Information Analysis Centre, Oak Ridge National Laboratory (2008), http://cdiac.ornl.gov/pns/current_ghg.html.

4. [illegible], Climatic Change, 68 (2005) 251–302.
5. [illegible], 85 (2008) 17–21.

Chapter 2

The Role of Widespread Surface Solar Radiation Trends in Climate Change: Dimming and Brightening

Shabtai Cohen
Department of Environmental Physics and Irrigation, Institute of Soil, Water and Environmental Sciences, Agricultural Research Organization, The Volcani Centre, P.O. Box 6, Bet Dagan 50250, Israel

1. Introduction
2. Solar Radiation and its Measurement
2.1. Top (TOA) and Bottom (BOA) of the Atmosphere Solar Radiation and Atmospheric Transmission
2.2. Earth's Albedo and Net TOA Solar Radiation
2.3. BOA Radiation
2.4. Measurement of Surface Radiation
2.5. Comparing $E_g{\downarrow}$ from Different Sites
2.6. Archives of Surface Solar Radiation Measurements
3. Trends in Surface Solar Radiation or Global Dimming and Brightening
3.1. Global Dimming Reports in the Twentieth Century
3.2. From Dimming to Brightening
3.3. $E_g{\downarrow}$ Prior to the 1950s
3.4. Regional Changes
3.5. Cloud Trends and their Influence on $E_g{\downarrow}$
4. The Causes of Dimming and Brightening
5. The Influence of Solar Radiation Changes (Dimming and Brightening) on Climate
5.1. The Evaporation Conundrum
5.2. Soil Moisture Trends
5.3. The Hydrological Cycle
5.4. Daily Temperature Range (DTR)
5.5. Wind Speed and the Monsoon System
6. Conclusions
References

1. INTRODUCTION

The flux density and wavelength of electro-magnetic radiation emitted from a body depend on its temperature. On the earth's surface the wavebands that contain the most energy, and are therefore of prime interest in the context of climate influences, are those emitted by the sun and the earth. The calculation of spectral distributions from Planck's law using their approximate temperatures of 5800 and 300 K, for sun and earth, shows that 97% of the energy of solar and >99% of that of terrestrial radiation fall within the wavebands of 0.29–3 and 3–100 μm, respectively. Those wavebands are referred to as short wave (or solar) and long wave (or terrestrial) radiation [1]. The problem with the current ubiquitous, steady increase in atmospheric carbon dioxide concentration stems not from its direct influence on climate, but rather from its absorption of radiation in the long wave band, which decreases long wave radiative losses from the earth. Since its absorption in the solar spectrum is small, CO_2 has a negligible influence on the earth's solar radiation balance.

Global radiation ($E_g\downarrow$) is the total solar radiation falling on a horizontal surface at the earth's surface, that is, at the bottom of the atmosphere (BOA). Precise wide-spread measurements of $E_g\downarrow$ began in the early twentieth century and although it was first assumed that no multi-annual trends in this quantity occurred, by the 1970s there was evidence of significant decreases at some sites. As the evidence for large multi-decadal trends in $E_g\downarrow$ grew, the relationship between decreasing solar radiation (or global dimming) and wide spread decreasing pan evaporation was noticed. The energetic similarity of these changes led to scientific recognition that changes in $E_g\downarrow$ were playing a significant role in climate change. Previous assumptions that other parts of the earth's radiation balance were unchanging, have subsequently come under scrutiny.

This paper provides some background material on solar radiation and reviews some of the work done on the changing $E_g\downarrow$ and its influences on earth's climate.

2. SOLAR RADIATION AND ITS MEASUREMENT

2.1. Top (TOA) and Bottom (BOA) of the Atmosphere Solar Radiation and Atmospheric Transmission

Several of the quantities encountered when studying the earth's short wave radiation balance are easily computed. Understanding these relationships can give the quantitatively minded reader more confidence about solar radiation and its trends.

Black-body radiation is described by the Stefan–Boltzman equation, that is,

$$B = \sigma T^4$$

where B is radiant flux density emitted from a black body of temperature T, and σ is the Stefan–Bolzmann constant, 5.67×10^{-8} W·m^{-2}·K^{-4}. Taking the sun's average surface temperature to be 5800 K, calculating solar output for a sphere of solar radius, 6.96×10^8 m, and irradiating a large sphere whose radius is the earth–sun distance, or one astronomical unit (1.5×10^{11} m), the radiant flux density reaching a surface normal to the sun's rays on the earth before it is influenced by the atmosphere, that is, the extra-terrestrial 'solar constant', is 1380 $W{\cdot}m^{-2}$, which is very close to the currently accepted value of 1366 $W{\cdot}m^{-2}$ [2]. The latter varies during the year by about 3.3% due to eccentricity of the earth's orbit. As long as the solar surface temperature and composition doesn't change, the yearly average will be constant. In fact, the solar constant has varied by much less than 1% over the past few centuries [3,4]. The ratio of the area of a sphere to that of a circle of the same radius is 4, so the mean solar radiant energy reaching the TOA is 342 $W{\cdot}m^{-2}$.

TOA (or extra-terrestrial) solar radiation on a plane parallel to the surface varies with the solar zenith angle, that is, the angle between the vertical and the solar vector. Calculation of solar angles and TOA solar radiation is straightforward and given elsewhere [1,5–7]. TOA values are used to compare with BOA measurements in order to determine atmospheric absorption of radiation, for example, atmospheric transmission and turbidity and aerosol optical depth.

2.2. Earth's Albedo and Net TOA Solar Radiation

The earth's planetary albedo depends mostly on cloudiness, but also on land use. There is no scientific theory to indicate that the planetary albedo has been and will remain constant, and a change of 1% in its value can have a large impact on the earth's climate system [8]. Accurate measurements of the albedo began in the 1980s. Satellite observations made continuously during the past twenty years indicate that it is relatively constant at $29 \pm 2\%$ [9,10]. These measurements are close to previous estimates of 30 [11] and 31% [12]. However, analyses of earthshine measurements suggest that it may have changed by as much as 5% during the past 15 a [13–15]. The earthshine measurements have met with some criticism [9], but they are based on sound theory. In the future, if additional sites are added to the earthshine observation network, these measurements may gain more acceptance and the differences between the earthshine and satellite measurements will have to be resolved. Taking the current earth albedo to be 29%, the net solar input into the planet is about 243 $W{\cdot}m^{-2}$ [16].

2.3. BOA Radiation

From this brief discussion of TOA solar radiation balance we jump to the situation at the surface below the atmosphere where the solar radiation balance is confounded by atmospheric transmissivity and surface albedo. The former depends mostly on cloudiness and cloud properties, but also on dust and other aerosols. The latter, which has a small influence on downward radiation, depends on surface properties, which are influenced by land use and climate.

As solar radiation traverses the atmosphere it is absorbed and reflected by gases and non-gaseous particles [17]. Ozone is responsible for absorption of most of the UV radiation, that is, the solar radiation at wavelengths below 0.29 μm; at larger wavelengths oxygen and ozone absorption is negligible. Water vapour is a significant absorber in the infra-red portion of the solar spectrum above 0.7 μm. Carbon dioxide absorption of solar radiation is negligible. Aerosols can scatter and reflect some of the radiation back to space. Clouds can reflect most of the radiation back to space. Radiation reflected from the earth's surface can be re-reflected back, and so surface albedo can influence the downward flux. Thus, BOA solar radiation is much less than that at TOA, and is commonly divided into two fluxes: direct radiation coming from a 2.5–5° angle centred in the direction of the sun, and diffuse radiation arriving from the rest of the sky hemisphere above the observer. The total of these two, that is, global radiation ($E_g\downarrow$), is the total solar energy available at the surface.

2.4. Measurement of Surface Radiation

Total short wave 'solar' radiant flux density on a horizontal surface on the earth's surface (BOA), that is, global radiation, $E_g\downarrow$, is measured with a pyranometer. First class pyranometers measure the temperature difference between an exposed optically black surface and either a white surface (in the older instruments) or the lower non-exposed surface using a thermopile. In order to exclude thermal radiation and advection of heat from the surroundings the black surface is covered with two quartz glass domes which transmit radiation between 200 and 4500 nm wavelength, and a temperature correction circuit is incorporated into the instrument. Another type of 'pyranometer' in common use, due to its lower cost, is based on a selenium cell which upon illumination causes an electrical current to flow. The sensor is covered with appropriate filters to measure solar radiation, but the maximum wavelength measured is 1100 nm, so total solar radiation is determined indirectly by assuming that the ratio of the full spectrum to that below 1100 nm is constant. In most outdoor conditions the assumption is good enough for many applications, for example, calculation of crop water requirements, but the non-thermopile pyranometers are not acceptable for first class meteorological measurement.

Frequent cleaning of the dome and yearly calibration of sensors is necessary in order to ensure the reliability of measurements. These and other constraints have led to sparse measurement networks producing reliable data for solar radiation as compared to those measuring air temperature. Most of the networks began to operate during the International Geophysical Year, 1957–1958.

A second widely used surface measure which has been of interest is sunshine duration (SSD), or the amount of time that direct solar radiation exceeds a threshold of 120 $W \cdot m^{-2}$, corresponding approximately to direct irradiance at 3° solar elevation under clear sky conditions [1]. This measure has been shown to be highly correlated with global radiation, both on a single day basis as well as for yearly totals [18,19]. Instruments measuring SSD came into use in the nineteenth century, and some of their history has been recently reviewed [20]. Many measurement series dating back to the nineteenth century are available in various forms, and analysis of these has enabled a rough view of variations in solar radiation for more than a century (e.g. [21,22]).

In addition to surface measurements, satellite based sensors have been monitoring earth radiance in different wavebands for more than two decades. Algorithms have been developed to use these measurements to calculate solar radiation at the surface. These measurements have the advantage of spatial averaging over an area several orders of magnitude larger than the few square centimetres measured by the surface based sensors, and the ongoing efforts to improve the reliability and accuracy of the satellite measurements has led to their increased acceptance.

2.5. Comparing $E_g\downarrow$ from Different Sites

When comparing sites it is convenient to consider annual totals of $E_g\downarrow$, since seasonal variations can be large and vary greatly areally. However, $E_g\downarrow$ varies with altitude and latitude. One way to normalize data from different sites is to determine the transmission of a unit atmosphere, which is similar to turbidity [6,23]. Yearly means of $E_g\downarrow$ are converted to atmospheric transmittance, τ_m, by dividing by integrated yearly extraterrestrial solar irradiance on a horizon tal surface (S_o) computed for the latitude of the measurements, that is,

$$\tau_m = \frac{\int E_g \downarrow \mathrm{dt}}{\int S_o \downarrow \mathrm{dt}} \tag{1}$$

Transmittance is also an exponential function of the optical thickness of the atmosphere k, and the vertical non-dimensional air mass, m, such that

$$\tau_m = \exp(-km)$$

or

$$k = -\ln(\tau_m)/m \tag{2}$$

For a unit air mass ($m = 1$) Eqn (2) yields

$$\tau_1 = \exp(-k) = \exp(\ln(\tau_m)/m) \tag{3}$$

Values of τ_1, which expresses the yearly average transmittance of a unit atmosphere at the site, are computed for each yearly mean of $E_g\downarrow$, where m is computed from site altitude using a simple altimetric relationship like:

$$m = \exp\frac{-A}{8200} \tag{4}$$

and A is site altitude (m) (after [6]). A second method to normalize data from different sites is multiple regression of $E_g\downarrow$ on time and site parameters, where the influence of altitude is taken as linear, but site latitude (Φ) is taken as $\cos^3(\Phi)$ [23].

2.6. Archives of Surface Solar Radiation Measurements

Solar radiation data measured by the different national weather services and conforming to WMO standards are collected in various national archives and are available from national weather services. Much of this data has also been collected in two archives – the Global Energy Balance Archive (GEBA) in Zurich, Switzerland [24], and the World Radiation Data Center (WRDC) archive in St. Petersberg, Russia, which was established by the WMO in 1964. GEBA has incorporated much of the data from the WRDC archive after strict quality control filtering, while the WRDC archive should be used with caution.

Data from the US is managed by the National Renewable Energy Laboratory's (NREL) Renewable Resource Data Center (RReDC, at website: www.nrel.gov/rredc). Although solar radiation has been measured in the US for about 75 a, first class long term data is available for only few of the stations in their network.

The World Radiation Monitoring Center (WRMC, http://www.bsrn.awi.de/) archives data from the Baseline Surface Radiation Network (BSRN, [25]), which is a small number of stations (currently about 40) in contrasting climatic zones, covering a latitude range from 80°N to 90°S, where solar and atmospheric radiation is measured with instruments of the highest available accuracy and with high time resolution (1–3 min). The BSRN program began in the late 1990s and is based on voluntary participation of organizations measuring radiation in different countries.

3. TRENDS IN SURFACE SOLAR RADIATION OR GLOBAL DIMMING AND BRIGHTENING

Significant multi-year trends in $E_g\downarrow$ during the first decades that measurements were made were reported by a few scientists during the twentieth century. Many of these decreasing trends, called 'global dimming' [23], were in excess of 1% per decade. They were viewed with considerable scepticism by the scientific community. The reasons for this scepticism are important because they reflect on the way current science is carried out. Here are some possibilities:

a. Previous texts, which were accepted as foundations of climate science, assumed that earth's solar radiation budget was constant on the short term time scale (i.e. hundreds of years [26]), although changes in solar activity and the solar constant were included as possible drivers for long term (i.e. 10^3–10^7 a) climate changes (see Ref. [27] for a review of climate change theories up to the mid 1960s).

b. Climate change science has been dominated by the influence of the ubiquitous and steadily increasing atmospheric greenhouse gases, and especially CO_2. A large effort has been made to establish that this change is large enough to warrant worldwide political action. The magnitude of 'global dimming' was clearly of the same order of magnitude as the greenhouse gas influence. If large changes were occurring un-noticed to the scientific community, how good was our understanding of climate and climate change? That question may have been viewed as a threat to the attempts to harness political action and the unprecedented funding that climate change science was receiving [28].

c. Climate change science has focused on TOA influences (e.g. TOA radiative forcing) and assumed that the distribution of energy within the system is less important.

d. Solar radiation is highly variable spatially and temporally and this high variability has hampered integration of worldwide trends. This is in sharp contrast with greenhouse gases which mix well in the atmosphere and whose rate of increase can be discerned within a few years.

3.1. Global Dimming Reports in the Twentieth Century

Suraqui et al. [29] reported 'severe changes over the years in solar radiation' and issued a call for 'a careful study of incoming radiation at different places throughout the world ... to determine the exact kind, order of magnitude and their causes ...'. The 'severe changes' referred to emerged from the measurements at the site of the Smithsonian Institution's former solar radiation monitoring station on Mt. St. Katherine in the southern Sinai peninsula (28°31′N, 33°56′E, 2643 m altitude). Measurements using modern radiometers as well as some of the original instruments employed between 1933 and 1937 showed a 12% loss in global radiation during the intervening four decade interval.

Atsumu Ohmura, whose background was in glaciology, and who headed the GEBA archive [24], reported at a conference that solar radiation was decreasing at many sites where it was being measured. His colleagues, who were highly sceptical of his findings, discouraged him from pursuing this, and the report was published (or temporarily buried) in a little known conference proceedings [30]. Russak [31] reported decreasing trends of 0.2–0.6 $W{\cdot}m^{-2}{\cdot}a^{-2}$ for a few stations in northern Europe. Gerald Stanhill, who used solar radiation measurements for determining evaporation and crop water use in arid environments, was intrigued by the decreasing trends in solar radiation that he found in radiation records. Stanhill and Moreshet [32] analyzed data from 45 stations for the years 1958, 1965, 1975 and 1985, and found a statistically significant average worldwide decrease of $E_g\downarrow$ totalling 5.3% (or 0.34 $W{\cdot}m^{-2}{\cdot}a^{-2}$) from 1958 to 1985. Decreasing trends of the same order of magnitude were found for sites in Australia [33], Japan [34], the arctic [35], Antarctica [36], Israel [37] and Ireland [38]. The largest decrease, found in Hong Kong, was 1.8 $W \cdot m^{-2} \cdot a^{-2}$, that is, a decrease in excess of 1% per year [39]. Other groups reported dimming for China [40], the former Soviet Union [41] and Germany [42,43]. Reductions in solar radiation were larger for urban industrial sites, but even at sites remote from pollution $E_g\downarrow$ was usually decreasing at a rapid rate.

Gilgen et al. [44] reviewed trends found in the GEBA archive. Their paper, entitled 'Means and trends of short wave irradiance at the surface estimated from GEBA Data', included analyses of accuracy and biases, and trends in $E_g\downarrow$ for different regions of the world. The final sentence of the abstract noted that 'on most continents, shortwave irradiance decreases significantly in large regions, and significant positive trends are observed only in four small regions'.

Stanhill and Cohen [23] tabulated the negative trends for different sites around the world. Of the 30 stations where detailed analyses of trends had been published, at 28 $E_g\downarrow$ had decreased and only at two, Dublin, Ireland and Griffith, Australia, had $E_g\downarrow$ increased (by 0.56 and 0.76 $W \cdot m^{-2} \cdot a^{-2}$, respectively). They also analysed solar radiation records from the geophysical year, 1958, and the years 1965, 1975, 1985 and 1992. These records were from between 145 (1958) and 303 (1992) stations whose measurements conformed to WMO standards. Average transmittance of a unit atmosphere for the northern hemisphere was 0.52 in 1957 and declined steadily to 0.44 in 1992 while that for the southern hemisphere averaged 0.57 until 1985 and declined between 1985 and 1992 to 0.52. A spline fit to the latitudinal distribution of $E_g\downarrow$ showed that the decrease during the 34 a period had been especially large in the industrialized region of the northern hemisphere with a centre at ~35°N and a width of ~20°. This feature and an analysis of the various possible reasons for the dimming phenomenon, led to the conclusion that particulate aerosols, and especially those from anthropogenic sources, were the cause of the changes. Similar conclusions were drawn at about the same time by Liepert and Lohmann [45].

Many subsequent studies have highlighted similar trends based on data collected from the mid twentieth century and onwards. Trends for individual sites are highly variable, and for some places and some parts of the world no change or increases in solar radiation have been found.

3.2. From Dimming to Brightening

Recent studies [46,47] have found evidence for a reversal in the negative trends in solar radiation, which, for many sites changed to positive trends in the late 1980s and early 1990s. The data sets analysed were from the GEBA archive [46] and, for the first time, long term trends in satellite data from 1983 to 2001 [47]. However, there is an inconsistency between the two studies, since the satellite data show brightening over the oceans and no trend over the land surfaces while the surface GEBA and BSRN measurements are mostly land based and show clear brightening during this period. The reversal in the trend is thought to be related to the decreases in air pollution in Europe and other parts of the western world following legislation that limited air pollution. The positive trend has not led to a full recovery in $E_g\downarrow$ and current levels of solar radiation in most places where dimming took place are still below the values measured during the 1950s. A selection of widespread trends reported for $E_g\downarrow$ is given in Table 2.

A list of the publications on global dimming, brightening and related topics was compiled by M. Roderick at ANU and is kept more or less up to date. It can be found on the web at http://www.rsbs.anu.edu.au/ResearchGroups/EBG/index.php. Several international meetings have been held to discuss these topics (Table 1).

TABLE 1 International meetings held on changing surface solar radiation and related changes in evaporation

Organizing Organization and event	Date	Session title	Location	Reference
AGU/CGU joint assembly	17–18 May 2004	Magnitude and Causes of Decreasing Surface Solar Radiation	Montreal, Canada	[91]
Australian Academy of Science International workshop	22–23 November 2004	Pan evaporation: An example of the detection and attribution of trends in climate variables	Canberra, Australia	[92]
EGU general assembly	15–20 April 2007	Surface Radiation Budget, Radiative Forcings and Climate Change	Vienna, Austria	
AGU fall meeting	10–14 December 2007	Pan Evaporation Trends: Observations, Interpretations, and the Ecohydrological Implications	San Francisco, CA, USA	
Israel Science Foundation international workshop	10–14 February 2008	Global dimming and brightening	Ein Gedi, Israel	[93]
EGU general assembly	13–18 April 2008	Surface Radiation Budget, Radiative Forcings and Climate Change	Vienna, Austria	

TABLE 2 **Selected estimates of widespread trends in surface solar radiation from surface measurements and satellite-based estimates. Based on Ref. [94]**

Study	Time period	Energy trend per decade/ ($W{\cdot}m^{-2}$)	Comments
Surface			
[23]	From mid-1950s to 1992	−3	Trend analysis of about 30 sites of various lengths, and data from five years from 1957 to 1992 for >145 stations
[95]	1960–1990	−2	Trend analysis of GEBA and US NREL data sets from 1960 to 1990
[44]	From mid-1950s to 1990	−3	Statistics of the GEBA data set based on about 300 sites of various length
[51]	From mid-1950s to 1990	−1.6	Analysis of GEBA data to constrain the "urbanization" effect. Separation of sparsely populated sites (<0.1 million inhabitants) and
		−4.1	populated sites (>0.1 million inhabitants)
[96]	1977–1990	−2	Trend analysis of five records of the GMD data set from remote sites from South Pole to Barrow, Alaska
[46]	1993–2004	4.7	Trend analysis of 18 BSRN records
[46]	1985–2005	2.2	Decadal change between (1985–1995) and (1995–2005) based on 320 GEBA sites
Satellite			
[47]	1983–2001	1.6	Global. University of Maryland algorithm with ISCCP Clouds – Global average
		2.4	Ocean surfaces
		−0.5	Land surfaces
[97]	1984–2000	2.4	Global. ISCCP Clouds with own RT model
[58]	1984–2000	0.4	Global (ISCCP FD)
		1	Ocean (ISCCP FD)
		−1	Land (ISCCP FD)

Notes: *GMD – Global monitoring division of NOAA, ISCCP FD – International Satellite Cloud Climatology Project result data sets.*

3.3. $E_g\downarrow$ Prior to the 1950s

Little is known about $E_g\downarrow$ prior to the 1950s and since temperature changes then are well documented, such information could be valuable for understanding the influences of $E_g\downarrow$ on climate. Stanhill and Cohen [18,19] used SSD data as proxies for $E_g\downarrow$ based on recent simultaneous measurements of both measures, in order to deduce trends of $E_g\downarrow$ from 1891 to 1987 for the US and from 1890 to 2004 for Japan. SSD was found to be well correlated with $E_g\downarrow$ and therefore can serve as a proxy. The data from the US and Japan were from 106 and 65 stations with at least 70 and 35 a of data each, respectively. In the US mean SSD increased from 1891 to the 1930s and then decreased until the mid-1940s. In Japan a similar increase was observed from 1900 to the mid-1940s. This was followed by a decline until the late 1950s. Palle and Butler [22] found a decrease in SSD for four stations in Ireland for the period from 1890 to the 1940s. Sanchez-Lorenzo et al. [48] analysed SSD for the Iberian Peninsula for 1931–2004 and found a dimming trend from the 1950s to the early 1980s followed by brightening, but the early data (1931–1950) showed no clear trend. Thus, it is possible to obtain estimates of $E_g\downarrow$ for the first half of the twentieth century and many SSD data sets exist, but more work is needed to understand this period.

3.4. Regional Changes

The areal extent of the changes in global radiation and their global impact has been the subject of much debate and some investigation. Significant rates of dimming and brightening have been observed at many sites remote from major sources of air pollution, for example the polar regions [35,36], and the largest trends have been observed in heavily polluted regions (e.g. Hong Kong [39], India [49] and China [50]), suggesting a significant relationship between pollution rates and global radiation trends. Alpert et al. [51] found that dimming from the 1950s to the 1980s averaged 0.41 $W{\cdot}m^{-2}{\cdot}a^{-2}$ for highly populated sites while for sparsely populated sites, that is, populations $<0.1 \times 10^6$ dimming was only 0.16 $W{\cdot}m^{-2}{\cdot}a^{-2}$. In equatorial locations with low population density there were slightly increasing trends. Since most of the globe is sparsely populated this implies that the spatially averaged changes in $E_g\downarrow$ are significant, but smaller than those obtained by averaging the data, which may be biased toward population centres. However, to date no model has been developed to integrate population density and its influence on $E_g\downarrow$ with the worldwide grid of $E_g\downarrow$ in order to update the estimates of dimming and brightening, and current estimates revolve around those given in Table 2. Trends observed from satellites are for wide regions ([47]; Table 2) and it is encouraging that those trends are similar to those computed by averaging data from surface stations.

3.5. Cloud Trends and their Influence on $E_g\downarrow$

Changes in cloudiness during parts of the dimming and brightening periods were studied by Joel Norris [52]. The data was from both surface data sets and satellite observations. The surface set, which was divided into $10° \times 10°$ cells, was from the Extended Edited Cloud Report Archive (EECRA), and included ground based cloud observations from land stations (1971–1996) and ship reports (1952–1997). These showed that zonal mean upper-level cloud cover at low and middle latitudes decreased by 1.5%-sky-cover between 1971 and 1996 over land and by about 1%-sky-cover between 1951 and 1997 over ocean. The upper level data were closely related to satellite (ISCCP) estimates for an overlapping period. Estimates of the cloud cover influence on solar radiation showed that between 1952 and 1997 over mid-latitude oceans cloud changes decreased $E_g\downarrow$ by about 1 $W{\cdot}m^2$, and over northern mid-latitude land areas cloud changes increased $E_g\downarrow$ slightly. For low-latitude land and ocean regions cloud changes increased $E_g\downarrow$ from the 1980s to the mid-1990s. These changes in cloudiness are relatively small, and although they probably played a significant part in global dimming and brightening, they could not be considered to be major players. Similar conclusions, that is, that cloud trend influences on short wave radiative forcing could not account for most of the global dimming and brightening, were made by Norris and Wild [53], who subtracted the estimated cloud cover influence on solar radiation from surface $E_g\downarrow$ data in the GEBA archive and found that dimming and brightening trends in the residual $E_g\downarrow$ were unchanged.

4. THE CAUSES OF DIMMING AND BRIGHTENING

Dimming and brightening are related to aerosol loading of the atmosphere and the influences of aerosols on atmospheric transmittance. The influence of natural aerosols from volcanic eruptions can be seen in the sharp declines in $E_g\downarrow$ for the year or two following the eruptions of El Chichon in 1983 and Pinatubo in 1991 [54]. Stanhill and Cohen [23] reviewed the possible causes for dimming in the context of a simplified expression:

$$E_g \downarrow = E_o \exp\left[-\left(\tau_r + \tau_g + \tau_w + \tau_a + \tau_c\right)\right] \tag{5}$$

where $E_g\downarrow$ is estimated from the extraterrestrial irradiance at the top of the atmosphere, E_o, modified by a chain of five transmissivities τ which quantify the solar scattering and absorbing properties of the different components of the atmosphere. These include τ_r, representing Rayleigh scattering; τ_g, permanent gas absorption; τ_w, absorption by water vapour; and τ_a and τ_c, the absorption and scattering by the aerosols and cloud components, respectively. The only factor whose known changes and influence on global radiation are large enough to cause changes of the magnitude observed is aerosol loading. Aerosol influences on radiation include direct effects, that is, absorption, reflection and scattering of radiation by aerosols, and indirect effects, referring to aerosol mediated changes in cloud albedo (the Twomey effect), rain suppression (the Albrecht effect), and cloud lifetime. The large changes in $E_g\downarrow$ can be pinned to some extent on anthropogenic pollution, as suggested by the large dimming in urban mega-cities and the industrialised zone of the Northern Hemisphere. The connection between dimming and aerosols has been clearly demonstrated (e.g. [55]), and known changes in aerosol loading of the atmosphere are well correlated with the transition from dimming to brightening in the 1980s [56].

Prior to the twenty-first century scientists studying aerosols had suspected that aerosol influences on climate were far larger than was being acknowledged and Satheesh and Ramanathan [57] demonstrated the large radiative forcing that can be caused by aerosols. As the evidence for worldwide dimming of a magnitude of several percent has mounted scientists who were studying aerosol influences have begun to implement the full extent of aerosol influences in models of earth's climate (e.g. [58,59]).

5. THE INFLUENCE OF SOLAR RADIATION CHANGES (DIMMING AND BRIGHTENING) ON CLIMATE

5.1. The Evaporation Conundrum

Potential evaporation rates in many places in the world decreased during the second half of the twentieth century. As with solar radiation measurements, a major client for these measurements is the agricultural community, where evaporation rates are used to determine irrigation scheduling and application rates. Measurement of evaporation is usually done with an evaporimeter of the evaporation pan type, for example, the US class-A pan and Russian GGI-3000 pan [60]. Specifications of pan size, deployment and exposure are given in the previous reference. Networks of pans have been established in many parts of the world.

Evaporation of water requires large quantities of energy. Therefore, one model of evaporation is the energy budget of the evaporating surface, that is,

$$R_n = \lambda E + C + G \quad \text{and} \quad \lambda E = R_n - C - G \tag{6}$$

where R_n is net radiation absorbed by the surface, λ is the latent heat of vaporisation, E is the evaporative flux, C is convective heat transfer with the environment and G is surface heat flux and/or energy storage. For annual totals, heat flux and energy storage can usually be ignored and evaporation depends only on net radiation and convection.

Evaporation from a wet surface (i.e. potential evaporation) can also be viewed as a diffusion process where water vapour is transported from the surface to the surrounding air, that is,

$$\lambda E = \rho c_p [e_s(T_s) - e_a]/(\gamma r) = \frac{\rho c_p}{\gamma (r_a + r_s)} [e_s(T_s) - e_a] \qquad (7)$$

where ρ and c_p are air density and heat capacity, respectively, $e_s(T_s)$ and e_a are water vapour pressure in air for saturation at surface temperature (T_s) and ambient conditions, respectively, and γ is the psychrometric constant. r is the resistance to vapour transport from the wet surface to the point of interest in the air where humidity is measured, which in turn can be separated into a bulk surface resistance (r_s) and boundary-layer aerodynamic resistance (r_a). This second description of evaporation emphasises that it is influenced not only by radiation, but also by aerodynamic parameters like air temperature, humidity and wind speed, as well as surface parameters like roughness. Viewing both the energy budget and diffusion models of evaporation together, it is clear that climate factors determine the partitioning of radiative energy absorbed by a surface between the energy dissipation processes, that is, evaporation and convection.

The two approaches [Eqns (6) and (7)] can be used to solve for evaporation from a wet surface with few assumptions, giving the Penman equation [61], that is,

$$\lambda E = \frac{\Delta}{\gamma^* + \Delta}(R_n - G) + \frac{\rho c_p}{r_a(\gamma^* + \Delta)}[e_s(T_a) - e_a] \qquad (8)$$

where T_a is air temperature, Δ is the slope of the relationship between saturation vapour pressure and temperature, and γ^* is a bulk psychrometric constant which depends on surface properties. The expression ($e_s(T_a) - e_a$) is the air vapour pressure deficit (VPD), which is a function of temperature and humidity. Thus, evaporation from a wet surface can be partitioned between radiative and aerodynamic influences on evaporation, where the radiative term (the left hand part of the Penman equation) is dominated by solar radiation and the aerodynamic term (the right hand part) depends on air temperature, humidity and wind speed. When analysing changes in potential (pan) evaporation Eqn (8) can help to determine which climatic factor has caused the change.

Widespread reductions in pan evaporation during the second half of the twentieth century were first reported for the former Soviet Union and much of the northern hemisphere [62,63]. These reports were considered evidence of global warming, which was thought to be increasing regional evaporation but decreasing pan evaporation due to a feedback influence of increasing regional humidity on local (or pan) potential evaporation [64] (see below). However, Stanhill and Cohen [23] considered decreasing evaporation to be evidence for decreasing solar radiation and Cohen et al. [65] showed that in Israel's arid conditions the overwhelming influence on evaporation is solar radiation. A full analysis of environmental factors showed that decreasing solar radiation was decreasing potential evaporation rates. Qian et al. [50] found a striking correspondence between decreasing $E_g\downarrow$ and pan evaporation in China.

Two Australian biologists, Roderick and Farquhar [66], analysed worldwide changes in temperature and humidity and their relationship to evaporation rates. If regional evaporation were increasing and causing local pan evaporation to decrease then VPD should be decreasing [see Eqn (8)]. However, there was no evidence that this was occurring worldwide. Daily minimum temperatures are closely related to the daily dew point temperature and air vapour pressure (e_a), since excess humidity precipitates as dew when the air is coolest in the early morning. Saturation vapour pressure (e_s) increases exponentially with increasing temperature, so if average and minimum temperatures increase at the same rates, VPD will increase and this should increase evaporation rates. However, worldwide minimum temperatures are increasing much faster than average temperatures and Roderick and Farquhar reasoned that this might be stabilizing VPD, as observed in climate data from the US. This implied that the aerodynamic term in the Penman equation [Eqn (8)] was stable; and if evaporation was decreasing it would have to be caused by decreasing net radiation, which is dominated by solar radiation. Roderick and Farquhar continued to develop a rigorous estimate of the evaporative equivalent to solar radiation. For a first order analysis the evaporative equivalent of radiative energy is expressed by λ, whose value is $\sim$2.4 MJ·kg^{-1} and 1 kg of water will cover a surface area of 1 m^2 to a depth of 1 mm. For the region of the FSU where both radiation and evaporation trends were available, solar radiation, which was in the range of 3000–4000 MJ·m^{-2}·a^{-1}, had declined by $\sim$9% or 315 MJ·m^{-2} in three decades, which is equivalent to 131 mm of water. This is similar to the average reported evaporation reduction during that period, $\sim$111 mm of water. Thus, the reported reductions in evaporation rates matched those for solar radiation, and the pan evaporation data set corroborated the reported dimming trends in $E_g\downarrow$. Roderick and Farquhar's analysis [66] convinced many scientists that dimming was real and was having a significant impact on earth's climate.

Evaporation at most sites in Australia has decreased significantly during the period on record, with no signs of recovery during the 'brightening' era [67]. The climate parameters that could be causing this were investigated by Roderick et al. [68] using a physical model similar to Eqn (8). They found that the primary cause for the reduction in evaporation in Australia was decreasing wind speed with some regional contributions from decreasing solar radiation.

The question as to whether changes in pan evaporation are similar or opposite to changes in regional evaporation involves the 'complementary' hypothesis [69], which hypothesises that when regional evaporation changes, air humidity changes in the same direction, and a feedback occurs which has an opposite effect on local evaporation. The hypothesis [70] considers the sum of regional and local (e.g. pan) evaporation to be equal to a constant value, making them 'complementary'. For example, in the Tibetian plateau, $E_g\downarrow$ and pan evaporation decreased from 1966 to 2003 [71], yet regional evaporation increased [72].

Since global radiation influences both local and regional evaporation similarly, when global radiation changes the constant of the complementary equation may also change. Nevertheless, when significant changes in air temperature occur, especially if accompanied by changes in wind speed, which have also been noted for many sites, changes in pan evaporation cannot be taken as unambiguous evidence for dimming, brightening or warming [73].

5.2. Soil Moisture Trends

Another line of evidence for changes in regional evaporation rates has come from the study of soil moisture data from an extensive network of stations in the Ukraine where plant available soil moisture for the top 1 m of soil is determined gravimetrically every 10 days from April to October at 141 stations from fields with either winter or spring cereals. The data, from 1958 to 2002 [74], shows that soil moisture increased until approximately 1980 and then levelled off. No trends in rainfall were observed for this region while air temperature increased slightly. As noted above, one of the first reports of dimming was from this region during the period in question [41]. The observed changes in soil moisture were opposite to the predictions that global warming would lead to soil desiccation [75,76]. Thus, Robock and Li [74] concluded that the changes in soil moisture were evidence of dimming and its reduction of regional evaporation rates. Subsequent modelling with a sophisticated land surface model, which included a decreasing trend of solar radiation along with increasing CO_2 and global warming, demonstrated similar increases in soil moisture [77].

5.3. The Hydrological Cycle

Regional evaporation rates are a central part of the hydrological cycle, and so the question as to whether decreases in pan evaporation indicate decreasing or increasing regional evaporation is of great importance. An increasing hydrological cycle with increased regional evaporation would lead to increased rainfall rates. However, it would also increase cloudiness whose feedback influence would cause a decrease in $E_g\downarrow$. As noted above, cloud changes have been relatively small.

Prior to the twenty-first century, it was assumed that global warming would enhance evaporation and lead to an enhancement (or spinning up) of the hydrological cycle. Ramanathan et al. [49] evaluated the influences of anthropogenic aerosols on solar and thermal radiation balances, atmospheric temperature profiles and climate. They found that 'aerosols enhance scattering and absorption of solar radiation and produce brighter clouds that are less efficient at releasing precipitation. These in turn lead to large reductions in the amount of solar irradiance reaching Earth's surface, a corresponding increase in solar heating of the atmosphere, changes in the atmospheric temperature structure, suppression of rainfall, and less efficient removal of pollutants. Thus, these aerosol effects can lead to a weaker hydrological cycle'. A case in point is the Indian sub-continent where anthropogenic aerosol 'brown clouds' can reduce $E_g\downarrow$ by more than 10% and change the regional hydrological cycle. In particular, dark aerosols absorb solar radiation and cause enhanced atmospheric warming and decreased $E_g\downarrow$, which decreases surface temperatures and evaporation rates. Together, these enhance atmospheric stability and spin down the hydrological cycle [78].

Liepert et al. [79] and Wild et al. [80] also considered that a reduction of $E_g\downarrow$ and related reductions in evaporation rates could be 'spinning down' the hydrological cycle. They argued that reductions in surface solar radiation were only partly offset by enhanced down-welling longwave radiation from the warmer and moister atmosphere and that the radiative imbalance at the surface leads to weaker latent and sensible heat fluxes and hence to reductions in evaporation and precipitation despite global warming. This is in line with experimental evidence of the influence of aerosols on climate [81].

5.4. Daily Temperature Range (DTR)

$E_g\downarrow$ is directly related to maximum mid-day temperatures since it heats the surface. The same factors that reduce $E_g\downarrow$, that is, clouds, haze and aerosols, increase downwelling long-wave radiation at night leading to higher night-time, or minimum daily temperatures. Therefore, it is no surprise that $E_g\downarrow$ is significantly correlated with daily temperature range (DTR, [82]). Various episodes of temperature changes that correspond to sudden changes in atmospheric aerosol loading have been reported. One dramatic demonstration of the influence of aerosol on DTR was shown by Travis et al. [83], who studied climate data for the period of the World Trade Centre tragedy in September 2001. During the three days that air traffic in the US was grounded there were no atmospheric contrails, leading to increased $E_g\downarrow$ and an increase of ~1 °C in DTR. Stanhill and Moreshet [34] found an average 18% increase in $E_g\downarrow$ during Yom Kippur (the Day of Atonement) in Israel, which is a one day Jewish holiday in the fall when industries close and car use is minimal. Analysis of data from 1963–2003 shows that average daily total DTR increased on Yom Kippur by 0.31 °C (Stanhill and Cohen, unpublished data). Robock and Mass [84] and Mass and Robock [85] showed that tropospheric aerosol loading from the 1980 Mt. St. Helens volcanic eruption strongly reduced the diurnal temperature range for several days in the region with the volcanic dust, and surface temperature effects under smoke from forest fires was correlated with a reduction in daytime temperatures [86,87].

Global surface temperatures have been increasing since the beginning of the industrial era. As noted by Roderick and Farquhar [66] minimum temperatures have been increasing faster than maximum temperatures and thus DTR has been decreasing. This may also be related to decreasing surface radiation.

Wild et al. [88] used DTR to analyse the influence of changes in $E_g\downarrow$ on global temperatures. They contend that global dimming masked global warming until the 1980s and that during the global brightening era the accelerating temperature increases demonstrate the full (unmasked) global warming that is caused by greenhouse gases.

5.5. Wind Speed and the Monsoon System

Another mechanism for the influence of changes in $E_g\downarrow$ on climate is sea warming and its influence on wind speed and the monsoon rain system [89]. Xu et al. [90] showed that wind speeds over China have decreased because of dimming. This is related to the increased atmospheric stability caused by aerosol mediated warming of the atmosphere as surface radiation decreases. Thus, aerosols over China changed the land-ocean temperature contrast, affecting monsoon winds.

6. CONCLUSIONS

Global radiation $E_g\downarrow$ decreased significantly (i.e. dimming) from the beginning of widespread measurements in the 1950s to the late 1980s over large parts of the globe and then partly recovered (i.e. brightening) in many places. The areal extent of these changes is not certain because of the large spatial variability, but the mean trends are evident in satellite estimates of global radiation. The trends are apparently caused by anthropogenic aerosols which reduce surface short wave radiation directly and indirectly through their influence on cloud properties. Changes in $E_g\downarrow$ have played a part in regional and global changes in DTR (positively correlated) as well as soil moisture (negatively correlated) and potential evaporation rates (positively correlated), but in some cases potential evaporation has changed due to other factors. Dimming may have offset global warming between the 1950s and 1980s while the more recent brightening may have unmasked the full extent of global warming, as seen in the accelerated temperature increase since the early 1990s.

REFERENCES

1. WMO. Measurement of Radiation. Chapter 7, in: Guide to Meteorological Instruments and Methods of Observation. WMO-No. 8 (Draft Seventh edition). World Meteorological Organization, Geneva, 2006.
2. J.L. Monteith, M.H. Unsworth. Principles of Environmental Physics, 2nd ed., Arnold publishers, London, 1990.
3. P.R. Goode, E. Pallé, J. Atmos. Solar-Terrestrial Phys. 69 (2007) 1556–1568.
4. C. Frohlich, Space Sci. Rev. 125 (2006) 53.
5. R.J. List (Ed.), Smithsonian Meteorological Tables, sixth ed., Smithsonian Institute, Washington, DC, 1966.

6. G.S. Campbell, J.M. Norman, An Introduction to Environmental Biophysics, second ed., Springer-Verlag, NewYork, 1998.
7. G.W. Paltridge, C.M.R. Platt, Radiative Processes in Meteorology and Climatology, Elsevier, New York, 1976.
8. A. Raval, V. Ramanathan, Nature 342 (1989) 758–761.
9. B.A. Wielicki, T. Wong, N. Loeb, P. Minnis, K. Priestley, R. Kandel, Science 308 (2005) 825.
10. V. Ramanathan, iLEAPS Newsletter 5 (2008) 18–20.
11. B.R. Barkstrom, in: Bruce R. Barkstrom (Ed.), Earth radiation budget measurements: pre-ERBE, ERBE, and CERES. Long-Term Monitoring of the Earth's Radiation Budget, Proc. SPIE 1299, pp. 52–60, 1990.
12. W.B. Rossow, Y.-C. Zhang, J. Geophys. Res. 100 (1995) 1167–1197, doi:10.1029/94JD02746.
13. E. Pallé, P.R. Goode, P. Montanes-Rodriguez, S.E. Koonin, Science 304 (2004) 1299–1301.
14. E. Pallé, P. Montañés-Rodriguez, P.R. Goode, S.E. Koonin, M. Wild, S. Casadio, Geophys. Res. Lett., 32, (2005) L21702, doi:10.1029/2005GL023847.
15. P.R. Goode, E. Palle', J. Atmos Solar-Terrestrial Phys., 69 (2007) 1556–1568.
16. A. Ohmura, New radiation and energy balance of the world and its variability, in: H. Fischer, B. Sohn (Eds.), IRS 2004: Current Problems in Atmospheric Radiation, pp. 327–330, 2006.
17. J.T. Kiehl, K.E. Trenberth, Bull. Amer. Meteor. Soc. 78 (1997) 197–208.
18. G. Stanhill, S. Cohen, J. Clim. 18 (2005) 1503–1512.
19. G. Stanhill, S. Cohen, J. Met. Soc. Jpn. 86 (2008) 57–67.
20. G. Stanhill, Weather 58 (2003) 3–11.
21. D.J. Hatch, J. Meteorol. U.K. 6 (1981) 101–113.
22. E. Pallé, C.J. Butler, Int. J. Climatol. 21 (2001) 709–729.
23. G. Stanhill, S. Cohen, Agric. For. Meteorol. 107 (2001) 255–278.
24. A. Ohmura, H. Gilgen, M. Wild, Global Energy Balance Archive GEBA, World Climate Program—Water Project A7, Report 1: Introduction. Zuercher Geografische Schriften Nr. 34, Verlag der Fachvereine, Zurich, 1989.
25. A. Ohmura, H. Gilgen, H. Hegner, G. Muller, M. Wild, E.G. Dutton, B. Forgan, C. Frohlich, R. Philipona, A. Heimo, G. Konig-Langlo, B. McArthur, R. Pinker, C.H. Whitlock, K. Dehne, Bull. Amer. Meteor. Soc. 79 (1998) 2115–2136.
26. M. Budyko, The heat balance of the Earth's surface. (translated by Nina A. Stepnova from: Teplovoi balans zemnoi poverkhnosti; Gidrometeorologicheskoe iz datel'stovo. Leningrad) U.S. Dept. Commerce, Washington, DC, 1956.
27. W.D. Sellers, Physical Climatology, The University of Chicago Press, Chicago, 1965.
28. G. Stanhill, EOS, 80 (1999) 396–397.
29. S. Suraqui, H. Tabor, W.H. Klein, B. Goldberg, Solar Energy 16 (1974) 155–158.
30. A. Ohmura, H. Lang, Secular variation of global radiation over Europe, in: J. Lenoble, J.F. Geleyn (Eds.), Current Problems in Atmospheric Radiation, pp. 98–301, Deepak, Hampton, VA, 1989.
31. V. Russak, Tellus 42 B (1990) 206–210.
32. G. Stanhill, S. Moreshet, Clim. Change 21 (1992) 57–75.
33. G. Stanhill, J.D. Kalma, Austr. Met. Mag. 43 (1994) 81–86.
34. G. Stanhill, S. Moreshet, Clim. Change 26 (1994) 89–103.
35. G. Stanhill, Phil. Trans. R. Soc. A 352 (1995) 247–258.
36. G. Stanhill, S. Cohen, J. Clim. 10 (1997) 2078–2086.
37. G. Stanhill, A. Ianitz, Tellus 49 B (1997) 112–122.

38. G. Stanhill, Int. J. Climatol. 18 (1998) 1015–1030.
39. G. Stanhill, J.D. Kalma, Int. J. Climatol. 15 (1995) 933–941.
40. X. Li, X. Zhou, W. Li, Acta. Met. Sin. 9 (1995) 57–68.
41. G.M. Abakumova, E.M. Feigelson, V. Bussak, V.V. Stadnik, J. Clim. 9 (1996) 1319–1327.
42. B.G. Liepert, Int. J. Climatol. 17 (1997) 1581–1593.
43. B.G. Liepert, G.J. Kukla, J. Clim. 10 (1997) 2391–2400.
44. H. Gilgen, M. Wild, A. Ohmura, J. Clim. 11 (1998) 2042–2061.
45. B.G. Liepert, U. Lohmann, J. Clim. 14 (2001) 1078–1091.
46. M. Wild, H. Gilgen, A. Roesch, A. Ohmura, C.N. Long, E.G. Dutton, B. Forgan, A. Kallis, V. Russak, A. Tsvetkov, Science 308 (2005) 847–850.
47. R.T. Pinker, B. Zhang, E.G. Dutton, Science 308 (2005) 850–854.
48. A. Sanchez-Lorenzo, M. Brunetti, J. Calbó, J. Martin-Vide, J. Geophys. Res. 112 (2007) D20115, doi:10.1029/2007JD008677.
49. V. Ramanathan, P.J. Crutzen, J.T. Kiehl, D. Rosenfeld, Science 294 (2001) 2119.
50. Y. Qian, D.P. Kaiser, L.R. Leung, M. Xu, Geophys. Res. Lett. 33 (2006) L01812, doi:10.1029/2005GL024586.
51. P. Alpert, P. Kishcha, Y.J. Kaufman, R. Schwarzbard, Geophys. Res. Lett. 32 (2005) L17802, doi:10.1029/2005GL023320.
52. J.R. Norris, J. Geophys. Res. 110 (2005) D08206, doi:10.1029/2004JD005600.
53. J.R. Norris, M. Wild, J. Geophys. Res. 112 (2007) D08214, doi:10.1029/2006JD007794.
54. M. Wild, H. Gilgen, A. Roesch, A. Ohmura, C.N. Long, E.G. Dutton, B. Forgan, A. Kallis, V. Russak, A. Tsvetkov, Science 308 (2005) Supporting online material. http://www.sciencemag.org/cgi/data/308/5723/847/DC1/1.
55. B.G. Liepert, I. Tegen. J. Geophys. Res. -Atm. 107 (D12) (2002) 4153–4168.
56. D.G. Streets, Y. Wu, M. Chin, Geophys. Res. Lett. 33 (2006) L15806, doi:10.1029/2006GL026471.
57. S.K. Satheesh, V. Ramanathan, Nature 405 (2000) 60–63.
58. A. Romanou, B. Liepert, G.A. Schmidt, W.B. Rossow, R.A. Ruedy, Y.C. Zhang, Geophys. Res. Lett. 34 (2007) L05713, doi:10.1029/2006GL028356.
59. M.M. Kvalevag, G. Myhre, J. Clim. 20 (2007) 4874–4883.
60. WMO, Measurement of evaporation. Chapter 10. In: Guide to Meteorological Instruments and Methods of Observation. WMO-No. 8 (Draft Seventh edition). World Meteorological Organization, Geneva. 2006.
61. J.L. Monteith. Evaporation and environment, in: G.E. Fogg (Ed.), The State and Movement of Water in Living Organisms, pp. 205–234, Cambridge University Press. Cambridge, U.K., 1965.
62. T.C. Peterson, V.S. Golubev, P.Y. Groisman, Nature 377 (1995) 687–688.
63. V.S. Golubev, J.H. Lawrimore, P. Ya. Groisman, N.A. Speranskaya, S.A. Zhuravin, M.J. Menne, T.C. Peterson, R.W. Malone, Geophys. Res. Lett. 28 (2001) 2665–2668.
64. W. Brutsaert, M.B. Parlange, Nature 396 (1998) 30.
65. S. Cohen, A.Ianitz, G.Stanhill, Agric. For. Meteorol. 111 (2002) 83–91.
66. M.L. Roderick, G.D. Farquhar, Science 298 (2002) 1410–1411.
67. M.L. Roderick, G.D. Farquhar, Int. J. Climatol., 24 (2004) 1077–1090.
68. M.L. Roderick, L.D. Rotstayn, G.D. Farquhar, M.T. Hobbins, Geophys. Res. Lett. 34 (2007) L17403, doi:10.1029/2007GL031166.
69. R.J. Bouchet, Evapotranspiration re'elle evapotranspiration potentielle, signification climatique, Symp. Publ. 62, Int. Assoc. Sci. Hydrol., Berkeley, CA, pp. 134–142, 1963.

70. W. Brutsaert, Evaporation into the Atmosphere: Theory, History and Applications, D. Reidel Publishing Co., Dordrecht, Holland, 1982.
71. S.B. Chen, Y.F. Liu, A. Thomas, Clim. Change 76 (2006) 291–319.
72. Y. Zhang, C. Liu, Y. Tang, Y. Yang, J. Geophys. Res. 112 (2007), D12110, doi:10.1029/2006JD008161.
73. W. Brutsaert, Geophys. Res. Lett. 33 (2006) L20403, doi:10.1029/2006GL027532.
74. A. Robock, M. Mu, K. Vinnikov, I.V. Trofimova, T.I. Adamenko, Geophys. Res. Lett. 32 (2005) L03401 doi:10.1029/2004GL021914.
75. S. Manabe, R.T. Wetherald, J. Atmos. Sci. 44 (1987) 1211–1235.
76. J.M. Gregory, J.F.B. Mitchell, A.J. Brady, J. Clim. 10 (1997) 662–686.
77. A. Robock, H. Li, Geophys. Res. Lett. 33 (2006) L20708, doi:10.1029/2006GL027585.
78. V. Ramanathan, C. Chung, D. Kim, T. Bettge, L. Buja, J.T. Kiehl, W.M. Washington, Q. Fu, D.R. Sikka, M. Wild, Proc. Natl. Acad. Sci. USA 102 (2005) 5326–5333.
79. B.G. Liepert, J. Feichter, U. Lohmann, E. Roeckner, Geophys. Res. Lett. 31 (2004) L06207, doi:10.1029/2003GL019060.
80. M. Wild, A. Ohmura, H. Gilgen, D. Rosenfeld, Geophys. Res. Lett. 31 (2004) L11201, doi:10.1029/2003GL019188.
81. D. Rosenfeld, Science, 287 (2000) 1796–2793.
82. K.L. Bristow, G.S. Campbell, Agric. For. Meteorol. 31 (1984) 159.
83. D.J. Travis, A.M. Carleton, R.G. Lauritsen, Nature 418 (2002) 601.
84. A. Robock, C. Mass, Science 216 (1982) 628–630.
85. C. Mass, A. Robock, Mon. Wea. Rev. 110 (1982) 614–622.
86. A. Robock, Science 242 (1988) 911–913.
87. A. Robock, J. Geophys. Res. 96 (D11) (1991) 20,869–20,878.
88. M. Wild, A. Ohmura, K. Makowski, Geophys. Res. Lett. 34 (2007) L04702, doi:10.1029/2006GL028031.
89. L.D. Rotstayn, U. Lohmann, J. Clim. 15 (2002) 2103–2116.
90. M. Xu, C.P. Chang, C. Fu, Y. Qi, A. Robock, D. Robinson, H. Zhang, J. Geophys. Res. 111 (2006) D24111, doi:10.1029/2006JD007337.
91. S. Cohen, B. Liepert, G. Stanhill, EOS 85 (2004) 362.
92. R. Gifford (Ed.) Pan evaporation: An example of the detection and attribution of trends in climate variables. Australian Academy of Science, National Committee for Earth System Science. Proceedings of a workshop, Canberra, 22–23 November 2004, 2005.
93. G. Ohring, S. Cohen, J. Norris, A. Robock, Y. Rudich, M. Wild, W. Wiscombe, EOS, 89 (2008) 212 and supplemental material at http://www.agu.org/eos_elec/2008/ohring_89_23.html
94. M. Wild, N. Loeb, G. Stanhill, B. Liepert, P. Alpert, J. Calbo, C. Long, G. Ohring, E. Palle, P. Kishcha, Measurements of GDB. Workgroup 1 report at the ISF International Workshop on GDB, Ein Gedi, Israel. 2008.
95. B.G. Liepert, Geophys. Res. Lett. 29 (2002) 1421, doi:10.1029/2002GL014910.
96. E.G. Dutton, D.W. Nelson, R.S. Stone, D. Longenecker, G. Carbaugh, J.M. Harris, J. Wendell, J. Geophys. Res. 111 (2006) D19101, doi:10.1029/2005JD006901.
97. N. Hatzianastassiou, C. Matsoukas, A. Fotiadi, K.G. Pavlakis, E. Drakakis, D. Hatzidimitriou, I. Vardavas, Atmos. Chem. Phys. 5 (2005) 2847–2867.

Chapter 3

The Role of Space Weather and Cosmic Ray Effects in Climate Change

Lev I. Dorman
Head of Cosmic Ray and Space Weather Center with Emilio Sègre Observatory, affiliated to Tel Aviv University, TECHNION and Israel Space Agency, P.O. Box 2217, Qazrin 12900, Israel
Chief Scientist of Cosmic Ray Department of IZMIRAN Russian Academy of Science, Troitsk 142090, Moscow Region, Russia

1. **Introduction**
2. **Solar Activity, Cosmic Rays and Climate Change**
   - 2.1. Long-Term Cosmic Ray Intensity Variations and Climate Change
   - 2.2. The Possible Role of Solar Activity and Solar Irradiance in Climate Change
   - 2.3. Cosmic Rays as an Important Link between Solar Activity and Climate Change
   - 2.4. The Connection between Galactic Cosmic Ray Solar Cycles and the Earth's Cloud Coverage
   - 2.5. The Influence of Cosmic Rays on the Earth's Temperature
   - 2.6. Cosmic Ray Influence on Weather during Maunder Minimum
   - 2.7. The Influence of Long-Term Variations of Cosmic Ray Intensity on Wheat Prices (Related to Climate Change) in Medieval England and Modern USA
   - 2.8. The Connection between Ion Generation in the Atmosphere by Cosmic Rays and Total Surface of Clouds
   - 2.9. The Influence of Big Magnetic Storms (Forbush Decreases) and Solar Cosmic Ray Events on Rainfall
   - 2.10. The Influence of Geomagnetic Disturbances and Solar Activity on the Climate through Energetic Particle Precipitation from Inner Radiation Belt

2.11. On the Possible Influence of Galactic Cosmic Rays on Formation of Cirrus Hole and Global Warming
2.12. Description of Long-Term Galactic Cosmic Ray Variation by both Convection-Diffusion and Drift Mechanisms with Possibility of Forecasting of Some Part of Climate Change in Near Future Caused by Cosmic Rays
2.13. Influence of Long-Term Variation of Main Geomagnetic Field on Global Climate Change through Cosmic Ray Cutoff Rigidity Variation
2.14. Atmospheric Ionisation by Cosmic Rays: The Altitude Dependence and Planetary Distribution
2.15. Project 'Cloud' as an Important Step in Understanding the Link between Cosmic Rays and Cloud Formation
3. The Influence on the Earth's Climate of the Solar System Moving Around the Galactic Centre and Crossing Galaxy Arms
4. The Influence of Molecular-dust Galactic Clouds on the Earth's Climate
5. The Influence of Interplanetary Dust Sources on the Earth's Climate
6. Space Factors and Global Warming
7. The Influence of Asteroids on the Earth's Climate
8. The Influence of Nearby Supernova on the Earth's Climate
9. Discussion and Conclusions
Acknowledgments
References

1. INTRODUCTION

There are a number of space phenomena that influence the Earth's climate and determined its long-term and short-term changes. These include:

- the variability of the Sun's irradiation flux energy;
- the variations of the Earth's orbital characteristics;
- the variable solar activity (with periods of 8–15 a (year), average period of about 11 a), general solar magnetic field (average period of 22 a) together with the related phenomena of variable solar wind, coronal mass ejections and shocks in the Heliosphere and modulated galactic cosmic rays (CR) – see Section 2;
- the solar CR generated during great solar flares – see Section 2.9;
- the precipitation of energetic electrons and protons from the Earth's magnetosphere during magnetic disturbances – see Section 2.10;
- the variable Earth's magnetic field's influence on CR cutoff rigidity and changed galactic and solar cosmic ray intensity in the Earth's atmosphere – see Section 2.12;

- the moving of the solar system around the galactic centre and crossing the Galaxy arms – see Section 3;
- the impacts of the solar system with galactic molecular dust cloud – see Section 4;
- the impacts of the solar system with interplanetary zodiac dust cloud – see Section 5;
- asteroid impacts – see Section 7;
- nearby supernova explosions – see Section 8.

The first phenomenon is the subject of Chapter 2 by Shabtai Cohen, and the second is dealt with by Lucas Lourens in Chapter 5. In this Chapter the other phenomena are discussed and compared to anthropogenic induced changes. Details on CR behaviour in the Earth's atmosphere, magnetosphere and in space are the subject of recent publications by the author [1–3]. The role of these factors in our present climate change will be discussed in the final section of this chapter.

2. SOLAR ACTIVITY, COSMIC RAYS AND CLIMATE CHANGE

2.1. Long-Term Cosmic Ray Intensity Variations and Climate Change

About 200 a ago the famous astronomer William Herschel [4] suggested that the price of wheat in England was directly related to the number of sunspots. He noticed that less rain fell when the number of sunspots was small (Joseph in the Bible, recognised a similar periodicity in food production in Egypt, about 4000 a ago). The solar activity level is known from direct observations over the past 450 a, and from data of cosmogenic nuclides (through CR intensity variations) for more than 10 000 a [1,5]. Over this period there is a striking qualitative correlation between cold and warm climate periods and high and low levels of galactic CR intensity (low and high solar activity). As an example, Fig. 1 shows the change in the concentration of radiocarbon during the last millennium (a higher concentration of ^{14}C corresponds to a higher intensity of galactic CR and to lower solar activity). It can be seen from Fig. 1 that during 1000–1300 AD the CR intensity was low and solar activity high, which coincided with the warm medieval period (during this period Vikings settled in Greenland). After 1300 AD solar activity decreased and CR intensity increased, and a long cold period followed (the so-called Little Ice Age, which included the Maunder minimum 1645–1715 AD and lasted until the middle of nineteenth century).

2.2. The Possible Role of Solar Activity and Solar Irradiance in Climate Change

Friis-Christiansen and Lassen [7,8] found, from 400 a of data, that the filtered solar activity cycle length is closely connected to variations of the average surface temperature in the northern hemisphere. Labitzke and Van Loon [9]

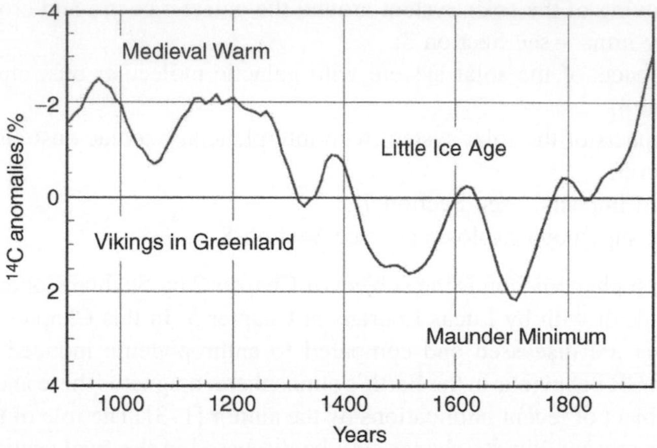

FIGURE 1 The change of CR intensity reflected in radiocarbon concentration during the last millennium. The Maunder minimum refers to the period 1645–1715, when sun spots were rare. From Ref. [6].

showed, from solar cycle data, that the air temperature increases with increasing levels of solar activity. Swensmark [6] also discussed the problem of the possible influence of solar activity on the Earth's climate through changes in solar irradiance. But the direct satellite measurements of the solar irradiance during the last two solar cycles showed that the variations during a solar cycle was only about 0.1%, corresponding to about 0.3 $W \cdot m^{-2}$. This value is too small to explain the present observed climate changes [10]). Much bigger changes during a solar cycle occur in UV radiation (about 10%, which is important in the formation of the ozone layer). High [11] and Shindell et al. [12] suggested that the heating of the stratosphere by UV radiation can be dynamically transported into the troposphere. This effect might be responsible for small contributions towards 11 and 22 a cycle modulation of climate but not to the 100 a of climate change that we are presently experiencing.

2.3. Cosmic Rays as an Important Link between Solar Activity and Climate Change

Many authors have considered the influence of galactic and solar CR on the Earth's climate. Cosmic Radiation is the main source of air ionisation below 40–35 km (only near the ground level, lower than 1 km, are radioactive gases from the soil also important in air ionisation) [1]. The first to suggest a possible influence of air ionisation by CR on the climate was Ney [13]. Swensmark [6] noted that the variation in air ionisation caused by CR could potentially influence the optical transparency of the atmosphere, by either a change in aerosol formation or influence the transition between the different phases of water.

Many other authors considered these possibilities [13–22]. The possible statistical connections between the solar activity cycle and the corresponding long-term CR intensity variations with characteristics of climate change were considered in Dorman et al. [23–25]. Dorman et al. [26] reconstructed CR intensity variations over the last 400 a on the basis of solar activity data and compared the results with radiocarbon and climate change data.

Cosmic radiation plays a key role in the formation of thunderstorms and lightnings [1]. Many authors [27–32] have considered atmospheric electric field phenomena as a possible link between solar activity and the Earth's climate. Also important in the relationship between CR and climate, is the influence of long-term changes in the geomagnetic field on CR intensity through the changes of cutoff rigidity [2]. One can consider the general hierarchical relationship to be: (solar activity cycles + long-term changes in the geomagnetic field) → (CR long-term modulation in the Heliosphere + long-term variation of cutoff rigidity) → (long-term variation of clouds covering + atmospheric electric field effects) → climate change.

2.4. The Connection between Galactic Cosmic Ray Solar Cycles and the Earth's Cloud Coverage

Recent research has shown that the Earth's cloud coverage (observed by satellites) is strongly influenced by CR intensity [6,18,20–22]. Clouds influence the irradiative properties of the atmosphere by both cooling through reflection of incoming short wave solar radiation, and heating through trapping of outgoing long wave radiation (the greenhouse effect). The overall result depends largely on the height of the clouds. According to Hartmann [33], high optically thin clouds tend to heat while low optically thick clouds tend to cool (see Table 1).

From Table 1 it can be seen that low clouds result in a cooling effect of about 17 $W{\cdot}m^{-2}$, which means that they play an important role in the Earth's radiation budget [34–36]). The important issue is that even small changes in the lower cloud coverage can result in important changes in the radiation budget and hence has a considerable influence on the Earth's climate (let us remember that the solar irradiance changes during solar cycles is only about 0.3 $W{\cdot}m^{-2}$).

Figure 2 shows a comparison of the Earth's total cloud coverage (from satellite observations) with CR intensities (from the Climax neutron monitor (NM)) and solar activity data over 20 a.

From Fig. 2 it can be seen that the correlation of global cloud coverage with CR intensity is much better than with solar activity. Marsh and Swensmark [21] came to conclusion that CR intensity relates well with low global cloud coverage, but not with high and middle clouds (see Fig. 3).

It is important to note that low clouds lead, as rule, to the cooling of the atmosphere. It means that with increasing CR intensity and cloud coverage

TABLE 1 Global annual mean forcing due to various types of clouds, from the Earth Radiation Budget Experiment (ERBE), according to Hartmann [33]

Parameter	High clouds Thin	High clouds Thick	Middle clouds Thin	Middle clouds Thick	Low clouds All	Total
Global fraction /(%)	10.1	8.6	10.7	7.3	26.6	63.3
Forcing (relative to clear sky):						
Albedo (SW radiation)/($W{\cdot}m^{-2}$)	−4.1	−15.6	−3.7	−9.9	−20.2	−53.5
Outgoing LW radiation /($W{\cdot}m^{-2}$)	6.5	8.6	4.8	2.4	3.5	25.8
Net forcing /($W{\cdot}m^{-2}$)	2.4	−7.0	1.1	−7.5	−16.7	−27.7

The positive forcing increases the net radiation budget of the Earth and leads to a warming; negative forcing decreases the net radiation and causes a cooling. (Note that the global fraction implies that 36.7% of the Earth is cloud free.)

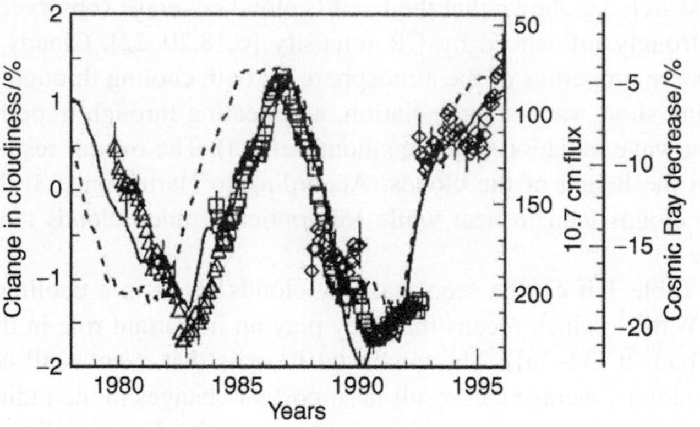

FIGURE 2 Changes in the Earth's cloud coverage: triangles – from satellite Nimbus 7, CMATRIX project [37]; squares – from the International Satellite Cloud Climatology Project, ISCCP [38]; diamonds – from the Defence Meteorological Satellite Program, DMSP [39,40]. Solid curve – CR intensity variation according to Climax NM, normalized to May 1965. Broken curve – solar radio flux at 10.7 cm. All data are smoothed using twelve months running mean. From Ref. [6].

(see Fig. 2), we can expect the surface temperature to decrease. It is in good agreement with the situation shown in Fig. 1 for the last 1000 a, and with direct measurements of the surface temperature over the last four solar cycles (see Section 2.5, below).

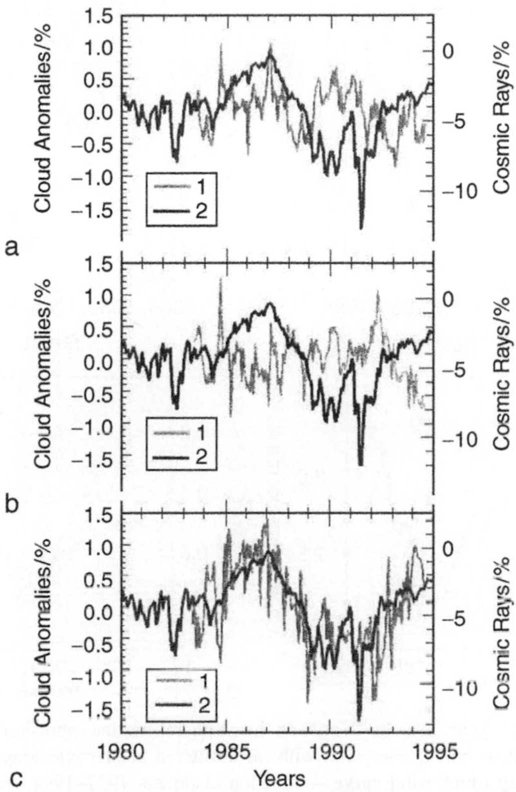

FIGURE 3 CR intensity obtained at the Huancayo/Haleakala NM (normalised to October 1965, curve 2) in comparison with global average monthly cloud coverage anomalies (curves 1) at heights, H, for: a – high clouds, $H > 6.5$ km, b – middle clouds, 6.5 km $> H > 3.2$ km, and c – low clouds, $H < 3.2$ km. From Ref. [21].

2.5. The Influence of Cosmic Rays on the Earth's Temperature

Figure 4 shows a comparison of 11 year moving average Northern Hemisphere marine and land air temperature anomalies for 1935–1995 with CR intensity (constructed for Cheltenham/Fredericksburg for 1937–1975 and Yakutsk for 1953–1994, [41]) and Climax NM data, as well as with other parameters (unfiltered solar cycle length, sunspot numbers and reconstructed solar irradiance).

From Fig. 4 one can see that the best correlation of global air temperature is with CR intensity, in accordance with the results described in Sections 2.1–2.4 above. According to Swensmark [6], the comparison of Fig. 4 with Fig. 2 shows that the increase of air temperature by 0.3 °C corresponds to a decrease of CR intensity of 3.5% and a decrease of global cloudiness of 3%; this is equivalent to an increase of solar irradiance on the Earth's surface of about 1.5 W·m^{-2} [42] and is about 5 times bigger than the solar cycle change of solar irradiance, which as we have seen, is only 0.3 W·m^{-2}).

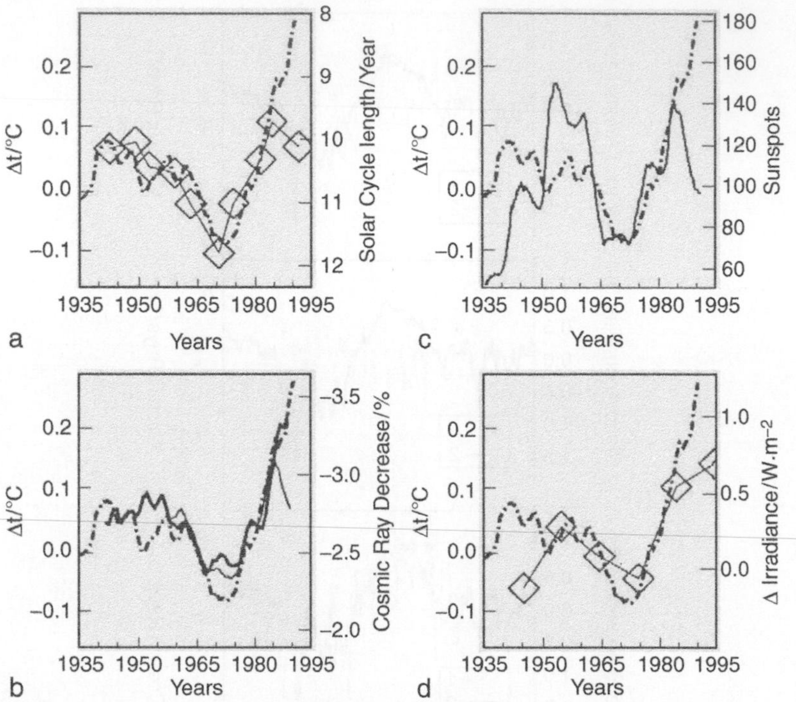

FIGURE 4 Eleven year average Northern hemisphere marine and land air temperature anomalies, Δt, (broken curve) compared with: a, unfiltered solar cycle length; b, Eleven year average CR intensity (thick solid curve – from ion chambers 1937–1994, normalized to 1965, and thin solid curve – from Climax NM, normalized to ion chambers); c, eleven-year average of sunspot numbers; and d, decade variation in reconstructed solar irradiance from Ref. [10] (zero level corresponds to 1367 $W{\cdot}m^{-2}$). From Ref. [6].

2.6. Cosmic Ray Influence on Weather during Maunder Minimum

Figure 5 shows the situation in the Maunder minimum (a time when sunspots were rare) for: solar irradiance [10,43]); concentration of the cosmogenic isotope ^{10}Be [44] – a measure of CR intensity [1]); and reconstructed air surface temperature for the northern hemisphere [45]).

The solar irradiance is almost constant during the Maunder minimum and about 0.24% (or about 0.82 $W{\cdot}m^{-2}$) lower than the present value (see Panel a in Fig. 5), but CR intensity and air surface temperature vary in a similar manner – see above sections; with increasing CR intensity there is a decrease in air surface temperature (see Panels b and c in Fig. 5). The highest level of CR intensity was between 1690–1700, which corresponds to the minimum of air surface temperature [46] and also to the coldest decade (1690–1700).

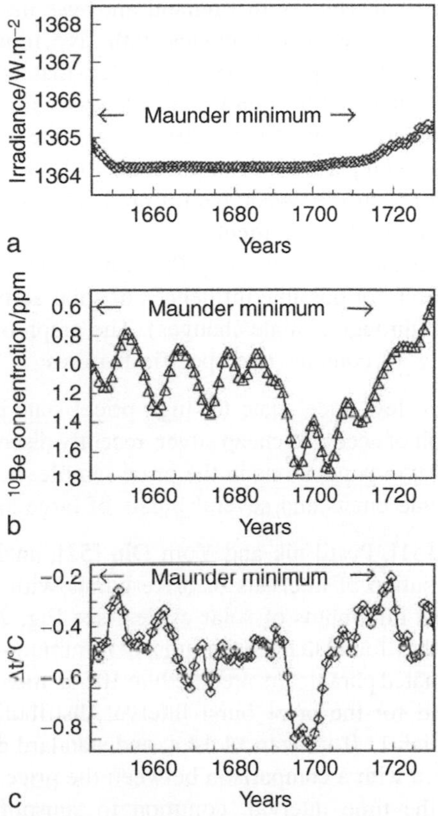

FIGURE 5 Situation in the Maunder minimum: a – reconstructed solar irradiance [9]; b – cosmogenic ^{10}Be concentration [44]; c – reconstructed relative change of air surface temperature, Δt, for the northern hemisphere [45]. From Swensmark [6].

2.7. The Influence of Long-Term Variations of Cosmic Ray Intensity on Wheat Prices (Related to Climate Change) in Medieval England and Modern USA

Herschel's observations [4] mentioned in Section 2.1, were based on the published wheat prices [47], and showed that five prolonged periods of sunspot numbers correlated with costly wheat. This idea was taken up by the English economist and logician William Stanley Jevons [48]. He directed his attention to the wheat prices from 1259 to 1400 and showed that the time intervals between high prices were close to 10–11 a. This work was later published by Rogers [49]. The coincidence of these intervals with the period of the recently discovered 11 year cycle of solar activity led him to suggest that the solar activity cycle was a 'synchronisation' factor in the fluctuations of wheat prices (Jevons [50]). As a next step, he extrapolated his theory to stock markets of the nineteenth century in England and was impressed by a close coincidence of five stock exchange panics with five minima in solar spot numbers that preceded these panics. He suggested that both solar and economic activities are subjected to a harmonic process with the same constant period of 11 a. However, the subsequent discovery of the non-harmonic behaviour of solar cycles, with periods varying from 8 to 15 a, and the later observation of lack of coincidence between panics predicted by Jevons [48,50] and the actual ones, destroyed his argument.

The Rogers [49] database was used by Pustil'nik et al. [51], Pustil'nik and Yom Din [52] to search for possible influences of solar activity and CR intensity on wheat prices (through climate changes). The graph of wheat prices as a function of time (Fig. 6) contains two specific features:

1. A transition from 'low price' state to 'high price' state during 1530–1630, possibly as a result of access to cheap silver, recently discovered New World.
2. The existence of two populations in the price sample: noise-like variations with low amplitude bursts and several bursts of large amplitude.

Pustil'nik et al. [51], Pustil'nik and Yom Din [52], analysed the data and compared the distribution of intervals of price bursts with the distribution of the intervals between minimums of solar cycles (see Fig. 7).

In their analysis they found that for the sunspot minimum–minimum interval distribution the estimated parameters are: median 10.7 a; mean 11.02 a; standard deviation 1.53 a and for the price burst interval distribution, the estimated parameters are: median 11.0 a; mean 11.14 a; and standard deviation 1.44 a.

The main problem with a comparison between the price and solar activity, is the absence of the time interval, common to sunspot observation data

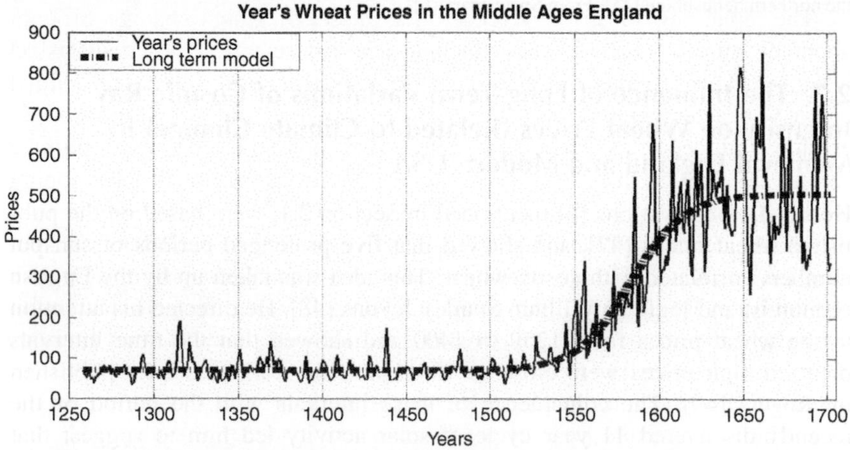

FIGURE 6 Wheat prices in England during 1259–1702 with a price transition at 1530–1630. From Refs. [51,52].

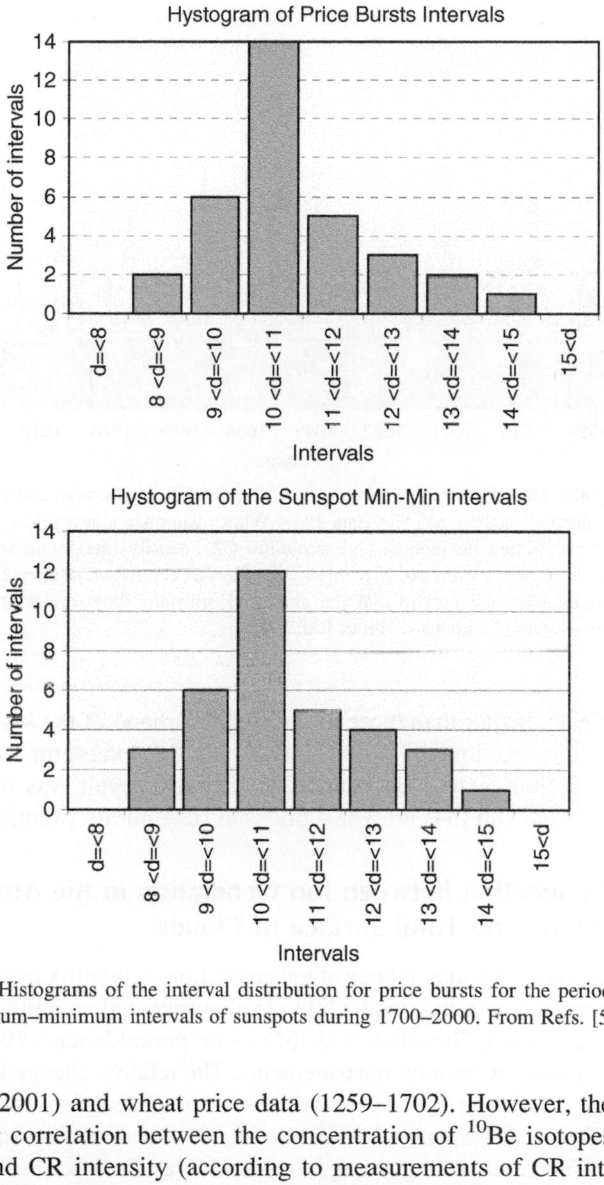

FIGURE 7 Histograms of the interval distribution for price bursts for the period, 1249–1702, and of minimum–minimum intervals of sunspots during 1700–2000. From Refs. [51,52].

(for 1700–2001) and wheat price data (1259–1702). However, the discovery of a strong correlation between the concentration of ^{10}Be isotopes in Greenland ice and CR intensity (according to measurements of CR intensity over the last 60 a [1]) sheds a new light on the problem. In Fig. 8, the wheat prices for 1600–1702 are shown and compared to ^{10}Be data [53]. White marks show prices, averaged for three-year intervals centred on moments of minimum CR intensity. Black marks correspond to average prices in three-year intervals for maximum CR intensities.

As can be seen from Fig. 8, all prices in the neighbourhoods of the seven maxima of CR intensity (correspond approximately to minima of solar activity)

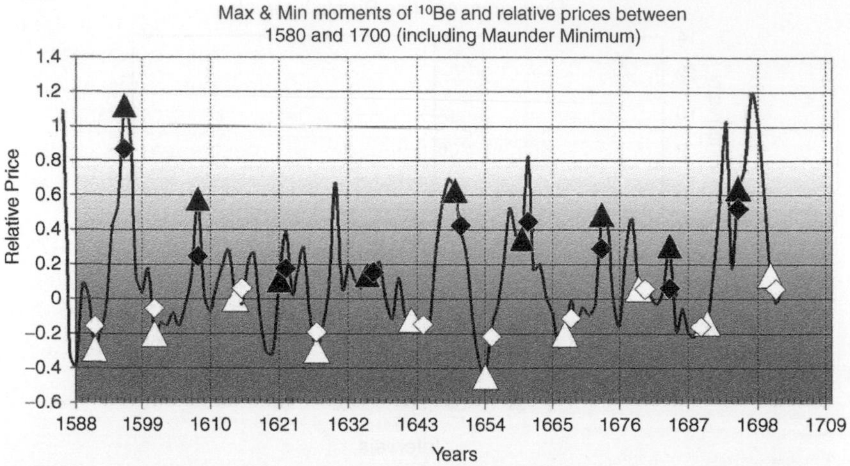

FIGURE 8 Systematic differences in wheat prices at moments of minimum and maximum CR intensity determined according to ^{10}Be data [53]. White diamonds show prices averaged for three-year intervals centred on moments of minimum CR intensity (maximum solar activity); black rectangles show prices averaged over three-year intervals centred on moments of maximum CR intensity (minimum solar activity). White and black triangles show prices at moments of minimum and maximum CR intensity. From Ref. [52].

are systematically higher than those in the neighbourhood of the seven minima of CR intensity (maxima of solar activity) in the long-term variation of CR intensity according to ^{10}Be data [53]. A similar result was obtained by Pustil'nik and Yom Din [54] for wheat prices in USA during twentieth century.

2.8. The Connection between Ion Generation in the Atmosphere by Cosmic Rays and Total Surface of Clouds

The time variation of the integral rate of ion generation, q, (approximately proportional to CR intensity) in the middle latitude atmosphere at an altitude between 2 and 5 km was found by Stozhkov et al. [55] for the period January 1984–August 1990 using regular CR balloon measurements. The relative change in q, $\Delta q/q$, have been compared with the relative changes of the total surface of clouds over the Atlantic Ocean, $\Delta S/S$, and are shown in Fig. 9: the correlation coefficient is 0.91 ± 0.04. This result is in good agreement with results described above (see Panel b in Fig. 4 and Panel c in Fig. 5) and shows that there is a direct correlation between cloud cover and CR generated ions.

2.9. The Influence of Big Magnetic Storms (Forbush Decreases) and Solar Cosmic Ray Events on Rainfall

A decrease of atmospheric ionisation leads to a decrease in the concentration of charge condensation centres. In these periods, a decrease of total cloudiness

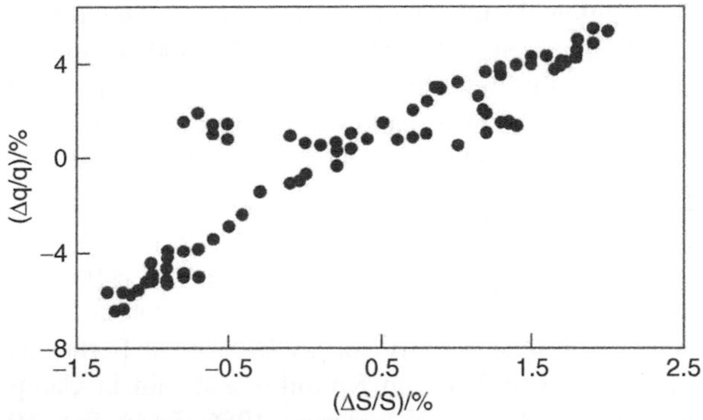

FIGURE 9 The positive relationship between the relative changes of total clouds covering surface over Atlantic Ocean, $\Delta S/S$, in the period January 1984–August 1990 [19] and the relative changes of integral rate of ion generation $\Delta q/q$ in the middle latitude atmosphere in the altitude interval 2–5 km. From Ref. [55].

and atmosphere turbulence together with an increase in isobaric levels is observed [56]). As a result, a decrease of rainfall is also expected. Stozhkov et al. [57–59] and Stozhkov [60] analysed 70 events of Forbush decreases (defined as a rapid decrease in observed galactic CR intensity, and caused by big geomagnetic storms) observed in 1956–1993 and compared these events with rainfall data over the former USSR. It was found that during the main phase of the Forbush decrease, the daily rainfall levels decreases by about 17%. Similarly, Todd and Kniveton [61,62] investigating 32 Forbush decreases events over the period 1983–2000, found reduced cloud cover of 18% [61] and 12% [62].

During big solar CR events, when CR intensity and ionisation in the atmosphere significantly increases, an inverse situation is expected and the increase in cloudiness leads to an increase in rainfall. A study [57–60] involving 53 events of solar CR enhancements, between 1942–1993, showed a positive increase of about 13% in the total rainfall over the former USSR.

2.10. The Influence of Geomagnetic Disturbances and Solar Activity on the Climate through Energetic Particle Precipitation from Inner Radiation Belt

The relationship between solar and geomagnetic activity and climate parameters (cloudiness, temperature, rainfall, etc.) was considered above and is the subject of much ongoing research. The clearly pronounced relationship observed at high and middle latitudes, is explained by the decrease of galactic CR intensity (energies in the range of MeV and GeV) with increasing solar and geomagnetic activity, and by the appearance of solar CR fluxes ionising the atmosphere [63]). This mechanism works efficiently at high latitudes, because CR particles with energy up to 1 GeV penetrate this region more easily due to its very low cutoff rigidity. Near the equator, in the Brazilian Magnetic

Anomaly (BMA) region, the main part of galactic and solar CR is shielded by a geomagnetic field. This field is at an altitude of 200–300 km and contains large fluxes of energetic protons and electrons trapped in the inner radiation belt. Significant magnetic disturbances can produce precipitation of these particles and subsequent ionisation of the atmosphere. The influence of solar-terrestrial connections on climate in the BMA region was studied by Pugacheva et al. [64]. Two types of correlations were observed: (1) a significant short and long time scale correlation between the index of geomagnetic activity Kp and rainfall in Sao Paulo State; (2) the correlation-anti-correlation of rainfalls with the 11 and 22 a cycles of solar activity for 1860–1990 in Fortaleza. Figure 10 shows the time relationship between Kp-index and rain in Campinas (23°S, 47°W) and in Ubajara (3°S, 41°W), during 1986. From Fig. 10, it can be seen that, with a delay of 5–11 days, almost every significant (>3.0) increase of the Kp-index is accompanied by an increase in rainfall. The effect is most noticeable at the time of the great geomagnetic storm of 8 February 1986, when the electron fluxes of inner radiation belt reached the atmosphere between 18 and 21 February [65]) and the greatest rainfall of the 1986 was recorded on 19 February. Again, after a series of solar flares, great magnetic disturbances were registered between 19 and 22 March 1991. On 22 March, a Sao Paolo station showed the greatest rainfall of the year.

The relationship between long-term variations of annual rainfall at Campinas, the Kp-index and sunspot numbers are shown in Figs. 11 and 12.

Figures 11 and 12 show the double peak structure of rainfall variation compared to the Kp-index. Only during the 20th solar cycle (1964–1975), weakest of the shown 6 cycles, an anti-correlation between rainfalls and sunspot numbers is observed in most of Brazil. The Kp – rainfall correlation is more pronounced in the regions connected with magnetic lines occupied by trapped particles.

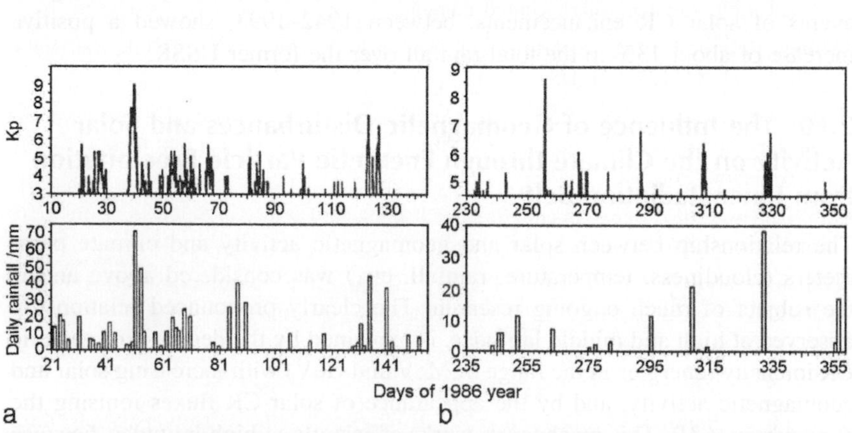

FIGURE 10 The Kp-index of geomagnetic activity (top panels) and rainfall level (bottom panels) in Campinas (left panels a) and in Ubajara (right panels b) in 1986. According to Pugacheva et al. [64].

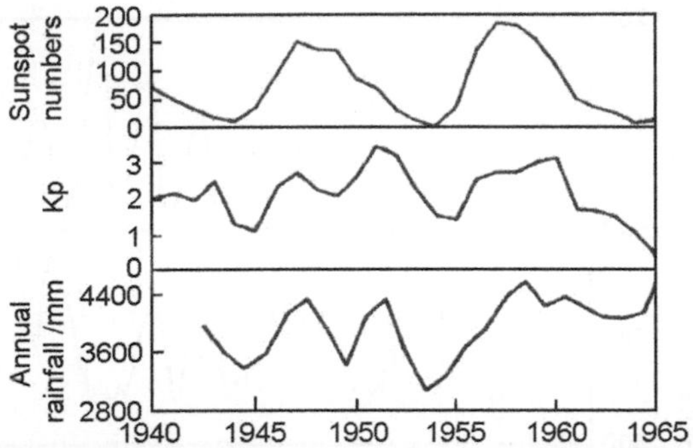

FIGURE 11 Long-term variations of rainfalls (Campinas, the bottom panel) in comparison with variations of solar and geomagnetic activity (the top and middle panels, respectively) for 1940–1965. From Ref. [64].

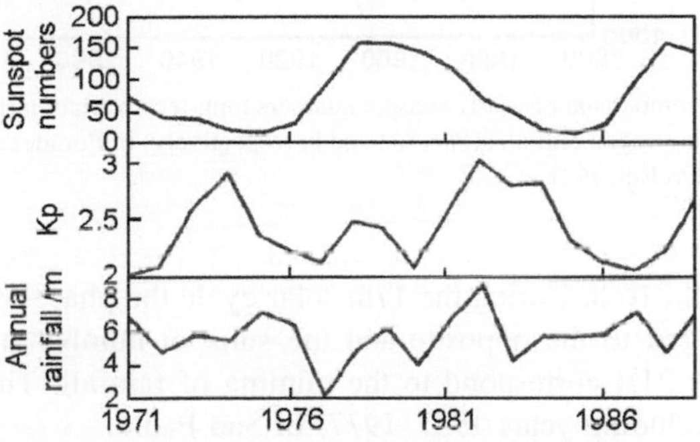

FIGURE 12 The same as in Fig. 11, but for 1971–1990. From Ref. [64].

In Fortaleza (4°S, 39°W), located in an empty magnetic tube ($L = 1.054$), it is the other kind of correlation (see Fig. 13).

From Fig. 13 it can be seen that a correlation exists between sunspot numbers and rainfall between 1860–1900 (11th–13th solar cycles) and 1933–1954 (17th and 18th cycles). The anti-correlation was observed during 1900–1933 (cycles 14th–16th) and during 1954–1990 (cycles 19th–21th). As far as sunspot numbers mainly anti-correlate with the galactic CR flux, an anti-correlation of sunspot numbers with rainfalls could be interpreted as a correlation of rainfalls with the CR. The positive and negative phases of the correlation interchange several times during the long time interval 1860–1990, that was observed earlier in North America (King [66]). Some climate events have a 22 a periodicity similar to the 22 a solar magnetic cycle. Panel b in Fig. 13 demonstrate 22 a periodicity of 11 a running averaged rainfalls in Fortaleza. The phenomenon is observed during 5 periods from 1860 to 1990. During the 11th–16th solar cycles (from 1860 until 1930), the maxima of rainfalls correspond to the maxima of sunspot numbers of odd solar cycles 11th, 13th, 15th and minima of rainfalls correspond to maxima of even solar

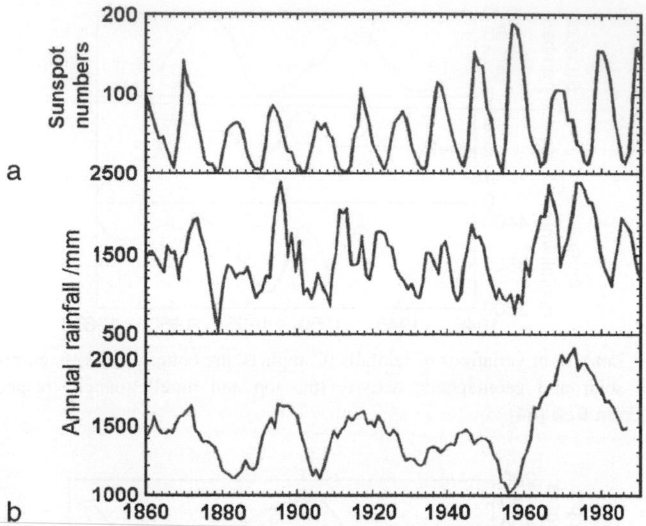

FIGURE 13 The comparison of yearly sunspot numbers long-term variation (the top panel) with 3 and 11 a running averaged rainfalls (Panels a and b, respectively) in Fortaleza (4°S, 39°W) during 1860–1990. From Ref. [64].

cycles 12th, 14th, 16th. During the 17th solar cycle the phase of the 22 a periodicity is changed to the opposite and the sunspot number maxima of odd cycles 19th and 21st correspond to the minima of rainfall. The effect is not pronounced (excluding years 1957–1977) in Sao Paolo.

The difference in results obtained in Refs. [60–62, 64] can be easily understood if we take into account the large value of the cutoff rigidity in the BMA region. This is the reason why the variations in galactic and solar CR intensity in the BMA region, are not reflected in the ionisation of the air and hence do not influenced the climate. However, in the BMA region other mechanism of solar and magnetic activity can influence climatic parameters such as energetic particle precipitation coming from the inner radiation belt.

2.11. On the Possible Influence of Galactic Cosmic Rays on Formation of Cirrus Hole and Global Warming

According to Ely and Huang [67] and Ely et al. [68], there are expected variations of upper tropospheric ionisation caused by long-term variations of galactic CR intensity. These variations have resulted in the formation of the cirrus hole (a strong latitude dependent modulation of cirrus clouds). The upper tropospheric ionisation is caused, largely, by particles with energy smaller than 1 GeV but bigger than about 500 MeV. In Fig. 14 is shown the long-term modulation of the difference between Mt. Washington and Durham for protons with kinetic energy 650–850 MeV.

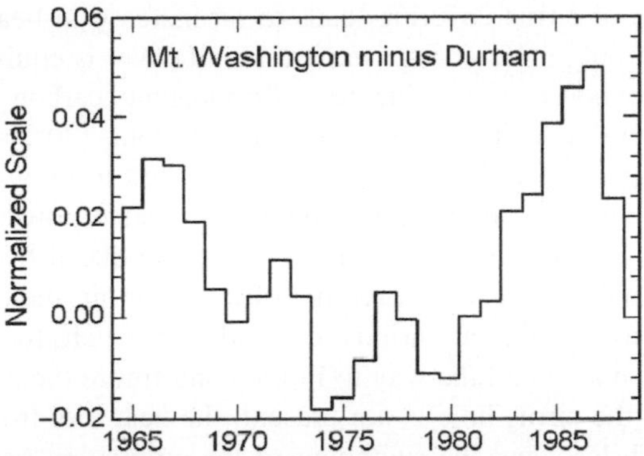

FIGURE 14 The observed 22 a modulation of galactic CR between 1.24 GV and 1.41 GV rigidity (i.e. protons with kinetic energy between 650–850 MeV, ionising heavily in the layer 200–300 g/cm^2). From Ely et al. [68].

Figure 14 clearly shows the 22 a modulation of galactic CR intensity in the range 650–850 MeV with an amplitude of more than 3%. Variations of upper tropospheric ionisation do have some influence on the cirrus covering and the 'cirrus hole' is expected to correspond to a decrease in CR intensity.

According to Ely et al. [68], the 'cirrus hole' was observed in different latitude zones over the whole world between 1962 and 1971, centred at 1966 (see Fig. 15).

Figure 15 gives the cirrus cloud cover data over a 25 a period, for the whole world, the equatorial zone (30°S–30°N) and the northern zone (30°N–90°N), showing fractional decreases in cirrus coverage of 7%, 4% and 17%, respectively.

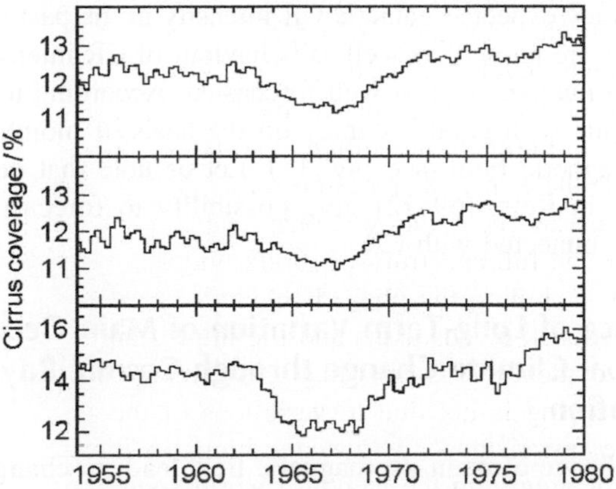

FIGURE 15 The 'cirrus hole' of the 1960s for: the whole world (the top panel); the equatorial zone (30°S–30°N; middle panel); the northern zone (bottom panel) From Ely et al. [68].

The decrease of cirrus covering leads to an increase in heat loss to outer space (note, that only a 4% change in total cloud cover is equivalent to twice the present greenhouse effect due to anthropogenic carbon dioxide). The influence of cirrus hole in the northern latitude zone (30°N–90°N), where the cirrus covering was reduced by 17%, is expected to be great (this effect of the cirrus hole is reduced in summer by the increase of lower clouds resulting in enhanced insulation) The low temperatures produced from mid to high latitude significantly increase the pressure of the polar air mass and cause frequent 'polar break troughs' at various longitudes in which, for example, cold air from Canada may go all the way to Florida and freeze the grapefruit [68]). However, when the cirrus hole is not present, the heat loss from mid to high latitudes is much less, and the switching of the circulation patterns (Rossby waves) is much less frequent.

2.12. Description of Long-Term Galactic Cosmic Ray Variation by both Convection-Diffusion and Drift Mechanisms with Possibility of Forecasting of Some Part of Climate Change in Near Future Caused by Cosmic Rays

It was shown in previous Sections that CR may be considered as sufficient links determined some part of space weather influence on the climate change. From this point of view it is important to understand mechanisms of galactic CR long-term variations and on this basis to forecast expected CR intensity in near future. In Dorman [69–71] it was made on basis of monthly sunspot numbers with taking into account time-lag between processes on the Sun and situation in the interplanetary space as well as the sign of general magnetic field (see Fig. 16); in Belov et al. [72] – mainly on basis of monthly data of solar general magnetic field (see Fig. 17). From Fig. 16 follows that in the frame of used in [69–71] convection-diffusion and drift models can be determined with very good accuracy expected galactic CR intensity in the past (when monthly sunspot numbers are known) as well as behaviour of CR intensity in future if monthly sunspot numbers can be well forecasted. According to Ref. [72], the same can be made with good accuracy on the basis of monthly data on the solar general magnetic field (see Fig. 17). Let us note that described above results obtained in Refs. [69–72] give possibility to forecast some part of climate change connected with CR.

2.13. Influence of Long-Term Variation of Main Geomagnetic Field on Global Climate Change through Cosmic Ray Cutoff Rigidity Variation

The sufficient change of main geomagnetic field leads to change of planetary distribution of cutoff rigidities R_c and to corresponding change of the i-th component of CR intensity $N_i(R_c, h_o)$ at some level h_o in the Earth's

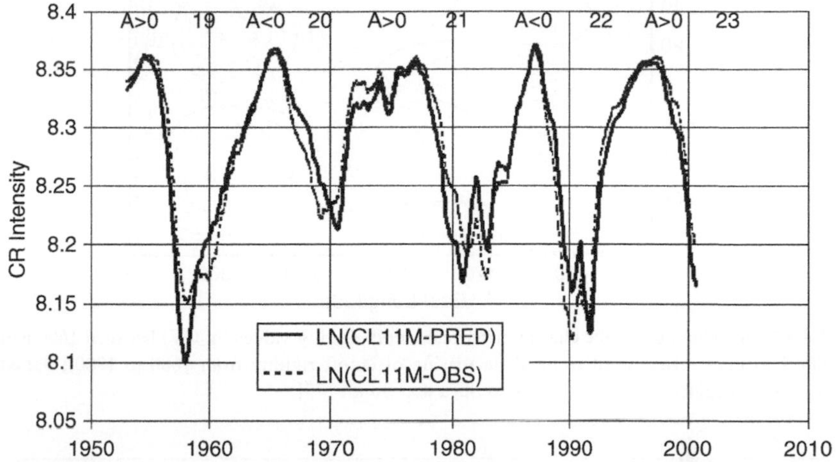

FIGURE 16 Comparison of observed by Climax neutron monitor CR intensity averaging with moving period eleven month LN(CL11M-OBS) with predicted on the basis of monthly sunspot numbers from model of convection-diffusion modulation, corrected on drift effects LN(CL11M-PRED). Correlation coefficient between both curves 0.97. From Dorman [71].

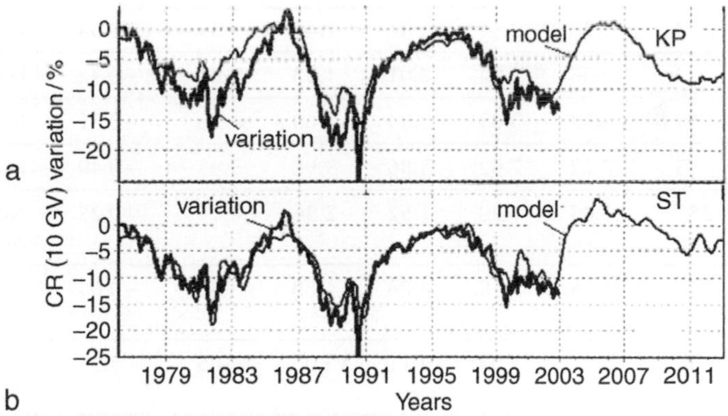

FIGURE 17 The forecast of galactic CR behaviour based on the predicted values of the global characteristics of the solar magnetic field, thick line – data of CR intensity observations (Moscow neutron monitor), thin line – the predicted CR variation up to 2013 based on data of Kitt Peak Observatory (upper panel) and based on data of Stanford Observatory (bottom panel). From Belov et al. [72].

atmosphere $\Delta N_i(R_c, h_o)/N_{io} = -\Delta R_c W_i(R_c, h_o)$, where $W_i(R_c, h_o)$ is the coupling function (see details in Chapter 3 of Ref. [1]). Variations of CR intensity caused by change of R_c are described in detail in Ref. [2], and here we will demonstrate results of Shea and Smart [73] on R_c changing for the last 300 and 400 a (see Fig. 18 and Table 2, correspondingly).

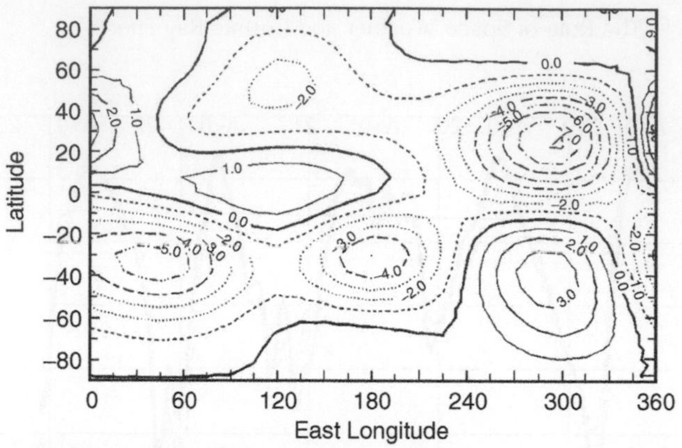

FIGURE 18 Contours of the change in vertical cutoff rigidity values (in GV) between 1600 and 1900. Full lines reflect positive trend (increasing of cutoff rigidity from 1600 to 1900); dotted lines reflect negative trend. According to Shea and Smart [73].

TABLE 2 Vertical cutoff rigidities (in GV) for various epochs 1600, 1700, 1800, 1900 and 2000, as well as change from 1900 to 2000 owed to changes of geomagnetic field. According to Shea and Smart [73]

Lat.	Long. (E)	Epoch 2000	Epoch 1900	Epoch 1800	Epoch 1700	Epoch 1600	Change 1900–2000	Region
55	30	2.30	2.84	2.31	1.49	1.31	−0.54	Europe
50	0	3.36	2.94	2.01	1.33	1.81	+0.42	Europe
50	15	3.52	3.83	2.85	1.69	1.76	−0.31	Europe
40	15	7.22	7.62	5.86	3.98	3.97	−0.40	Europe
45	285	1.45	1.20	1.52	2.36	4.1	+0.25	North America
40	255	2.55	3.18	4.08	4.88	5.89	−0.63	North America
20	255	8.67	12.02	14.11	15.05	16.85	−3.35	North America
20	300	10.01	7.36	9.24	12.31	15.41	+2.65	North America
50	105	4.25	4.65	5.08	5.79	8.60	−0.40	Asia
40	120	9.25	9.48	10.24	11.28	13.88	−0.23	Asia
35	135	11.79	11.68	12.40	13.13	14.39	+0.11	Japan
−25	150	8.56	9.75	10.41	11.54	11.35	−1.19	Australia
−35	15	4.40	5.93	8.41	11.29	12.19	−1.53	South Africa
−35	300	8.94	12.07	13.09	10.84	8.10	−3.13	South America

Table 2 shows that the change of geomagnetic cutoffs, in the period 1600–1900, is not homogeneous: of the 14 selected regions, 5 showed increasing cutoffs with decreasing CR intensity, and 9 regions showed decreasing cutoffs with increasing CR intensity. From Table 2 it can also be seen that at present time (from 1900 to 2000) there are sufficient change in cutoff rigidities: decreasing (with corresponding increasing of CR intensity) in 10 regions, and increasing (with corresponding decreasing of CR intensity) in 3 regions. These changes give trend in CR intensity change what we need to take into account together with CR 11 and 22 a modulation by solar activity, considered in Section 2.12.

2.14. Atmospheric Ionisation by Cosmic Rays: The Altitude Dependence and Planetary Distribution

The main process in the link between CR and cloudiness is the air ionisation which triggers chemical processes in the atmosphere. Figure 19 shows experimental data [74] of the galactic CR generation of secondary particles and absorption at different cutoff rigidities. Figure 20 illustrates the total ionisation of atmosphere by galactic CR (primary and secondary) as a function of altitude.

The planetary distribution of ionisation at the altitude of 3 km [75], is shown in Fig. 21 for the year 2000, and its time variation during 1950–2000 is presented in Fig. 22.

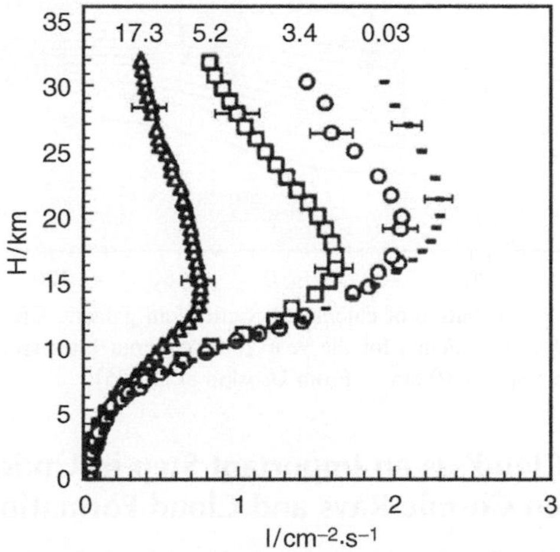

FIGURE 19 The absorption, *I*, curves of CR in the atmosphere at different cutoff rigidities (numbers at the top in units of GV) as a function of altitude, *H*. The horizontal bars indicate the standard deviations. From Ermakov et al. [74].

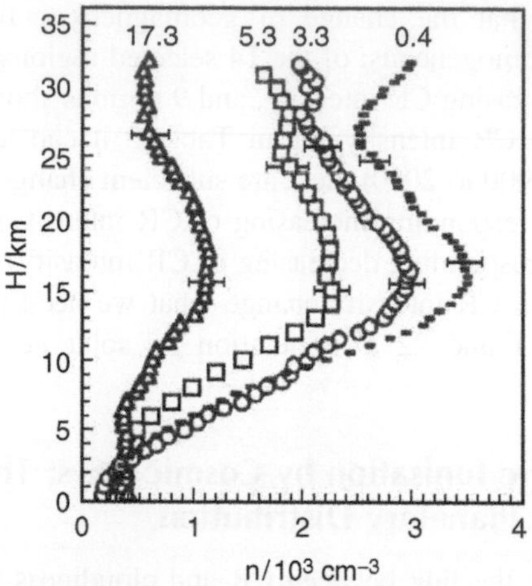

FIGURE 20 The ion concentration, n, profiles as a function of altitude, H, for different geomagnetic cutoff rigidities (numbers at the top are in units of GV). The horizontal bars indicate the standard deviations. From Ermakov et al. [74].

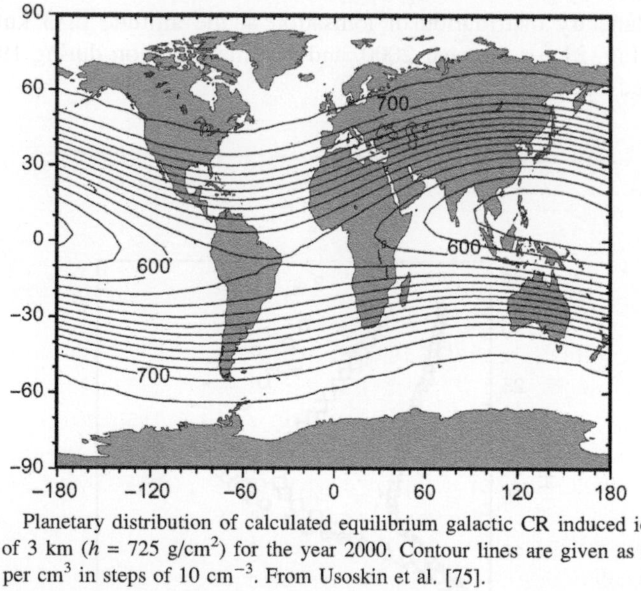

FIGURE 21 Planetary distribution of calculated equilibrium galactic CR induced ionisation at the altitude of 3 km (h = 725 g/cm^2) for the year 2000. Contour lines are given as the number of ion pairs per cm^3 in steps of 10 cm^{-3}. From Usoskin et al. [75].

2.15. Project 'Cloud' as an Important Step in Understanding the Link between Cosmic Rays and Cloud Formation

The many unanswered questions in understanding the relationship between CR and cloud formation is being investigated by a special collaboration, within the framework of European Organization for Nuclear Research,

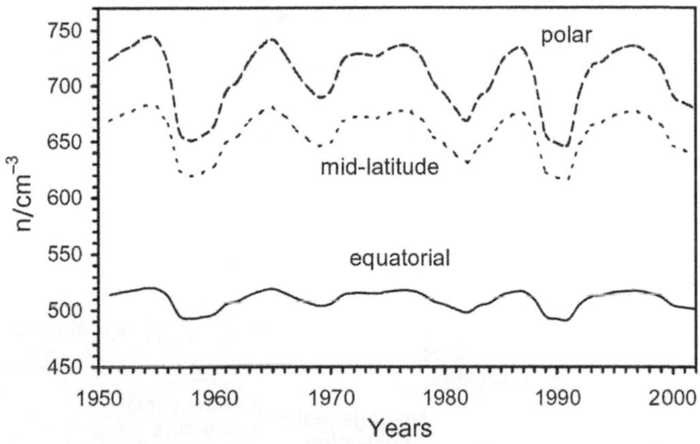

FIGURE 22 Calculated time profiles of the annual ionisation, n, at altitude of 3 km (h = 725 g/cm^2), induced by galactic CR, for three regions: polar (cutoff rigidity $R_c < 1$ GV), mid-latitudes ($R_c \approx 6$ GV) and equatorial ($R_c \approx 15$ GV) regions. From Ref. [75].

involving 17 Institutes and Universities [76]. The experiment, which is named 'CLOUD', is based on a cloud chamber (which is designed to duplicate the conditions prevailing in the atmosphere) and 'CRs' from CERN Proton Synchrotron. The Project will consider possible links between CR, variable Sun intensities and the Earth's climate change (see Fig. 23).

3. THE INFLUENCE ON THE EARTH'S CLIMATE OF THE SOLAR SYSTEM MOVING AROUND THE GALACTIC CENTRE AND CROSSING GALAXY ARMS

The influence of space dust on the Earth's climate has been reviewed [77]. Figure 24 shows the changes of planetary surface temperature for the last 520 Ma according [78]. These data were obtained from the paleoenvironmental records. During this period the solar system crossed Galaxy arms four times. In doing so, there were four alternating warming and cooling periods with temperature changes of more than 5 °C.

The amount of matter inside the galactic arms is more than on the outside. The gravitation influence of this matter attracts the inflow of comets from Oort's cloud to solar system [79,80]. It results in an increase in concentration of interplanetary dust in zodiac cloud and a cooling of the Earth's climate [81].

4. THE INFLUENCE OF MOLECULAR-DUST GALACTIC CLOUDS ON THE EARTH'S CLIMATE

The solar system moves relative to interstellar matter with a velocity about 30 km s^{-1} and sometimes passes through molecular-dust clouds. During these periods we can expect a decrease in sea level air temperature. According to

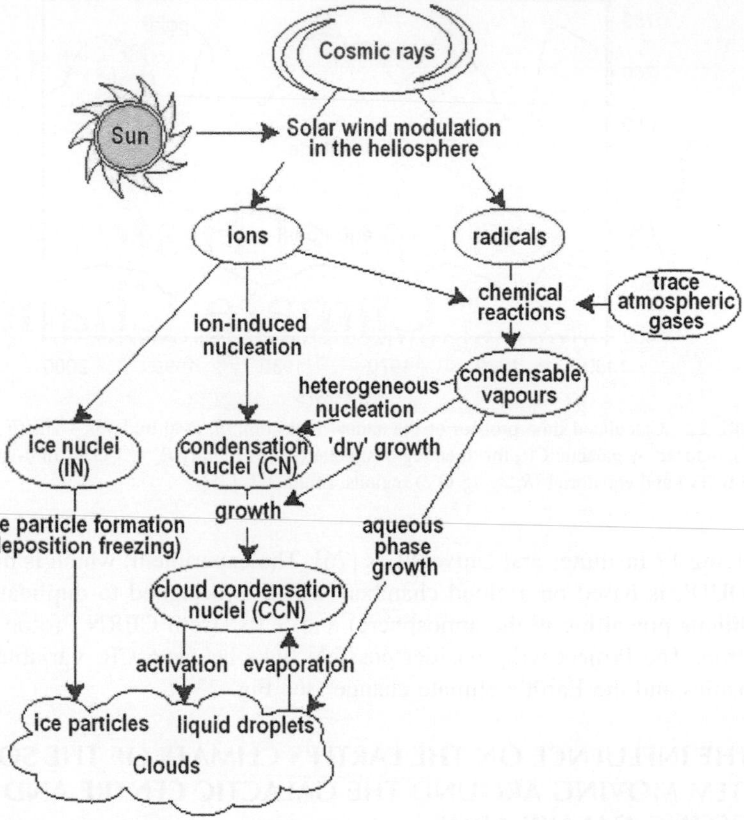

FIGURE 23 Possible paths of solar modulated CR influence on different processes in the atmosphere leading to the formation of clouds and their influence on climate. From Ref. [76].

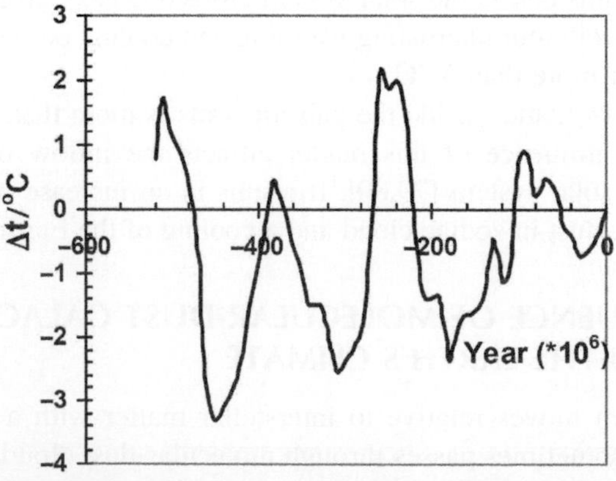

FIGURE 24 Changes of air temperature, Δt, near the Earth's surface for the last 520 Ma according to the paleoenvironmental records [78]. From Ref. [77].

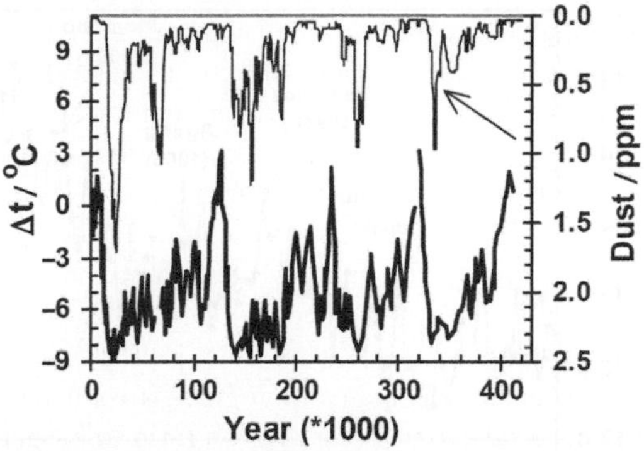

FIGURE 25 Changes of temperature, Δt, relative to modern epoch (bottom thick curve) and dust concentration (upper thin curve) over the last 420 000 a [83]. From Ref. [77].

Dorman [82], the prediction of the interaction of a dust-molecular cloud with the solar system can be performed by measurements of changes in the galactic CR distribution function. From the past we know that the dust between the Sun and the Earth has led to decreases of solar irradiation flux resulting in reduced global planetary temperatures (by 5–7 °C in comparison with the 0.8 °C increase due to the present greenhouse effect). The plasma in a moving molecular dust cloud contains a frozen-in magnetic field; this moving field can modify the stationary galactic CR distribution outside the Heliosphere. The change in the distribution function can be significant, and it should be possible to identify these changes when the distance between the cloud and the Sun becomes comparable with the dimension of the cloud. The continuous observation of the time variation of CR distribution function for many years should make it possible to determining the direction, geometry and the speed of the dust-molecular cloud relative to the Sun. Therefore, it should, in future, be possible to forecast climatic changes caused by this molecular-dust cloud.

Figure 25 shows the temperature changes at the Antarctic station Vostok (bottom curve), which took place over the last 420 000 a according to Petit et al. [83]. These data were obtained from isotopic analysis of O and H extracted from the ice cores at a depth 3300 m. It is seen from Fig. 25 that during this time the warming and cooling periods changed many times and that the temperature changes amounted up to 9 °C. Data obtained from isotope analysis of ice cores in Greenland, which cover the last 100 000 a [79], confirm the existence of large changes in climate.

5. THE INFLUENCE OF INTERPLANETARY DUST SOURCES ON THE EARTH'S CLIMATE

According to Ermakov et al. [77], the dust of zodiac cloud is a major contributory factor to climate changes in the past and at the present time. The proposed mechanism of cosmic dust influence is as follows: dust from

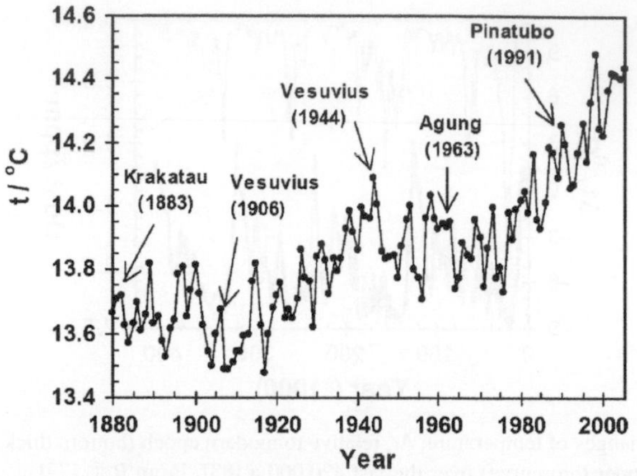

FIGURE 26 Yearly average values of the global air temperature, *t*, near the Earth's surface for the period from 1880 to 2005 [81]. Arrows show the dates of the volcano eruptions with the dust emission to the stratosphere and short times cooling after eruptions. From Ref. [77].

interplanetary space enters the Earth's atmosphere during the yearly rotation of the Earth around the Sun. The space dust participates in the processes of cloud formation. The clouds reflect some part of solar irradiance back to space. In this way the dust influences climate. The main sources of interplanetary dust are comets, asteroids and meteor fluxes. The rate of dust production is continually changing. The effect of volcanic dust on the Earth's air temperature is illustrated in Fig. 26 [81]. (Note air temperature can be found at ftp://ftp.ncdc.noaa.gov/pub/data/anomalies/global_meanT_C.all)

According to Ermakov et al. [77], the spectral analysis of global surface temperature during 1880–2005 shows the presence of several spectral lines that can be identified with the periods of meteor fluxes, comets and asteroids. The results of analysis have been used [77,84] to predict changes in climate over the next half-century: the interplanetary dust factor of cooling in the next few decades will be more important than the warming from greenhouse effect.

6. SPACE FACTORS AND GLOBAL WARMING

It is now commonly thought of that the current trend of the global warming is causally related to the accelerating consumption of fossil fuels by the industrial nations. However, it has been suggested that this warming is a result of a gradual increase of solar and magnetic activity over the last 100 a. According to Pulkkinen et al. [85], as shown in Figs. 27 and 28, the solar and magnetic activity has been increasing since the year 1900 with decreases in 1970 and post 1980. Figures 27 and 28, show that that the aa index of geomagnetic activity, (a measure of the variability of the interplanetary magnetic field,

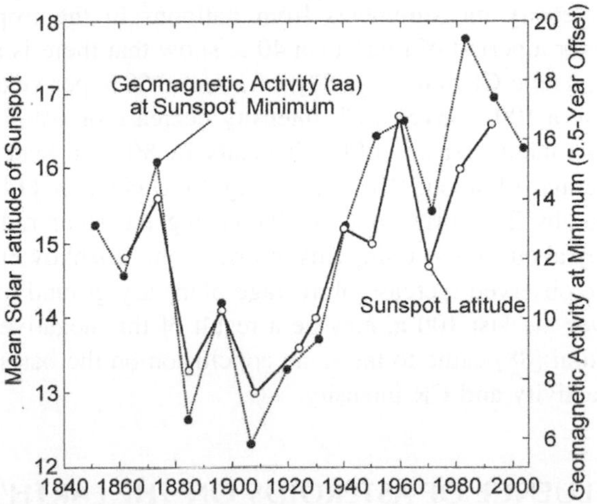

FIGURE 27 The geomagnetic activity (index aa) at the minimum of solar activity and the mean sunspots latitude, from 1840 to 2000. From Ref. [85].

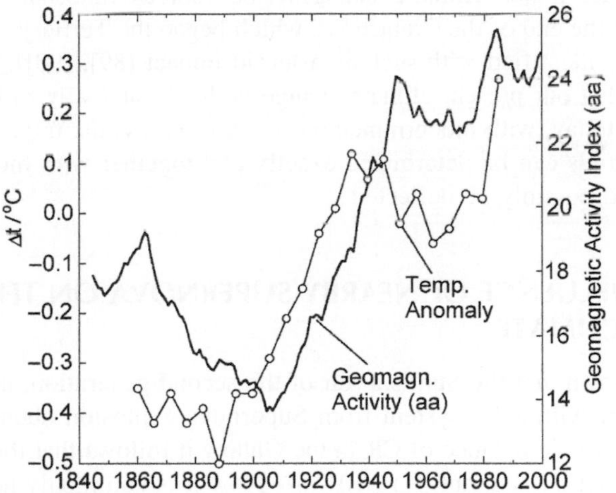

FIGURE 28 The geomagnetic activity (index aa) at the minimum of solar activity variation of the global temperature anomaly, Δt, from 1840 to 2000. From Ref. [85].

IMF), varies, almost in parallel, with the sunspot activity and with the global temperature anomaly.

It has been well established that the brightness of the Sun varies in proportion to solar activity. The brightness changes are very small and cannot explain all of the present global warming. However, the gradual increase of solar activity over the last hundred years has been accompanied by a gradual decrease of CR intensity in interplanetary space [86]. The direct measurements of CR intensity on the ground by the global network of NM as well as regular CR intensity measurements from balloons in the troposphere and stratosphere over a period of more than 40 a, show that there is a small negative trend of galactic CR intensity [87]) of about 0.08% per year. Extrapolating this trend to a 100 a, gives a CR intensity decrease on 8%. From Fig. 2 it can be seen that the decreasing of CR intensity by 8% will lead to a decrease of cloud coverage of about 2%. According to Dickinson [14], decreasing cloud coverage by 2% corresponds to increasing the solar radiation falling on the Earth by about 0.5%. Using this information, Stozhkov et al. [55] concluded that the observed increase of average planetary ground temperature of 0.4–0.8 °C over the last 100 a, may be a result of this negative trend of CR intensity. Sakurai [88] came to the same conclusion on the basis of analysing data of solar activity and CR intensity.

7. THE INFLUENCE OF ASTEROIDS ON THE EARTH'S CLIMATE

It is well known that asteroids have in the past, struck the Earth with sufficient force to make major climatic changes (the famous dinosaur-killing mass extinction at the end of the Cretaceous, which began the Tertiary era, has been convincingly identified with such an asteroid impact [89], [90]). However, it is unlikely that our present climate change is due in any way to such events. Fortunately today, with modern methods of Astronomy, the trajectory of dangerous asteroids can be determined exactly and together with modern rocket power, could possibly be deflected.

8. THE INFLUENCE OF NEARBY SUPERNOVA ON THE EARTH'S CLIMATE

It is well known that the Sun is a star of the second generation, in that it was born together with solar system from Supernova explosion about 5 Ga ago. From the energetic balance of CR in the Galaxy it follows that the full power for CR production is about 3×10^{33} W. Now it is commonly accepted that the Supernova explosions are the main source of galactic CR. At each explosion the average energy transferred to CR is about 10^{43}–10^{44} J. From this quantity we can determine the expected frequency of Supernova explosions in our Galaxy and in vicinity of the Sun, and estimate: the probability of Supernova explosions at different distances from the Sun; the expected UV radiation flux (destroyer of our ozone layer and hence a significant player in our Earth's climate), and the expected CR flux. It has been estimated in Dorman et al. [91] and Dorman [82] that if such an event does take place, the levels of CR radiation reaching our Earth could reach levels extremely dangerous to our civilisation and biosphere. Such an event is unlikely to be responsible for our present climate changes.

9. DISCUSSION AND CONCLUSIONS

Many factors from space and from anthropogenic activities can influence the Earth's climate. The initial response is that space factors are unlikely to be responsible for most of our present climate change. However, it is important that all possible space factors be considered, and from an analysis of past climate changes, we can identify our present phase and can predict future climates. During the last several hundred million years the Sun has moved through the galactic arms several times with resultant climate changes. For example, considering the effects due to galactic molecular-dust concentrated in the galactic arms, as given in Fig. 24, we can see that during the past 520×10^6 a, there were four periods with surface temperatures lower than what we are presently experiencing and four periods with higher temperatures. On the other hand, during the past 420 000 a (Fig. 25) there were four decreases of temperature (the last one was about 20 000–40 000 a ago: the so-called big ice period), and five increases of temperature, the last of which happened few thousand years ago. At present the Earth is in a slight cooling phase (of the order of one degree centigrade over several thousand years.

When considering CR variations as one of the possible causes of long-term climate change (see Section 2) we need to take into account not only CR modulation by solar activity but also the change of geomagnetic cutoff rigidities (see Table 2). It is especially important when we consider climate change on a scale of between 10^3 and 10^6 a: paleomagnetic investigations show that during the last 3.6×10^6 a the magnetic field of the Earth changed sign nine times, and the Earth's magnetic moment changed – sometimes having a value of only one-fifth of its present value [92] – corresponding to increases of CR intensity and decreases of the surface temperature.

The effects of space factors on our climate can be divided into two types:

- the 'gradual' type, related to changes on time scales ranging from 10^8 a to 11–22 a, producing effects which could be greater than that produced from anthropogenic factors, and
- the 'sudden' type, coming from Supernova explosions and asteroid impacts, for example, and which may indeed be catastrophic to our civilisation. Volcanic and anthropogenic factors are also in a sense, 'sudden' factors in their effect on climate change.

It is necessary to investigate all of the possible 'sudden' factors and to develop methods of forecasting and also for protecting the biosphere and the Earth's civilization from big changes in climate and environment. We cannot completely exclude the possibility that a Supernova explosion, for example, took place 20 a ago at a distance of say 30 light years away. In this case its influence on our climate and environment will be felt in 10 a time. According to Ellis and Schramm [93], in this case, UV radiation would destroy the Earth's ozone layer over a period of about 300 a. The recent observations of Geminga, PSR J0437–4715, and SN 1987A strengthen the case for one or more supernova extinctions having taken place during the Phanerozoic era. In this case a nearby supernova explosion would have depleted the ozone layer, exposing both marine and terrestrial organisms to potentially lethal solar UV radiation. In particular, photosynthesising organisms including phytoplankton and reef communities would most likely have been badly affected.

As Quante [94] noted, clouds play a key role in our climate system. They strongly modulate the energy budget of the Earth and are a vital factor in the global water cycle. Furthermore, clouds significantly affect the vertical transport in the atmosphere and determine, in a major way, the redistribution of trace gases and aerosols through precipitation. In our present-day climate, on average, clouds cool our planet; the net cloud radiative forcing at the top of the atmosphere is about $-20\ \mathrm{W{\cdot}m^{-2}}$. Any change in the amount of cloud or a shift in the vertical distribution of clouds, can lead to considerable changes in the global energy budget and thus affect climate [94].

Many of the 'gradual' types of space factors are linked to cloud formation. Quante [94] noted that galactic CR [19–22] was an important link between solar activity and low cloud cover (see Figs. 2 and 3). However, new data after 1995 shows that the problem is more complicated and the correlation no longer holds [95]. Kristjánsson et al. [96] pointed out that still many details are missing for a complete analysis, but a cosmic ray modulation of the low cloud cover seems less likely to be the major factor in our present climate change, but its role in future climate changes must not be ruled out.

In this Chapter much emphasis has been given to the formation of clouds and the influence CR plays (through ionisation and influence on chemical processes in atmosphere) in their formation. This does not imply that CR is the only factor in their formation; dust, aerosols, precipitation of energetic particles from radiation belts and greenhouse gases, all play their part. However, the influence of CR is important and has been demonstrated here through:

- a direct correlation during one solar cycle (Figs. 2 and 3) and also for much longer periods,
- the correlation of CR intensity with the planetary surface temperature (Figs. 1, 4 and 5),
- by the direct relationship between CR intensity and wheat prices in medieval England (Fig. 8),
- by the direct relationship between cloud formation and CR air ionisation (Fig. 9),
- by the relationship between geomagnetic activity and rainfall through precipitation of energetic particles from radiation belts (Figs. 10–13) and
- by linking CR intensities with the cirrus holes (Figs. 14 and 15).

The importance of CR cannot be stressed highly enough and it is important to develop methods for determining, with high accuracy, galactic CR intensity variations for the past, the present and for the near future.

In this Chapter several attempts have been made to explain the present climate change (the relatively rapid warming of the Earth discussed in Section 6 for 1937–1994), using space factors:

- through 11 a average CR intensities discussed in Section 2.5,
- by the increasing geomagnetic activity,
- by the decreasing CR intensity (of 8% over the past 100 a) and
- by relating the spectral analysis of Ermakov et al. [77] to global temperature during 1880–2005.

Their results show the presence of several spectral lines that can be identified with the periods of meteor fluxes, comets and asteroids. On the basis of this work, Ermakov et al. [77,84] has predicted a cooling of the Earth's climate over the next half-century which they believe will be more important than warming from greenhouse effect.

Finally, it appears that our present climate change (including a rapid warming of about 0.8 °C over the past 100 a, see Fig. 26) is caused by a collective action of several space factors, volcano activities (with the dust emission rising to the stratosphere, resulting in short term cooling after eruptions), as well as by anthropogenic factors with their own cooling and warming contributions. The relation between these contributions will determine the final outcome. At present the warming effect is stronger than the cooling effect. It is also very possible that the present dominant influence is anthropogenic in origin.

From Fig. 25 can be seen that now we are near the maximum global temperature reached over the past 400 000 a, so an additional rapid increase of even a few degree celsius could lead to an unprecedented and catastrophic situation. It is necessary that urgent and collective action be taken now by the main industrial countries and by the UN, to minimise the anthropogenic influence on our climate before it is too late. On the other hand, in future, if the natural change of climate results in a cooling of the planet (see Figs. 24 and 25), then special man-made factors, resulting in warming, may have to be used to compensate.

ACKNOWLEDGMENTS

My great gratitude to Abraham Sternlieb, Isaac Ben Israel, and Zvi Kaplan for constancy attention and support of work of Israel Cosmic Ray and Space Weather Center and Emilio Ségre Observatory; to Lev Pustil'nik, Yurii Stozhkov and Igor Zukerman for interesting discussions, and to Trevor Letcher for useful comments.

REFERENCES

1. L.I. Dorman, Cosmic Rays in the Earth's Atmosphere and Underground, Kluwer Academic Publishers, Dordrecht/Boston/London, 2004.
2. L.I. Dorman, Cosmic Rays in Magnetospheres of the Earth and other Planets, Springer, Netherlands, 2009.
3. L.I. Dorman, Cosmic Ray Interactions, Propagation, and Acceleration in Space Plasmas, Springer, Netherlands, 2006.
4. H. Herschel, Philosophical Transactions of the Royal Society, London, 91 (1801) 265–318.
5. J.A. Eddy, Science 192 (1976) 1189–1202.
6. H. Swensmark, Space Sci. Rev. 93 (2000) 175–185.
7. E. Friis-Christiansen, K. Lassen, Science 254 (1991) 698–700.
8. K. Lassen, E. Friis-Christiansen, J. Atmos. Solar-Terr. Phys. 57 (1995) 835–845.
9. K. Labitzke, H. van Loon, Ann. Geophys. 11 (1993) 1084–1094.
10. J. Lean, J. Beer, R. Breadley, Geophys. Res. Lett. 22 (1995) 3195–3198.
11. J.D. Haigh, Science 272 (1996) 981–984.
12. D. Shindell, D. Rind, N. Balabhandran, J. Lean, P. Lonengran, Science 284 (1999) 305–308.
13. E.R. Ney, Nature 183 (1959) 451–452.
14. R.E. Dickinson, Bull. Am. Met. Soc. 56 (1975) 1240–1248.
15. M.I. Pudovkin, O.M. Raspopov, Geomagn. Aeronomy 32 (1992) 593–608.
16. M. Pudovkin, S. Veretenenko, J. Atmos. Solar-Terr. Phys. 57 (1995) 1349–1355.
17. M. Pudovkin, S. Veretenenko, Adv. Space Res. 17(11) (1996) 161–164.
18. B.A. Tinsley, J. Geomagn. Geoelectr. 48 (1996) 165–175.
19. H. Swensmark, E. Friis-Christiansen, J. Atmos. Solar-Terr. Phys. 59 (1997) 1225–1232.
20. H. Swensmark, Phys. Rev. Lett. 81 (1998) 5027–5030.
21. N.D. Marsh, H. Swensmark, Phys. Rev. Lett. 85 (2000) 5004–5007.
22. N. Marsh, H. Swensmark, Space Sci. Rev. 94 (2000) 215–230.
23. L.I. Dorman, I.Ya. Libin, M.M. Mikalayunas, K.F. Yudakhin, Geomagn. Aeronomy 27 (1987) 303–305.
24. L.I. Dorman, I.Ya. Libin, M.M. Mikalajunas, Regional Hydrometeorol. (Vilnius), 12 (1988) 119–134.
25. L.I. Dorman, I.Ya. Libin, M.M. Mikalajunas, Regional Hydrometeorol. (Vilnius), 12 (1988) 135–143.
26. L.I. Dorman, G. Villoresi, I.V. Dorman, N. Iucci, M. Parisi, Proc. 25-th Intern. Cosmic Ray Conference, Durban (South Africa), 7 (1997) 345–348.
27. R. Markson, Nature 273 (1978) 103–109.
28. C. Price, Nature 406 (2000) 290–293.
29. B.A. Tinsley, Space Sci. Rev. 94 (2000) 231–258.
30. K. Schlegel, G. Diendorfer, S. Them, M. Schmidt, J. Atmos. Solar-Terr. Phys. 63 (2001) 1705–1713.
31. L.I. Dorman, I.V. Dorman, N. Iucci, M. Parisi, Y. Ne'eman, L.A. Pustil'nik, F. Signoretti, A. Sternlieb, G. Villoresi, I.G. Zukerman, J. Geophys. Res. 108(A5) (2003) 1181, SSH 2_1–8.
32. L.I. Dorman, I.V. Dorman, Adv. Space Res. 35 (2005) 476–483.
33. D.L. Hartmann, in: P.V. Hobbs (Ed.), Aerosol–Cloud–Climate Interactions, Academic Press (1993) p. 151.
34. G. Ohring, P.F. Clapp, J. Atmospheric Sci. 37 (1980) 447–454.
35. V. Ramanathan, R.D. Cess, E.F. Harrison, P. Minnis, B.R. Barkstrom, E. Ahmad, D. Hartmann, Science, 243(4887) (1989) 57–63.

36. P.E. Ardanuy, L.L. Stowe, A. Gruber, M. Weiss, J. Geophys. Res. 96 (1991) 18537–18549.
37. L.L. Stowe, C.G. Wellemayer, T.F. Eck, H.Y.M. Yeh, and the Nimbus-7 Team, J. Clim. 1 (1988) 445–470.
38. W.B. Rossow, R. Shiffer, Bull. Am. Met. Soc. 72 (1991) 2–20.
39. F. Weng, N.C. Grody, J. Geophys. Res. 99 (1994) 25535–25551.
40. R.R. Ferraro, F. Weng, N.C. Grody, A. Basist, Bull. Am. Met. Soc. 77 (1996) 891–905.
41. H.S. Ahluwalia, Proc. 25th Int. Cosmic Ray Conf., Durban, 2 (1997) 109–112.*
42. W.B. Rossow, B. Cairns, J. Clim., 31 (1995) 305–347.
43. J. Lean, A. Skumanich, O. White, Geophys. Res. Lett. 19 (1992) 1591–1594.
44. J. Beer, G.M. Raisbeck, F. Yiou, in: C.P. Sonett, M.S. Giampapa, M.S. Matthews (Eds.), The Sun in Time, University of Arizona Press (1991) 343–359.
45. P.D. Jones, K.R. Briffa, T.P. Barnett, S.F.B. Tett, The Holocene, 8 (1998) 455–471.
46. M.E. Mann, R.S. Bradley, M.K. Hughes, Nature 392 (1998) 779–787.
47. A. Smith, An Inquiry into the Nature and Causes of the Wealth of Nations, W. Strahan & T. Cadell, London, 1776.
48. W.S. Jevons, Nature 19 (1875) 33–37.
49. J.E.T. Rogers, Agriculture and Prices in England, vols. 1–8, Oxford, Clarendon Press, 1887.
50. W.S. Jevons, Nature 26 (1882) 226–228.
51. L. Pustil'nik, G. Yom Din, L. Dorman, Proc. 28th Intern. Cosmic Ray Conf. Tsukuba, 7 (2003) 4131–4134.
52. L. Pustil'nik, G. Yom Din, Solar Phys. 223 (2004) 335–356.
53. J. Beer, S. Tobias, N. Weiss, Solar Phys. 181 (1998) 237–249.
54. L. Pustil'nik, G. Yom Din, Solar Phys. 224 (2004) 473–481.
55. Yu.I. Stozhkov, V.I. Ermakov, P.E. Pokrevsky, Izvestia Russian Academy. Sci. Ser. Phys., 65(3) (2001) 406–410.
56. S.V. Veretenenko, M.I. Pudovkin, Geomagn. Aeronomy, 34 (1994) 38–44.
57. Yu. I. Stozhkov, P.E. Pokrevsky, I.M. Martin et al., Proc. 24th Int. Cosmic Ray Conf. Rome, 4 (1995) 1122–1125.
58. Yu.I. Stozhkov, J. Zullo, I.M. Martin et al., Nuovo Cimento, C18 (1995) 335–341.
59. Yu.I. Stozhkov, P.E. Pokrevsky, J. Jr. Zullo et al., Geomagn. Aeronomy 36 (1996) 211–216.
60. Yu. I. Stozhkov, J. Phys. G: Nucl. Part. Phys. 28 (2002) 1–11.
61. M.C. Todd, D.R. Kniveton, J. Geophys. Res. 106(D23) (2001) 32031–32042.
62. M.C. Todd, D.R. Kniveton, J. Atmos. and Solar-Terr. Phys. 66 (2004) 1205–1211.
63. B.A. Tinsley, G.W. Deen, J. Geophys. Res. 96(D12) (1991) 22283–22296.
64. G.I. Pugacheva, A.A. Gusev, I.M. Martin et al., Proc. 24th Intern. Cosmic Ray Conf., Rome, 4 (1995) 1110–1113.
65. I.M. Martin, A.A. Gusev, G.I. Pugacheva, A. Turtelli, Y.V. Mineev, J. Atmos. and Terr. Phys. 57(2) (1995) 201–204.
66. J.W. King, Astronautics and Aeronautics 13(4) (1975) 10–19.
67. J.T.A. Ely, T.C. Huang, Geophys. Res. Lett. 14(1) (1987) 72–75.
68. J.T.A. Ely, J.J. Lord, F.D. Lind, Proc. 24th Intern. Cosmic Ray Conf., Rome, 4 (1995) 1137–1140.
69. L.I. Dorman, Adv. Space Res. 35 (2005) 496–503.
70. L.I. Dorman, Annales Geophysicae 23(9) (2005) 3003–3007.
71. L.I. Dorman, Adv. Space Res. 37 (2006) 1621–1628.
72. A.V. Belov, L.I. Dorman, R.T. Gushchina, V.N. Obridko, B.D. Shelting, V.G. Yanke, Adv. Space Res. 35(3) (2005) 491–495.
73. M.A. Shea, D.F. Smart, Proc. 28th Intern. Cosmic Ray Conf., Tsukuba, 7 (2003) 4205–4208.

74. V.I. Ermakov, G.A. Bazilevskaya, P.E. Pokrevsky, Yu. I. Stozhkov, Proc. 25th Intern. Cosmic Ray Conf., Durbin, 7 (1997) 317–320.
75. I.G. Usoskin, O.G. Gladysheva, G.A. Kovaltsov, J. Atmos. Solar-Terr. Phys. 66(18) (2004) 1791–1796.
76. B. Fastrup, E. Pedersen, E. Lillestol et al. (Collaboration CLOUD), Proposal CLOUD, CERN/SPSC (2000) 2000–2021.
77. V. Ermakov, V. Okhlopkov, Yu. Stozhkov, Proc. European Cosmic Ray Symposium, Lesboa (2006) Paper 1–72.
78. J. Veizer, Y. Godderis, I.M. Francois, Nature 408 (2000) 698–701.
79. K. Fuhrer, E.W. Wolff, S.J. Johnsen, J. Geophys. Res. 104(D24) (1999) 31043–31052.
80. O.A. Maseeva, Astronomichesky vestnik 38(4) (2004) 372–382.
81. J. Hansen, R. Ruedy, J. Glascoe, and M. Sato, J. Geophys. Res. 104(D24) (1999) 30997–31022.
82. L.I. Dorman, Adv. Geosci. 14 (2008) 281–286.
83. J.R. Petit, J. Jouzel, D. Raunaud, et al., Nature, 399 (1999) 429–436.
84. V.I. Ermakov, V.P. Okhlopkov, Y.I. Stozhkov, Bull. Lebedev Phys. Inst. Russian Acad. Sci. Moscow 3 (2006) 41–51.
85. T.I. Pulkkinen, H. Nevanlinna, P.J. Pulkkinen, M. Lockwood, Space Sci. Rev. 95 (2001) 625–637.
86. M. Lockwood, R. Stamper, M.N. Wild, Nature 399(6735) (1999) 437–439.
87. Yu.I. Stozhkov, P.E. Pokrevsky, V.P. Okhlopkov, J. Geophys. Res. 105(A1) (2000) 9–17.
88. K. Sakurai, Proc. 28th Intern. Cosmic Ray Conf., Tsukuba, 7 (2003) 4209–4212.
89. L. Alvarez, W. Alvarez, F. Asaro, H. Michel, Science 208 (1980) 1095–1108.
90. V. Sharpton, K. Burke, A. Camargo-Zaroguera et al., Science 261 (1993) 1564–1567.
91. L.I. Dorman, N. Iucci, G. Villoresi, Astrophys. Space Sci., 208 (1993) 55–68.
92. A. Cox, G.B. Dalrymple, R.R. Doedl, Sci. Am. 216(2) (1967) 44–54.
93. J. Ellis, D.N. Schramm, Proc. Nat. Acad. Sci. USA Astron. 92 (1995) 235–238.*
94. M. Quante, J. Phys. IV France 121 (2004) 61–86.
95. J.E. Kristjánsson, A. Staple, J. Kristiansen, Geophys. Res. Lett. 23 (2002) 2107–2110.
96. J.E. Kristjánsson, J. Kristiansen, E. Kaas, Adv. Space Res. 34(2) (2004) 407–415.

Chapter 4

The Role of Volcanic Activity in Climate and Global Change

Georgiy Stenchikov
Department of Environmental Sciences, Rutgers University, New Brunswick, New Jersey 08901-855

1. Introduction
2. Aerosol Loading, Spatial Distribution and Radiative Effect
3. Volcanoes and Climate
3.1. Tropospheric Cooling and Stratospheric Warming
3.2. Effect on Hydrological Cycle
3.3. Volcanic Effect on Atmospheric Circulation
3.4. Volcanic Impact on Ocean Heat Content and Sea Level
3.5. Strengthening of Overturning Circulation
3.6. Volcanic Impact on Sea Ice
4. Summary
Acknowledgements
References

1. INTRODUCTION

Volcanic activity is an important natural cause of climate variations because tracer constituents of volcanic origin impact the atmospheric chemical composition and optical properties. This study focuses on the recent period of the Earth's history and does not consider a cumulative effect of the ancient volcanic degassing that formed the core of the Earth's atmosphere billions of years ago. At present, a weak volcanic activity results in gas and particle effusions in the troposphere (lower part of atmosphere), which constitute, on an average, the larger portion of volcanic mass flux into the atmosphere. However, the products of tropospheric volcanic emissions are short-lived and contribute only moderately to the emissions from large anthropogenic and natural tropospheric sources. This study focuses instead on the effects on climate of the Earth's explosive volcanism. Strong volcanic eruptions with a volcanic explosivity index (VEI) [1] equal to or greater than 4 could inject volcanic ash and sulfur-rich gases into the clean lower stratosphere at an altitude about 25–30 km, increasing their concentration thereby two to three orders of magnitude in comparison with the background level. Chemical transformations and gas-to-particle conversion of volcanic tracers form a volcanic aerosol layer that remains in the stratosphere for 2–3 years after an eruption, thereby impacting the Earth's climate because volcanic aerosols cool the surface and the troposphere by reflecting solar radiation, and warm the lower stratosphere, absorbing thermal IR and solar near-IR radiation [2]. Figure 1 shows stratospheric optical depth for the visible wavelength of 0.55 μm. It roughly characterises the portion of scattered solar light. Three major explosive eruptions occurred in the second part of the twentieth century, as depicted in Fig. 1: Agung of 1963, El Chichon of 1982, and Pinatubo of 1991.

Volcanic eruptions, like the Mt. Pinatubo eruption in 1991, with global visible optical depth maximizing at about 0.15, cause perturbation of the globally averaged radiative balance at the top of the atmosphere reaching $-3\ \mathrm{W{\cdot}m^{-2}}$ and cause a decrease of global surface air temperature by 0.5 K. Radiative impact of volcanic aerosols also produces changes in atmospheric circulation, forcing a positive phase of the Arctic Oscillation (AO) and counterintuitive

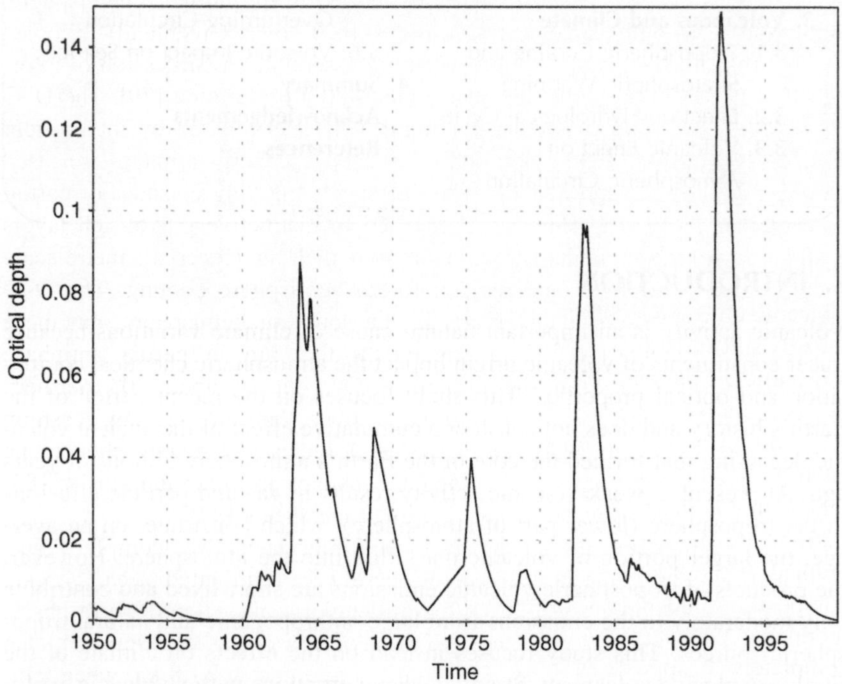

FIGURE 1 The total global mean normal optical depth τ of stratospheric aerosols for the Pinatubo period for the visible wavelength of 0.55 μm as a function of time. It causes attenuation of direct solar visible light with a factor of $\exp(-\tau/\cos\zeta)$, where $\cos\zeta$ is a cosine of zenith angle.

boreal winter warming in middle and high latitudes over Eurasia and North America [3–8]. In addition, stratospheric aerosols affect stratospheric chemistry serving as surfaces for heterogeneous reactions liberating anthropogenic chlorine and causing ozone depletion.

It was traditionally believed that volcanic impacts produced mainly short-term transient climate perturbations. However, the ocean integrates volcanic radiative cooling, and different components of the ocean respond over a wide range of time scales. Volcanically induced tropospheric temperature anomalies vanish in about 7 years, while volcanically induced sea ice extent and volume changes have a relaxation time scale closer to a decade. Volcanically induced changes in interior ocean temperature, the meridional overturning circulation (MOC), and steric height, have even longer relaxation times, from several decades to a century. Because of their various impacts on climate systems, volcanic eruptions play a role of natural tests, providing an independent means of assessing multiple climate feedback mechanisms and climate sensitivity [7–11].

There are several excellent reviews devoted to volcanic impacts on climate and weather [12–19]. The present study provides an overview of available observations of volcanic aerosols and discusses their radiative forcing and large-scale effects on climate. It focuses on recently discovered forced stratosphere troposphere dynamic interaction and long-term ocean response to volcanic forcing, and aims to add information to that already presented in the previous reviews.

2. AEROSOL LOADING, SPATIAL DISTRIBUTION AND RADIATIVE EFFECT

Volcanic emissions comprised of gases (H_2O, CO_2, N_2, SO_2, H_2S) and solid (mostly silicate) particles, that are usually referred to as volcanic ash. Volcanic ash particles are relatively large, exceeding 2 μm in diameter, and therefore deposit relatively quickly, that is, within a few weeks. They are responsible for short-term regional-to-continental perturbations of the Earth's radiative balance and meteorological parameters. H_2O, CO_2 and N_2 are abundant in the Earth's atmosphere, so individual volcanic perturbations of their concentrations are negligible. But SO_2 and H_2S, which quickly oxidize to SO_2 if erupted in the stratosphere, could significantly affect stratospheric chemical composition and optical properties. SO_2 gas absorbs UV and IR radiation, producing very strong localized stratospheric heating [20–22]. However, it completely disappears in about half a year and the major long-term impact of volcanic eruptions on climate is due to long-lived sulfate aerosols formed by oxidizing of SO_2 with a characteristic conversion time of about 35 days. Sulfate volcanic aerosols (submicron droplets of highly concentrated sulfuric acid) are transported globally by the Brewer–Dobson stratospheric circulation and eventually fall out in 2–3 years. A significant amount of volcanic aerosols that penetrate to the troposphere through the tropause folds is washed out in storm tracks. Aerosols deposited in downward branches of the Brewer–Dobson circulation in the Polar Regions are preserved in the polar ice sheets, recording the history of the Earth's explosive volcanism for thousands of years [23–25]. However, the atmospheric loadings calculated using volcanic time series from high-latitude ice records, suffer from uncertainties in observation data and poor understanding of atmospheric transport and deposition processes. The global instrumental observations of volcanic aerosols have been conducted during the last 25 a (years) by a number of remote sensing platforms. Total Ozone Mapping Spectrometer (TOMS) instrumentation onboard the Nimbus-7 provided SO_2 loadings from November 1978 until 6 May 1993 [26]. Prata et al. [27] recently developed a new retrieval technique to obtain SO_2 loadings from TOMS data. The Advanced Very High Resolution Radiometer (AVHRR) provides aerosol optical depth over oceans with 1 km spatial resolution in several visible and near-IR wavebands. However, column observations are not sufficient to reliably separate tropospheric and stratospheric aerosols.

The Stratospheric Aerosol and Gas Experiment (SAGE) and Stratospheric Aerosol Measurement (SAM) projects have provided more than 20 a of vertically resolved stratospheric aerosol spectral extinction, the longest such record. The 3-D observations are most valuable to understand stratospheric aerosols transformations and transport. However, there are significant gaps in the temporal-spatial coverage, for example, the eruption of El Chichón in 1982 (the second most important in the twentieth century after Mt. Pinatubo) is not covered by SAGE observations because the SAGE I instrument failed in 1981, and SAGE II was only launched in 1984. Fortunately, instruments aboard the Stratosphere Mesosphere Explorer (SME) filled the gap of 1982–1984 in 3-D aerosol observations. The saturation periods when the SAGE instrument could not see the direct sun light through the dense areas of aerosol cloud also could be partially reconstructed using lidar and mission observations [28,29]. It is important to utilize observations from the multiple platforms to improve data coverage, for example, combining SAGE II and Polar Ozone and Aerosol Measurement (POAM) data could help to fill in the polar regions. Randall et al. [30,31] have extensively intercompared the POAM and SAGE data and normalized them, combining them into a consistent data set.

Cryogenic Limb Array Etalon Spectrometer (CLAES), Improved Stratospheric and Mesospheric Sounder (ISAMS) and Halogen Occultation Experiment (HALOE) instruments launched on the Upper Atmosphere Research Satellite (UARS) provide additional information for the post-Pinatubo period. These instruments measure the aerosol volume extinction (HALOE) and volume emission (CLAES, ISAMS) in the near IR and IR bands. These three infrared instruments provide better horizontal coverage than SAGE, but do not penetrate lower than the 100 hPa level. They started operating in September 1991. CLAES and ISAMS stopped working after 20 months. The SAGE III instrument aboard the Russian Meteor III-3M satellite continued the outstanding SAGE aerosol data record [32,33] from 2001 to 2007. The new Moderate Resolution Imaging Spectroradiometer (MODIS) and Multiangle Imaging SpectroRadiometer (MISR) instruments have superior spatial and spectral resolutions, but mostly focus on the tropospheric aerosols and surface characteristics, providing column average observations.

Available satellite and ground-based observations were used to construct volcanic aerosol spatial-temporal distribution and optical properties [2,34–39]. Hansen et al. [37] improved a Goddard Institute for Space Studies (GISS) volcanic aerosols data set for 1850–1999, providing zonal mean vertically resolved aerosol optical depth for visible wavelength and column average effective radii. Amman et al. [34] developed a similar data set of total aerosol optical depth based on evaluated atmospheric loadings distributed employing a seasonally varying diffusion-type parameterisation that could also be used for paleoclimate applications (if aerosol loadings are available). Amman et al. [34], however, used a fixed effective radius of 0.42 μm for calculating aerosol optical properties and, in general, provided higher values of optical depth than in Hansen et al. [37]. Stenchikov et al. [39] used UARS observations to modified effective radii from Hansen et al. [37] implementing its variations with altitude, especially at the top of the aerosol layer where particles became very small. They conducted Mie calculations for the entire period since 1850 and implemented these aerosol characteristics in the new Geophysical Fluid Dynamics Laboratory (GFDL) climate model. The sensitivity calculations with different effective radii show that total optical depth vary as much as 20% when effective radius changes are in the reasonable range. The study of Bauman et al. [35,36] provides a new approach for calculating aerosol optical characteristics using SAGE and UARS data. Bingen et al. [40,41] have calculated stratospheric aerosols size distribution parameters using SAGE II data. A new partly reconstructed and partly hypothesized climate forcing time series for 500 years, that includes greenhouse gas (GHG) and volcanic effects, was developed by Robertson et al. [42].

Aerosol optical properties include aerosol optical depth (see Fig. 1), single scattering albedo (to characterize aerosol absorptivity) and asymmetry parameter (to define the directionality of scattering). Using these aerosol radiative characteristics one can evaluate aerosol radiative effect on climate system – aerosol radiative forcing at the top of the atmosphere. Figure 2 shows the total forcing and its short wave (SW) and long wave (LW) components. Increase of reflected SW radiation ranges from 3 to 5 $W{\cdot}m^{-2}$ but is compensated by aerosol absorption of outgoing LW radiation, so total maximum cooling of the system ranges from 2 to 3 $W{\cdot}m^{-2}$.

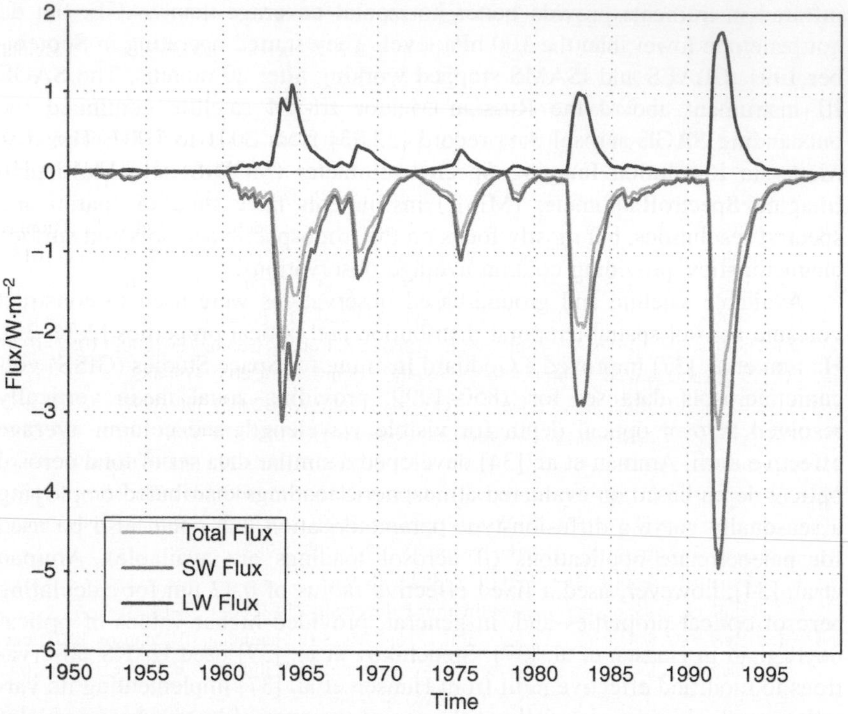

FIGURE 2 Volcanic aerosol total, short wave (SW) and long wave (LW) radiative forcing ($W \cdot m^{-2}$) at the top of the atmosphere for All-Sky conditions. Positive sign of the forcing corresponds to heating of climate system.

3. VOLCANOES AND CLIMATE

The perturbations of the Earth's radiative balance caused by strong volcanic eruptions dominate other forcings for 2–3 years. Their effect is seen in the atmosphere for about 5–7 years, and, as was recently discovered, for much longer in oceans [43–47]. Volcanic perturbations have been used for years as natural experiments to test models and to study climate sensitivity and feedback mechanisms. Many of these studies have focused on simulating the aftermath of the Mt. Pinatubo eruption in the Philippines at 15.1°N, 120.4°E in June 1991, which was both the largest eruption of the twentieth century and the eruption for which the stratospheric aerosol has been best observed [48–54]. During this eruption about 17 Tg (1 Tg = 10^{12} g) of SO_2 were injected into the lower stratosphere and subsequently converted into sulfate aerosols. There are three main foci of such studies addressed in the present study: analysis of the simulation of atmospheric temperature and precipitation response; simulation of the response of the extratropical circulation in the NH winter to season; and, recently emerged, analysis of volcanic impact on ocean.

The use of volcanic simulations as tests of model climate feedback and sensitivity is somewhat hampered by weather and climate fluctuations because any climate anomalies observed in the aftermath of these eruptions will also reflect other internally generated variability in the atmosphere–ocean system (e.g., El Niño/Southern Oscillation (ENSO), quasi-biennial oscillation (QBO) or chaotic weather changes). Due to limited observations one has to use models to better understand the physical processes forced in the climate system by volcanic impacts. With model simulations, one can perform multiple realizations to clearly isolate the volcanic climate signal, but the real world data are limited to the single realization during the period since quasi-global instrumental records have been available.

Models of different complexity were traditionally used to analyse volcanic climate impacts. Those models might simplify description of atmospheric and/or ocean processes [55], or mimic radiative effect of volcanic aerosol by decreasing of solar constant [56]. In the present study, to illustrate mechanisms of volcanic impacts on climate, a comprehensive coupled climate model, CM2.1, is used. Developed at the National Oceanic and Atmospheric Administration's (NOAA) Geophysical Fluid Dynamic Laboratory (GFDL), CM2.1 was used in the IPCC AR4 study [44,57]. This model calculates both atmosphere and ocean, and accounts interactively for volcanic aerosol radiative forcing. It is composed of four component models: atmosphere, land, sea ice and ocean. The coupling between the component models occurs at 2-h intervals. The atmospheric model has a grid spacing of 2.5° longitude by 2° latitude and 24 vertical levels. The dynamical core is based on the finite volume scheme of Lin [58]. The model contains a completely updated suite of model physics compared to the previous GFDL climate model, including new cloud prediction and boundary layer schemes, and diurnally varying solar insolation. The radiation code allows for explicit treatment of numerous radiatively important trace gases (including tropospheric and stratospheric ozone, halocarbons, etc.), a variety of natural and anthropogenic aerosols (including black carbon, organic carbon, tropospheric sulfate aerosols and volcanic aerosols), and dust particles. Aerosols in the model do not interact with the cloud scheme, so that indirect aerosol effects on climate are not considered. The land model is described in Milly and Shmakin [59]. Surface water is routed instantaneously to ocean destination points on the basis of specified drainage basins. The land cover type in the model uses a classification scheme with 10 different land cover types. The ocean model [60,61] has a nominal grid spacing of 1° in latitude and longitude, with meridional grid spacing decreasing in the tropics to 1/3° near the equator, and uses a tripolar grid to avoid polar filtering over the Arctic. The model has 50 vertical levels, including 22 levels with 10 m thickness each in the top 220 m. A novel aspect is the use of a true fresh-water-flux boundary condition. The sea ice model is a dynamical model with three vertical layers and five ice thickness categories. It uses the elastic-viscous-plastic rheology to calculate ice internal stresses, and a modified Semtner three-layer scheme for thermodynamics [62]. The aerosol optical characteristics were calculated following Stenchikov et al. [2] using optical depth from Sato et al. [38] and Hansen et al. [37]. The aerosol size distribution was assumed log-normal with fixed width of 1.8 μm [39].

In this study, various volcanic impacts on climate are illustrated using results from the model experiments and available observations. In each case, twin ensembles of volcano and control runs are conducted, and the response of the climate system calculated to volcanic forcing as the ensemble mean over the volcano runs minus the ensemble mean over the control runs. The variability within ensembles is used to estimate the statistical significance of climate signals.

3.1. Tropospheric Cooling and Stratospheric Warming

The analysis for the Pinatubo case is easier than for other big eruptions because aerosols were well observed and the climate responses were relatively well documented. However, Pinatubo erupted in an *El Niño* year and both volcanic and sea surface temperature (SST) effects overlapped at least in the troposphere. ENSO events that occurred near the times of volcanic eruptions could either mask or enhance the volcanic signal. Adams et al. [63] even argued that changing atmospheric circulation caused by volcanic eruptions could cause El Nino. Santer et al. [64] conducted a comprehensive analysis of the ENSO effect on the modelled and observed global temperature trends. Shindell et al. [4] addressed the issue of interfering volcanic and ENSO signals by specific sampling of eruptions so as SST signal will average out in the composite. Yang and Schlesinger [65,66] used Singular Value Decomposition (SVD) analysis to separate spatial patterns of the ENSO and volcanic signals in the model simulations and observations. They showed that ENSO signal is relatively weak over Eurasia but strong over North America contributing about 50% of the responses after the 1991 Mt. Pinatubo eruption.

The ENSO variability issue is addressed in the present study by comparing simulated and observed responses after extracting the *El Niño* contribution from the tropospheric temperature. Santer et al. [64] developed an iterative regression procedure to separate a volcanic effect from an *El Niño* signal using Microwave Sounding Unit (MSU) brightness temperature observations from the lower tropospheric channel 2LT [67]. The globally averaged synthetic 2LT temperature for the Pinatubo ensemble runs is calculated using model output and compared with the response from Santer et al. [64]. The simulated anomaly is calculated with respect to the mean over the corresponding control segments that have the same developing *El Niños* as in the perturbed runs. It is probably an ideal way to remove the *El Niño* effect from the simulations because the exact *El Niño* signal which would have developed in the model if the volcanic eruption did not occur is subtracted. This procedure, however, only works well for the initial *El Niño* when perturbed runs 'remember' their oceanic initial conditions. Figure 3 shows a comparison of synthetic ENSO-subtracted anomaly with the observed

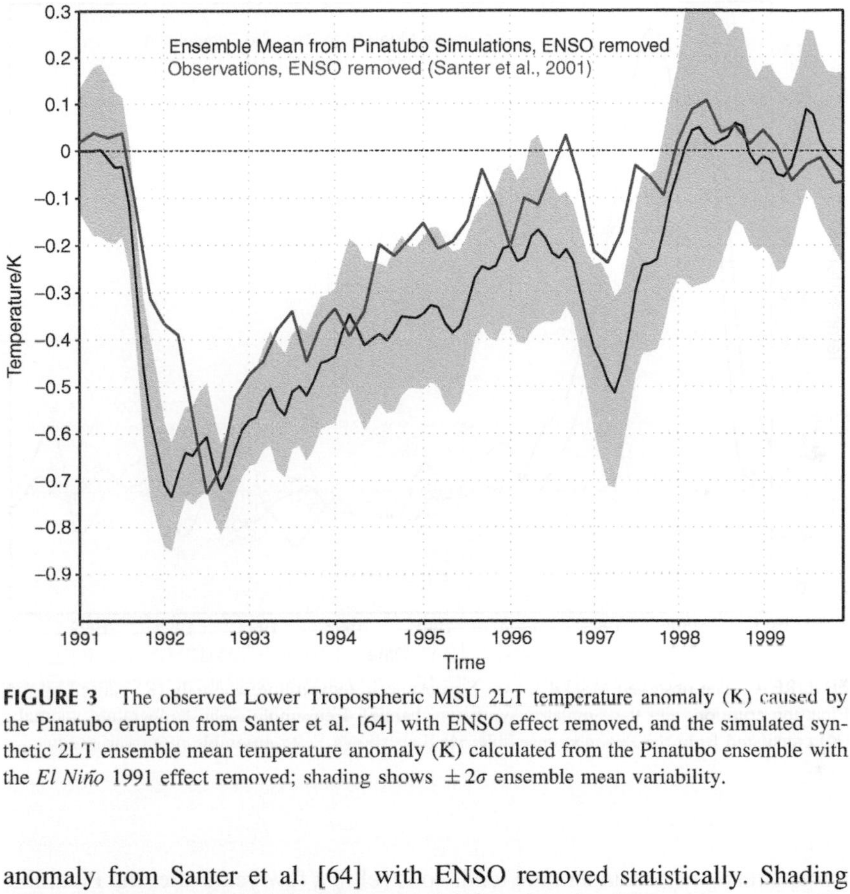

FIGURE 3 The observed Lower Tropospheric MSU 2LT temperature anomaly (K) caused by the Pinatubo eruption from Santer et al. [64] with ENSO effect removed, and the simulated synthetic 2LT ensemble mean temperature anomaly (K) calculated from the Pinatubo ensemble with the *El Niño* 1991 effect removed; shading shows $\pm 2\sigma$ ensemble mean variability.

anomaly from Santer et al. [64] with ENSO removed statistically. Shading shows doubled standard deviation variability for the 10-member ensemble mean. The observed MSU 2LT anomaly itself has much higher variability (not shown) because there is only one natural realization. Thus, the simulated Pinatubo signal in the lower tropospheric temperature reaches −0.7 K; it is statistically significant at 99% confidence level and the difference between simulated and observed responses is below the variability range. The lower tropospheric temperature anomaly reduces below the noise level in about 7 years, which corresponds approximately to the thermal response time of the ocean mixed layer [68].

For the lower stratosphere, a similar comparison was conducted as for the lower troposphere, but without removing ENSO because its effect in the lower stratosphere is fairly small. However, the stratospheric response to volcanic forcing might be affected by the phase of a QBO [7,69]. Figure 4 compares the simulated synthetic MSU channel 4 temperature for the lower stratosphere with the MSU 4 observations. The stratospheric warming is produced by aerosol IR and near-IR absorption. Ramaswamy et al. [70] discussed that the MSU

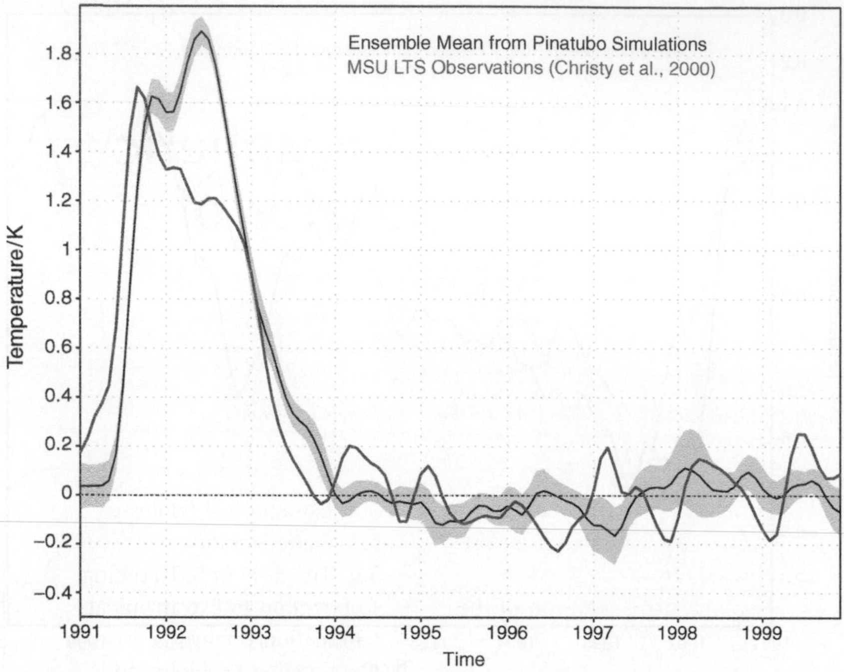

FIGURE 4 The observed MSU 4 Lower Stratospheric temperature anomaly (K) caused by the Pinatubo eruption, and the simulated synthetic channel 4 ensemble mean temperature anomaly (K) calculated from the Pinatubo ensemble; shading shows $\pm 2\sigma$ ensemble mean variability.

lower stratospheric temperature tends to level in a few years after the Pinatubo eruption; therefore, we calculate the anomalies in Fig. 4 with respect to the 1994–1999 mean both in the model and in the observation. The yellow shading shows the $\pm\ 2\sigma$ ensemble mean variability. The simulated signal compares well with the observation albeit slightly overestimates the stratospheric warming in the second year after the eruption. In the real world, the observed signal could be offset by the easterly phase of QBO in 1992/1993 but not in the model, which lacks QBO. The atmospheric response in the lower stratosphere follows the volcanic forcing and disappears in 3 years, as expected, when volcanic radiative forcing vanishes.

3.2. Effect on Hydrological Cycle

Precipitation is more sensitive to variations of Solar SW Radiation than Thermal IR Radiation because SW radiation directly affects the surface energy budget and links to global precipitation changes through evaporation. Therefore, one could expect that volcanic aerosols might decrease global

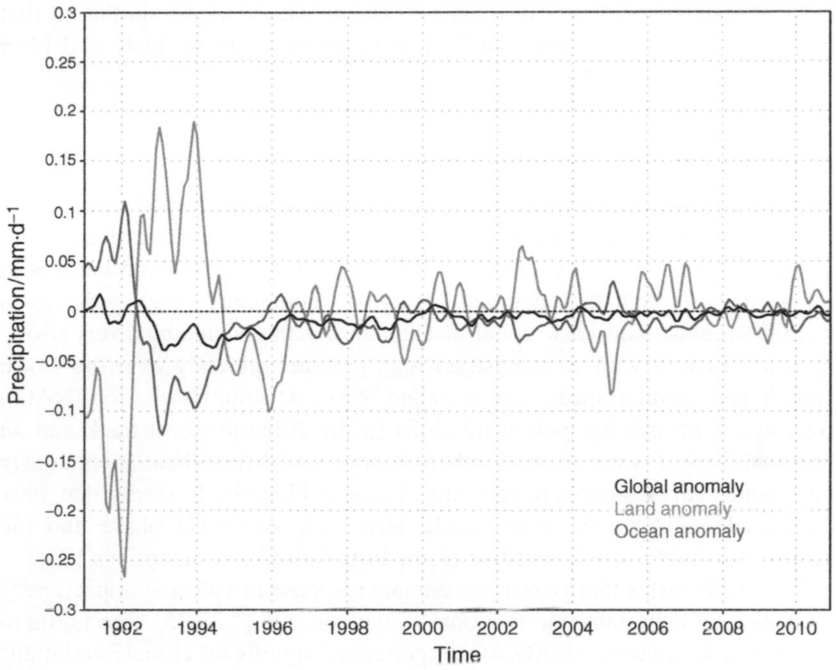

FIGURE 5 Time evolution of ensemble mean precipitation anomaly ($mm \cdot d^{-1}$) caused by Pinatubo eruption averaged over ocean, land, and globally, calculated with respect to a climatological mean.

precipitation for the period of 2–3 years when volcanic SW radiative forcing remains significant. This effect was detected in observations [71] and in model simulations [72,73]. The Pinatubo case-study analysis shows that in the ensemble mean results the global precipitation anomalies (Fig. 5) could be seen for almost 5–6 years because ocean cools and SST relaxes for about 7 years and affects the global hydrologic cycle. The precipitation anomalies over land and over ocean have different dynamics. The land precipitation drops during the first year because of rapid land radiative cooling. The ocean cooling and decrease of precipitation over ocean are delayed and reach maximum values in 3–4 years after the eruption when the sea surface temperature is coldest. The cold SST tends to shift precipitation over land and the land precipitation goes up compensating in part the decrease of precipitation over ocean. Geographically the precipitation anomalies are located in low latitude monsoon regions and could cause significant disruptions of food production in those regions with very high population density. It must be emphasized that ENSO contributes significantly to the observed precipitation anomalies. When ENSO signal is removed in the model results the amplitude of the precipitation anomalies significantly decreases although temporal behaviour does not change qualitatively. This suggests that in data analysis similar to that conducted by Trenberth and Dai [71], it is important to evaluate and filter out the ENSO contribution.

3.3. Volcanic Effect on Atmospheric Circulation

In the 2 years following major eruptions, the NH winter tropospheric circulation has typically been observed to display features characteristic of an anomalously positive AO index situation. This has a zonal-mean expression with low-pressure at high latitudes and a ring of anomalously high pressure in the mid-latitudes. This basic zonal-mean pattern is modulated by a very strong regional structure with an intensified high pressure anomaly over the North Atlantic and Mediterranean sectors called North Atlantic Oscillation (NAO). Consistent with this are pole-ward shifts in the Atlantic storm track and an increased flow of warm air to Northern Europe and Asia, where anomalously high winter surface temperatures are observed [7,8,39]. It seems that only low-latitude volcanic eruptions could affect the AO/NAO phase and the AO/NAO remain fairly insensitive to the high-latitude NH eruptions [74].

The mechanisms that govern the climate response to volcanic impacts very likely play an important role in global climate change [4,39,75]. The northern polar circulation modes (NAO/AO) experienced significant climate variations during the recent two decades and also are sensitive to volcanic forcing. The southern annular mode (SAM similar to AO) shows recently a very significant climate trend but is not sensitive to volcanic forcing [76]. Supposedly dynamics of the annular mode in the Southern Hemisphere are different than in the Northern Hemisphere and, to a great extent, are controlled by ocean processes and stratospheric ozone variations. In the present study, discussion is limited to NAO/AO.

The most robust effect on atmospheric temperature produced by volcanic aerosols is in the lower stratosphere. It is known that low-latitude explosive eruptions produce anomalously warm tropical lower stratospheric conditions and, in the NH winter, an anomalously cold and intense polar vortex. The tropical temperature anomalies at 50 hPa (Fig. 6) are a direct response to the enhanced absorption of terrestrial IR and solar near-IR radiation by the aerosols. The high-latitude winter perturbations at 50 hPa are a dynamical response to the strengthening of the polar vortex or polar night jet. This is due to stronger thermal wind produced by increasing of the equator-to-pole temperature gradient in the lower stratosphere [6,69,77–81].

The strengthening of the polar jet is amplified by a positive feedback between the polar NH winter vortex and vertical propagation of planetary waves. The stronger vortex reflects planetary waves decreasing deceleration and preserving axial symmetry of the flow. Stenchikov et al. [8] also found that tropospheric cooling caused by volcanic aerosols can affect storminess and generation of planetary waves in the troposphere. This tends to decrease

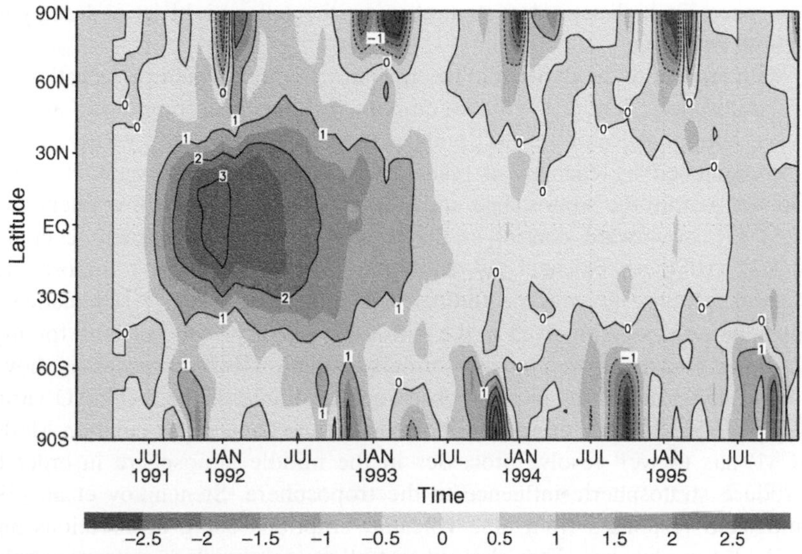

FIGURE 6 Zonal and ensemble mean stratospheric temperature anomaly (K) at 50 hPa (at about 25 km) calculated with respect to control experiment. (See Color Plate 4).

the flux of wave activity and negative angular momentum from the troposphere into the polar stratosphere, reducing wave drag on the vortex. To show this, Stenchikov et al. [8] conducted experiments with only solar, mostly tropospheric and surface cooling (no stratospheric warming). In these experiments, a positive phase of the AO was also produced because aerosol-induced tropospheric cooling in the subtropics decreases the meridional temperature gradient in the winter troposphere between 30°N and 60°N. The corresponding reduction of mean zonal energy and amplitudes of planetary waves in the troposphere decreases wave activity flux into the lower stratosphere. The resulting strengthening of the polar vortex forces a positive phase of the AO.

The high-latitude eruptions cannot warm lower stratosphere, and cannot cool subtropics as much as can low-latitude eruptions. Oman et al. [74] used GISS Model-E to simulate a climate impact of the 1912 Katmai eruption in Alaska. They calculated a 20-member ensemble of simulations and found that the volcanic aerosol cloud spread mostly north of 30 N could not produce a significant winter warming pattern even if it produced a higher hemispheric optical depth than that of the Pinatubo eruption in 1991.

Stenchikov et al. [8] also partitioned the dynamic effect of polar stratospheric ozone loss, caused by heterogeneous chemistry initiated by volcanic aerosols in the post-Pinatubo period. They found that ozone depletion caused a positive phase of the AO in late winter and early spring by cooling the lower stratosphere in high latitudes, strengthening the polar night jet and delaying the final warming.

With respect to the dynamical mechanisms through which perturbations of the stratospheric annular circulation can influence tropospheric annular modes, Song and Robinson [82] pointed out that tropospheric westerlies can be strengthened by changes of planetary wave vertical propagation and/or reflection within the stratosphere and associated wave–zonal flow interaction [77,81,83], downward control or the nonlinear effect of baroclinic eddies [84–87]. All these mechanisms could play a role in shaping tropospheric dynamic response to volcanic forcing. The diagram in Fig. 7 schematically shows the processes involved in the AO/NAO sensitivity to volcanic forcing.

The up-to-date climate models formally include all those processes shown in Fig. 7 but can not produce the observed amplitude of the AO/NAO variability [39,75]. Shindell et al. [6] reported that the General Circulation Model (GCM) has to well resolve processes in the middle atmosphere in order to reproduce stratospheric influence to the troposphere. Stenchikov et al. [39] composited responses from nine volcanic eruptions using observations and IPCC AR4 model runs. They showed that all models produce a stronger polar vortex in the Northern Hemisphere as a response to volcanic forcing but the dynamic signal penetrated to the troposphere is much weaker in the models than in observations. Figure 8 shows simulations by different models in the

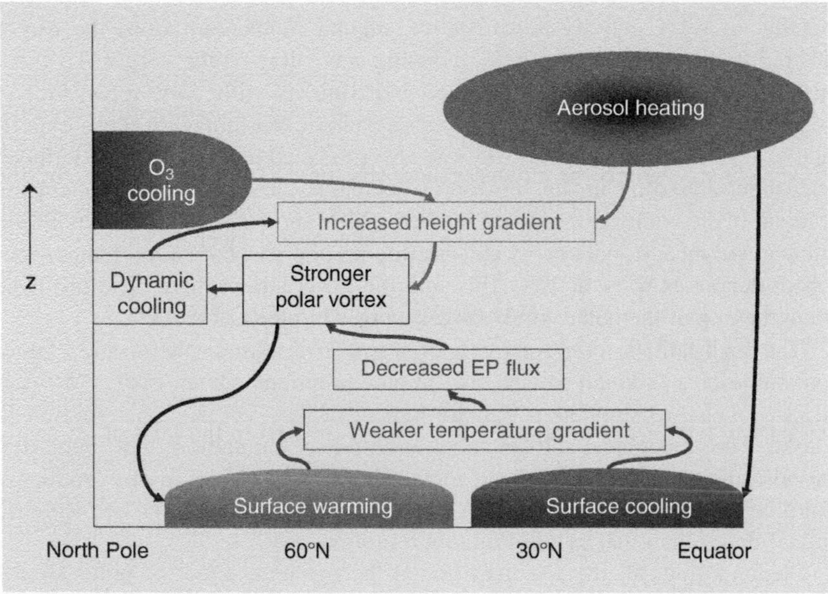

FIGURE 7 Schematic diagram depicting how the stratospheric and tropospheric gradient mechanisms are triggered by volcanic aerosol clouds in the tropical stratosphere. The wave feedback mechanism amplifies the response.

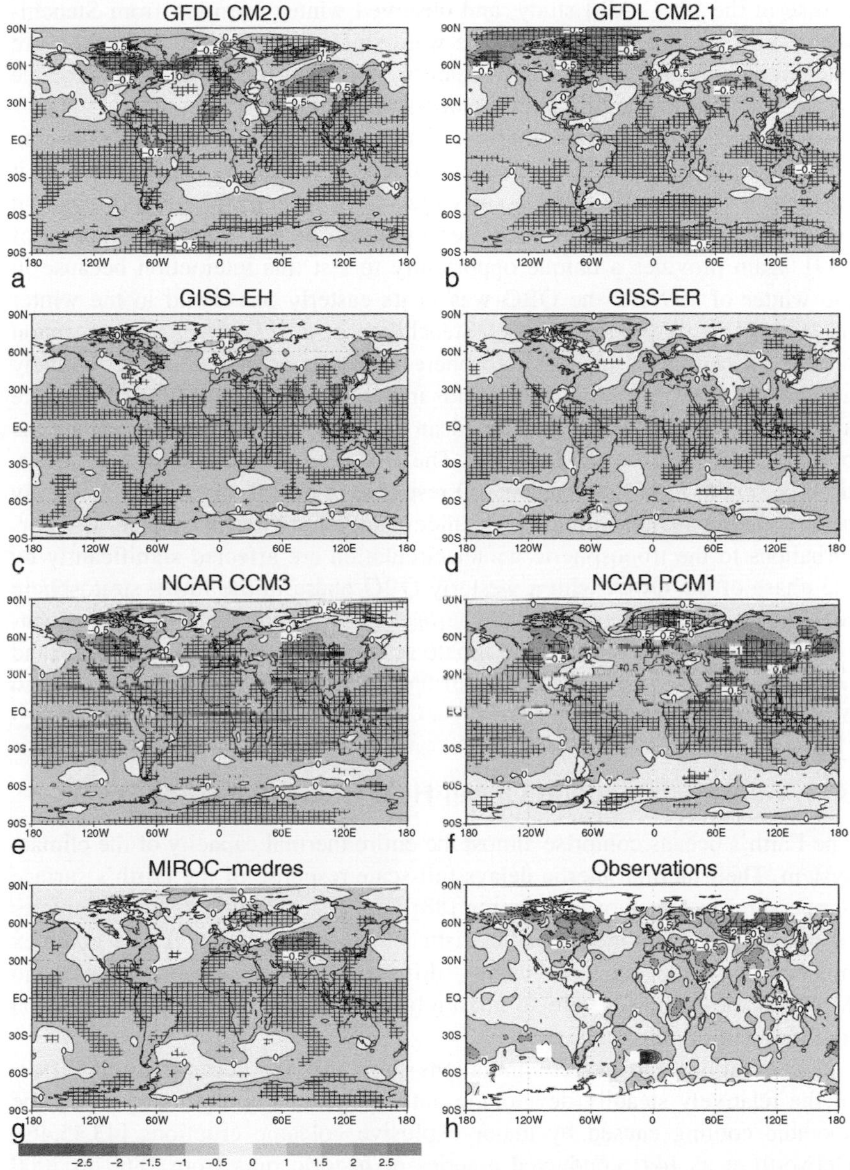

FIGURE 8 Surface winter (DJF) air temperature anomalies (K) composited for nine major volcanic eruptions from 1883 until present and averaged for two seasons and all available ensemble members: IPCC AR4 model simulations (a–g); observations from HadCRUT2v dataset (h). Hatching shows the areas with at least 90% confidence level calculated using a two-tailed local *t*-test. (See Color Plate 5).

course of the IPCC AR4 study, and observed winter warming from Stenchikov et al. [39] that caused by a pole-ward shift of tropospheric jet and more intensive transport of heat from ocean to land. The model tends to produce winter warming but significantly underestimates it.

It should be mentioned that the dynamic response to volcanic forcing could interact with the QBO that modulates the strength of polar vortex: it weakens and destabilizes the polar vortex in its easterly phase and makes it stronger and more stable in its westerly phase. The Mt. Pinatubo eruption of 1991 again provides a unique opportunity to test this interaction because in the winter of 1991/92 the QBO was in its easterly phase and in the winter of 1992/93 in its westerly phase. Stenchikov et al. [7] developed a version of the SKYHI troposphere–stratosphere–mesosphere model that effectively assimilates observed zonal mean winds in the tropical stratosphere to simulate a very realistic QBO and performed an ensemble of 24 simulations for the period 1 June 1991 to 31 May 1993. The model produced a reasonably realistic representation of the positive AO response in boreal winter that is usually observed after major eruptions. Detailed analysis shows that the aerosol perturbations to the tropospheric winter circulation are affected significantly by the phase of the QBO, with a westerly QBO phase in the lower stratosphere resulting in an enhancement of the aerosol effect on the AO. Improved quantification of the QBO effect on climate sensitivity helps to better understand mechanisms of the stratospheric contribution to natural and externally forced climate variability.

3.4. Volcanic Impact on Ocean Heat Content and Sea Level

The Earth's oceans comprise almost the entire thermal capacity of the climate system. Their thermal inertia delays full-scale response of the Earth's surface temperature to greenhouse warming [88]. The rate at which heat accumulates in oceans is an important characteristic of global warming. It is a complex process that involves slow energy diffusion and large-scale transport in MOC, as well as faster vertical mixing by seasonal thermo-haline convection and by wind-driven gyres.

Observations and model simulations show that the ocean warming effect of the relatively steadily developing anthropogenic forcings is offset by the sporadic cooling caused by major explosive volcanic eruptions [43,45,46]. Delworth et al. [44] conducted a series of historic runs from 1860 to 2000 in the framework of the IPCC AR4 study using GFDL CM2.1, and partitioning contributions of different forcings. Figure 9 shows the ensemble mean ocean heat content anomalies in the 0–3000 m depth range for a subset of the runs from Delworth et al. [44], calculated accounting for all the time varying forcing agents ('ALL') and for volcanic and solar forcings only ('NATURAL'). However, the solar effect for this period is small compared to the volcanic effect. The 'ALL' compares well with the Levitus et al. [89]

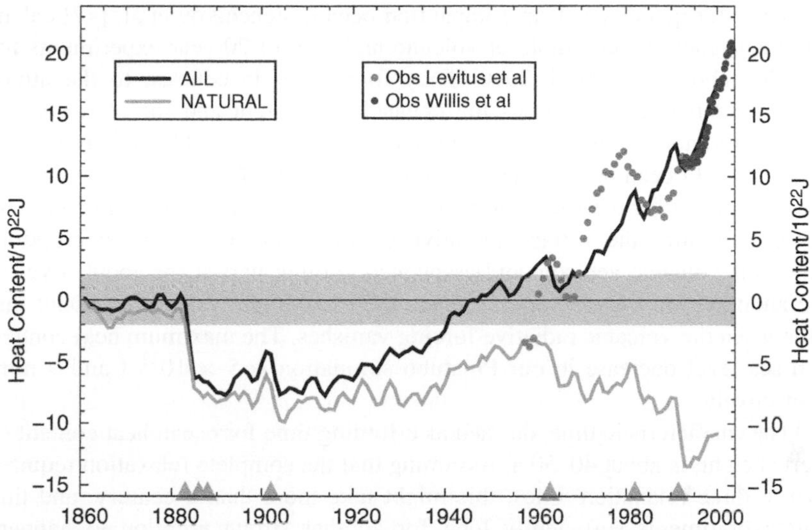

FIGURE 9 The ensemble mean ocean heat content anomalies in the 0–3000 m ocean layer. 'ALL' refers to the ensemble mean calculated with all the time varying forcing agents: well mixed greenhouse gases, anthropogenic aerosols, stratospheric and tropospheric ozone, land use, solar irradiance and volcanic aerosols. 'NATURAL' refers to the ensemble calculated accounting for volcanic and solar forcings only. The red and purple circles depict observational estimates based on, respectively, over 0–3000 m layer [89] and over 0–750 m layer [90]. Constant offsets have been added to the observed data so that their means were the same as the model data over the period of overlap. The shaded triangles along the time-axes denote the times of major volcanic eruptions. The shading shows plus or minus 2-standard deviations of ocean heat content estimated from a 2000 a control run of the climate model with forcings fixed at the 1860 level.

and Willis et al. [90] observations shown in Fig. 9, and, even better, with the improved analyses from Carton et al. [91] and Dominigues et al. [92] (not shown). Both 'ALL' and 'NATURAL' anomalies are highly statistically significant and far exceed the 'CONTROL' variability shown by shading. The cumulative cooling effect of natural forcings reaches 10^{23} J by year 2000, which is right between the estimates obtained by Church et al. [43] and Gleckler et al. [45] who conducted similar analysis, and offsets about one third of 'ALL' minus 'NATURAL' ocean warming. The volcanic signal exceeds two standard deviations of unforced variability throughout the entire run since the Krakatau eruption in 1883. This result suggests that the observed frequency and strength of the Earth's explosive volcanism in the nineteenth and twentieth centuries [1] was sufficient to produce a 'quasi-permanent' signature in the global oceans. Also, ocean warming (cooling) causes expansion (contraction) of water and therefore affects sea level or, so called, thermosteric height. This effect comprises a significant portion of the observed contemporary sea level rise.

To better quantify volcanic impact on ocean, Stenchikov et al. [47] calculated a 10-member ensemble of volcano and control 20-year experiments for the Pinatubo period 1991–2010. They found that in contrast to the atmospheric temperature responses the ocean heat content and the steric height remain well above noise level for decades. Figures 10 and 11 show anomalies of the global ocean heat content and the steric height for the Pinatubo ensembles calculated for the whole-depth ocean and for the upper 300 m layer. The ocean integrates the surface radiative cooling from the volcanic eruption. Since the volcanic aerosols and associated cooling persist for about 3 years, the anomalies in Figs. 10 and 11 reach their maximum value after about this time when the volcanic radiative forcing vanishes. The maximum heat content and sea level decrease in our Pinatubo simulation is 5×10^{22} J and 9 mm, respectively.

The characteristic time, defined as e-folding time for ocean heat content or steric height, is about 40–50 a. Assuming that the complete relaxation requires two to three relaxation times, this might take more than a century, and that length of time is sufficiently long for another strong eruption to happen. Therefore, the 'volcanic' cold anomaly in the ocean never disappears at the present frequency of the Earth's explosive volcanism.

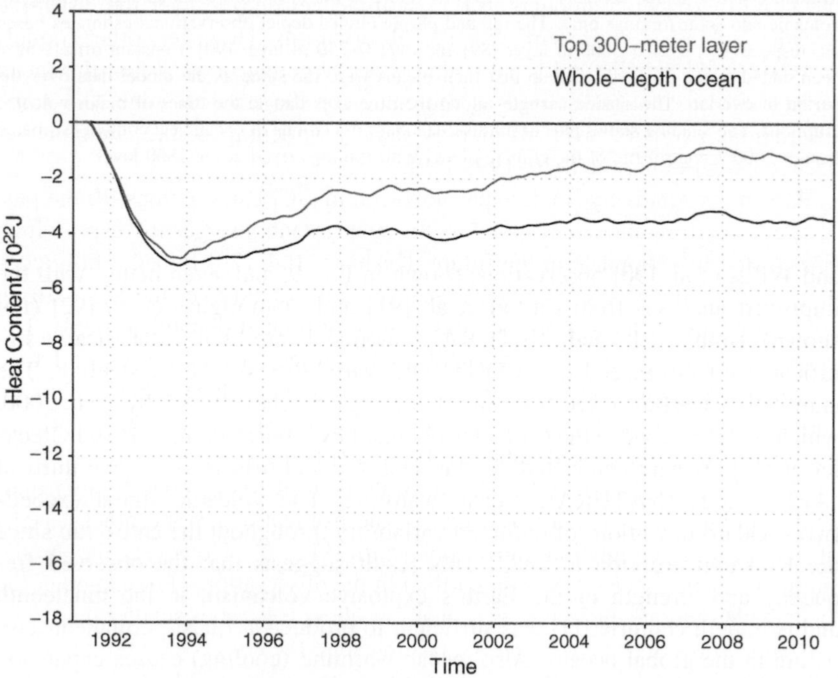

FIGURE 10 The global and ensemble mean ocean heat content (10^{22} J) anomaly for 300-m and whole depth ocean for the Pinatubo ensemble calculated with respect to ensemble control.

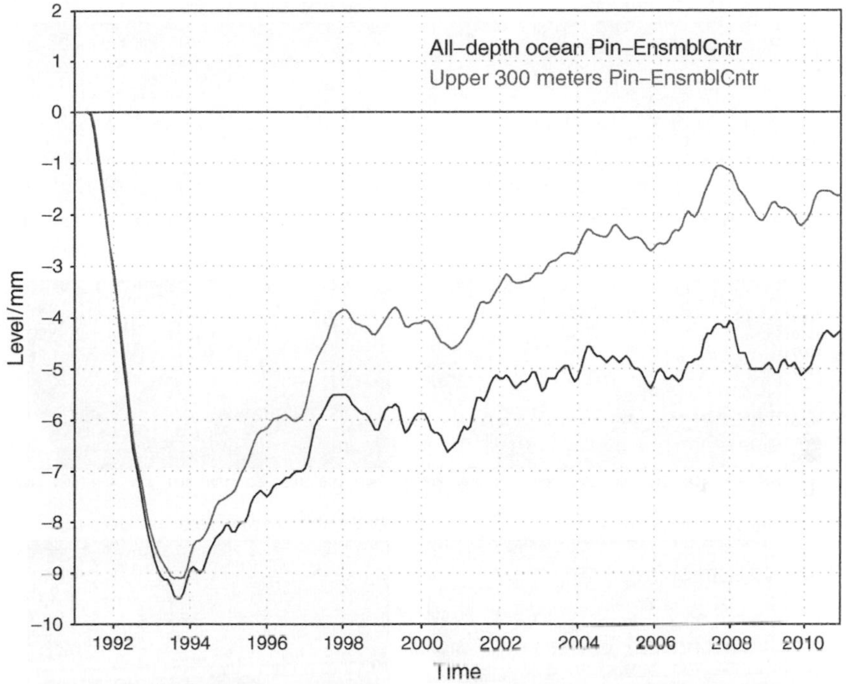

FIGURE 11 The global and ensemble mean thermosteric height anomalies (mm) for 300-m and whole depth ocean for the Pinatubo ensemble calculated with respect to ensemble control.

3.5. Strengthening of Overturning Circulation

The short-wave cooling from volcanic aerosol results in a cold surface temperature anomaly that develops during the first 3 years until volcanic aerosols vanish. Cold surface water is gradually transferred into the deeper ocean layers. A volcanically induced cooling leads to reduced precipitation and river runoff at high latitudes of the Northern Hemisphere, thereby leading to more saline (and hence denser) upper ocean conditions in the higher latitudes of the Northern Hemisphere. Both these factors (colder ocean temperature and enhanced salinity) destabilise the water column, making them more prone to ocean convection. The increased ocean convection tends to enhance the MOC. Further, an enhanced positive phase of the AO also leads to an MOC increase [93].

As a result, the MOC increases in response to the volcanic forcing (see Fig. 12). The maximum increase is 1.8 Sverdrups or about 9% (Sv; 1 Sv = 10^6 $m^3 \cdot s^{-1}$). The MOC has inherent decadal time scales of adjustment, and is thus maximum at some 5–15 a after the volcanic eruptions. An increase in

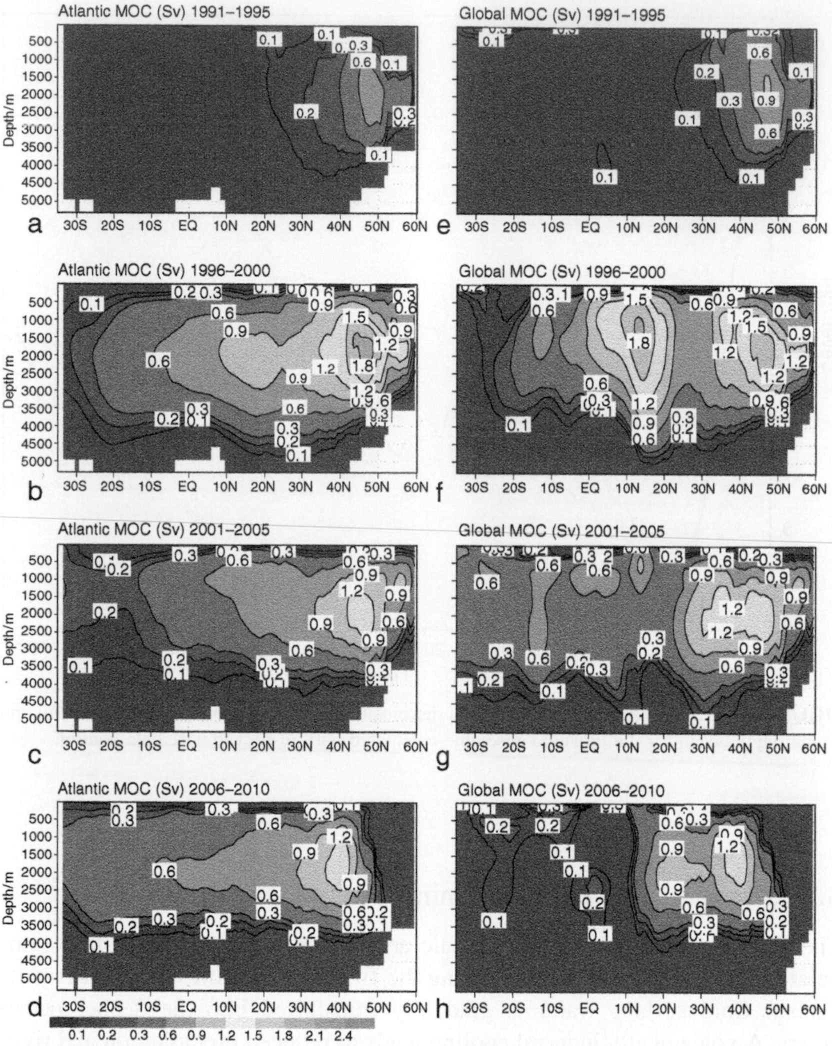

FIGURE 12 The 5-year means MOC anomalies (Sv) from the Pinatubo ensemble averaged zonally over Atlantic basin (a–d) and over the globe (e–h). (See Color Plate 6).

MOC also could cause in part the asymmetry of the ocean temperature response in the high northern and southern latitudes.

The simulations show a tendency for cooling of the deep waters in the Southern Ocean and warming in the deep waters of the Northern Ocean. This asymmetry could also be caused in part by the redistribution of ocean salinity, the forced positive phase of the AO during a few years following a volcanic eruption, and by a significant increase of sea ice extent and volume in the Northern Hemisphere.

3.6. Volcanic Impact on Sea Ice

The effect of volcanic forcing on the sea ice extent in the Northern Hemisphere is of great interest because significant loss of perennial sea ice under global warming is occurring in the Northern Hemisphere. Therefore, it is very important to better understand what factors could affect them the most. Figures 13 and 14 show the anomaly of the northern hemispheric mean annual maximum and minimum ice extent and mass for the Pinatubo runs. The Maximum Sea Ice Extent anomalies reach 0.6×10^6 km^2 in the Pinatubo run – it takes at least 5 years to develop. So, sea ice extent responds more strongly not to the radiative forcing but to ocean temperature and circulation. The sea ice extent relaxes to zero for a decade. It must be mentioned that both observed and simulated ice extent anomalies are not statistically significant, though in simulations they exceed one standard deviation. The minimum ice extent is more sensitive to radiative cooling and ocean temperature; therefore, its anomaly is stronger than the anomaly of the maximum ice extent reaching 0.9×10^6 km^2. It builds up in 3 years when the strongest ocean cooling develops and then declines for about 10 years.

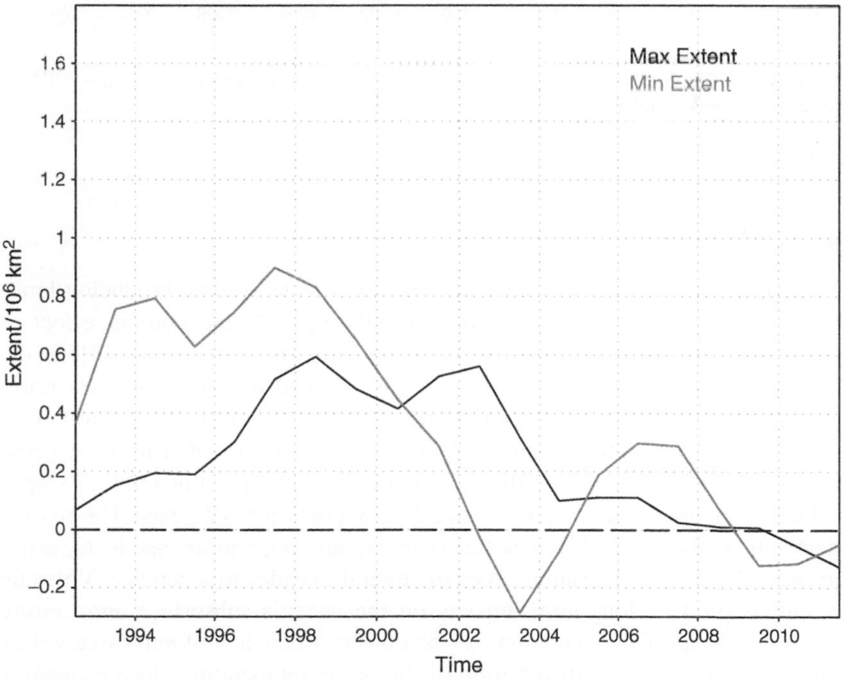

FIGURE 13 The Northern Hemisphere anomalies of maximum and minimum ice extent (10^6 km^2) for the Pinatubo ensemble.

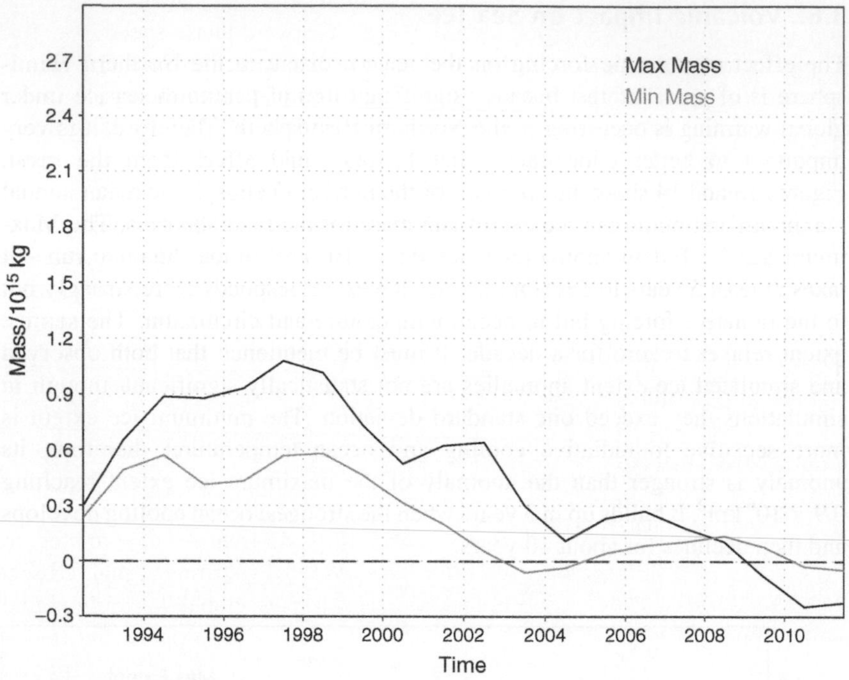

FIGURE 14 The Northern Hemisphere anomalies of maximum and minimum ice mass (10^{15} kg) for the Pinatubo ensemble.

4. SUMMARY

Volcanic eruptions force all elements of the climate system, producing long-term climate signals in ocean. The cumulative volcanic cooling effect at present offsets about one third of anthropogenic ocean warming [44]. In the atmosphere, however, volcanic signals are masked by meteorological noise in about 7 years in the model ensembles and much sooner in the real world. Radiative forcing produced by explosive volcanic events that have occurred in historic periods lasts for about 3 years. The volcanically induced tropospheric temperature anomalies reduce below noise for ~7 years. The sea ice responds on the decadal time scale. Deep ocean temperature, sea level, salinity and MOC have relaxation times of several decades to a century. Volcanic eruptions produce long-term impacts on the ocean's subsurface temperature and steric height that accumulate at the current frequency of explosive volcanic events. The vertical distribution of the ocean temperature change signal is asymmetric at high latitudes. A cooling signal penetrates to depth at high Southern latitudes, while a warming signal penetrates to depth at high latitudes of the Northern Hemisphere. This asymmetry is caused in part by an increase in MOC. The decrease of ocean steric height in our simulations, caused by the Pinatubo eruption, reaches 9 mm in comparison with 5 mm estimated by Church et al. [43] from observations. The ocean heat content decreases by 5×10^{22} J. The maximum sea ice extent and ice mass increase in the Pinatubo runs by 0.5×10^{6} km^2 and 1.0×10^{15} kg, respectively. This corresponds to 3% and 5% of the model 'control' maximum extent and mass in the Pinatubo runs. The simulated minimum ice extent is more sensitive to volcanic forcing than the maximum ice extent. The Atlantic MOC strengthens in the Pinatubo runs very significantly by 1.8 Sv or 9% of its maximum value.

Atmospheric temperature anomalies forced by the Pinatubo eruption in the troposphere and lower stratosphere are well reproduced by the models. However, forced AO/NAO responses are underestimated and observed sea level and ocean heat content anomalies are overestimated by all models. Nevertheless, all model results and observations suggest that volcanoes could produce long-lasting impact on ocean heat content and thermo-steric level that, in fact, could affect estimates of current climate trends. Quasi-periodic nature of volcanic cooling facilitates ocean vertical mixing and might have an important effect on the thermal structure of the deep ocean. Therefore, it has to be realistically implemented in climate models for calculating 'quasi-equilibrium' initial conditions, climate reconstructions and for future climate projections.

ACKNOWLEDGEMENTS

This work was supported in part by NASA grant NNG05GB06G, NSF grantATM-0351280, by UCAR Visiting Scientist Program and by the NOAA Geophysical Fluid Dynamics Laboratory (GFDL).

REFERENCES

1. T. Simkin, Annu. Rev. Earth Planet Sci. 21 (1993) 427–452.
2. G.L. Stenchikov, I. Kirchner, A. Robock, H.F. Graf, J.C. Antuña, R.G. Grainger, A. Lambert, L. Thomason, J. Geophys. Res. 103 (1998) 13837–13857.
3. M. Collins, in: A. Robock, C. Oppenheimer (Eds.), Predictions of Climate Following Volcanic Eruptions, American Geophysical Union, Washington, DC, 2003, pp. 283–300.
4. D. Shindell, G. Schmidt, M. Mann, G. Faluvegi, J. Geophys. Res. 109 (2004) D05104.
5. D. Shindell, G. Schmidt, R. Miller, M. Mann, J. Clim. 16 (2003) 4094–4107.
6. D.T. Shindell, G.A. Schmidt, R.L. Miller, D. Rind, J. Geophys. Res. 106 (2001) 7193–7210.
7. G. Stenchikov, K. Hamilton, A. Robock, V. Ramaswamy, M.D. Schwarzkopf, J. Geophys. Res. 109 (2004) D03112.
8. G. Stenchikov, A. Robock, V. Ramaswamy, M.D. Schwarzkopf, K. Hamilton, S. Ramachandran, J. Geophys. Res. 107 (2002) 4803.
9. G.J. Boer, M. Stowasser, K. Hamilton, Clim. Dyn. 28 (2007) 481–502.
10. B.J. Soden, R.T. Wetherald, G.L. Stenchikov, A. Robock, Science 296 (2002) 727–730.
11. T.M.L. Wigley, C.M. Ammann, B.D. Santer, S.C.B. Raper, J. Geophys. Res. 110 (2005) D09107.

12. M.L. Asaturov, M.I. Budyko, K.Y. Vinnikov, P.Y. Groisman, A.S. Kabanov, I.L. Karol, M.P. Kolomeev, Z.I. Pivovarova, E.V. Rozanov, S.S. Khmelevtsov, Volcanic Stratospheric Aerosols and Climate (in Russian), St. Petersburg, Russia, Gidrometeoizdat, 1986, 256pp.
13. H.W. Elsaesser, Isolating the climatologic effects of volcanoes. Report UCRL-89161. Lawrence Livermore National Laboratory, Livermore, CA, 1983, 29pp.
14. K.Y. Kondratyev, Volcanoes and Climate. WCP-54, WMO/TD-166, World Meteorological Organisation, Geneva, 1988, 103pp.
15. K.Y. Kondratyev, I. Galindo, Volcanic Activity and Climate, A. Deepak, Hampton, VA, 1997, 382pp.
16. H.H. Lamb, Philos. Trans. R. Soc. Lond. Ser. A 266 (1970) 425–533.
17. A. Robock, Rev. Geophys. 38 (2000) 191–219.
18. O.B. Toon, in: A. Deepak (Ed.), Atmospheric Effects and Potential Climatic Impact of the 1980 Eruptions of Mount St. Helems, NASA Conference Publication 2240, 1982, pp. 15–36.
19. O.B. Toon, J.B. Pollack, Am. Sci. 68 (1980) 268–278.
20. M.F. Gerstell, J. Crisp, D. Crisp, J. Clim. 8 (1995) 1060–1070.
21. D.J. Lary, M. Balluch, S. Bekki, Q J R Meteorol. Soc. 120 (1994) 1683–1688.
22. W. Zhong, J. Haigh, R. Toumi and S. Bekki, Q. J. R. Meteorol. Soc. 122 (1996) 1459–1466.
23. D. Budner, J. Cole-Dai, in: A. Robock, C. Oppenheimer (Eds.), The Number and Magnitudes of Large Explosive Volcanic Eruptions Between 904 and 1865 A.D.: Quanitative Evidence From A New South Pole Ice Core, American Geophysical Union, Washington, DC, 2003, pp. 165–176.
24. E. Mosley-Thompson, T.A. Mashiotta, L.G. Thompson, in: A. Robock, C. Oppenheimer (Eds.), High Resolution Ice Core Records of Late Holocene Volcanism: Current and Future Contributions From The Greenland PARCA Cores, American Geophysical Union, Washington, DC, 2003, pp. 153–164.
25. K. Yalcin, C. Wake, M. Germani, J. Geophys. Res. 107 (2002) 4012.
26. A.J. Krueger, S.J. Schaefer, N.A. Krotkov, G. Bluth, S. Barker, Geophys. Monograph 116 (2000) 25–43.
27. A. Prata, W. Rose, S. Self, D. O'Brien, in: A. Robock, C. Oppenheimer (Eds.), Global, Long-Term Sulphur Dioxide Measurements From TOVS Data: A New Tool For Studying Explosive Volcanism and Climate, American Geophysical Union, Washington, DC, 2003, pp. 75–92.
28. J.C. Antuña, A. Robock, G.L. Stenchikov, L.W. Thomason, J.E. Barnes, J. Geophys. Res.-Atm. 107 (2002) pp. 4194.
29. J.C. Antuña, A. Robock, G.L. Stenchikov, J. Zhou, C. David, J. Barnes, L. Thomason, J. Geophys. Res. 108 (2003) 4624.
30. C.E. Randall, R.M. Bevilacqua, J.D. Lumpe, K.W. Hoppel, J. Geophys. Res. 106 (2001) 27525–27536.
31. C.E. Randall, R.M. Bevilacqua, J.D. Lumpe, K.W. Hoppel, D.W. Rusch, E.P. Shettle, J. Geophys. Res. 105 (2000) 3929–3942.
32. L. Thomason, T. Peter, Assessment of Stratospheric Aerosol Properties. WCRP-124, WMO/TD-1295, SPARC Report 4. World Climate Research Program. (2006) 318pp.
33. L.W. Thomason, G. Taha, Geophys. Res. Lett. 30 (2003) 1631.
34. C. Amman, G. Meehl, W. Washington, C. Zender, Geophys. Res. Lett. 30 (2003) 1657.
35. J. Bauman, P. Russell, M. Geller, P. Hamill, J. Geophys. Res. 108 (2003) 4382.
36. J. Bauman, P. Russell, M. Geller, P. Hamill, J. Geophys. Res. 108 (2003) 4383.
37. J. Hansen, J. Geophys. Res. 107 (2002) 4347.
38. M. Sato, J. Hansen, M.P. McCormick, J. Pollack, J. Geophys. Res. 98 (1993) 22987–22994.

39. G. Stenchikov, K. Hamilton, R.J. Stouffer, A. Robock, V. Ramaswamy, B. Santer, H.F. Graf, J. Geophys. Res. 111 (2006) D07107.
40. C. Bingen, D. Fussen, F. Vanhellemont, J. Geophys. Res. 109 (2004) D06201.
41. C. Bingen, D. Fussen, F. Vanhellemont, J. Geophys. Res. 109 (2004) D06202.
42. A. Robertson, J. Overpeck, D. Rind, E. Mosley-Thomson, G. Zelinski, J. Lean, D. Koch, J. Penner, I. Tegen, R. Healy, J. Geophys. Res. 106 (2001) 14783–14803.
43. J. Church, N. White, J. Arblaster, Nature 438 (2005) 74–77.
44. T.L. Delworth, V. Ramaswamy, G.L. Stenchikov, Geophys. Res. Lett. 32 (2005) L24709.
45. P.J. Gleckler, T.M.L. Wigley, B.D. Santer, J.M. Gregory, K. AchutaRao, K.E. Taylor, Nature 439 (2006) 675.
46. J. Gregory, J. Lowe, S. Tett, J. Clim. 19 (2006) 4576–4591.
47. G. Stenchikov, V. Ramaswamy, T. Delworth, Impact of Big Tambora Eruption on ENSO, Ocean Heat Uptake, and Sea Level, PP31E-07, Presented at 2007 Fall AGU Meeting, San Francisco, CA, 2007.
48. A.J. Baran, J.S. Foot, J. Geophys. Res. 99 (1994) 25673–25679.
49. J.E. Barnes, D.J. Hoffman, Geophys. Res. Lett. 24 (1997) 1923–1926.
50. G.J.S. Bluth, S.D. Doiron, A.J. Krueger, L.S. Walter, C.C. Schnetzler, Geophys. Res. Lett. 19 (1992) 151–154.
51. G.J.S. Bluth, W.I. Rose, I.E. Sprod, A.J. Krueger, J. Geol. 105 (1997) 671–683.
52. A. Lambert, R.G. Grainger, J.J. Remedios, C.D. Rogers, M. Corney, F.W. Taylor, Geophys. Res. Lett. 20 (1993) 1287–1290.
53. W.G. Read, L. Froidevaux, J.W. Waters, Geophys. Res. Lett. 20 (1993) 1299–1302.
54. P. Minnis, E.F. Harrison, L.L. Stowe, G.G. Gibson, F.M. Denn, D.R. Doelling, W.L. Smith Jr., Science 259 (1993) 1411–1415.
55. T. Crowley, Science 289 (2000) 270–277.
56. A. Broccoli, K. Dixon, T. Delworth, T. Knutson, R. Stouffer, J. Geophys. Res. 108 (2003) 4798.
57. GFDL. CM2.x References, 2006, Available from: http://nomads.gfdl.noaa.gov/CM2.X/references.
58. S.-J. Lin, Monthly Weather Rev. 132 (2004) 2293–2307.
59. P.C.D. Milly, A.B. Shmakin, J. Hydrometeor. 3 (2002) 283–299.
60. A. Gnanadesikan, K.W. Dixon, S.M. Griffies, V. Balaji, M. Barreiro, J.A. Beesley, W.F. Cooke, T.L. Delworth, R. Gerdes, M.J. Harrison, I.M. Held, W.J. Hurlin, H.-C. Lee, Z. Liang, G. Nong, R.C. Pacanowski, A. Rosati, J. Russell, B.L. Samuels, Q. Song, M.J. Spelman, R.J. Stouffer, C.O. Sweeney, G. Vecchi, M. Winton, A.T. Wittenberg, F. Zeng, R. Zhang, J.P. Dunne, J. Clim. 19 (2006) 675–697.
61. S.M. Griffies, A. Gnanadesikan, K.W. Dixon, J.P. Dunne, R. Gerdes, M.J. Harrison, A. Rosati, J.L. Russell, B.L. Samuels, M.J. Spelman, M. Winton, R. Zhang, Ocean Sci. 1 (2005) 45–79.
62. M. Winton, J. Atmos. Oceanic Technol. 17 (2000) 525–531.
63. J.B. Adams, M.E. Mann, C.M. Ammann, Nature 426 (2003) 274–278.
64. B.D. Santer, T.M.L. Wigley, C. Doutriaux, J.S. Boyle, J.E. Hansen, P.D. Jones, G.A. Meehl, E. Roeckner, S. Sengupta, K.E. Taylor, J. Geophys. Res. 106 (2001) 28033–28059.
65. F. Yang, M. Schlesinger, J. Geophys. Res. 106 (2001) 14757–14770.
66. F. Yang, M. Schlesinger, J. Geophys. Res. 107 (2002) 4073.
67. J.R. Christy, R.W. Spencer, W.D. Braswell, J. Atmos. Oceanic Technol. 17 (2000) 1153–1170.
68. S. Manabe, R.J. Stouffer, J. Geophys. Res. 85 (1980) 5529–5554.

69. I. Kirchner, G. Stenchikov, H. Graf, A. Robock, J. Antuña, J. Geophys. Res. 104 (1999) 19039–19055.
70. V. Ramaswamy, M.D. Schwarzkopf, W. Randel, B. Santer, B.J. Soden, G. Stenchikov, Science 311 (2006) 1138–1141.
71. K. Trenberth, A. Dai, Geophys. Res. Lett. 34 (2007) L15702.
72. L. Oman, A. Robock, G.L. Stenchikov, T. Thordarson, Geophys. Res. Lett. 33 (2006) L18711.
73. A. Robock, Y. Liu, J. Clim. 7 (1994) 44–55.
74. L. Oman, A. Robock, G. Stenchikov, G.A. Schmidt, R. Ruedy, J. Geophys. Res. 110 (2005) D13103.
75. R.L. Miller, G.A. Schmidt, D. Shindell, J. Geophys. Res. 111 (2006) D18101.
76. A. Robock, T. Adams, M. Moore, L. Oman, G. Stenchikov, Geophys. Res. Lett. 34 (2007) L23710.
77. K. Kodera, J. Geophys. Res. 99 (1994) 1273–1282.
78. K. Kodera, Y. Kuroda, Geophys. Res. Lett. 27 (2000) 3349–3352.
79. K. Kodera, Y. Kuroda, J. Geophys. Res. 105 (2000) 12361–12370.
80. Y. Ohhashi, K. Yamazaki, J. Meteorol. Soc. Jpn. 77 (1999) 495–511.
81. J. Perlwitz, H.-F. Graf, J. Clim. 8 (1995) 2281–2295.
82. Y. Song, W.A. Robinson, J. Atmos. Sci. 61 (2004) 1711–1725.
83. J. Perlwitz, N. Harnik, J. Clim. 16 (2003) 3011–3026.
84. R. Black, J. Clim. 15 (2002) 268–277.
85. R.X. Black, B.A. McDaniel, J. Clim. 17 (2004) 3990–4004.
86. P.H. Haynes, M.E. McIntyre, T.G. Shepherd, C.J. Marks, K.P. Shine, J. Atmos. Sci. 48 (1991) 651–680.
87. V. Limpasuvan, D.J.W. Thompson, D.L. Hartmann, J. Clim. 17 (2004) 2584–2596.
88. G. Meehl, W. Washington, W. Collins, J. Arblaster, A. Hu, L. Buja, W. Strand, H. Teng, Science 307 (2005) 1769–1772.
89. S. Levitus, J. Antonov, T. Boyer, Geophys. Res. Lett. 32 (2005) L02604.
90. J.K. Willis, D. Roemmich, B. Cornuelle, J. Geophys. Res. 109 (2004) C12036.
91. J. Carton, A. Santorelli, J. Clim. 2008, 21 (22), 6015–6035.
92. C. Dominigues, J. Church, N. White, P. Gleckler, S. Wijffels, P. Barker, J. Dunn, Nature 453 (2008) 1090–1093.
93. T.L. Delworth, K.W. Dixon, Geophys. Res. Lett. 33 (2006) L02606.

Chapter 5

The Role of Variations of the Earth's Orbital Characteristics in Climate Change

Lucas J. Lourens
Department of Earth Sciences, Faculty of Geosciences, Utrecht University, Budapestlaan 4, 3584 CD Utrecht, The Netherlands

Erik Tuenter
Institute for Marine and Atmospheric Research Utrecht (IMAU), P.O. Box 80 000, 3508 TA Utrecht, The Netherlands

1. Introduction
2. Astronomical Parameters
2.1. Eccentricity
2.2. Precession and Obliquity
2.3. Insolation
3. Orbital-Induced Climate Change
3.1. Ice Ages
3.2. Low and Mid-Latitude Climate Changes
3.3. Greenhouse World
4. Conclusion
References

1. INTRODUCTION

The climate of the Earth is characterised by trends, aberrations and quasi-periodic oscillations varying over a broad range of time-scales [1]. The trends are largely controlled by plate tectonics, and thus tend to change gradually on a million year (Ma) time scale. Aberrations occur when certain thresholds are passed and are manifested in the geological record as unusually rapid (less than a few thousand of years) or extreme changes in climate. The quasi-periodic oscillations are mostly astronomically paced; they are driven by astronomical perturbations that affect the Earth's orbit around the Sun and the orientation of the Earth's rotation axis with respect to its orbital plane. These perturbations are described by three main astronomical cycles: eccentricity (shape of the Earth's orbit), precession (date of perihelion) and obliquity (angle between the equator and orbital plane), which together determine the spatial and seasonal pattern of insolation received by the Earth, eventually resulting in climatic oscillations of tens to hundreds of thousands of years. The expression of these astronomical-induced climate oscillations is found in geological archives of widely different ages and environments.

Computation of the orbital solution of the Earth is complex because the Earth's motion is perturbed by our Moon and all the other planets of the Solar System. Much of our knowledge of the planetary orbits dates back to the investigations of Johannes Kepler (1571–1630) and the universal gravitational theory of Sir Isaac Newton (1643–1727). The first approximate solutions were established by Lagrange [2,3] and Pontécoulant [4], but it was Louis Agassiz [5], who formulated a sweeping theory of Ice Ages that triggered the search for a correlation between large-scale climatic changes and variations of the Earth's astronomical parameters. Shortly after, Adhémar [6] proposed that glaciations originated from the precession of the Earth's rotation axis that alters the lengths of the seasons. He suggested that when the lengths of the winters last longer a glaciation would occur. According to his theory, the Northern Hemisphere (NH) and the Southern Hemisphere (SH) would be glaciated during the opposing phases of the precession cycle. He evidenced his idea with the present Antarctic ice sheet and the fact that the NH is essentially not glaciated.

After the publication of a more precise solution of the Earth by Le Verrier [7], Croll [8] proposed that the variation of the Earth's eccentricity was also an important parameter for understanding past climates through its modulation of precession. He elaborated Adhémar's idea that winter insolation is critical for glaciation, but argued that the large continental areas covered with snow would turn into ice sheets because of a positive ice-albedo feedback.

The first computations of the variations of obliquity due to secular changes in the motion of the Earth's orbital plane are due to Pilgrim [9]. His computations were later used by Milankovitch [10] to establish his mathematical basis for the theory of the Ice Ages. Since then the understanding of the climate response to orbital forcing has evolved and is the subject of this chapter. However, all the necessary elements for the insolation computations were present in Milankovitch's work.

2. ASTRONOMICAL PARAMETERS

Bretagnon [11] made an important improvement to the orbital solution by computing terms of second order and third degree in the secular (mean) equations. His solution was used by Berger [12,13] for the computation of the precession and insolation quantities for the Earth following Sharav and Boudnikova [14,15]. Berger's publications have since been extensively used for paleoclimate reconstructions and climate modelling under the acronym Ber78.

Laskar [16–18] computed in an extensive way the secular equations giving the mean motion of the whole Solar System. It was clear from his computations that the traditional perturbation theory could not be used for the integration of the secular equations, due to strong divergences that became apparent in the system of the inner planets [16]. This difficulty was overcome by switching to a numerical integration of the secular equations with steps of 0.5 ka. These computations provided a much more accurate solution for the orbital motion of the Solar System over 10 Ma [18,19].

Extending his integration to 200 Ma, Laskar [20,21] demonstrated that the orbital motion of the planets, and especially that of the terrestrial planets, is chaotic, with an exponential divergence corresponding to an increase of the error by a factor 10 every 10 Ma. It seems, therefore, almost impossible to obtain a precise astronomical solution for paleoclimate studies over more than a few tens of millions of years [22].

A comparison between the La90 solution and the first direct numerical integration of the Solar System by Quinn and coworkers [23] revealed that the main obliquity and precession periods of the two solutions diverge with time over the past 3 Ma [24]. In the QTD91 solution, a term was introduced which describes the change in the speed of rotation of the Earth as a result of the dissipation of energy by the tides. If the same, present-day value is used in the La90 solution the discrepancy with QTD91 is almost completely removed. The resulting La90 solution with tidal dissipation set to the present-day value is now generally termed as the La93 solution. In this solution, also a second term can be modified. This term refers to the change in the dynamical ellipticity of the Earth, which may strongly depend on the build-up and retreat of large ice caps [25–28] and/or on long-term mantle convection processes [29]. Similar to the tidal dissipation term, a small change in the dynamical ellipticity of the Earth will change the main precession and obliquity frequencies.

The uncertain values of the tidal dissipation and dynamical ellipticity of the Earth are considered as the most limiting factors to obtain accurate solutions for the precession and obliquity time series of the Earth over a time span of millions of years, while the orbital part of the La93 solution was considered to be reliable over 10–20 Ma [24]. At present, there exists only one possible way to test the extent of change of both parameters in the (geological) past. This test involves a statistical comparison between the obliquity–precession interference patterns in the insolation time series and those observed in geological records [30]. Lourens and coworkers [31] showed for instance by using a record of climate change from the eastern Mediterranean, that over the past 3 Ma the decline in the speed of rotation was on average smaller than the average value obtained for the present day; this is probably a result of the large ice caps that dominated Earth's climate from the Late Pliocene to present.

In 2004, Laskar and coworkers [32] presented a new numerical solution from −250 to 250 Ma, which has been used for the direct calibration of the youngest geological period, the Neogene, spanning the last 23 Ma [33], and of the early Paleogene [34–36]. Beyond 40–50 Ma the chaotic evolution of the orbits still prevents a precise determination of the Earth motion. However, the most regular component of the orbital solution (i.e., the 405 ka period in eccentricity) could still be used over the last 250 Ma or full Mesozoic era.

2.1. Eccentricity

The Earth's orbit around the Sun is an ellipse. The plane in which the Earth moves around the Sun is called the Ecliptic of date, Ec_t (Fig. 1). The Sun is roughly located in one of its two foci. The eccentricity (e) of the Earth's orbit is defined as:

$$e = \frac{\sqrt{a^2 - b^2}}{a^2} \tag{1}$$

where a is the ellipse semi-major axis and b the semi-minor axis. The current eccentricity is 0.0167 but in the past hundred millions of years eccentricity has varied from about 0.0669 to almost 0.0001; that is, a near-circular orbit [32]. For the past 15 Ma, the three most important periods in the series expansion for eccentricity are about 405, 124 and 95 ka (Fig. 2e).

2.2. Precession and Obliquity

The locations along the Earth's orbit where the Sun is perpendicular to the equator at noon are called equinoxes (Fig. 1a). Then the night lasts as long as the day at all latitudes. Today this occurs on 20 March (vernal equinox, NH spring) and on 23 September (autumnal equinox, NH autumn). The summer (winter) solstice is defined as the location of the Earth when the Sun appears directly overhead at noon at its northernmost (southernmost) latitude, that is, the tropic of Cancer at 23.44°N (Capricorn at 23.44°S), which occurs on 21 June (22 December).

The Earth's rotational axis (φ) revolves around the normal (n) to the orbital plane like a spinning top (ψ in Fig. 1). This rotation causes a clockwise movement of the equinoxes and solstices along the Earth's orbit, called precession. The quasi-period of precession is 25 672 a relative to the stars, but because the Earth's orbit rotates in a counter clockwise direction with respect to the reference fixed Ecliptic (Ec_0) at Julian date J2000, the net period of precession is about 21.7 ka. The general precession in longitude ψ is thus defined by $\psi = \Lambda - \Omega$, where Ω is the longitude of the ascending node (N), and Λ the inclination of Ec_t (Fig. 1). The angle between the Earth's equatorial plane (Eq_t, Fig. 1) and Ec_t is the obliquity (ε). The current value for ε is 23.44° but it varied from about 22 to 24.5° during the past 15 Ma with a main period of about 41 ka (Fig. 2c). In the earlier episodes of Earth's history, obliquity oscillated at a much shorter period (i.e., ~29 ka at 500 Ma [37]). This is because the Earth's rate of rotation has declined with time due to tidal friction.

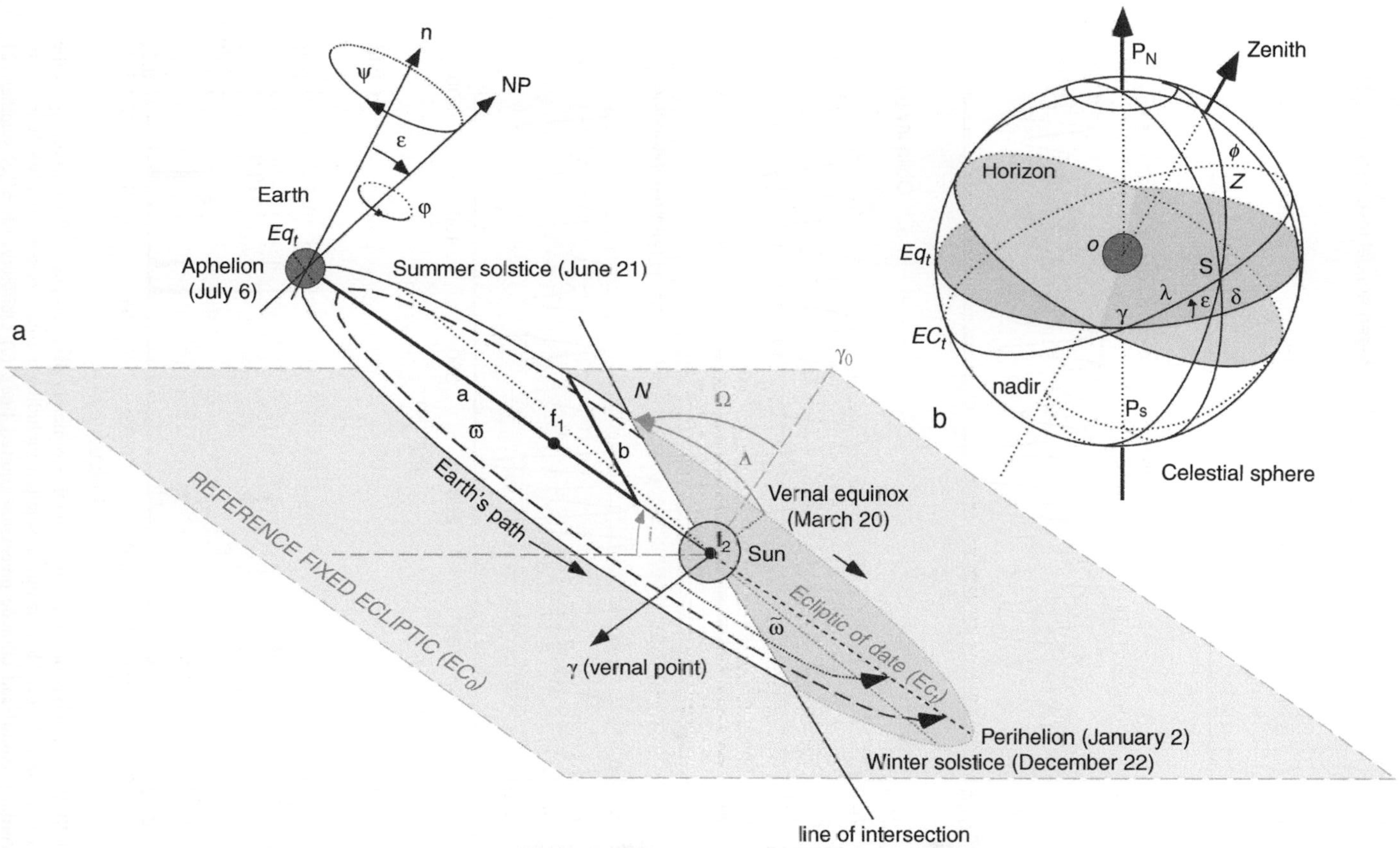

FIGURE 1 Astronomical configuration of the Earth. (a) Elements of the Earth's orbital parameters (modified after [101]). (b) Position of a point (S) on the celestial sphere (modified after [38]). See text for explanations. (See Color Plate 7).

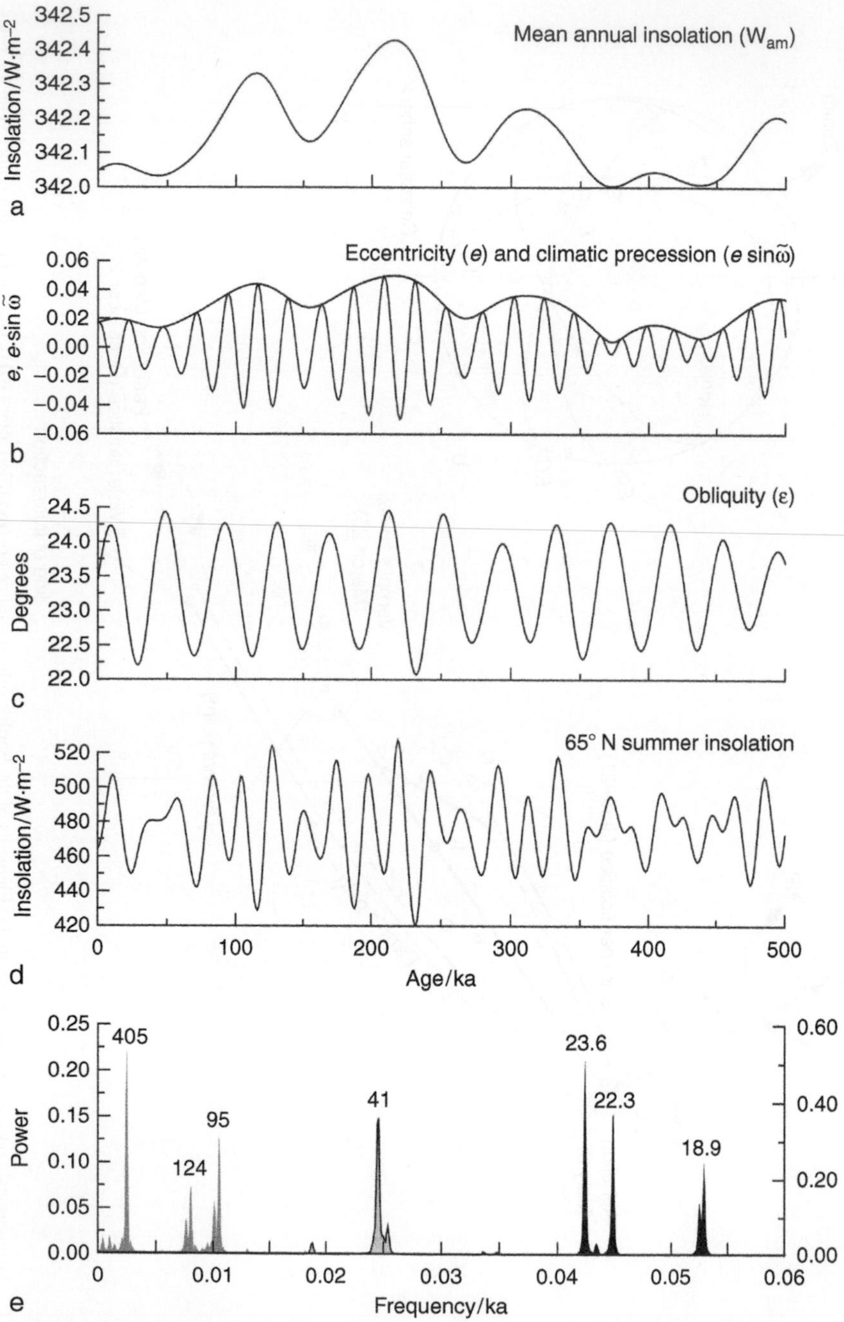

FIGURE 2 Variations of the Earth's orbital parameters over the past 500 ka according to the La04 solution [32]. (a) Mean annual insolation including a solar constant of 1368 W·m^{-2}. (b) Eccentricity (solid) and climatic precession (dotted line). (c) Obliquity. (d) 65°N summer (21 May–20 July) insolation. (e) Combined power spectrum of eccentricity (grey), obliquity (grey plus solid line) and precession (black) for the past 15 Ma.

For paleoclimate studies, the usual quantity that relates more to insolation is the climatic precession index $e \cdot \sin \overline{\omega}$, where $\overline{\omega} = \varpi + \psi$. ϖ is the approximation of the longitude of perihelion of the Earth from the fixed J2000 and $\overline{\omega}$ the resulting longitude of perihelion from the moving vernal equinox (Fig. 1a). In practice, however, the numerical calculations are done using the direction in which the Sun is seen from the Earth at the beginning of the spring, the so-called vernal point γ (Fig. 1) as reference. In most cases the climatic precession index is, therefore, given by $e \cdot \sin \tilde{\omega}$, where $\tilde{\omega} = \overline{\omega} - 180°$ [13]. This implies that climatic precession is at a minimum when NH summer solstice is in perihelion, so that when $\tilde{\omega}$ is 270° (or $\overline{\omega}$ is 90°).

The eccentricity term in the climatic precession index is operating as a modulator of the precession-related insolation changes (Fig. 2b). In case of a circular orbit (eccentricity is zero), perihelion is undefined and there is no climatic effect associated with precession, while in case the Earth's orbit is strongly elongated the effect of precession on insolation is at a maximum. The three most important periods of the climatic precession parameter over the past 15 Ma are about 23.6, 22.3 and 18.9 ka (Fig. 2e). Just as for obliquity, the periods of precession shorten back in time due to tidal dissipation with ~3–4 ka over the past 500 Ma [37].

2.3. Insolation

If the orbital parameters are known, the insolation for any latitude and at any time of the year can be computed. The mean annual insolation at the surface of the Earth depends only on the eccentricity and is represented by the following equation [24]:

$$W_{am} = \frac{S_0}{4\sqrt{(1 - e^2)}} \tag{2}$$

S_0 is called the 'solar constant'. In fact, the intensity of the Sun varies along with the number of sunspots. Recent observations have shown that when sunspots are numerous (scarce) the solar constant is about 1368 $W \cdot m^{-2}$ (1365 $W \cdot m^{-2}$). In literature, the various astronomical computations include values for S_0 ranging from 1350 $W \cdot m^{-2}$ [24] to 1360 $W \cdot m^{-2}$ [38]. The variations in mean annual insolation are very small, as they depend on the square of the eccentricity, with the largest mean annual insolation values reached during eccentricity maxima (Fig. 2a). Orbital-induced mean annual insolation changes are, therefore, not seen as the primary cause of past climate changes. On the contrary, according to Milankovitch's theory [10] summer insolation at high northern latitudes (Fig. 2d) played in particular a crucial role on the waxing and waning of the ice sheets. The theory states that in case insolation in summer was not high enough, ice sheets could expand. It is, therefore, important to compute the daily (or monthly) insolation at any given point on the Earth.

Following Berger and Loutre [38], insolation W received on a horizontal surface at latitude ϕ and a given time (H) during the course of the year (λ) is described by:

$$W(\phi, \lambda, H) = S_0\left(\frac{a}{r}\right)^2 \cos z \tag{3}$$

where r is the distance to the Sun and z the solar zenith angle (or zenith distance). The horizontal surface of the position of the observer (o) refers to the plane perpendicular to the direction of the local gravity, while the zenith is the point vertically upward (Fig. 1b). The zenith distance z of any point S (i.e., the position of the Sun seen in the sky from the observer at time H) on the celestial sphere is the angular distance from the zenith measured along the vertical circle through the given point (Fig. 1b). It varies from 0 to 180°.

The point S can also be calculated from the angle between the meridian (great circle through the celestial poles P_n and P_s, the zenith and the nadir) and the secondary great circle through the point and the poles (Fig. 1b). This angle is called the hour angle H. This gives the following relationship:

$$\cos z = \sin\phi \sin\delta + \cos\phi \cos\delta \cos H \tag{4}$$

The declination δ is the angular distance of point S measured from the equator on the secondary great circle. The latitude ϕ is the angular distance from the equator to the zenith measured on the meridian. The declination δ is related to the true longitude λ of the Earth by:

$$\sin\delta = \sin\lambda \sin\varepsilon \tag{5}$$

Over one year, λ varies from 0 to 360° while δ varies between $-\varepsilon$ and $+\varepsilon$. The Earth–Sun distance r is given by the ellipse equation:

$$r = \frac{a(1-e^2)}{1 + e\cos v} \tag{6}$$

with v being the true anomaly related to the true longitude λ of the Earth by:

$$v = \lambda - \varpi \tag{7}$$

Combining Eqns (4)–(7), Eqn (3) can be rewritten as

$$W(\phi, \lambda, H) = S_0\frac{(1 + e\cos(\lambda - \varpi))^2}{(1-e^2)^2}(\sin\phi \sin\lambda \sin\varepsilon + \cos\phi \cos\delta \cos H) \tag{8}$$

Over one year, ε, ϖ and e are assumed to be constant. Over a given day λ and δ are assumed to be constant, while H varies from 0 at solar noon to 24 h (0–360°). The long-term behaviour of each factor in Eqn (8) is thus governed by a different orbital parameter. The obliquity ε drives $\cos z$, the precession ϖ drives $(1 + e\cos(\lambda - \varpi))^2$ and the eccentricity e drives $(1-e^2)^{-2}$. Note that the eccentricity appears as $(1-e^2)^{-2}$ while in the mean annual insolation it appears as $(1-e^2)^{1/2}$ (see Eqn (2)).

To illustrate the influence of precession on insolation, we plotted the monthly averaged zonal insolation difference between a climatic precession minimum and maximum situation (Fig. 3a). This comparison shows that high northern (southern) latitudes receive (dispatch) more than 100 W·m^{-2} of additional insolation during summer (~20%) when they occur in perihelion

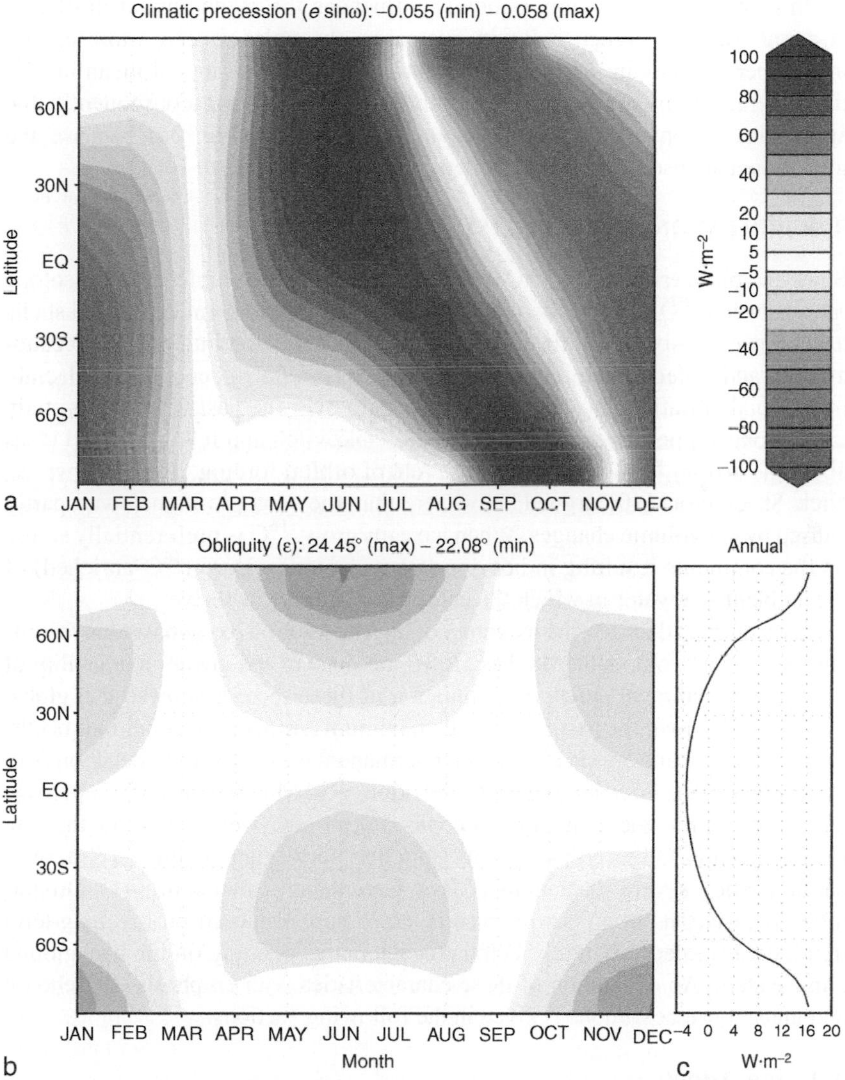

FIGURE 3 Monthly incoming differences in W·m^{-2} at the top of the atmosphere. (a) Insolation differences between a minimum (−0.055) and a maximum (+0.058) climatic precession configuration. (b) Insolation differences between a maximum obliquity (Tilt = 24.45°) and a minimum obliquity (Tilt = 22.08°) configuration with zero eccentricity. (c) As in (b), but now plotted the annual incoming insolation differences. Solar constant = 1360 W·m^{-2}. (See Color Plate 8).

(aphelion). In contrast, the accompanying NH (SH) winters receive less (more) insolation. Thus during a precession minimum seasonal contrasts at the NH increase, whereas they decrease at the SH. In contrast, a change in obliquity causes a simultaneous shift in seasonal contrasts at both hemispheres (Fig. 3b). From an obliquity minimum to maximum situation high-latitudes receive more than 50 $W \cdot m^{-2}$ of additional insolation during summer, while winters gain considerable ($\sim$15 $W \cdot m^{-2}$) less insolation.

In contrast to precession, obliquity influences the mean annual insolation at certain latitude. When obliquity increases, the poles receive more energy in summer but stay in the polar night during winter. The annual mean insolation, therefore, increases symmetrically at the poles and consequently the annual insolation must decrease around the equator (Fig. 3c) because the global annual insolation does not depend on obliquity (Eqn (2)).

3. ORBITAL-INDUCED CLIMATE CHANGE

Since the pioneering work of Cesare Emiliani [39], the stable oxygen isotope ratio between ^{16}O and ^{18}O (denoted by $\delta^{18}O$) of calcareous (micro) fossil shells has been extensively studied to improve our understanding of paleoceanographic and paleoclimate changes. He used this ratio to reconstruct glacial–interglacial variations in sea water temperature over the past 500 ka. His study gave strong support to the hypothesis of Milankovitch and revolutionised ideas about the history of the oceans and the role of orbital forcing. Soon afterwards, Nick Shackleton [40] argued, however, that the isotopic signal was partly caused by ice volume changes. When ice caps grow, ^{16}O is preferentially stored on the continents resulting in heavier oxygen isotope values (^{18}O-enriched) of the ambient sea water in which the calcareous organisms thrive.

Over the past decades, the inventory of high-resolution oxygen isotope records across the Cenozoic, 0–65 Ma, has grown, because of the greater availability of high-quality sediment cores. A compilation of these records showed that global climate cooled over the past 50 Ma with maximum temperature conditions occurring between 50 and 55 Ma [1]. The first permanent ice caps start to occur on Antarctica around the Eocene-Oligocene transition, $\sim$34 Ma. Also recently, extensive ice-rafted debris, including macroscopic dropstones, were found in the late Eocene to early Oligocene sediments from the Norwegian-Greenland Sea, indicating already severe glaciations of East Greenland at that time [41]. Orbital-induced variations in $\delta^{18}O$ were also detected superimposed on this long-term trend, but revealed different spectral characteristics pending on the background climate state. An evaluation of these characteristics with emphasis on icehouse and greenhouse conditions is given in the following sections.

3.1. Ice Ages

Through the development of radio-isotopic dating methods, power spectra could be obtained from the oxygen isotope records in the time domain. These methods clearly demonstrate that for the past 700 ka, major climate cycles have followed variations in obliquity and precession, although the dominant spectral power occurs at ~100 ka [42,43]. Understanding the mechanisms which control this long-term variability remains an outstanding question in climate sciences [44]. The most widely adopted explanation is that it originates from a nonlinear response to the precession forcing [43,45]. Other theories relate the 100 ka glacial rhythm directly to eccentricity [46], but the insolation changes that may have caused this are probably too small to be of much climatic relevance (Fig. 2a).

A third category of theories attribute the 100 ka glaciations to an internal oscillation of the Earth's ice–atmosphere–ocean climate system [47], which is nonlinear phase-locked to the external Milankovitch forcing [48,49]. An historically important example of a nonlinear oscillator is the model of Imbrie and Imbrie [45] which may be written as:

$$\frac{dV}{dt} = \frac{1 \pm b}{T_m}(X - V) \tag{9}$$

where X is the model's forcing function (i.e., the 65°N summer insolation), T_m is a mean time constant of the ice sheet response and b a nonlinearity coefficient which switches sign depending on whether ice volume is increasing or decreasing. For the late Pleistocene a T_m of 17 ka and b of 0.6 were estimated, which result in a 4–5 ka lag (ice-sheet response time) for the precession components and an 8 ka lag for obliquity [50]. In the latest marine benthic oxygen isotope stack of Lisiecki and Raymo [51], the same model has been applied. A plot of the LR04 record for the last 350 000 a is presented as overlay of the model's output in Fig. 4a. Evidently, changes in the marine $\delta^{18}O$ record lag 65°N summer insolation (Fig. 4d) with a few thousands of years as a result of the adopted values for T_m and b. The $\delta^{18}O$ record, however, preserves not only an ice volume signal, but also a deep-water temperature component, which should be taken into account for a more accurate estimate of the $\delta^{18}O$ response to orbital forcing [52].

Using an inverse modelling technique, Bintanja and coworkers [53] seperated the LR04 $\delta^{18}O$ record into an ice volume and a temperature component. The ice volume component is expressed in terms of sea level equivalent (Fig. 4b), while the temperature component is translated into the annual surface air temperature (T_{air}) over the continents north of 40°N (Fig. 4c). Clearly, T_{air} leads ice volume increases up to a few thousands of years, because ice sheets will only start to grow (inception) below a certain temperature threshold (−5 °C), and they can not expand faster than the rate at which mass is gained through snow accumulation [53]. During deglaciations, surface air temperature and sea-level increase almost in-concurrence, presumably the result of the rapid melt-down of the large ice sheets, with sea-level rises of over 1.5 cm·a^{-1} during the major terminations $T_{I–IV}$ (Fig. 4b).

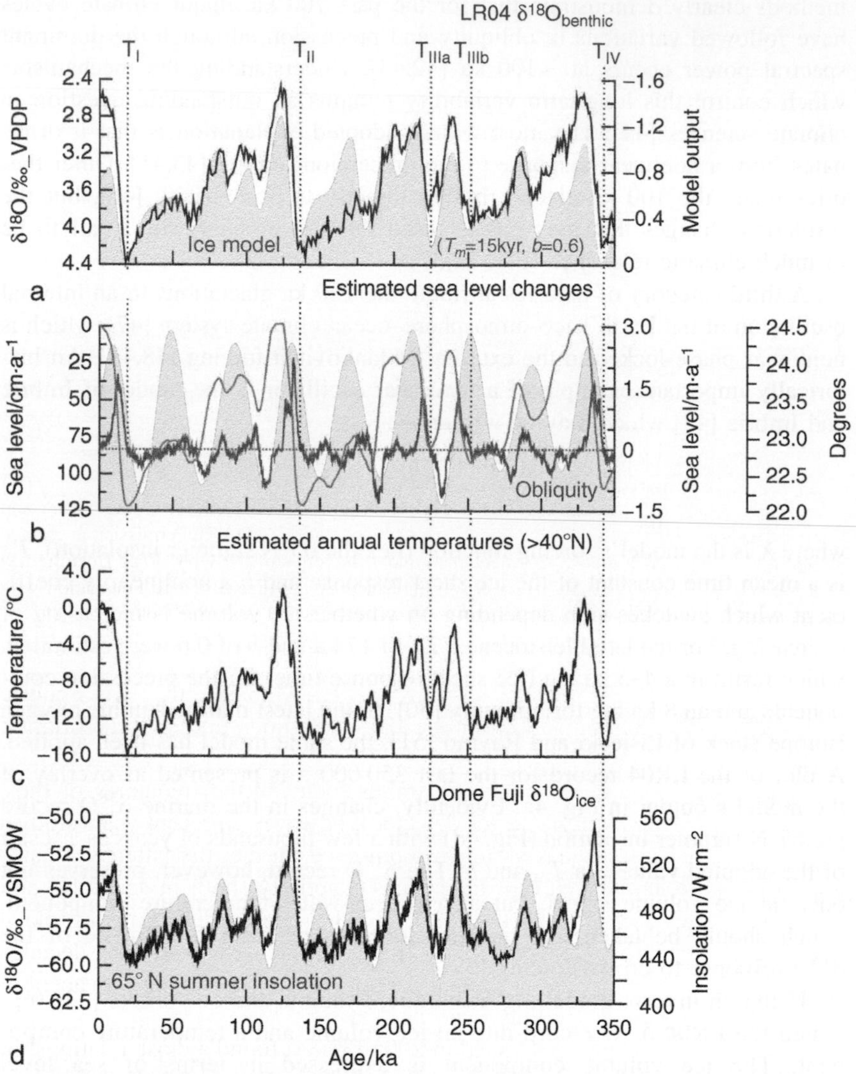

FIGURE 4 Comparison between orbital cycles and climatic proxy records for the past 350 ka (late Pleistocene). (a) LR04 $\delta^{18}O$ benthic stacked record (solid) plotted as overlay to its tuning target, represented by the outcome (thin line and grey area) of simple ice sheet model [45] including a mean time constant of the ice sheet response (T_m) of 15 ka and a nonlinearity coefficient (b) of 0.6 and the 65°N summer insolation as input [51]. (b) Reconstructed global sea level (thin line) and sea level change (solid line) derived from the LR04 $\delta^{18}O$ benthic stacked record and the use of an inverse modelling technique [53]. On the background (grey) is plotted the obliquity time series. Major terminations (I–IV) are defined by the first increase in sea level at the end of the glaciations. (c) The modelled surface air temperature deviation from present (mean over the continents between 40°N and 80°N) [53]. (d) Dome Fuji $\delta^{18}O_{ice}$ (solid) as overlay to the 65°N summer insolation curve [54].

The estimated T_{air} component of the marine $\delta^{18}O$ record resembles the Antarctic $\delta^{18}O_{ice}$ record of Dome Fuji [54], although the latter record tend to lead T_{air} in the order of a few thousands of years (Fig. 4d). There are several explanations which may account for the discrepancy between the insolation-induced response times of T_{air} and the Antarctic $\delta^{18}O_{ice}$ record. First of all, the $\delta^{18}O_{ice}$ record depends, besides local temperature, on a variety of factors such as sea water $\delta^{18}O$ and the temperature of the water vapour source area. A reconstruction of the $\delta^{18}O_{ice}$-derived local temperatures showed for example a slightly larger response time, of ~2 ka, to the insolation forcing [54].

Another part of the discrepancy may arise from uncertainties in the chronologies of either Dome Fuji or the LR04 $\delta^{18}O$ record. It should be noted that the LR04 $\delta^{18}O$ chronology is not directly constraint by radio-isotopic measurements of the marine cores that were incorporated in the LR04 stack, but it relies on the correlation between Thorium-230 and Protactinium-231 dated sea level reconstructions from coral terraces [55–58] and their signature in the $\delta^{18}O$ record, where no distinction has been made between the temperature and ice volume contribution of the $\delta^{18}O$ signal. Although the chronology of the last glacial cycles is well constraint there are conflicting estimates for the age of the penultimate and earlier deglaciations, which argue for [59] and against [60–62] the Milankovitch theory.

The Dome Fuji chronology, on the other hand, is based on tuning of the O_2/N_2 ratio of the trapped ice to the local variations in summer insolation (21 December at 77°S). The O_2/N_2 ratio lacks a strong 100 ka response, which makes this proxy more suitable for tuning than the $\delta^{18}O_{atm}$ record applied previously by Shackleton [52]. Dating uncertainties in this time scale range from 0.8 to 2.9 ka at the tie points [54]. Given these uncertainties, the increases in Antarctic temperature and atmospheric carbon dioxide concentration coincide with the rising phase of NH summer insolation during the last four terminations (Fig. 4d), thereby supporting the Milankovitch theory [54].

The role of obliquity is less highlighted in glacial theories despite the fact that from about 1 to 3 Ma and also during older geological periods, such as the Middle Miocene (14–15 Ma), smaller ice sheets varied at an almost metronomic 41 ka rhythm [63–66]. There are several mechanism proposed to explain the obliquity-dominated climate cycles (Fig. 5a). The most straightforward possibility is that because high-latitude (annual and summer) insolation declines with a reduced tilt of the Earth's axis (Figs. 3b and c), the ice caps will grow, Earth's albedo increases, and global mean temperatures decrease [10]. Another possibility is that during obliquity minima the meridional gradient of insolation during the summer half-year of both hemispheres increases, causing an increased moisture transport to the poles and hence the buildup of large ice caps [67]. Evidently, most periods of maximum sea level lowering or ice-sheet growth over the past 350 ka occur during obliquity minima (Fig. 4). In addition, Huybers and Wunsch [68] presented simple stochastic and deterministic models that describe the timing of the late Pleistocene glacial

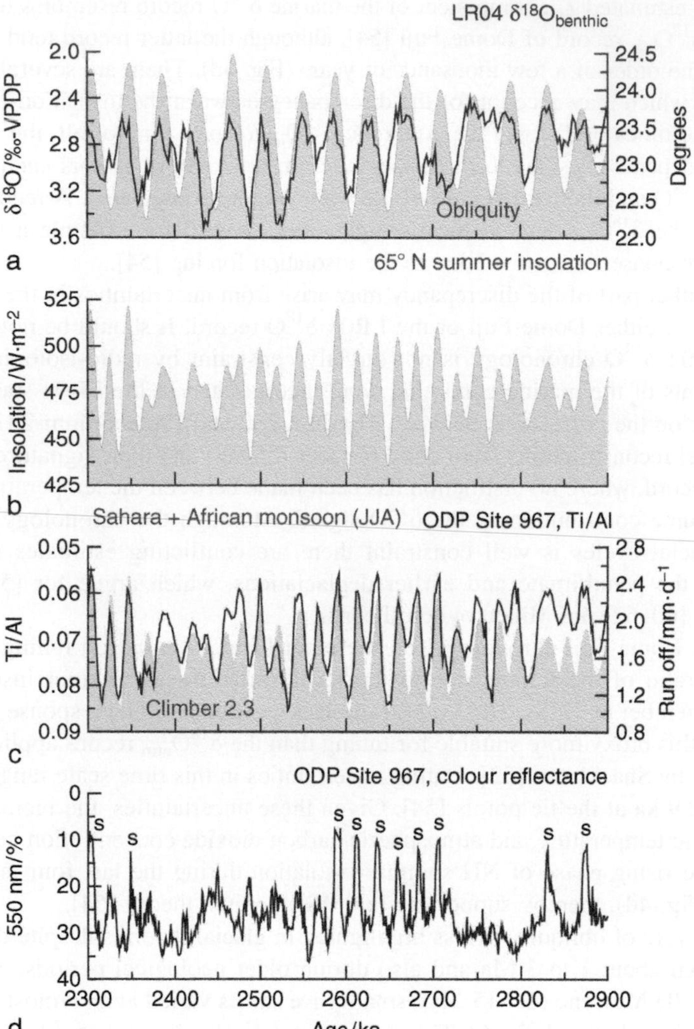

FIGURE 5 Comparison between orbital cycles and climatic proxy records for the late Pliocene time interval (2.3–2.9 Ma). (a) LR04 δ^{18}O benthic stacked record (solid) [51] plotted as overlay to the obliquity time series (thin line and grey area). (b) The 65°N summer insolation curve. (c) Reconstructed Ti/Al ratio of ODP Site 967 (solid) [31,102] as overlay to our modelled (see text for explanations) runoff associated with the Sahara and the African monsoon (blue line and grey area). (d) Colour reflectance data of ODP Site 967 (solid). Sapropels (s) are marked by very low (dark) values.

terminations purely in terms of obliquity forcing, although their findings were not yet confirmed by for instance the new results of Dome Fuji [54].

To summarise, the development of glacial-independent chronologies has become one of the major challenges in climate sciences to further unravel the Milankovitch theory of the Ice Ages. These chronologies could provide new insights to key issues such as the phase relation between climate change and the atmospheric carbon dioxide concentration, the feedback mechanisms associated with the buildup of large icecaps, or whether terminations are caused by internal or external processes.

3.2. Low and Mid-Latitude Climate Changes

The expression of orbital-induced climate oscillations is not restricted to glacial–interglacial variability. Data and models revealed that climate variations in the low and mid-latitudes are dominated by the precession cycle [69–73]. For instance, high-resolution absolute-dated oxygen isotope records from speleothems of central China have provided insights into the factors that control the strength of the East Asian Monsoon for the past 224 ka [74]. The record is dominated by precession cyclicity that is synchronous within dating errors with NH summer insolation. This supports the idea that on orbital timescales (sub)tropical monsoons respond dominantly and instantly to changes in NH summer insolation.

Another example is the cyclic occurrence of sapropels (organic-rich layers) in the marine sediments of the Mediterranean throughout the last 13 Ma [33,75]. The underlying mechanism that caused their formation gave rise to a contentious debate over the relative importance of anoxia caused by stable stratification [76] versus productivity [77]. The stratification hypothesis links the reduced oxygen conditions of the deep waters during sapropel formation to a weaker thermohaline circulation caused by lowered surface water density conditions in the eastern Mediterranean. Rossignol-Strick [78] proposed that these circumstances were triggered by the enhanced discharge of the river Nile during precession minima (NH summer insolation maxima, Fig. 2) when the strength of the African monsoon is at a maximum [70]. Climate modelling experiments, including a regional ocean model for the Mediterranean Sea, revealed however, that the precession-induced increase in net precipitation over the Mediterranean Sea itself is of equal or greater importance than the increase in runoff from the bordering continents [79]. Evidence for enhanced primary productivity has been gathered by a variety of geochemical and micropaleontological proxy records [77,80–84]. As possible causes for the enhanced nutrient supply to the mixed layer has been proposed a reversal in the flow directions of the nutrient-poor surface and nutrient-enriched deep waters, increased runoff, and the development of a Deep Chlorophyll Maximum (DCM).

To illustrate the different orbital characteristics of low and high-latitude driven climate signals a comparison is shown between the open ocean LR04 $\delta^{18}O$ record and the sapropel patterns of ODP (Ocean Drilling Program) Site 967 (Eratosthenes Seamount, eastern Mediterranean) for the time interval between 2.3 and 2.9 Ma are shown in Figs. 5a and d, respectively. Obliquity dominates the globally recorded high-latitude driven glacial–interglacial oscillations, whereas the climatic precession determines the circum-Mediterranean climate variability. The sapropels do not occur during all NH summer insolation maxima, indicating that their formation is bound to a threshold in the ocean–climate system. From the same core, changes in the Titanium to Aluminum ratio (Ti/Al), however, do show a striking correspondence with 65°N summer insolation throughout the studied interval, where high (low) Ti/Al values are interpreted to reflect a relative larger (lower) contribution of windblown (e.g., Sahara dust) versus river-transported (e.g., by the river Nile) derived terrigenous material [31] (Fig. 5c). Clearly, changes in northern African aridity conditions respond almost linearly to the orbital forcing, that is, containing both an obliquity and a precession signal.

To further unravel the link between the Ti/Al index and northern African aridity we simulated changes in runoff derived from the northern part of the African continent using an atmosphere–ocean–vegetation model of intermediate complexity, CLIMBER-2.3 [85]. The atmospheric model is a statistical–dynamical model with a resolution of 10° in latitude and ~51° in longitude. The terrestrial vegetation model, VECODE (VEgetation COntinuous Description), computes the fraction of the potential vegetation (i.e., grass, trees and bare soil) from the annual sum of positive day-temperatures and the annual precipitation [86]. These vegetation changes affect the land-surface albedo and the hydrological cycle. The ocean model [87] computes the zonally averaged temperature, salinity and velocity for three separate basins (Atlantic, Indian and Pacific oceans). The latitudinal resolution is 2.5° and the vertical resolution is 20 unequal levels. We have run a transient simulation for the time-interval between 2.25 and 3.0 Ma in which the only forcing is variations in insolation induced by the orbital parameters. Boundary conditions like orography, land–sea configuration, ice sheets and concentration of trace gasses were kept constant at pre-industrial values. As an indicator for northern African aridity, we extracted the total amount of runoff for the months June, July and August of the Sahara (20°N–30°N, 11°W–40°E) and African monsoon (10°N–20°N, 11°W–40°E) grid boxes [88]. This transient experiment shows that within a precession period runoff fluctuated between 1 and 2.6 $mm \cdot d^{-1}$, with the lowest values being associated with minima in 65°N summer insolation (Figs. 5b and c). The Ti/Al record and the modelled runoff shows a very good similarity revealing that the Ti/Al index reflects northern African aridity.

Spectral analyses of the Ti/Al index and simulated runoff show that African runoff is determined by precession and obliquity of which the first dominates the signal. Similar results have been found using several generic radiation patterns with an AGCM (Atmospheric General Circulation Model) in permanent July mode [89] or with time slice experiments of orbital extremes using the intermediate complexity model ECBilt [90]. These experiments show that during precession minimum or obliquity maximum configurations the African monsoon intensifies and extends further northwards.

In, contrast to obliquity, precession also influences the seasonal timing of the occurrence of the maximum precipitation [90]. The influence of obliquity could be due to stronger wind from the Atlantic Ocean into southern North Africa forced by deepening of the convergence zone over southern Asia during maximum summer insolation at high latitudes.

As for the origin of Ice Ages and the timing of major terminations, uncertainties exist in our knowledge of the exact phase relation between astronomical forcing and low-latitude climate changes. Where most model experiments suggest that tropical monsoons respond instantly to changes in NH summer insolation [70,71,88], reconstructions of Indian monsoon variability from the Arabian Sea have proposed a long response time of up to 8.0 ka after the inferred precession minimum configuration [91–95]. Also radiometric dates of the youngest sapropel in the Mediterranean (S1) suggest a time lag of ~3.0 ka between the last precession minimum at 11.5 ka and the midpoint of the S1 dated at 8.5 ka [30]. The Chinese speleothem records, on the other hand, support an in-phase relationship of the East-Asian summer monsoon with NH July insolation [74]. Several scenarios have been proposed to explain the long phase lag of the marine records, ranging from the influence of glacial–interglacial variability on the monsoon to a SH forcing through latent heat transport [92].

3.3. Greenhouse World

During the late Oligocene and early Miocene (~18–27 Ma), when the Polar regions were only partially ice-covered, benthic isotope records exhibit, besides a dominant obliquity component, a strong response to eccentricity forcing [96]. In the absence of permanent ice caps between 35 and 65 Ma, the imprint of eccentricity seems even more prominent, although the benthic isotope records currently available for the early Cenozoic lack adequate resolution to fully characterise obliquity variance [1]. The pronounced eccentricity imprint can be explained by filtering effects of the precession forcing due to continental geography and differences in land–sea heating, especially in the tropics [73]. A variety of processes have been suggested for exporting the signals to higher latitudes, including changes in ocean and atmospheric circulation, heat-transport, precipitation or the global carbon cycle and pCO2. Evidence of changes in the carbon cycle are given for instance by the Oligocene–Miocene carbon isotope ($\delta^{13}C$) records, which exhibit pervasive large-amplitude 100 and 400 ka oscillations that are highly coherent with the benthic oxygen isotope records [96,97].

Also during the late Paleocene and early Eocene (~60–50 Ma), eccentricity has significantly modulated the carbon isotope records of the Atlantic and Pacific oceans [98]. Cramer and coworkers [98] identified several short-lived $\delta^{13}C$ depletions, which they linked to maxima in the Earth's orbital eccentricity cycle. They linked, however, the much larger Carbon Isotope Excursion (CIE) that marks the Paleocene/Eocene boundary to a minimum in the

400 ka eccentricity cycle, thereby excluding orbital-forcing as triggering mechanism for the Paleocene/Eocene Thermal Maximum (PETM). This is in contrast to the more recent findings of Lourens and coworkers [34], showing on basis of the more complete successions from the southern Atlantic Walvis Ridge depth transect [99] that the PETM and Eocene Thermal Maximum 2 do correspond to 400 and 100 ka eccentricity maxima. They suggested that the critical conjunction of short, long and very long eccentricity cycles and the long-term late Palaeocene to early Eocene warming trend may have favoured the build-up of a significant methane hydrate reservoir before its release during both hyperthermal events.

Although the proposed orbital control as forcing mechanism of Paleogene hyperthermal events should be confirmed, it is evident that eccentricity has left its mark on the global carbon cycle. Moreover, the appearance of this modulation became more visible in the geological archives when the impact of the obliquity-dominated glacial cycles is at a minimum. Evidence that these long-term changes in the carbon cycle determined global climate has not yet been solved. In particular, cross-spectral comparison between the Oligocene–Miocene $\delta^{18}O$ and $\delta^{13}C$ records revealed a time lag of more than 20 ka in the 405 ka eccentricity band, suggesting a response rather than a forcing to global climate change [96]. In addition, the conspicuous absence of the long eccentricity signal in the Pleistocene glacial cyclicity raised the so-called '400 ka problem' [45]. On the other hand, the covariance between light $\delta^{13}C$ values and severe dissolution horizons in the deep sea during the greenhouse conditions of the late Paleocene and early Eocene indicate that changes in the carbon cycle through orbital forcing has had an important impact on ocean acidification and the position of the lysocline and calcite compensation depth [100].

4. CONCLUSION

The role of orbital forcing in climate change has been unequivocally shown by their characteristic patterns in sedimentary archives, ice cores and proxy records. Although our knowledge of orbital forcing is concerned with long-term natural climate cycles, it is of fundamental importance to assess and remediate global climate change problems on short-term periods. In particular, the integration of climate modelling experiments with geological observations will provide these insights required for a better understanding of climate change in the past and near future. Considerable challenges will have to be addressed before the full spectrum of orbital-induced climatic variability has been unravelled, including the phase behaviour of different parts of the climate system, feedback mechanisms and the impact on ecosystem dynamics.

From all the evidence, it is most likely that the climate change that we are currently experiencing is not due to variations of the Earth's orbital movements. With the fast rising CO_2 concentrations in the atmosphere, general orbital theories dealing with the icehouse world conditions will probably not account for future predictions. Integrating our knowledge of geological times when greenhouse gas conditions were those as being predicted, we might be able to decipher the role of orbital forcing in future climate change scenarios.

REFERENCES

1. J.C. Zachos, M. Pagani, L. Sloan, E. Thomas, K. Billups, Science 292 (2001) 686–693.
2. J.L. Lagrange, Memoirs of Berlin Academy, Oeuvre complètes t. V, vol. 1781, Gauthier-Villars, Paris, 1870, 125–207.
3. J.L. Lagrange, Memoirs of Berlin Academy, Oeuvre complètes t. V, vol. 1782, Gauthier-Villars, Paris, 1870, 211–344.
4. G.D. Pontécoulant, Théorie Analytique of Système of Monde, t III, Bachelier, Paris, 1834.
5. L. Agassiz, Étude sur les glacier, Neuchâtel (1840).
6. J.A. Adhémar, Révolutions de la mer, privately published, Paris, 1842.
7. U. Le Verrier, Ann. Obs. Paris II, Mallet-Bachelet, Paris, 1856.
8. J. Croll, Philos. Mag. 28 (1864) 121–137.
9. L. Pilgrim, Versuch einer rechnerischen Behandlung des Eiszeitproblems. Jahreshefte des Vereins für vaterländische Naturkunde in Württemberg, Stuttgart 60: 26–117 (1904).
10. M. Milankovitch, R. Serb. Acad. Spec. Publ. 133 (1941) 1–633.
11. P. Bretagnon, Astron. Astrophys. 30 (1974) 141–154.
12. A. Berger, Astron. Astrophys. 51 (1976) 127–135.
13. A. Berger, J. Atmos. Sci. 35 (1978) 2362–2367.
14. S.G. Sharav, N.A. Boudnikova, Bull. I.T.A XI-4(127) (1967) 231–265.
15. S.G. Sharav, N.A. Boudnikova, Trud. I.T.A XIV (1967) 48–84.
16. J. Laskar, Théorie générale planétaire: Eléments orbitaux des planètes sur 1 million d'années, Thèse, Observatoire de Paris (1984).
17. J. Laskar, Astron. Astrophys. 144 (1985) 133–146.
18. J. Laskar, Astron. Astrophys. 157 (1986) 59–70.
19. J. Laskar, Astron. Astrophys. 198 (1988) 341–362.
20. J. Laskar, Nature 338 (1989) 237–238.
21. J. Laskar, Icarus 88 (1990) 266–291.
22. J. Laskar, Philos. Trans. R. Soc. Lond. A 357 (1999) 1735–1759.
23. T.R. Quinn, S. Tremaine, M. Duncan, Astron. J. 101 (1991) 2287–2305.
24. J. Laskar, F. Joutel, F. Boudin, Astron. Astrophys. 270 (1993) 522–533.
25. K. Lambeck, The earth's variable rotation: Geophysical causes and consequences, Cambridge University Press, 458 p. (1980).
26. W.R. Peltier, X. Jiang, Geophys. Res. Lett. 21 (1994) 2299–2302.
27. J.X. Mitrovica, A.M. Forte, Geophys. J. Int. 121 (1995) 21–32.
28. J.X. Mitrovica, A.M. Forte, R. Pan, Geophys. J. Int. 128 (1997) 270–284.
29. A.M. Forte, J.X. Mitrovica, Nature 390 (1997) 676–680.
30. L.J. Lourens, A. Antonarakou, F.J. Hilgen, A.A.M.v. Hoof, C. Vergnaud-Grazzini, W.J. Zachariasse, Paleoceanography 11 (1996) 391–413.
31. L.J. Lourens, R. Wehausen, H.J. Brumsack, Nature 409 (2001) 1029–1032.
32. J. Laskar, P. Robutel, F. Joutel, M. Gastineau, A.C.M. Correia, B. Levrard, Astron. Astrophys. 428 (2004) 261–285.
33. L.J. Lourens, F.J. Hilgen, N.J. Shackleton, J. Laskar, D. Wilson, in: F. Gradstein, J. Ogg, A. Smith (Eds.), A Geologic Time Scale 2004, Cambridge University Press, UK, 2004, pp. 409–440.

34. L.J. Lourens, A. Sluijs, D. Kroon, J.C. Zachos, E. Thomas, U. Röhl, J. Bowles, I. Raffi, Nature 435 (2005) 1083–1087.
35. T. Westerhold, U. Rohl, J. Laskar, I. Raffi, J. Bowles, L.J. Lourens, J.C. Zachos, Paleoceanography 22 (2007) doi:10.1029/2006PA001322.
36. K.F. Kuiper, A. Deino, F.J. Hilgen, W. Krijgsman, P.R. Renne, J.R. Wijbrans, Science 320 (2008) 500–504.
37. A. Berger, M.F. Loutre, J. Laskar, Science 255 (1992) 560–565.
38. A. Berger, M.F. Loutre, in: J.-C. Duplessy, M.-T. Spyridakis (Eds.), Long-Term Climatic Variations, vol. I22, Springer-Verlag, Berlin Heidelberg, 1994, pp. 107–151.
39. C. Emiliani, J. Geol. 63 (1955) 538–578.
40. N.J. Shackleton, Nature 215 (1967) 15–17.
41. J.S. Eldrett, I.C. Harding, P.A. Wilson, E. Butler, A.P. Roberts, Nature 446 (2007) 176–179.
42. J.D. Hays, J. Imbrie, N.J. Shackleton, Science 194 (1976) 1121–1132.
43. J. Imbrie, A. Berger, E.A. Boyle, S.C. Clemens, A. Duffy, W.R. Howard, G. Kukla, J. Kutzbach, D.G. Martinson, A. McIntyre, A. Mix, B. Molfino, J.J. Morley, L.C. Peterson, N.G. Pisias, W.L. Prell, M.E. Raymo, N.J. Shackleton, J.R. Toggweiler, Paleoceanography 8 (1993) 699–735.
44. B. Saltzman, Dynamical Paleoclimatology: Generalised Theory of Global Climate Change, vol. 80, Elsevier, New York, 2001, pp. 320.
45. J. Imbrie, J.Z. Imbrie, Science 207 (1980) 943–952.
46. R. Benzi, G. Parisi, A. Sutera, A. Vulpiani, Tellus 34 (1982) 10–16.
47. H. Gildor, E. Tziperman, Paleoceanography 15 (2000) 605–615.
48. K.A. Maasch, B. Saltzman, J. Geophys. Res. 95 (1990) 1955–1963.
49. E. Tziperman, M.E. Raymo, P. Huybers, C. Wunsch, Paleoceanography 21 (2006) PA4206.
50. J. Imbrie, J.D. Hays, D.G. Martinson, A. McIntyre, A.C. Mix, J.J. Morley, N.G. Pisias, W.L. Prell, N.J. Shackleton, in: A. Berger, J. Imbrie, J. Hays, G. Kukla, B. Saltzman (Eds.), Milankovitch and Climate, Understanding the Response to Astronomical Forcing, Part I, D. Reidel Publishing Company, Dordrecht/Boston/Lancaster, 1984, pp. 269–305.
51. L.E. Lisiecki, M.E. Raymo, Paleoceanography 20 (2005) PA1003.
52. N.J. Shackleton, Science 289 (2000) 1897–1902.
53. R. Bintanja, R.S.W.v.d. Wal, J. Oerlemans, Nature 437 (2005) 125–128.
54. K. Kawamura, F. Parrenin, L.E. Lisiecki, R. Uemura, F. Vimeux, J.P. Severinghaus, M.A. Hutterli, T. Nakazawa, S. Aoki, J. Jouzel, M.E. Raymo, K. Matsumoto, H. Nakata, H. Motoyama, S. Fujita, K. Goto-Azuma, Y. Fujii, O. Watanabe, Nature 448 (2007) 912–916.
55. W. Broecker, D.L. Thurber, J. Goddard, T.L. Ku, R.K. Matthews, K.J. Mesolella, Science 159 (1968) 297–300.
56. R.L. Edwards, J.H. Chen, T.-L. Ku, G.J. Wasserburg, Science 236 (1987) 1547–1553.
57. R.L. Edwards, H. Cheng, M.T. Murrell, S.J. Goldstein, Science 276 (1997) 782–786.
58. E. Bard, B. Hamelin, R.G. Fairbanks, Nature 346 (1990) 456–458.
59. T.D. Herbert, J.D. Schuffert, D. Andreasen, L. Heusser, M. Lyle, A. Mix, A.C. Ravelo, L.D. Stott, J.C. Herguera, Science 293 (2001) 71–76.
60. G.M. Henderson, N.C. Slowey, Nature 404 (2000) 61–66.
61. C.D. Gallup, H. Cheng, F.W. Taylor, R.L. Edward, Science 295 (2002) 310–313.
62. I.J. Winograd, J.M. Landwehr, K.R. Ludwig, T.B. Coplen, A.C. Riggs. Q. Res. 48 (1997) 141–154.
63. N.J. Shackleton, J. Backman, H. Zimmerman, D.V. Kent, M.A. Hall, D.G. Roberts, D. Schitker, J. Baldauf, Nature 307 (1984) 620–623.
64. W.F. Ruddiman, M.E. Raymo, D.G. Martinson, B.M. Clement, J. Backman, Paleoceanography 4 (1989) 353–412.

65. M.E. Raymo, W.F. Ruddiman, J. Backman, B.M. Clement, D.G. Martinson, Paleoceanography 4 (1989) 413–446.
66. A. Holbourn, W. Kuhnt, M. Schulz, H. Erlenkeuser, Nature 438 (2005) 483–487.
67. M.E. Raymo, K. Nisancioglu, Paleoceanography 18 (2003) doi:10.1029/2002PA000791.
68. P. Huybers, C. Wunsch, Nature 434 (2005) 491–494.
69. E.M. Pokras, A.C. Mix, Nature 326 (1987) 486–487.
70. J.E. Kutzbach, Science 214 (1981) 59–61.
71. J.E. Kutzbach, X. Liu, Z. Liu, G. Chen, Clim. Dyn. 30 (2008) 567–579.
72. D.A. Short, J.G. Mengel, Nature 323 (1986) 48–50.
73. D.A. Short, J.G. Mengel, T.J. Crowley, W.T. Hyde, G.R. North, Q. Res. 35 (1991) 157–173.
74. Y. Wang, H. Cheng, R.L. Edwards, X. Kong, X. Shao, S. Chen, J. Wu, X. Jiang, X. Wang, Z. An, Nature (2008), 451, 1090.
75. F. Hilgen, Newslett. Stratigr. 17 (1987) 109–127.
76. E. Olausson, Reports of the Swedish Deep Sea Expedition 1947–1948, vol. 8, 1961, pp. 353–391.
77. S.E. Calvert, Oceanol. Acta 6 (1983) 255–267.
78. M. Rossignol-Strick, Nature 304 (1983) 46–49.
79. P.T. Meijer, E. Tuenter, J. Mar. Syst. 68 (2007) 349–365.
80. E.J. Rohling, W.W.C. Gieskes, Paleoceanography 5 (1989) 531–545.
81. D. Castradori, Paleoceanography 8 (1993) 459–471.
82. J.P. Sachs, D.J. Repeta, Science 286 (1999) 2485–2488.
83. A.E.S. Kemp, R.B. Pearce, I. Koizumi, J. Pike, S. Jea Rance, Nature 398 (1999) 57–61.
84. S.J. Schenau, A. Antonarakou, F.J. Hilgen, L.J. Lourens, I.A. Nijenhuis, C.H. van der Weijden, W.J. Zachariasse, Mar. Geol. 153 (1999) 117–135.
85. V. Petoukhov, A. Ganopolski, V. Brovkin, M. Claussen, A. Eliseev, C. Kubatzki, S. Rahmstorf, Clim. Dyn. 16 (2001) 1–17.
86. V. Brovkin, A. Ganopolski, Y. Svirezhev, Ecol. Modell. 101 (1997) 251–261.
87. T.F. Stocker, L.A. Mysak, Clim. Change 20 (1992) 227–250.
88. E. Tuenter, S.L. Weber, F.J. Hilgen, L.J. Lourens, A. Ganopolski, Clim. Dyn. 24 (2005) 279–295.
89. W.L. Prell, J.E. Kutzbach, J. Geophys. Res. 92 (1987) 8411–8425.
90. E. Tuenter, S.L. Weber, F.J. Hilgen, L.J. Lourens, Glob. Planet. Change 36 (2003) 219–235.
91. S. Clemens, W. Prell, D. Murray, G. Shimmield, G. Wedon, Nature 353 (1991) 720–725.
92. S.C. Clemens, D.W. Murray, W.L. Prell, Science 274 (1996) 943–948.
93. G.J. Reichart, L.J. Lourens, J.W. Zachariasse, Paleoceanography 13 (1998) 607–621.
94. M.A. Altabet, D.W. Murray, W.L. Prell, Paleoceanography 14 (1999) 732–743.
95. P. Wang, S. Clemens, L. Beaufort, P. Braconnot, G. Ganssen, Z. Jian, P. Kershaw, M. Sarnthein, Quat. Sci. Rev. 24 (2005) 595–629.
96. J.C. Zachos, N.J. Shackleton, J.S. Revenaugh, H. Pälike, B.P. Flower, Science 292 (2001) 274–278.
97. H.A. Paul, J.C. Zachos, B.P. Flower, A. Tripati, Paleoceanography 15 (2000) 471–485.
98. B.S. Cramer, J.D. Wright, D.V. Kent, M.-P. Aubry, Paleoceanography 18 (2003), doi: 10.1029/2003PA000909.
99. J.C. Zachos, D. Kroon, P. Blum, J. Bowles, P. Gaillot, T. Hasegawa, E.C. Hathorne, D.A. Hodell, D.C. Kelly, J.-H. Jung, S.M. Keller, Y.-S. Lee, D.C. Leuschner, Z. Liu, K.C. Lohmann, L. Lourens, S. Monechi, M. Nicolo, I. Raffi, C. Riesselman, U. Röhl, S.A. Schellenberg, D. Schmidt, A. Sluijs, D. Thomas, E. Thomas, H. Vallius. (2004), *Proc. ODP, Init. Repts.*, Vol. 208: College Station, TX (Ocean Drilling Program).
100. J.C. Zachos, G.R. Dickens, R.E. Zeebe, Nature 451 (2008) 279–283.
101. L. Hinnov, in: F. Gradstein, J. Ogg, A.Smith (Eds.), A Geologic Time Scale 2004, Cambridge University Press, UK, 2004, pp. 55–62.
102. R. Wehausen, H.-H. Brumsack, Mar. Geol. 153 (1999) 161–176.

Part II

A Geological History of Climate Change

Chapter 6

A Geological History of Climate Change

Jan Zalasiewicz and Mark Williams
Department of Geology, University of Leicester, University Road, Leicester LE1 7RH, UK

1. Introduction
2. Climate Models
3. Long-Term Climate Trends
4. Early Climate History
5. Phanerozoic Glaciations
6. The Mesozoic–Early Cenozoic Greenhouse
7. Development of the Quaternary Icehouse
8. Astronomical Modulation of Climate
9. Milankovitch Cyclicity in Quaternary (Pleistocene) Climate History
10. Quaternary Sub-Milankovitch Cyclicity
11. The Holocene
12. Climate of the Anthropocene
13. Conclusions
Acknowledgement
References

1. INTRODUCTION

Earth's climate is now changing in response to an array of anthropogenic perturbations, notably the release of greenhouse gases; an understanding of the rate, mode and scale of this change is now of literally vital importance to society. There is presently intense study of current and historical (i.e. measured) changes in both perceived climate drivers and the Earth system response. Such studies typically lead to climate models that, in linking proposed causes and effects, are aimed at allowing prediction of climate evolution over an annual to centennial scale.

However, the Earth system is complex and imperfectly understood, not least as regards resolving the effect of multiple feedbacks in the system, and of assessing the scale and importance of leads, lags and thresholds ('tipping points') in climate change. There is thus a need to set modern climate studies within a realistic context. This is most effectively done by examining the preserved history of the Earth's climate. Such study cannot provide precise replicas of the unplanned global experiment that is now underway (for the sum of human actions represents a geological novelty). However, it is providing an increasingly detailed picture of the nature, scale, rate and causes of past climate change and of its wider effects, as regards for instance sea level and biota. Imperfect as it is, it provides an indispensable context for modern climate studies, not least as a provision of ground truth for computer models (see below) of former and present climate.

Aspects of climate that are recorded in strata include temperature and seasonality [1,2], humidity/aridity [3], and wind direction and intensity [4]. Classical palaeoenvironmental indicators such as glacial tills, reef limestones and desert dune sandstones have in recent years been joined by a plethora of other proxy indicators. These include many biological (fossilised pollen, insects, marine algae) and chemical proxies (e.g. Mg/Ca ratio in biogenic carbonates). Others are isotopic: oxygen isotopes provide information on temperature and ice volume; carbon isotopes reflect global biomass and inputs (of methane or carbon dioxide) into the ocean/atmosphere system; strontium and osmium are proxies for weathering, and the latter, with molybdenum also, for oceanic oxygenation levels. Other proxies include recalcitrant organic molecules: long-chain algal-derived alkenones as sea temperature indicators [5] and isorenieratane as a specific indicator of photic zone anoxia [6]. These and many other proxies are listed in Ref. [7]. Levels of greenhouse gases such as carbon dioxide and methane going back to 800 ka can be measured in ice cores [8]. For older times, (somewhat imprecise) proxies have been used, such as leaf stomata densities [9,10], palaeosol chemistry [11] or boron isotopes [12]; estimates of greenhouse gas concentrations in the atmosphere have also been arrived at by modelling [13,14].

2. CLIMATE MODELS

Since the 1960s, computer models of climate have been developed that provide detailed global and regional projections of future climate and reconstructions of deep time climate. Some of these models are used to simulate conditions during icehouse climates, for example, of the Late Proterozoic [15], whilst others simulate warm intervals of global climate, such as during the Mesozoic greenhouse [16]. The most widely applied computer simulations of palaeoclimate are general circulation models (GCMs). The increasing complexity of these models has followed the exponential growth in computer power.

GCMs divide the Earth into a series of grid boxes. Within each of these grid boxes, variables important for the prediction of climate are calculated, based upon the laws of thermodynamics and Newton's laws of motion. At progressive time steps of the model the reaction between the individual grid boxes is calculated. GCM simulations rely on establishing key boundary conditions. These conditions include solar intensity, atmospheric composition (e.g. level of greenhouse gases), surface albedo, ocean heat transport, geography, orography, vegetation cover and orbital parameters. Whilst solar luminosity can be estimated with a high degree of confidence for different time periods, some of the other boundary conditions are much more difficult to establish, and the magnitude of the problem increases with greater age. Thus, models of Late Proterozoic climate can establish solar luminosity as 93% of present, but the geography of Proterozoic palaeocontinents is much more controversial [15]. It depends on geological data, in this example from remnant magnetism, preserved within rocks and placing the continents in their ancient position according to the Earth's magnetic field.

Geological data (e.g. sedimentology, palaeontology) are essential to 'ground truth' climate models, to establish whether they are providing a realistic reconstruction of the ancient world, and also to provide data for calibrating boundary conditions for the models. Of major importance for GCM palaeoclimate reconstructions is accurate information about sea surface temperatures (SSTs), as this provides a strong indication of how ocean circulation was working. The most extensive deep time reconstruction of SSTs is that of the United States Geological Survey PRISM Group, based on a global dataset of planktonic foraminifera [17]. This dataset has been used for calibrating a range of climate model scenarios for the 'mid Pliocene warm period' and also includes an extensive catalogue of terrestrial data [18]. This time interval is used for potential comparison with the path of future global warming [19].

3. LONG-TERM CLIMATE TRENDS

Earth's known climate history, as decipherable through forensic examination of sedimentary strata, spans some 3.8 Ga (billion years), to the beginning of the Archaean (Fig. 1). The previous history, now generally assigned to the Hadean Eon, is only fragmentarily recorded as occasional ancient mineral fragments contained within younger rocks – particularly of highly resistant zircon dated to nearly 4.4 Ga ago [20] and thus stretching back to very nearly the beginning of the Earth at 4.56 Ga ago [21]. The chemistry of these very ancient fragments hints at the presence of a hydrosphere even at that early date, though one almost certainly disrupted by massive meteorite impacts [22]. Certainly, by the beginning of the Archaean, oceans had developed, and an atmosphere sufficiently reducing to allow the preservation of detrital minerals such as pyrite and uraninite that would not survive in the presence of free oxygen [23].

From then until the present, Earth's climate has remained within narrow temperature limits that have allowed the presence of abundant liquid water, water vapour and variable amounts of water ice, the last of these (when present) generally accumulating at high latitudes and/or high altitudes. This is despite widely accepted astrophysical models suggesting that the sun has

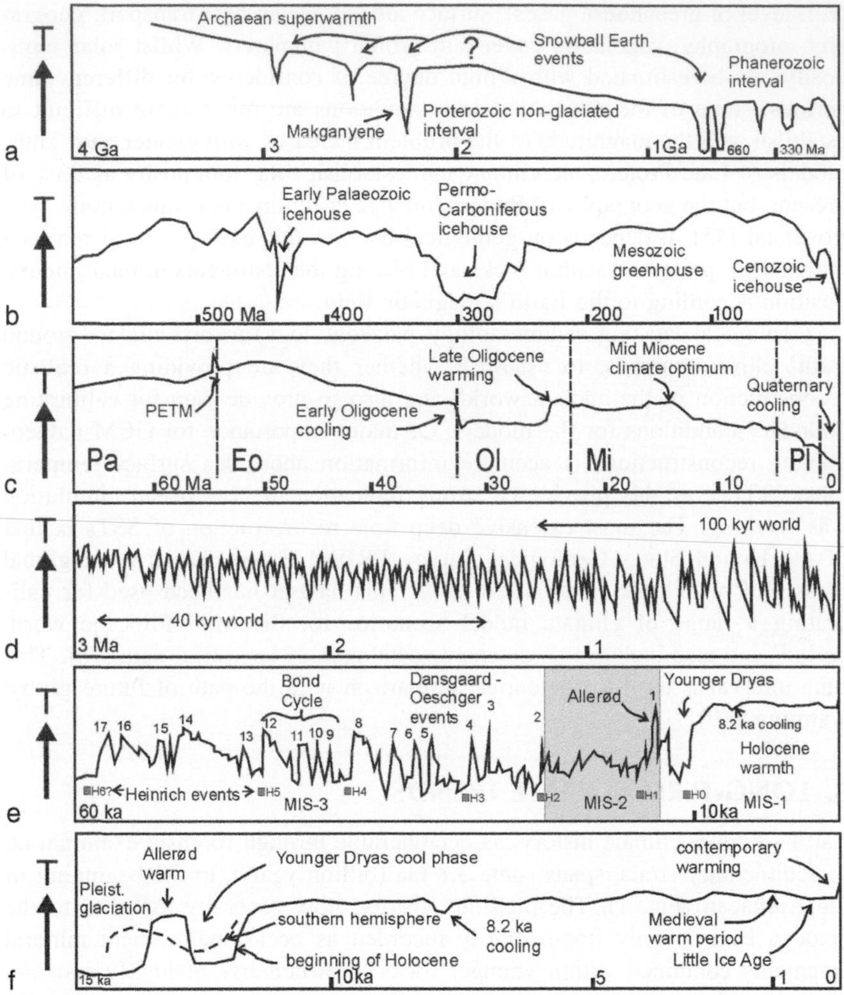

FIGURE 1 Global climate variation at six different timescales. Data adapted from sources including [7,27,48,71,95,100]. On the left side of the figure, the figure 'T' denotes relative temperature. Note that the line denoting 'T' is derived from $\delta^{18}O$ from benthic foraminifera for the Cenozoic time slices (c–e), but for the intervals with polar ice this line will record a combination of ice volume and temperature change.

increased its luminosity by some 20% since the early Archaean, and contrasts sharply with the history of our planetary neighbours: Venus now having a surface temperature of ca. 400 °C with a dense anhydrous atmosphere dominated by carbon dioxide (representing approximately the amount of carbon that on Earth is bound up in rock form as carbonates and hydrocarbons); and Mars with an early history of running surface water (roughly during the Earth's Hadean Eon) and subsequently being essentially freeze-dried.

Hypotheses to explain the Earth's climate stability (that has allowed *inter alia* a continuous lineage of living organisms) have included such as the Gaia hypothesis [24], in which the totality of the Earth's biota operate to maintain optimum conditions for their existence (via feedback mechanisms that involve such factors as albedo and atmospheric composition). Currently, it is thought that terrestrial silicate weathering (a largely abiotic mechanism) is an important factor in Earthly homeostasis [25]. Thus, as temperatures rise through an increase in greenhouse gases, increased reaction rates of rainwater (i.e. dilute carbonic acid) with rock – allied to increased humidity from enhanced evaporation rates – will cause drawdown of carbon dioxide, thus lowering temperatures [26]. Similarly, as greenhouse gas levels and temperatures fall, diminished rates of weathering will allow carbon dioxide levels to rise, and so warm the Earth's climate. The silicate weathering mechanism operates on timescales of hundreds of thousands to millions of years, with greenhouse gas levels having fallen throughout Earth history as the sun's luminosity has increased. At shorter timescales, this mechanism may be over-ridden by other factors, to allow the production of climate states that are hotter or colder than the long-term average.

4. EARLY CLIMATE HISTORY

At long time scales, Earth's (post-Hadean) climate history can be broadly divided into: *greenhouse* (or *hothouse*) states, when the Earth's climate was generally warm, with little or no polar ice; and *icehouse* states with substantial high/mid (and sometimes low latitude) ice masses over land and ocean. Ability to resolve the duration and timing of these states becomes increasingly better as the geological record becomes younger, with a gulf, in particular, between a Phanerozoic record (from 0.542 Ga) that is highly resolved because of an abundant fossil content and a Precambrian record in which dating and correlation are based upon sporadic radiometric dates and, increasingly, chemical and event stratigraphy. Similarly, the Quaternary glaciation is much better resolved than previous Phanerozoic glaciations.

The earliest reasonable indications of climate in the Archaean hint at a very warm world (Fig. 1a): silicon and oxygen isotopes in Archaean and early Proterozoic rocks suggest temperatures of some 50–80 °C, before temperatures declined to 20–30 °C by 1.5 Ga [27,28]. Most of the post-Hadean Precambrian seems to have roughly equated to a greenhouse world in general, and high-carbon dioxide and high-methane atmospheres have both been suggested as a means of maintaining high temperatures in the face of a faint early sun [29,30].

There were, though, the striking exceptions of the 'Snowball Earth' glaciations (Fig. 1a). There are possible representatives (certainly glacial if not 'Snowball') in the early Proterozoic at ca. 2.5–2.0 Ga [31]. But, they are most typical of the late Proterozoic within the 'Cryogenian Period' (now widely used as a geological time period, but not yet properly defined and ratified). Stratigraphic and palaeomagnetic evidence suggests widespread icesheets in at least two pulses (Sturtian 740–660 Ma ago and Marinoan 660–635 Ma ago [32] that reached into tropical latitudes, with ice present on all main continents. Budyko [33] suggested a theoretical basis for a snowball glaciation, showing that if ice extended to within 30° latitude of the equator, the ice albedo effect would produce a positive feedback mechanism allowing ice sheets to grow to the equator. It has been proposed, controversially, that ice encased the entire globe (the 'hard snowball' variant: [34]), preventing exchange between land/oceans and atmosphere. This has been disputed, with opponents preferring 'slush ball', 'zipper-rift' or 'high tilt' Earth models [32] leaving significant areas of ocean ice-free.

Whichever version is nearer the truth, these appear to have been extreme excursions of the Earth system, with deglaciation being rapid, perhaps 'catastrophic', and marked by the deposition of unique 'cap carbonate' deposits – dolomites and limestones that, worldwide, immediately overlie the glacial deposits. Deglaciation mechanisms commonly involve crossing thresholds in greenhouse gas concentrations. In the 'hard snowball' model this takes the form of volcanic carbon dioxide being prevented from dissolving in the ocean or reacting with rock (because of their carapace of ice), and hence building up to levels high enough to cause rapid ice melt, with acid rain then reacting rapidly with newly exposed bedrock to generate alkalinity that precipitated as carbonates. In the 'slush ball' model, deglaciation hypotheses include massive release of methane, with at least local isotopic evidence of methane release accompanied by ice-melt [35]. Perhaps in support of a 'slush ball' or alternative glacial hypothesis, some GCMs do not replicate the conditions in which a 'hard' Snowball Earth could develop even with very low levels of atmospheric carbon dioxide prescribed [15].

5. PHANEROZOIC GLACIATIONS

Phanerozoic time has also been dominated overall by 'greenhouse' conditions [7]. Glaciations during the Phanerozoic were less extreme, neither reaching the equator nor being associated with post-glacial cap carbonates. Three main glaciations took place (Fig. 1b): a late Ordovician/early Silurian 'Early Palaeozoic Icehouse' (ca. 455–425 Ma) [36], with an end-Ordovician glacial maximum [37] that collapsed in a rapid deglaciation; a long-lived Permo-Carboniferous glaciation (ca. 325–270 Ma) [7] with ice covering much of the palaeocontinent Gondwana (leaving widespread traces in South America and Africa, then over the South Pole); and the current glaciation, that began in the southern hemisphere through the Eocene–Oligocene Epoch boundary interval (ca. 35 Ma) with ice growing on Antarctica [38], and developed into a full-scale bipolar glaciation around the beginning of the Quaternary Period, at ca. 2.6 Ma, with the significant expansion of northern hemisphere ice.

Each of these glaciations took place in different contexts, particularly as regards the carbon cycle. The Early Palaeozoic Icehouse took place in the effective absence of either a terrestrial flora or of widespread well-developed (and hence carbon-rich) soils. Hence, the oceans and marine sediments were of prime importance in carbon storage, with the intermittent anoxia of those oceans perhaps playing a key role as thermostat, episodically enhancing carbon sequestration that led to cooling [36]. In the Carboniferous, the explosive growth and widespread burial of plants on deltaic/coastal plain sediments (subsequently becoming coal) has long been considered key in driving down atmospheric carbon dioxide and leading to glaciation [39]. Other mechanisms have been invoked, such as continental rearrangement to alter patterns of ocean currents and hence global heat transport [40].

6. THE MESOZOIC–EARLY CENOZOIC GREENHOUSE

These early Phanerozoic switches between greenhouse and icehouse give invaluable (and increasingly well resolved) information on the mode and rate of climate change. However, it is the temporal background to, and the development of, the current glaciation that offers the most resolved history and the best clue to causal and controlling mechanisms. This is partly because of a biota that is closer to the present one and hence more interpretable, but crucially because there is a widespread oceanic record (buried under the present ocean floors) to accompany that from land and continental seas; Palaeozoic ocean deposits, by contrast, have almost all been obliterated through subduction, with only rare fragments being preserved by obduction on to destructive continental margins.

Mesozoic and early Tertiary climate was generally in 'greenhouse' mode with little (but generally some) polar ice, widespread epicontinental seas and ocean circulation driven by salinity rather than temperature differences (and hence more sluggish than today's, with a tendency to anoxia). Within this broad pattern, there were warmer and colder intervals [7]. Fossil evidence shows that high latitudes, in particular were considerably warmer during this interval, with extensive near-polar forests [16].

This interval includes brief (0.1–0.2 Ma) climate 'spikes' in which sudden temperature rises were accompanied by biotic changes and marked changes in carbon isotopes. These changes suggest massive (thousands of gigatonnes) transfer of carbon from rock reservoirs to the atmosphere/ocean system with the consequence of ocean acidification as well as warming [41]. The best-known of these [42,43] were in the Toarcian Age of the Jurassic Period (ca. 183 Ma) and at the boundary of the Paleocene and Eocene epochs (ca. 55 Ma). The most likely mechanism seems to be some initial warming (perhaps from volcanic carbon dioxide) that triggered large-scale dissociation of methane hydrates from the sea floor [44], although the baking of coal basins by igneous intrusions [45] may also be implicated. By whichever mechanism, the relevance for contemporary global warming is clear as, while humankind has not yet released as much carbon (ca. 600 Gt (gigatonnes)), it has done it much more quickly [46]. Re-equilibration of climate following the spikes was likely achieved via silicate weathering [26,47].

7. DEVELOPMENT OF THE QUATERNARY ICEHOUSE

The development of the Tertiary/Quaternary icehouse took place as a series of steps (Fig. 1c), with relatively rapid transitions between one climate state and the next, strongly suggesting the common operation of thresholds or 'tipping points' [48]. The early Oligocene inception is clearly seen as an isotopic and Mg/Ca signal, in benthic foraminifera [49], of ocean cooling and de-acidification [50] linked to the growth of substantial ice on Antarctica. Two mechanisms have been invoked, that in reality were likely inter-related: the separation of South America from Antarctica to open the Drake Passage and hence to allow a continuous circum-Antarctic cold current [51]; and a steep drop in carbon dioxide levels from about $\times 4$ to $\times 2$ present-day levels [38].

Subsequent Tertiary history includes Mid Miocene warming, possibly associated with release of carbon dioxide to the atmosphere via volcanism or meteorite impact (see Ref. [52] for an overview of possible causes) during which tundra conditions were developed at high southern latitudes within 1500 km of the South Pole [53], and late mid-Miocene cooling (often termed the 'Monterey event' [54]), which may have been influenced by drawdown of carbon dioxide from the atmosphere or by changes to ocean heat transport that triggered ice sheet growth and cooling [55].

The subsequent Pliocene Epoch marks the final phase of 'late Tertiary' climate. The Early and Mid Pliocene represent conditions that overall were somewhat warmer than present, with global ice volumes smaller, and global surface temperatures perhaps 2–3 °C warmer [56]. The last phase of this warmer world was the 'mid Pliocene warm period' some three million years ago [17]. Following this interval, global temperatures decreased, ice volumes increased, and the amplitude of glacial-interglacial oscillation also increased [57] heralding the intensification of Northern Hemisphere Glaciation (NHG). As the last interval of warmth, the 'mid Pliocene warm period' has received growing attention as a possible comparison for the path of future global warming [58].

The intensification of NHG that is characteristic of the Quaternary Period (sensu [59]) was marked by the growth of substantial ice in the northern polar region [60]. It is associated with ice-rafted debris appearing in North Atlantic Ocean floor deposits, together with the beginning of substantial loess accumulation in central Europe and China, the drying of Africa to create extensive savannah areas, and other global phenomena. This event may partly reflect a further carbon dioxide threshold [61], with strontium isotope evidence of increased rock weathering, not least from uplift phases of the Himalayas [62]. However, there is strong evidence to suggest the importance of enhanced ice growth rather than simply temperature, with the development of the 'snow gun' hypothesis [63] in which the bringing of a warm moisture-laden ocean current against a cold north American continent led to increased snow precipitation and ice formation on that continent, and hence (via increased albedo and other feedbacks) to further cooling.

8. ASTRONOMICAL MODULATION OF CLIMATE

Over the last 40 years, an astronomical pacemaker for the Quaternary 'Ice Age' has been established beyond doubt, comprising variations in orbital eccentricity ('stretch'), axial tilt and precession ('wobble') with dominant periodicities of roughly 100, 40 and 20 ka, respectively [48]. These produced small variations in the amount and seasonal distribution of sunlight reaching the Earth that, when amplified by various feedback mechanisms – notably via variations in atmospheric greenhouse gas concentrations – led to the well-established pattern of Quaternary glacial/interglacial and stadial/interstadial changes. This mechanism was famously championed in the early twentieth century by Milutin Milankovitch [64], fell out of favour because the timing of individual glaciations as deduced from the fragmentary terrestrial record did not seem to fit, and then was triumphantly vindicated by analysis using oxygen isotopes from fossil foraminifera, that reflected temporal variations in ambient temperature and ice volume of the more complete ocean record [65,66].

The exploitation of Milankovitch cycles has subsequently developed in various directions. It has become a stratigraphic tool for dating and correlation, not only in the Quaternary, but in Tertiary and yet older strata [67], where a longer, 400 ka, orbital 'stretch' cycle is used as a more or less invariant 'pulse' that can be exploited stratigraphically and even quasi-formalised [68]. This in turn has led to the realisation that climate in greenhouse as well as icehouse times was modulated by astronomical forcing, with variations in humidity/aridity and biological productivity producing patterns that, although more subtle than those produced by large ice volume changes, are nonetheless recognisable.

Also, the detailed expression of Milankovitch cycles has come under scrutiny. Astronomical calculation can precisely reveal insolation variations and hence predict the climate patterns that should result. The observed patterns from the stratal record depart from this in several ways. Firstly, they typically show a 'sawtooth' pattern rather than the predicted temporally symmetrical one: thus, individual glacial phases tend to develop slowly but finish abruptly. Secondly, the periodicity that is expected to be dominant is not always so, as will be seen below. Thirdly, and particularly in cold phases, there are marked, higher-frequency 'sub-Milankovitch' climate cycles that have been well-described (also see below) but have not yet had adequate explanation.

9. MILANKOVITCH CYCLICITY IN QUATERNARY (PLEISTOCENE) CLIMATE HISTORY

The Quaternary displays a marked progression of overall climate state that may be regarded as an intensification of the glacial signature through time. The early Quaternary is dominated by the 40 ka axial tilt signal. About a million years ago, this gave way to dominance by the 100 ka orbital eccentricity cycle that has persisted to the present (Fig. 1d). This dominance has yet to be explained satisfactorily, for it would not be predicted from consideration of calculated insolation patterns over this interval, in which the eccentricity signal should be small. Suggested explanations have included the evolution of the ice-sheet/substrate system to resonate (i.e. most easily grow and decay) to a 100 ka periodicity [69,70]; these explanations are tentative, for detailed models linking ice volume to insolation remain elusive [71]. The dominance by eccentricity has been accompanied by colder glacial maxima and warmer interglacial peaks, and it is this interval that has seen the greatest advances of ice, and in general represents the 'ice ages' of vernacular usage.

The past million years includes a detailed record of atmospheric composition as well as temperature, in the form of the ice core data extracted from Greenland and Antarctica (with some ice core data of shorter duration from mountain glacier ice elsewhere) [72]. The longest current record is from Antarctica, extending to ca. 800 ka [73,74] and planned drilling is aimed at extending the record to beyond a million years ago, and so into the 'forty kiloyear world'. The Greenland record goes back to little more than 130 ka, and so just into the last interglacial phase; but it is of high-resolution, because of a greater rate of snowfall, and is of great value in also allowing detailed comparison with the southern hemisphere, given the different climate behaviour of the hemispheres at short time scales (discussed below).

The combination of atmospheric composition records with climate proxy records (through hydrogen and oxygen isotope data, dust concentrations and so on) is extremely powerful (indeed, unique in the geological record); but, it is not precisely calibrated because ice data directly relates to deposition, while the gas data relates to the time of final closure of air bubbles in the ice, some distance down in the snow pack. The uncertainty that stems from this is small but important, because the correlation of carbon dioxide and methane levels with temperature is so close that questions of cause-and-effect have arisen. The consensus now is that astronomically driven insolation thresholds lead to small temperature rises, leading to carbon dioxide/methane increases that then strongly amplify the temperature rises [72].

The glacial-to-interglacial difference seen in the ice core records is about 100 ppm (from ca. 180 to ca. 280 ppm pCO_2, respectively), representing several hundred gigatonnes of carbon that must be stored somewhere during glacial phases. Terrestrial storage via increased plant growth is unlikely, given the diminution of vegetated land during glacials, though storage in carbon-rich permafrost soils ('yedoma') has been mooted [75]. Ocean storage is generally considered more likely, and it is tempting to link this with the enhanced dust supply noted in the ice core records, that would fertilise open ocean waters and enhance carbon drawdown via increased plankton growth. However, ocean sediment records of barium (a proxy for plankton productivity) do not generally show increases during glacial phases. One means of combining low plankton productivity and increased trapping of carbon dioxide is to have a more stratified glacial ocean, limiting nutrient supply from below because of a stronger surface water 'lid' and also storing more dissolved carbon dioxide at depth [76]. There is evidence for such a model in the form of glacial-phase benthic foraminifer tests containing excess 'old' (i.e. radiocarbon-poor) carbon [77].

10. QUATERNARY SUB-MILANKOVITCH CYCLICITY

Examination of high-resolution Quaternary records suggests significant climate variability that takes place on a sub-Milankovitch scale, a variability that is particularly marked in the cold phases that make up the bulk of the record (Fig. 1e). Thus, the cold phase that separates the present interglacial and the preceding (Eemian) one comprises not only five precession cycles, but also 26 well-marked temperature oscillations, termed Dansgaard–Oeschger (D–O) cycles. These are most clearly expressed in the northern hemisphere, where they comprise rapid warming (of 8–16 °C over Greenland) followed by slower cooling [78], to produce what are essentially a succession of inter-stadial and stadial units that average some 1470 a in duration [79]. The D–O cycles may be grouped into larger Bond cycles, terminated by intermittent (every several 1000 years) Heinrich events [80]: iceberg 'armadas' released from the Laurentide and Scandinavian ice sheets marking episodes of partial collapse (Fig. 1e). The Heinrich events led to distinctive gravel-rich layers within sea floor sediments (brought in from melting icebergs), metre-scale rises in global sea level and rapid northern hemisphere cooling. The D–O cycles have one-to-one counterparts in the southern hemisphere, but more muted ones (about 1–3 °C in Antarctica) that are in partial antiphase (Fig. 1f), being offset from the northern D–O events by about 90° (northern cold coinciding with southern warming) rather than in 'see-saw' fashion [81,82]. The causal mechanisms of the D–O cycles and related phenomena remain unclear, having been ascribed to changes in solar luminosity [83] and also to 'binge-purge' cycles of the great ice sheets [84].

The transition into the current Holocene interglacial was complex: thus, glacial conditions in the northern hemisphere were terminated at ca. 14.5 ka, with rapid deglaciation ushering in the millennial-scale Allerød warm phase, itself terminated by rapid cooling into the Younger Dryas cold interval, also lasting about a thousand years. This finished abruptly at 11.8 ka, when temperatures in the northern hemisphere rose by ca. 5 °C in about a decade, ushering in the warm and relatively stable conditions of the Holocene.

The reversal into the Younger Dryas has been ascribed to a major meltwater flood from the Laurentide ice-sheet into the north Atlantic, putting a low-salinity 'lid' on the north Atlantic, hence stopping the formation of the cold dense (high-salinity) North Atlantic Deep Water current and its ultimate return flow, the north Atlantic Drift ('Gulf Stream'); eventual re-start of this oceanic circulation pattern brought warmth once more back to the region. As with the D–O cycles, correlation with the southern hemisphere was complex, partly out-of-phase, and it is debated whether the climate changes were driven from the north or the south [85,86].

11. THE HOLOCENE

The Holocene is simply the latest of the many interglacial phases of the Quaternary; it is now longer than the preceding three interglacials by some 2000–3000 a [73], but only one-third of the length of the preceding one, OIS 11 [87], that lasted one-and-a-half, rather than half a precession cycle; it is still unclear to which style of interglacial the Holocene 'naturally' belongs to on astronomical grounds (and thus what its 'natural' duration might be). Its duration to date has also been linked with the slow rise in atmospheric carbon dioxide levels from 260 to 280 ppm, ascribed (controversially) to pre-industrial forest clearance by humans [88].

To date, though, other than a brief northern-hemisphere cooling event at 8.2 ka (also ascribed to a meltwater pulse from the decaying ice-sheets: [89], the Holocene has seen remarkable stability of temperature and sea level, even when compared with other interglacials. Climate variation within it includes subdued, millennial-scale temperature oscillations of 1 °C or so, examples being the 'Medieval warm period' and succeeding 'Little Ice Age', with sea level variations of perhaps 1 or 1–2 m [90]. As with the D–O cycles, their cause is obscure. Other shorter-period variations include the ENSO/El Nino events and the North Atlantic Oscillation (NAO) (of a few years periodicity each); as with the millennial scale variations, these have far-reaching global impacts on such factors as regional rainfall patterns via a series of global teleconnections [91].

12. CLIMATE OF THE ANTHROPOCENE

About two centuries ago, human population rose above a billion (it is now over six billion). Widespread industrialisation, powered by fossil fuels, also started then and continues to this day – indeed, is currently accelerating [92,93]. The sum total of physical, chemical and biological changes associated with this has led to the concept of the Anthropocene, a geological interval dominated by human activity [94]; if considered as a formal stratigraphic unit at an Epoch level [95], it follows that the Holocene has terminated.

Climate drivers of the Anthropocene are already well outside Holocene norms, for instance in: the marked increase in greenhouse gas levels (now higher than in pre-industrial times by the amount separating glacial and interglacial phases of the past, the change being considerably more rapid than either glacial-to-interglacial changes [96] or those associated with, say, the Toarcian event [46]; the changing nature of carbon sinks associated with land-use changes; and, as we write, the diminishing albedo associated with rapidly waning Arctic sea ice. The current greenhouse warming is acting on an already warm phase, and hence is bringing in a novel environmental state. The long-term effect, if median predictions of the Inter-Governmental Panel on Climate Change [97] come to pass, may resemble the brief 'super-interglacial' suggested by Broecker [98], the normal Quaternary Milankovitch cyclical climate changes subsequently resuming. Alternatively, modelled changes to the long-term carbon balance, together with threshold effects, suggest perturbation to at least several glacial cycles [99]. In whichever scenario (but particularly in the latter), the effects of the current warming will have geologically long-lasting effects.

13. CONCLUSIONS

The history of Earth's climate system, as deduced from forensic examination of strata, has shown a general very long-term stability, which has probably been maintained by a complex interaction between the biosphere, atmosphere, hydrosphere, cryosphere and lithosphere. Superimposed on this overall stability has been a variety of climate perturbations on timescales ranging from multi-million year to sub-decadal, inferred to have been driven, amongst others, by variations in palaeogeography, greenhouse gas concentrations, astronomically forced insolation and inter-regional heat transport. Current anthropogenic changes to the Earth system, particularly as regards changes to the carbon cycle, are geologically significant. Their effects may likely include the onset of climate conditions of broadly pre-Quaternary style such as those of the 'mid-Pliocene warm period', with higher temperatures (particularly at high latitudes), substantially reduced polar ice cover, and modified precipitation and biotic patterns.

ACKNOWLEDGEMENT

We thank Philip Gibbard, Alan Haywood and Euan Nisbet for their comments on earlier drafts of this paper.

REFERENCES

1. A.L.A. Johnson, J.A. Hickson, J. Swan, M.R. Brown, T.H.E. Heaton, S. Chenery, P.S. Balson, in: E.M. Harper, J.D. Taylor, J.A. Crame (Eds.), The Evolutionary Biology of the Bivalvia, vol.177, Geological Society, London, 2000, pp. 425–439, Special Publications.

2. D.S. Jones, W.D. Allmon, Lethaia 28 (1995) 61–74.
3. B. Jenny, B.L. Valero-Garcés, R. Villa-Martínez, R. Urrutia, M. Geyh, H. Veit, Quat. Res. 58 (2002) 160–170.
4. C.M. Rowe, D.B. Loope, R.J. Oglesby, R. Van der Voo, C.E. Broadwater, Science 318 (2007) 1284.
5. K.T. Lawrence, T.D. Herbert, P.S. Dekens, A.C. Ravelo, in: M. Williams, A. Haywood, F.J. Gregory, D.N. Schmidt (Eds.), The Micropalaeontological Society, The Geological Society, London, 2007, pp. 539–562, Special Publication.
6. F. Kenig, J.D. Hudson, J.S. Sinninghe Damsté, B.N. Popp, Geology 32 (2004) 421–424.
7. A.P.M. Vaughan, in: M. Williams, A. Haywood, F.J. Gregory, D.N. Schmidt (Eds.), The Micropalaeontological Society, The Geological Society, London, 2007, pp. 5–59, Special Publication.
8. E. Brook, Nature 453 (2008) 291–2.
9. J.C. McElwain, W.G. Chaloner, Ann.Bot. 76 (1995) 389–395.
10. W.M. Kürschner, J. van der Burgh, H. Visscher, D.L. Dilcher, Mar. Micropaleontol. 27(1996) 299–331.
11. G.J. Retallack, Science 276 (1997) 583–585.
12. P. Pearson, M. Palmer, Nature 406 (2000) 695–699.
13. R.A. Berner, Geochimica et Cosmochimica Acta 70 (2006) 5653–5664.
14. D.L. Royer, R.A. Berner, J. Park, Nature 446 (2007) 530–32.
15. L.E. Sohl, M.A. Chandler, in: M. Williams, A. Haywood, F.J. Gregory, D.N. Schmidt (Eds.), The Micropalaeontological Society, The Geological Society, London, 2007, pp. 61–80, Special Publication.
16. B. Sellwood, P. Valdes, in: M. Williams, A. Haywood, F.J. Gregory, D.N. Schmidt (Eds.), The Micropalaeontological Society, London, 2007, pp. 201–224, Special Publication.
17. H.J. Dowsett, in: M. Williams, A. Haywood, F.J. Gregory, D.N. Schmidt (Eds.), The Micropalaeontological Society, The Geological Society, London, 2007, pp. 459–480, Special Publication.
18. U. Salzmann, A. Haywood, D. Lunt, P. Valdes, D. Hill, Global Ecol. Biogeogr. 17 (2008) 432–447.
19. E.E. Jansen, J. Overpeck, K.R. Briffa, J.-C. Duplessy, F. Joos, V. Masson-Delmotte, D. Olago, B. Otto-Bliesner, W.R. Peltier, S. Rahmstorf, R. Ramesh, D. Raynaud, D. Rind, O. Solomina, R. Villalba, D. Zhang, in: S. Solomon, D. Qin, M. Manning, Z. Chen, M. Marquis, K.B. Averyt, M. Tignor, H.L. Miller (Eds.), Cambridge University Press, Cambridge, UK and New York, NY, USA, 2007.
20. S.E. Wilde, J.W. Valley, W.H. Peck, C.M. Graham, Nature 409 (2001) 175–178.
21. J. Baker, M. Bizzarro, N. Wittig, J. Connelly, H. Haack, Nature 436 (2005) 1127–1131.
22. E.G. Nisbet, N.H. Sleep, Nature 409 (2001) 1083–1091.
23. B. Rasmussen, R. Buick, Geology 27 (1999) 115–118.
24. J.E. Lovelock, Nature 344 (1990) 100–102.
25. M.J. Bickle, Terra Nova 8 (1996) 270–276.
26. A.S. Cohen, A.L. Coe, S.M. Harding, L. Schwark, Geology 32 (2004) 157–160.
27. F. Robert, M. Chaussidon, Nature 443 (2006) 969–972.
28. L.P. Knauth, Palaeogeogr. Palaeoclimatol. Palaeoceanogr. 219 (2005) 53–69.
29. H. Ohmototo, Y. Watanabe, K. Kumazawa, Nature 429 (2004) 395–399.
30. T.W. Lyons, Nature 429 (2004) 359–360.
31. I.A. Hilburn, J.L. Kirschvink, E. Tajika, R. Tada, Y. Hamano, S. Yamamoto, Earth Planet. Sci. Lett. 409 (2005) 175–178.
32. I.J. Fairchild, M.J. Kennedy, J. Geol. Soc. Lond. 164 (2007) 895–921.
33. M.I. Budyko, Tellus 21 (1969) 611–619.
34. P.F. Hoffman, A.J. Kaufman, G.P. Halverson, D.P. Schrag, Science 281 (1998) 1342.
35. M. Kennedy, D. Mrofka, C. von der Borch, Nature 453 (2008) 642–645.

36. A.A. Page, J.A. Zalasiewicz, M. Williams, L.E. Popov, in: M. Williams, A. Haywood, F.J. Gregory, D.N. Schmidt (Eds.), The Micropalaeontological Society, The Geological Society, London, 2007, pp. 123–156, Special Publication.
37. P.J. Brenchley, J.D. Marshall, G.A.F. Carden, D.B.R. Robertson, D.G.F. Long, T. Meidla, L. Hints, T.F. Anderson, Geology 22 (1994) 295–298.
38. R.M. DeConto, D. Pollard, 2003, Nature 421 (2007) 245–249.
39. T.D. Frank, L.P. Birgenheier, I.P. Montañez, C.R. Fielding, M.C. Rygel, in: C.R. Fielding, T.D. Frank, J.L. Isbell (Eds.), Geological Society of America Special, 2008, pp. 331–342, Paper 441.
40. A.G. Smith, K.T. Pickering, J. Geol. Soc., Lond. 160 (2003) 337–340.
41. J. Zachos, U. Röhl, S.A. Schellenberg, A. Sluijs, D.A. Hodell, D.C. Kelly, E. Thomas, M. Nicolo, I. Raffi, L.J. Lourens, H. McCarren, D. Kroon, Science 308 (2005) 1611–1615.
42. A.S. Cohen, A.L. Coe, D.B. Kemp, J. Geol. Soc., Lond. 164 (2007) 1093–1108.
43. J.C. Zachos, G.R. Dickens, R.E. Zeebe, Nature 451 (2008) 279–283.
44. D.J. Thomas, J.C. Zachos, T.J. Bralower, E. Thomas, S. Bohaty, Geology 30 (2002) 1067–1070.
45. J.C. McElwain, J. Wade-Murphy, S.P. Hesselbo, Nature 435 (2005) 479–482.
46. D.B. Kemp, A.L. Coe, A.S. Cohen, L. Schwark, Nature 437 (2005) 396–399.
47. M.E. Smith, A.R. Carroll, E.R. Mueller, Nat. Geosci. 1 (2008), 370–374.
48. J. Zachos, M. Pagani, L. Sloan, E. Thomas, K. Billups, Science 292 (2001) 686–693.
49. C. Lear, in: M. Williams, A. Haywood, F.J. Gregory, D.N. Schmidt (Eds.), The Micropalaeontological Society, The Geological Society, London, 2007, pp. 313–322, Special Publication.
50. A. Merico, T. Tyrrell, P.A. Wilson, Nature 452 (2008) 979–982.
51. J.P. Kennett, J. Geophys. Res. 82 (1977) 3843–3859.
52. G.J. Retallack, Palaeogeogr. Palaeoclimatol. Palaeoecol. 214 (2004) 97–123.
53. A.R. Lewis, D.R. Marchant, A.C. Ashworth, L. Hedenäs, S.R. Hemming, J.V. Johnson, M.J. Leng, M.L. Machlus, A.E. Newton, J.I. Raine, J.K. Willenbring, M. Williams, A.P. Wolfe, PNAS 105.
54. B.P. Flower, J.P. Kennett, Geology 21 (1993) 877–880.
55. A.E. Shevenell, J.P. Kennett, D.W. Lea, Science 305 (2004) 1766–1770.
56. A.M. Haywood, P. Dekens, A.C. Ravelo, M. Williams, Geochem. Geophys. Geosyst. 6 (2005) 1–20.
57. L. Liesecki, M.E. Raymo, Paleoceanography 20 (2005), 17, PA1003.
58. A.M. Haywood, H.J. Dowsett, P.J. Valdes (Eds.), Phil. Trans. R. Soc. A 367 (2009), 3–204.
59. P.L. Gibbard, A.G. Smith, J.A. Zalasiewicz, T.L. Barry, D. Cantrill, A.L. Coe, J.C.W. Cope, A.S. Galp, F.J. Gregory, J.H. Powell, P.F. Rawson, P. Stone, C.N. Waters, Boreas 34 (2005) 1–6.
60. M. Mudelsee, M.E. Raymo, Paleoceanography 20 (2005) PA-4022.
61. D.J. Lunt, G.L. Foster, A.M. Haywood, E.J. Stone, Nature 454 (2008) 1102–1105.
62. J.M. Edmond, Science 258 (1992) 1594–1597.
63. G.H. Haug, A. Ganopolski, D.M. Sigman, A. Rosell-Mele, G.E.A. Swann, R. Tiedemann, S.L. Jaccard, J. Bollman, M.A. Maslin, M.J. Leng, G. Eglinton, Nature 433 (2005) 821–825.
64. M. Milankovitch, Kanon der Erdbestrahlungen und seine Anwendung auf das Eiszeitenproblem, R. Serbian Academy Belgrade, 1941 Special Publication 132.
65. N. Shackleton, Nature 215 (1967) 15–17.
66. N.J. Shackleton, N.D. Opdyke, Quat. Res. 3 (1973) 39–55.
67. A. Gale, J. Hardenbol, B. Hathway, W.J. Kennedy, J.R. Young, V. Phansalkar, Geology 30 (2002) 291–294.
68. B.S. Wade, H. Pälike, Paleoceanography 19 (2004) PA001042.
69. P.U. Clark, R.B. Alley, D. Pollard, Science 286 (1999) 1104–1111.
70. R. Bintanja, R.S. van der Wal, Nature 454 (2008) 869–872.
71. M. Raymo, P. Huybers, Nature 451 (2008) 284–285.
72. E.W. Wolff, Episodes 31 (2008) 219–221.

73. J.R. Petit, J. Jouzel, D. Raynaud, N.I. Barkov, J.M. Barnola, I. Basile, M. Bender, J. Chapellaz, M. Davis, G. Delaygue, M. Delmotte, V.M. Kotlyakov, M. Legrand, V.Y. Lipenkov, C. Lorius, L. Pépin, C. Ritz, E. Salzmann, M. Stievenard, Nature 399 (1999) 429–436.
74. E. Brook, Nature 453 (2008) 291–2.
75. S.A. Zimov, E.A.G. Schuur, F.C. Chapin III, Science 312 1612–1613.
76. S.L. Jaccard, G.H. Haug, D.M. Sigman, T.F. Pedersen, H.R. Thierstein, U. Röhl, Science 308 (2005) 1003–1006.
77. T.M. Marchitto, S.J. Lehman, J.D. Ortiz, J. Flüchtiger, A. van Gen, Science 316 (2007), 1456–1459.
78. C. Huber, M. Leuenberger, R. Spahni, J. Flückiger, J. Schwander, T.F. Stocker, S. Johnsen, A. Landais, J. Jouzel, Earth Planet. Sci. Lett. 245 (2006) 504–519.
79. G. Bond, W. Showers, M. Cheseby, R. Lotti, P. Almosi, P. deMenocal, P. Priore, H. Cullen, I. Hajdas, G. Bonani, Science 278 (1997) 1257–1266.
80. G. Bond, H. Heinrich, W. Broecker, L. Labeyrie, J. Mcmanus, J. Andrews, S. Huon, R. Jantschik, S. Clasen, C. Simet, K. Tedesco, M. Klas, G. Bonani, S. Ivy, Nature 360 (1992) 245–249.
81. EPICA community members, Nature 444 (2006) 195–198.
82. E.J. Steig, Nature 444 (2006) 152–153.
83. H. Braun, M. Christl, S. Rahmnstorf, A. Ganopolski, A. Mangini, C. Kubatzki, K. Roth, B. Kromer. Nature 438 (2005) 208–211.
84. D.R. MacAyeal, Paleoceanography 8 (1993) 775–784.
85. T.F. Stocker, Science 297 (2002) 1814–1815.
86. A.J. Weaver, O.A. Saenko, P.U. Clark, J.X. Mitrovica, Science 299 (2003) 1709–1713.
87. EPICA community members. Nature 429 (2004) 623–628.
88. W.F. Ruddiman, Clim. Change 61 (2003) 261–293.
89. J.D. Marshall, B. Lang, S.F. Crowley, G.P. Weedon, P. van Calsteren, E.H. Fisher, R. Holme, J.A. Holmes, R.T. Jones, A. Bedford, S.J. Brooks, J. Bloemendal, K. Kiriakovlakis, J.D. Ball. Geology 207 (2007) 639–642.
90. K-E. Behre, Boreas 36 (2007) 82–102.
91. T.M. Rittenour, J. Brigham-Grett, M.E. Mann, Science 288 (2000) 1039–1042.
92. W. Steffen, A. Sanderson, P.D. Tyson, J. Jäger, P.A. Matson, B. Moore III, F. Oldfield, K. Richardson, H.J. Schnellhuber, B.L. Turner II, R.J. Wasson, Global Change and the Earth System: A Planet under Pressure, Springer-Verlag, Berlin, 2004.
93. W. Steffen, P.J. Crutzen, J.R. McNeill, Ambio 36 (2008) 614–621.
94. P.J. Crutzen, Nature 415 (2002) 23.
95. J. Zalasiewicz, M. Williams, A. Smith, T.L. Barry, A.L. Coe, P.R. Brown, P. Brenchley, D. Cantrill, A. Gale, P. Gibbard, F.J. Gregory, M.W. Hounslow, A.C. Kerr, P. Pearson, R. Knox, J. Powell, C. Waters, J. Marshall, M. Oates, P. Rawson, P. Stone, GSA Today 18 (2) (2008) 4–8.
96. E. Mannin, A. Indermühle, A. Dällenbach, J. Flückiger, B. Stauffer, T.S. Stocker, D. Reynaud, J.M. Barnola, 2001, Science 291 (2001) 112–114.
97. IPCC (Intergovernmental Panel on Climate Change, Climate change 2007: synthesis report. (2007) Summary for policy makers. Available at: http://www.ipcc.ch/SPM2feb07.pdf.
98. W.S. Broecker, How to Build a Habitable Planet, Eldigio Press, NewYork, 1987, pp. 291.
99. T. Tyrrell, J.G. Shepherd, S. Castle, Tellus B 59 (2007) 664–672.
100. S. Rahmstorf, Nature 419 (2002) 207–214.

Part III

Indicators of Climate and Global Change

Chapter 7

Changes in the Atmospheric Circulation as Indicator of Climate Change

Thomas Reichler
Department of Meteorology, University of Utah, Salt Lake City, Utah

1. **Introduction**
2. **The General Circulation of the Atmosphere**
3. **The Poleward Expansion of the Tropical Circulation**
   3.1. Observation-Based Evidence
   3.2. Model-Based Evidence
4. **The Decreasing Intensity of the Tropical Circulation**
5. **Emerging Mechanisms**
   5.1. Tropical Tropopause Heights
   5.2. Extratropical Eddies
   5.3. Static Stability
   5.4. The Role of SST Forcing
6. **Connection to Extratropical Circulation Change**
7. **Outstanding Problems and Conclusions**

Acknowledgments
Appendix: List of Abbreviations
References

1. INTRODUCTION

The strength, direction, and steadiness of the prevailing winds are crucial for climate. Winds associated with the atmospheric circulation lead to transports of heat and moisture from remote areas and thereby modify the local characteristics of climate in important ways. Specific names, such as extratropical Westerlies, tropical Trades, and equatorial Doldrums remind us of the significance of winds for the climate of a region and for the human societies living in it.

The purpose of this chapter is to discuss changes in the structure of the atmospheric circulation and its associated winds that have taken place during recent decades. These changes are best described as poleward displacements of major wind and pressure systems throughout the global three-dimensional atmosphere. The associated trends are important indicators of climate change and are likely to have profound influences on ecosystems and societies.

This review is focused on two important examples of such change: first, tropical circulation change related to a poleward expansion of the Hadley cell (HC) and second, extratropical circulation change, as manifested by a poleward shift of the zone of high westerly winds in the midlatitudes, also known as an enhanced positive phase of the annular modes (AMs). Although both changes are associated with similar poleward displacements, it still remains to be seen whether the two phenomena are directly connected.

As with most aspects of climate change, the circulation changes that occurred over the past are still relatively subtle, making it difficult to distinguish them from naturally occurring variations. The difficulty of reliably monitoring the global circulation is an additional complication. Long-term records of the atmosphere exist at few locations only, and most regions of the Earth are not observed. The problem of sparse observations can be partly overcome by utilizing meteorological reanalyses, which represent a combination of numerical weather predictions and available observations. In the present context, however, reanalyses are only of limited use, since changes in the mix of used observations over time create spurious trends in the data [1].

Because of the difficulty in observing the atmospheric circulation and its long-term trends, this review will not only rely on observation-based evidence but also include findings from general circulation models (GCMs). GCMs are certainly not perfect representations of the real system, but they are extremely valuable in situations where observations alone are not giving sufficient information. For example, they allow producing consistent time series of virtually any length, location, and quantity. GCMs can be used to perform actual experiments of the Earth's climate system in its full complexity, an undertaking that would be impossible in a laboratory setting. This makes GCMs indispensable research tools, in particular for the search for human influences on climate.

The most intriguing challenges regarding the atmospheric circulation and climate change are to understand what the nature of this change is, what the consequences for surface climate are, and what the underlying causes and mechanisms are. At the beginning of this review, we will develop some basic understanding of the nature of the atmospheric circulation to provide necessary information for the remainder of this chapter. Next, I will give an overview of observation and model-based evidence of past circulation change. This discussion will primarily focus on the tropical widening phenomenon. I will continue by presenting some of the mechanisms that have been put forward in the literature to explain the widening. Later, I will clarify the relationship to other important forms of climate change and in particular to extratropical circulation change. I will conclude by summarizing some outstanding research questions and by highlighting possible impacts of atmospheric circulation change on other components of the global climate system.

2. THE GENERAL CIRCULATION OF THE ATMOSPHERE

The general circulation of the atmosphere describes the global three-dimensional structure of atmospheric winds. Halley [2] was probably the first to realize that the sphericity of the Earth and the resulting spatially non-uniform distribution of solar heating are the basic drivers behind this circulation. The tropics absorb about twice the solar energy that the higher latitudes absorb, creating a meridional gradient in temperature and potential energy. Some of the potential energy is converted into kinetic energy [3], which is manifested as wind. The winds are then deflected under the influence of the rotating Earth, creating the complicated flow patterns of the general circulation.

Atmospheric flow leads to systematic transports and conversions of energy within the Earth climate system. The different forms of energy involved are sensible heat, latent heat, potential energy, and kinetic energy. Typically, the energy transports are directed against spatial gradients, thus reducing the contrasts between geographical regions. For example, the winds transport warm air from the tropics to the extratropics and cold air in the opposite direction, decreasing the temperature contrasts between low and high latitudes. Similarly, the general circulation redistributes water from the oceans to the continents and supplies land surfaces with life-bringing precipitation. In other words, the atmospheric circulation exerts a moderating influence on climate and reduces the extremes in weather elements. The atmospheric winds also help drive the oceans, which in turn redistribute heat from low to high latitudes, nutrients from the ocean interior to the surface, and carbon from the atmosphere to the ocean. Because of its important role in redistributing properties within the climate system the general circulation has also been dubbed the "great communicator" [4].

The distinction into tropical and extratropical regimes is fundamental for the Earth atmosphere. In the extratropics, large-scale motions are governed by quasi-geostrophic theory, a simple framework related to the near perfect balance between the pressure gradient force and the Coriolis force. The extratropical circulation is dominated by cyclones, which are also called storms, eddies, or simply waves. These cyclones are the product of baroclinic instability, which develops particularly strongly during winter as a consequence of the intense pole-to-equator temperature gradient during that season. The storm-track regions over the western parts of the Pacific and Atlantic oceans are the preferred locations for the development of such systems.

In the tropics, the Coriolis force is weak, and other effects such as friction, and diabatic and latent heating become important [5]. The resulting tropical circulation is very distinct from the extratropics. The HC [6] is the most prominent tropical circulation feature. It extends through the entire depth of the troposphere from the equator to the subtropics (ca. 30° latitude) over both hemispheres (Fig. 1). The cell develops in response to intense solar heating in the Inter Tropical Convergence Zone (ITCZ) near the equator. The moist tropical

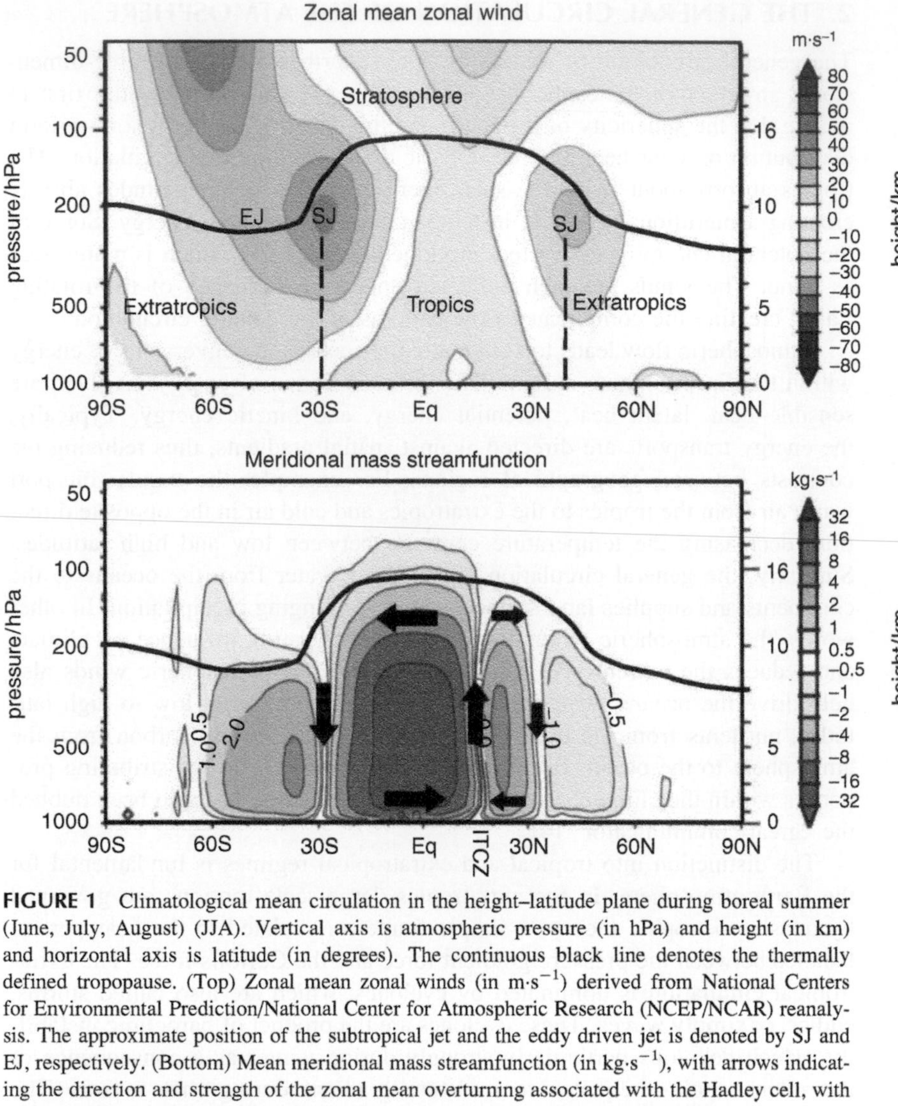

FIGURE 1 Climatological mean circulation in the height–latitude plane during boreal summer (June, July, August) (JJA). Vertical axis is atmospheric pressure (in hPa) and height (in km) and horizontal axis is latitude (in degrees). The continuous black line denotes the thermally defined tropopause. (Top) Zonal mean zonal winds (in m·s^{-1}) derived from National Centers for Environmental Prediction/National Center for Atmospheric Research (NCEP/NCAR) reanalysis. The approximate position of the subtropical jet and the eddy driven jet is denoted by SJ and EJ, respectively. (Bottom) Mean meridional mass streamfunction (in kg·s^{-1}), with arrows indicating the direction and strength of the zonal mean overturning associated with the Hadley cell, with a strong winter cell in the SH and a weak summer cell in the NH. (See Color Plate 9).

air warms, becomes buoyant, and rises towards the upper troposphere. The rising air cools adiabatically, leading to condensation, release of latent heat, and production of clouds and intense precipitation. In the upper troposphere, the air then diverges towards the poles and descends in the subtropics. The air is now dry and warm since it lost its moisture but retained much of the latent heat gained while rising. Consequently, the climate under the descending branch of the HC is characterized by dry conditions and relatively high pressure. The

HC is closed by the trade winds at the surface, which take up moisture from the oceans before they converge into the ITCZ.

The Walker circulation [7,8] is another important tropical circulation system, representing east-west oriented overturning of air across the equatorial Pacific. It is driven by low pressure and convection in the west, and high pressure and subsidence in the east. The pressure differences across the Pacific are due to warm sea surface temperatures (SSTs) over the west and rather cool SSTs over the east. Variations in these SSTs and the Walker circulation are closely related to the El Niño Southern Oscillation (ENSO) phenomenon, a naturally occurring instability of the coupled atmosphere–ocean system that has worldwide climate impacts [9].

The meridional overturning associated with the HC is also important for the extratropical circulation. For example, the poleward moving air in the upper branch of the HC tends to conserve angular momentum, spinning up a region of high zonal winds over the subsiding branch of the HC. This is the subtropical jet (Fig. 1). The jet, however, is not entirely angular momentum conserving, mainly because of the stirring action of the midlatitudes storms [10]. The stirring creates net fluxes of zonal momentum out of the jet and into the midlatitudes, which are so-called divergences and convergences of eddy-momentum. The consequence of these fluxes is a slowing of the subtropical jet and the creation of another wind maximum poleward of the subtropical jet. This second zone of high-wind speeds is the eddy-driven or polar-front jet [11]. This jet is often merged with the subtropical jet, giving the appearance of only one tropospheric jet centered at $\sim 30°$ latitude [12]. Only over the southern hemisphere (SH) and during winter are the two jet systems fairly well separated.

How does climate change impact the atmospheric circulation? Alterations of the radiative balance of the Earth due to climate change modify regional temperature and humidity structures. The winds respond to the resulting gradients and change the intensity and structure of the circulation. In the following sections, I will present evidence that such change is already taking place, and discuss some of the underlying theoretical mechanisms.

3. THE POLEWARD EXPANSION OF THE TROPICAL CIRCULATION

The location of the poleward boundaries of the tropics are not defined in unique and commonly accepted ways. This is related to the lack of an easily identifiable boundary between the extratropics and tropics. Atmospheric features undergo a more or less gradual transition between the two zones. The poleward extent of the tropics, therefore, depends on the definition of specific indicators of tropical width. Indicators that have been used in the past can be roughly divided into two categories. The first includes dynamical indicators, which focus on characteristic features in the atmospheric circulation at the outer edges of the tropics. Examples are the poleward boundary of the HC, the position of the subtropical jet cores, or the latitude where the surface winds change from westerly to easterly. The second consists of physical indicators, which utilize other aspects of the atmosphere that exhibit relatively sharp gradients at the tropical edge. These include the amount of outgoing longwave radiation, the concentration in stratospheric ozone, the height of the thermally defined tropopause, the relative humidity of the air, or the difference between precipitation and evaporation at the surface.

3.1. Observation-Based Evidence

Rosenlof [13] was probably the first to investigate long-term trends in the width of the tropics by studying the latitudinal extent of the upwelling branch of the Brewer–Dobson circulation in the lower stratosphere. This circulation represents a slow meridional overturning that extends through troposphere and stratosphere, with upwelling in the tropics and downwelling in higher latitudes. Rosenlof applied this indicator to reanalyses and found that the width of the tropics has increased by about 3° latitude per decade during the period 1992–2001. This rate is rather large and likely contains considerable observational uncertainty.

Continuing the pioneering work by Rosenlof, a subsequent study by Reichler and Held [14] focused on the structure of the global tropopause as another indicator of tropical width. This indictor is based on the well-known distinction between the tropics, where the tropopause is high, and the extratropics, where the tropopause is low (Fig. 1). The advantage of this method is that the tropopause is a relatively well observed atmospheric feature that can be easily derived from three-dimensional temperature fields. Using data from radiosondes (Fig. 2) and reanalyses (Fig. 3), it was found that the tropics have been expanding by about 0.4° latitude per decade since 1979. The same study arrived at very similar results by examining the separation distance between the two subtropical jets. Although the new widening figure was considerably smaller than what was found earlier [13], it confirmed the original result that the tropics were expanding.

The initial studies sparked a flurry of new research activity, aimed at better understanding the new phenomenon and its underlying cause. For instance, Fu et al. [15] examined long-term data (1979–2005) from the satellite-borne microwave sounding unit and found that the midtropospheric global warming signal was most pronounced in the subtropics (15–45°). It was argued that the enhanced warming was caused by a poleward shift of the subtropical jets. Hudson et al. [16] defined the location of the tropical edges from the characteristic distribution of total ozone between the tropics and the extratropics. They examined long-term records of total ozone from the Total Ozone Mapping Spectrometer instruments and found that the area over the northern hemisphere (NH) occupied by low ozone concentrations, which is indicative

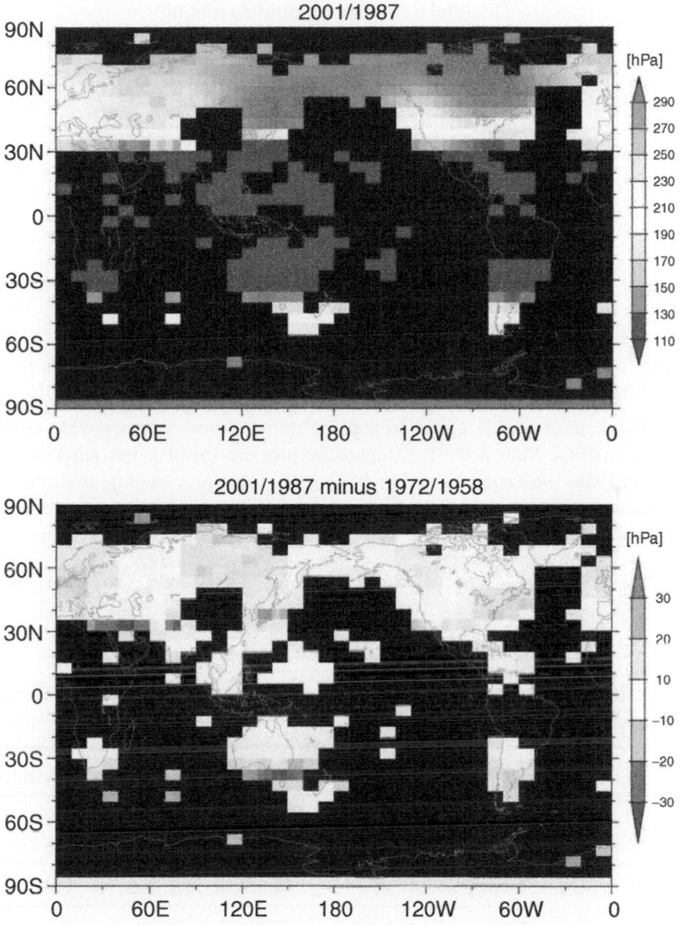

FIGURE 2 Changes in tropopause pressure during boreal winter (December, January, February) derived from gridded radiosonde data HADRT V2.1 [93]. (Top) Absolute tropopause pressure (in hPa) averaged over 1987–2001. (Bottom) Differences in tropopause pressure (in hPa) between the late period 1987–2001 and an early period 1958–1972. Bluish grid points indicate that tropopause pressure is decreasing and tropopause heights are increasing. The bluish banded structures over southern Australia and southern Europe indicate a trend toward tropical tropopause conditions and thus a widening of the tropics. Adapted from [14]. (See Color Plate 10).

for tropical regions, has increased over time. Seidel and Randel [17] also used the tropopause criterion to distinguish between the tropics and extratropics and examined the bimodal distribution of tropopause heights in the subtropics. Applying this measure to radiosonde and reanalysis data they again concluded that the tropics have been expanding.

Table 1 provides an overview of these and other relevant studies. Individual widening estimates range between 0.3° and 3° latitude per decade, with

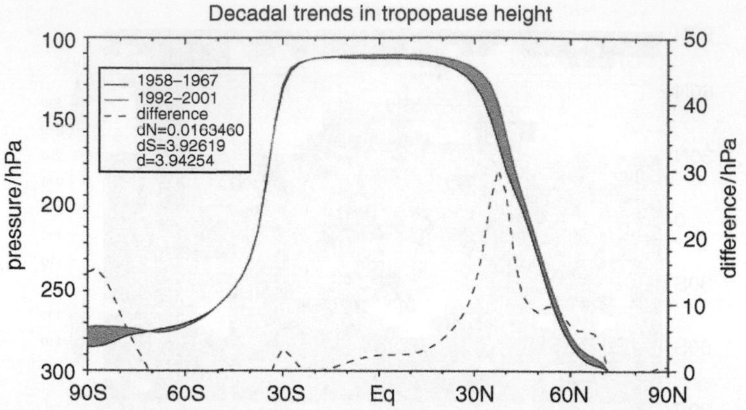

FIGURE 3 Zonal mean profile of the tropopause during boreal summer (JJA) derived from NCEP/NCAR reanalysis. Shown are the tropopause pressure (in hPa, left axis) averaged over the years (black) 1958–1967 and (dark grey) 1992–2001. Light grey shading indicates the differences in tropopause position between the two periods. The dashed curve shows the difference in tropopause pressure (in hPa, right axis). Horizontal axis shows latitude (in degrees). Adapted from [14].

TABLE 1 Estimates of tropical widening (in degrees latitude per decade) from observation-based studies

Study	Indicator	Data	Widening
Rosenlof [13]	Tropical upwelling (60 hPa)	Analyses	3.0
Reichler and Held [14]	Tropopause height	Radiosonde	0.4
	Tropopause height	Reanalyses	0.7
Fu et al. [15]	Tropospheric temperatures	MSU	0.7
Hudson et al. [16]	Total ozone	TOMS	1.0 (NH only)
Seidel and Randel [17]	Tropopause height	Radiosonde, reanalyses	1.8–3.1
Hu and Fu [59]	Outgoing longwave radiation	Various satellite sensors	1.5
	Mean meridional circulation	Reanalyses	1.0
Archer and Caldeira [92]	Jet stream separation	Reanalyses	0.3
Seidel et al. [23]	Jet stream separation	Reanalyses	1.0

a consensus widening of about 1.4°. The wide range of outcomes may be reconcilable in terms of observational uncertainties and methodological differences. However, some of the estimates are probably unrealistically large. For example, a sustained widening of 3° or more over the past three decades would have led to pronounced shifts in climate that have not been observed. Excluding some of the outliers, the most likely consensus estimate is, therefore, close to 1° latitude widening per decade over the recent decades.

Another important aspect of the observed tropical widening is its regional and seasonal structure. At least two studies suggest that the widening trend is strongest during summer of the respective hemisphere and that it is generally more pronounced over the SH than over the NH. In other words, the tropical expansion is largest over the SH during December, January, and February, and it is smallest over the NH during the same months.

3.2. Model-Based Evidence

The observed expansion is also reproduced by climate models that are driven with the observed history of forcings over the past decades. For example, most of the twentieth century scenario integrations of the Fourth Assessment Report of the Intergovernmental Panel on Climate Change (IPCC-AR4) [18] reproduce a widening of the tropopause [14]. The widening in the model with the largest expansion amounts to 0.7° latitude per decade over the last three decades, which is consistent with the observations. However, other models simulate much smaller rates, and some even negative ones. When the mean meridional circulation is used as indicator for the tropical edge, the same simulations averaged over all models show a widening of 0.2° latitude per decade over the period 1970–1999 [19].

Given the relatively small expansion seen in GCM simulations for the past one may ask how models respond to stronger greenhouse gas increase, which is expected to take place in the future. Kushner et al. [20] forced a fully coupled GCM with ~1% CO_2 increase per year and found over the SH a strong poleward shift of the westerly jet and of several related dynamical fields. The A2 scenario integrations of the IPCC-AR4 project, which correspond to a strong future increase in greenhouse gases, also reproduce robust poleward shifts of the jets [21], with an ensemble mean response of ~0.2° latitude per decade over the period 2000–2100 [19,22] (Fig. 4).

The aforementioned studies demonstrate that GCMs respond to anthropogenic forcings in expected ways, that is, the tropical edges and other aspects of the general circulation move poleward (Fig. 4). However, the model simulated trends seem to be smaller than in the observations. For example, the mean widening rate under the A2 scenario is about five times smaller than what was apparently observed during the past, despite the strong increase in greenhouse gas forcing under the A2 scenario. One might conclude that models have deficits in simulating the full extent of the widening. Seidel et al. [23],

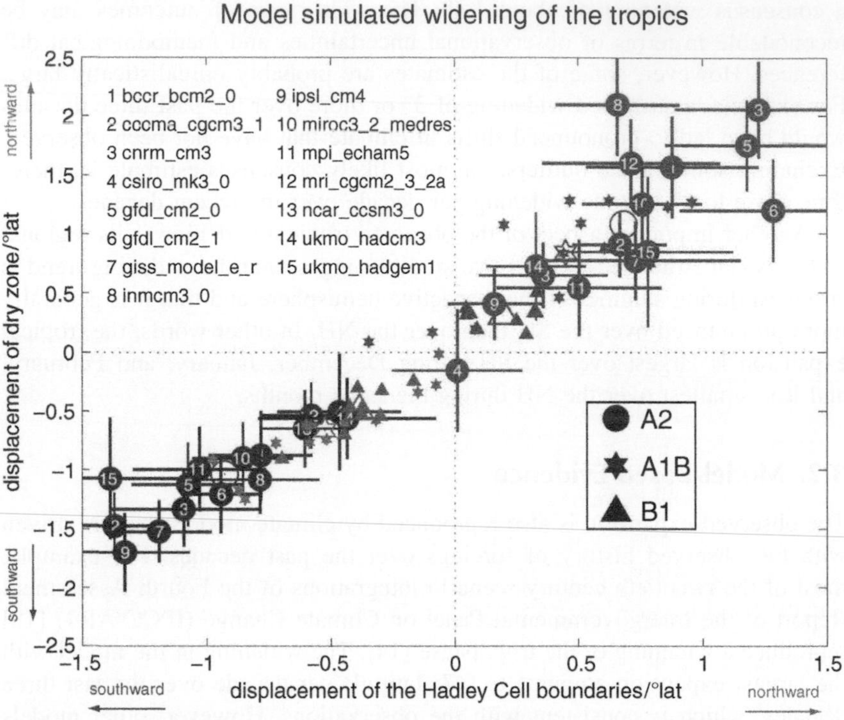

FIGURE 4 Displacements of (*x*-axis) the poleward boundaries of the tropical HC and (*y*-axis) the subtropical dry zone for different GCMs and scenarios. Reddish colors are for the NH and bluish colors are for the SH. The circles, hexagrams, and triangles denote the changes (2081–2100 minus 2001–2020, in degrees latitude) estimated from the A2, A1B, and B1 scenarios of the IPCC-AR4 simulations. The open symbols denote the multimodel ensemble mean values. The crosses centered on each circle show the 95% confidence intervals using a Student's *t*-test. Adapted from Ref. [22].

for example, raised the possibility that the poor representation stratospheric processes in the IPCC-AR4 models [24,25] may be in part responsible.

However, there are also reasons that may help to reconcile the discrepancies between observed and simulated trends. For example, the width of the tropics may undergo large natural swings on decadal and longer time scales; although model-derived estimates of this component of variability do not seem to support this explanation [19]. In addition, despite the increase in greenhouse gases, future trends may be smaller than past trends because of the expected recovery of stratospheric ozone [26]. And lastly, given the difficulty to make consistent long-term atmospheric observations, it is likely that the observed trends contain large uncertainties. Based on these limitations, it is currently impossible to say how realistic models simulate past and future widening trends.

4. THE DECREASING INTENSITY OF THE TROPICAL CIRCULATION

Another important aspect of the tropical overturning circulation is its strength. Theoretical considerations suggest that global warming weakens the strength of the mean tropical circulation [27,28]. This can be understood from the increasing moisture holding capacity of air in a warmer world, which is not followed along by an equivalent intensification of the hydrological cycle. A circulation slow-down is required to compensate for the difference.

Long-term observations of sea level pressure over the tropical Pacific reveal a weakening of the zonally asymmetric Walker circulation [29,30], which is consistent with the theoretical findings. Model simulations suggest that this downward trend is largely due to anthropogenic forcing and that this trend is going to continue in the future [28,31,32]. Warming of SSTs seems to be mostly responsible for the weakening [33].

For reasons yet to be understood, models also suggest that the weakening of the tropical circulation affects mostly the east–west oriented Walker circulation and not so much the zonal-mean HC [22,28]. This finding is also supported by the analysis of radiosonde data [34]. Atmospheric reanalyses give a somewhat mixed picture in this respect, with some indicating intensification and others showing no change [34,35]. This discrepancy may be related to well-known quality problems of the reanalysis [1]. Overall, the relative stability of the HC strength suggests that intensity changes in the tropical circulation are probably less important for the poleward movement of the HC and other elements of the general circulation.

5. EMERGING MECHANISMS

Understanding why the tropics are expanding under climate change is an area of active research. Several ideas have been put forward so far which, individually or together, may help to explain the phenomenon. Here, I will discuss three principal mechanisms that have been suggested in the literature: the changes in tropical tropopause heights, extratropical eddy activity, and static stability. In addition, I will explain what role recent and future SST changes have for the tropical width.

5.1. Tropical Tropopause Heights

Analysis of radiosonde [36] and reanalysis data [37,38] shows that the height of the global tropopause has increased over the past decades, and GCM experiments indicate that anthropogenic climate change is likely responsible for this increase [39]. This increase has been suggested as a possible reason for the poleward expansion of the tropical circulation. For example, nearly inviscid theory for axisymmetric circulations, proposed by Held and Hou [40], suggests that the meridional extent of the Hadley circulation varies proportionally with the square root of its vertical depth. However, applying this scaling to the past observed tropopause height increase of about 200 m [39] leads to a tropical expansion of only 0.1° latitude per decade, which is less than what is suggested by the observations and by most models. Analyses of idealized [41,42] and more complex climate models [22,43] also demonstrated that the Held and Hou theory does not provide a good explanation for the full parameter dependence of the meridional extent of the HC.

Other studies have suggested that changes in tropopause heights poleward of the jet are key to the poleward shift of the jet and the tropical edges [21,44,45]. These modeling studies have in common that the height of the tropopause is controlled by externally imposed temperature changes above or below the tropopause. However, this not only affects the height of the tropopause but also the meridional temperature gradients, the zonal winds, and the vertical wind shear by way of the thermal wind relationship. The additional circulation changes make it difficult to unequivocally assign the cause for the tropical widening to the lifting of the tropopause. In addition, none of the above studies puts forward a convincing physical mechanism by which tropopause height changes impact the position of the jets and the tropical edges.

5.2. Extratropical Eddies

The aforementioned recent increase in global tropopause heights is closely associated with systematic temperature changes below and above the tropopause [46]. Temperatures have been warming in the troposphere and cooling in the stratosphere, both of which have shown to be related to anthropogenic activity [47–49]. The pattern of warming and cooling also affects the zonal wind structure in the region of the subtropical upper troposphere and lower stratosphere (UTLS). This is related to the height structure of the tropopause. In the tropics, the tropopause is high and global warming reaches up to ~16 km. In the extratropics, the tropopause is low and warming reaches only up to ~12 km, followed by cooling in the stratosphere above. Thus, at intermediate heights of the UTLS region (~12–16 km) the tropics warm and the extratropics cool, leading to an increase in meridional temperature gradients, and, by the thermal wind relationship, to an increase of zonal wind speeds above.

Chen and Held [50] proposed a mechanism that establishes a connection between the pattern of warming and cooling in the UTLS region, the associated zonal wind anomalies, and the poleward movements of the jets. Extratropical tropospheric eddies play a central role in this mechanism. The eddies tend to move eastward with the zonal flow and equatorward toward the subtropics until they approach their critical latitudes, where their phase speed equals the speed of the background zonal flow. There, the waves grow in amplitude, break irreversibly, and decelerate the flow as a result of the absorbed wave activity. A key to understanding the mechanism is that the zonal wind in the UTLS region determines the eastward phase speed of extratropical tropospheric waves. Climate change related increases in UTLS winds increase the speed of the waves. According to critical layer theory, the now faster waves cannot penetrate as far equatorward into the regions of decreasing zonal winds. This in turn confines the zone of the eddy-driven jet more poleward and leads to a more positive state of the AMs.

Critical elements of this mechanism were identified in both model simulations and observations [50]. Later, this mechanism was extended by arguing that the poleward shift of the eddy-driven subsidence in the subtropics not only affects the AMs but also the HC related portion of the subsidence, which would move the boundaries of the HC and thus the tropical edges poleward [51].

5.3. Static Stability

Changes in the vertical temperature structure of the atmosphere provide yet another explanation for the tropical widening. Such changes are related to the vertical non-uniformity of the tropospheric global warming signal. Observations as well as model experiments indicate that the global warming signal in the upper troposphere is stronger than in the lower troposphere and that it maximizes in the tropical upper troposphere [52,53]. The upper tropospheric amplification is a well-established consequence of the quasi-moist adiabatic adjustment of the atmosphere, which leads to an increase in static stability in both the tropics [54,55] and extratropics [56–58].

Theory proposed by Held [10] establishes a connection between static stability and tropical width. The theory assumes that the upper, poleward moving branch of the HC is angular momentum conserving. The poleward moving air increases its zonal wind speed until it becomes baroclinically unstable and breaks down under the growing vertical wind shear. This marks the latitude of the outer boundary of the HC. Global warming related increases in static stability postpone the point where the atmosphere becomes baroclinically unstable. As a consequence, the HC expands towards higher latitudes.

The original theory was later refined by arguing that the poleward movement of the HC is intimately tied to the eddy-driven jet [51]. Global warming related reduction of baroclinicity at the equatorward flank of the eddy-driven jet stabilizes eddy growth and moves the jet and the associated subsidence toward the poles. The HC follows along since in the subtropics both HC and eddy-driven jet are associated with subsidence.

Independent of which interpretation is best, studies with both idealized models [41,43] and full global climate models [22,51] confirm that the Held [10] theory holds reasonably well in model simulated climates. For example, in idealized parameter sweep experiments, which were forced with prescribed

SSTs, Frierson et al. [43] find that the global mean warming is the primary reason for the expansion of the HC and that increases in meridional temperature gradients play only a secondary role. It is also noteworthy that the global warming related increase in static stability is expected to be particularly strong during summer and over the SH [57], which is consistent with the regional and seasonal patterns of the observed tropical widening [14,59].

5.4. The Role of SST Forcing

Surface temperatures over the tropical oceans undergo changes over time, which have been shown to have important consequences for the global atmospheric circulation [60,61]. These SST changes are primarily related to the natural ENSO phenomenon and to anthropogenic climate change. ENSO related SST fluctuations are periodic in nature and mainly affect the equatorial Pacific. Besides, global SSTs exhibit significant long-term trends that are associated with anthropogenic climate change [62]. The trends over the tropical Pacific resemble the SST pattern that exists during the warm phase of ENSO, which is related to the climate transition from 1976 to 1977 and the associated upward swing of the Pacific Decadal Oscillation (PDO) [63,64].

Various studies have demonstrated that the tropics are contracting during the warm phase of ENSO (El Niño), as indicated by equatorward displacements of the jet, storm track, eddy momentum divergence, and edge of the HC [33,51,65–68]. This equatorward shift is most pronounced over the SH, but it is also detectable over the NH. One way of understanding the contraction is the intensification of the thermally driven Hadley circulation as the equatorial SSTs become warmer [69]. The stronger HC leads to a westerly acceleration in its upper, poleward moving branch and thus to a strengthening of the subtropical jet. Invoking similar arguments as before, this moves the critical latitude for extratropical wave propagation equatorward, allowing the extratropical eddies to penetrate deeper into the tropics than during normal or cold ENSO conditions. As a result, elements of the circulation, including the tropical edge, shift equatorward.

An alternative explanation for the contraction of the tropics during El Niño is that the increased equatorial heating increases the pole-to-equator temperature gradient and draws the zone of maximum baroclinicity towards the equator. Consequently, the eddy-driven part of the circulation is shifted towards lower latitudes. There is also evidence that the increase in surface baroclinicity in association with El Niño impacts the type and number of non-linear wave breaking, which in turn may change the structure and position of the jet [68,70].

Lu et al. [71] found that the tropics are also contracting when a GCM is only forced by the observed history of SST and sea ice distribution but when atmospheric radiative effects due to natural and anthropogenic sources are excluded from the forcing. This can be understood from the El Niño-like pattern in the long-term trend of equatorial Pacific SSTs, which causes an equatorward shift of the circulation. This suggests that the tropical widening is largely a result of radiative effects (i.e., increased greenhouse gases and/or stratospheric ozone depletion), and that tropical effects are less important than extratropical effects. The finding by Lu et al. [71] also illustrates that climate change related shifts of the general circulation are complex and that the full response cannot be explained from one single mechanism alone.

6. CONNECTION TO EXTRATROPICAL CIRCULATION CHANGE

The extratropical circulation also undergoes important shifts under climate change, as revealed by observations and model data. The most prominent examples are the AMs, which are the dominant modes of large-scale extratropical variability [72]. The AMs are defined by shifts in sea level pressure between the high and middle latitudes, which are associated with equivalent barotropic changes in zonal winds, temperature, and geopotential height. There exists a tight relationship between AM variability and the position of the eddy-driven jet [65]; in other words, a positive AM is congruent with a poleward shift of the eddy-driven jet and its associated momentum fluxes.

The AMs have exhibited positive trends in both hemispheres in recent decades [73,74]. These were associated with lower than normal pressure over the poles and meridional shifts of the eddy-driven jet and the surface westerlies. The past trends were robust over the SH [15,74,75] but somewhat ambiguous over the NH [76–79]. Climate change simulations suggest that these trends are caused by increases in greenhouse gases and stratospheric ozone depletion [20,80–83].

Changes in the extratropical circulation are largely consistent with trends in the tropical circulation. For example, Previdi and Liepert [84] found in the IPCC-AR4 simulations a significant connection between variability in the AMs and the width of the HC: An increase in the AM index is accompanied by a poleward movement of the HC, and future upward trends in the AMs explain about half of the future expansion of the tropics. Consistent with the tendency toward more positive AMs, the twenty-first century simulations of the IPCC-AR4 [80] and other simulations [85] show that the extratropical storm tracks and the zone of the maximum surface westerlies move poleward and become more intense under global warming.

The observed and projected shifts of the AM under climate change can be explained, at least in part, by the eddy mechanisms [86] mentioned before: increased greenhouse gases and/or stratospheric ozone depletion lead to upper tropospheric warming and lower stratospheric cooling across the tropopause slope, which increases the UTLS winds and the phase speed of midlatitudes eddies. Invoking critical layer arguments, this results in a poleward shift of the eddy momentum flux convergence in the midlatitudes (i.e., a more positive AM). At the same time, the eddy-driven subsidence in the subtropics is also shifting poleward, which helps to explain the phase variability between the AM and the tropical width.

Stratospheric climate change also seems to be connected to the widening of the circulation. Increases in greenhouse gases and ozone depleting substances over the past have led to a substantial cooling of the stratosphere, especially over higher latitudes [48]. The resulting changes in zonal wind structure and subsequent dynamical interaction between the stratosphere and troposphere [87,88] may influence tropospheric climate [25]. The recent increase in the AM index over the SH has been linked to stratospheric ozone depletion over the Antarctic [74,81,82]. In the future, the additional build-up of greenhouse gases is expected to lead to a year-round positive shift of the AMs in both hemispheres. Over the SH, this positive trend is opposed by the expected recovery of stratospheric ozone over the SH [26]. Model simulations indeed demonstrate that ozone recovery has a seasonal effect that dominates and reverses the positive AM trend during summer [24,89].

7. OUTSTANDING PROBLEMS AND CONCLUSIONS

In summary, there exists considerable scientific evidence that key-elements of the atmospheric circulation have been moving poleward during the last few decades. Current theories as well as model experiments indicate that human activity in association with greenhouse gas increases and stratospheric ozone depletion is the most likely cause for the trends. However, it cannot be ruled out that natural climate variability also plays an important role, and there are many other aspects of these shifts that are not well understood.

The most notable reason for this deficit is the lack of climate-quality observations. Such observations are required to accurately characterize the nature of change, to validate climate models, and to falsify theories. However, as demonstrated by the wide range of outcomes from the tropical widening studies, the uncertainties in observing this phenomenon are large. Better observations are also needed to explore the regional and seasonal characteristics of the trends, and to decide how much of the observed change is due to natural low-frequency variability. It is hoped that some of these issues can be resolved in the near future, when satellite-based observation records will be long enough to be useful for climate studies. One of these records could come from the global positioning system radio occultation (GPS-RO) technique, which has great potential for monitoring the poleward edges of the tropics [90].

To some extent, model simulations can make up for what observations are lacking. But, as with the observations, the spread of outcomes in the current generation of models is still unsatisfactorily large. Although most models indicate a tropical expansion over the past, some actually simulate a contraction. These discrepancies may be related to differences in forcings, but important systemic inter-model differences are likely to be responsible as well. Understanding why the various models arrive at different answers and reconciling the differences would greatly aid in understanding the underlying causes and mechanisms for the widening and lead to more reliable simulations of the future widening.

There has been a marked improvement in our theoretical understanding of the circulation change. Three important mechanisms have been presented in this review, but it is still unclear which, if any, is correct. If several mechanisms are involved, then their relative contributions need to be understood. There is a strong indication that structural changes in extratropical eddies in relationship with lower stratospheric temperature change are behind some of the trends. However, increasing amounts of water vapor in a warmer climate may also modify eddy structure and thus lead to circulation change. This issue has not been adequately addressed, so far. A related question is whether the extratropical and tropical circulation trends have a common cause and whether they are connected to a similar mechanism.

What are the consequences of changes in the general circulation for other components of the Earth climate system? So far, the basic structure of the atmospheric circulation has remained unaltered and the position changes are only of modest amplitude. But even small shifts in the location of the HC, jets, and stormtracks can have important implications for regional climates by modifying patterns of storminess, temperature, and precipitation [85,91]. Particularly sensitive are regions with large spatial gradients in their normal distribution of precipitation, like the subtropical dry-zones (Fig. 4). There, even small trends decide whether there is a surplus or a deficit in overall rain. For example, the expansion of the HC may cause drier conditions over the subtropical semi-arid regions, including the Mediterranean, the southwestern United States, southern Australia, and southern Africa [23], and it was speculated that this process is already under way [91]. Atmospheric circulation change may also alter ocean currents. Because oceans are important regulators of climate, this may induce complicated and unexpected feedbacks, which either amplify or diminish the original cause for change.

Given the important role of the atmospheric general circulation for climate, any change in its structure is of concern. It may lead to profound changes in other parts of the global climate system with potentially important implications for natural ecosystems and human societies.

ACKNOWLEDGMENTS

I am grateful to Jian Lu for the many discussions we had on this and other topics. I thank Paul Staten for reviewing the manuscript and for offering helpful suggestions. I gratefully acknowledge financial support from the National Science Foundation under grant ATM 0532280.

APPENDIX: LIST OF ABBREVIATIONS

AM	annular mode
ENSO	El Niño Southern Oscillation
GCM	general circulation model
HC	Hadley cell
IPCC-AR4	fourth assessment report of the intergovernmental panel on climate change
ITCZ	inter tropical convergence zone
NH	northern hemisphere
PDO	Pacific decadal oscillation
SH	southern hemisphere
SST	sea surface temperature
UTLS	upper troposphere/lower stratosphere

REFERENCES

1. K.E. Trenberth, D.P. Stepaniak, J.W. Hurrell, M. Fiorino, J. Clim. 14 (2001) 1499.
2. E. Halley, Philos. Transact. 16 (1686) 153.
3. E.N. Lorenz, Tellus 7 (1955) 157.
4. D.L. Hartmann, Global Physical Climatology, vol. 56, International Geophysics Series Academic Press, San Diego, 1994.
5. T. Reichler, J.O. Roads, J. Clim. 18 (2005) 619.
6. G. Hadley, Philos. Transact. 39 (1735) 58.
7. G.T. Walker, Mem. Indian Meteor. Dept. 24 (1924) 275.
8. P.R. Julian, R.M. Chervin, Mon. Wea. Rev. 106 (1978) 1433.
9. K.E. Trenberth, G.W. Branstator, D. Karoly, A. Kumar, N.C. Lau, C. Ropelewski, J. Geophys. Res.-Oceans, 103 (1998) 14291.
10. I.M. Held, Woods Hole Geophysical Fluid Dynamics Program, Woods Hole Oceanographic Institute, Woods Hole, MA, 2000, 54pp.
11. G.K. Vallis, Atmospheric and Oceanic Fluid Dynamics: Fundamentals and Large-Scale Circulation, Cambrdige University Press, Cambridge, 2006.
12. S. Lee, H.K. Kim, J. Atmos. Sci. 60 (2003) 1490.
13. K.H. Rosenlof, J. Meteorol. Soc. Jpn. 80 (2002) 831.
14. T. Reichler, I. Held, Paper presented at the AMS Conference on Climate Variability and Change, Cambridge, MA, 2005.
15. Q. Fu, C. Johanson, J. Wallace, T. Reichler, Science 312 (2006) 1179.
16. R.D. Hudson, M.F. Andrade, M.B. Follette, A.D. Frolov, Atmos. Chem. Phys. 6 (2006) 5183.
17. D.J. Seidel, W.J. Randel, J. Geophys. Res 112 (2007) doi:10.1029/2007JD008861.
18. IPCC, Climate Change 2007: The physical science basis, Summary for policymakers. Intergovernmental Panel on Climate Change Secretariat, c/o WMO, Geneva, 2007.
19. C.M. Johanson, Q. Fu, J. Climate, (2009) (submitted). This paper is still under review, and is likely to appear in 2009 (not 2008).
20. P.J. Kushner, I.M. Held, T.L. Delworth, J. Clim. 14 (2001) 2238.
21. D.J. Lorenz, E.T. DeWeaver, J. Geophys. Res. 112 (2007) D10119.
22. J. Lu, G.A. Vecchi, T. Reichler, Geophys. Res. Lett. 34 (2007) doi:10.1029/ 2006GL028443.
23. D.J. Seidel, Q. Fu, W.J. Randel, T. Reichler, Nat. Geosci. 1 (2008) 21.

24. S.-W. Son, L.M. Polvani, D.W. Waugh, H. Akiyoshi, R. Garcia, D. Kinnison, S. Pawson, E. Rozanov, T.G. Shepherd, K. Shibata, Science, 320 (2008) 1486.
25. M.P. Baldwin, M. Dameris, T.G. Shepherd, Science 316 (2007) 1576.
26. V. Eyring, D.W. Waugh, G.E. Bodeker, E. Cordero, H. Akiyoshi, J. Austin, S.R. Beagley, B. Boville, P. Braesicke, C. Brühl, N. Butchart, M.P. Chipperfield, M. Dameris, R. Deckert, M. Deushi, S.M. Frith, R.R. Garcia, A. Gettelman, M. Giorgetta, D.E. Kinnison, E. Mancini, E. Manzini, D.R. Marsh, S. Matthes, T. Nagashima, P.A. Newman, J.E. Nielsen, S. Pawson, G. Pitari, D.A. Plummer, E. Rozanov, M. Schraner, J.F. Scinocca, K. Semeniuk, T.G. Shepherd, K. Shibata, B. Steil, R. Stolarski, W. Tian, and M. Yoshiki, J. Geophys. Res., 112 (2007) doi:10.1029/2006JD008332.
27. T.R. Knutson, S. Manabe, J. Clim. 8 (1995) 2181.
28. I.M. Held, B.J. Soden, J. Clim. 19 (2006) 5686.
29. G.A. Vecchi, et al., Nature 441 (2006) 73.
30. M. Zhang, H. Song, Geophys. Res. Lett. 33 (2006) doi:10.1029/ 2006GL025942.
31. G.A. Vecchi, B.J. Soden, J. Clim. 20 (2007) 4316.
32. H.L. Tanaka, N. Ishizaki, A. Kitoh, Tellus A 56 (2004) 250.
33. C. Deser, A.S. Philips, J. Climate, (2009) in press.
34. C.M. Mitas, A. Clement, Geophys. Res. Lett. 33 (2006) doi:10.1029/ 2005GL024406.
35. C. Mitas, A. Clement, Geophys. Res. Lett. 32 (2005) doi:10.1029/ 2004GL021765.
36. D.J. Seidel, W.J. Randel, J. Geophys. Res. 111 (2006) doi:10.1029/2006JD007363.
37. W. Randel, F. Wu, D. Gaffen, J. Geophys. Res. 105 (D12) (2000) 15509.
38. R. Sausen, B.D. Santer, Meteorol. Z. 12 (2003) 131.
39. B.D. Santer, M.F. Wehner, T.M.L. Wigley, R. Sausen, G.A. Meehl, K.E. Taylor, C. Ammann, J. Arblaster, W.M. Washington, J.S. Boyle, W. Brüggemann, Science, 301 (2003) 479.
40. I.M. Held, A.Y. Hou, J. Atmos. Sci. 37 (1980) 515.
41. C.C. Walker, T. Schneider, J. Atmos. Sci. 63 (2006) 3333.
42. T. Schneider, Annu. Rev. Earth Planet. Sci. 34 (2006) 655.
43. D. Frierson, J. Lu, G. Chen, Geophys. Res. Lett. 34 (2007) doi:10.1029/ 2007GL031115.
44. J.D. Haigh, M. Blackburn, R. Day, J. Clim. 18 (2005) 3672.
45. G.P. Williams, J. Atmos. Sci. 63 (2006) 1954.
46. J. Austin, T. Reichler, J. Geophys. Res., 113 (2008) D00B10.
47. IPCC, Climate Change 2007: The physical science basis, Contribution of Working Group I to the Fourth Assessment Report of the Intergovernmental Panel on Climate Change. S. Solomon *et al.*, Eds., Cambridge University Press, Cambridge, MA, 2007.
48. WMO, Scientific Assessment of Ozone Depletion: 2006 World Meteorological Organization, Geneva, Switzerland, 2007.
49. V. Ramaswamy, M.D. Schwarzkopf, W.J. Randel, B.D. Santer, B.J. Soden, G.L. Stenchikov, Science, 311 (2006) 1138.
50. G. Chen, I.M. Held, Geophys. Res. Lett. 34 (2007) doi:10.1029/ 2007GL031200.
51. J. Lu, G. Chen, D.M.W. Frierson, J. Climate, 21 (2008) 5835.
52. T.R. Karl, S.J. Hassol, C.D. Miller, W.L. Murray, Temperature Trends in the Lower Atmosphere: Steps for Understanding and Reconciling Differences U.S., Climate Change Science Program and the Subcommittee on Global Change Research. National Oceanic and Atmospheric Administration, National Climatic Data Center, Asheville, NC, 2006.
53. B.D. Santer, P.W. Thorne, L. Haimberger, K.E. Taylor, T.M.L. Wigley, J.R. Lanzante, S. Solomon, M. Free, P.J. Gleckler, P.D. Jones, T.R. Karl, S.A. Klein, C. Mears, D. Nychka,

G.A. Schmidt, S.C. Sherwood, and F.J. Wentz, Int. J. Climatol., (2008) doi:10.1002/joc.1756.
54. K.M. Xu, K.A. Emanuel, Mon. Wea. Rev. 117 (1989) 1471.
55. R.J. Allen, S.C. Sherwood, Nat. Geosci. 1 (2008) 399.
56. M.N. Juckes, J. Atmos. Sci. 57 (2000) 3050.
57. D.M.W. Frierson, Geophys. Res. Lett. 33 (2006) doi:10.1029/ 2006GL027504.
58. D.M.W. Frierson, J. Atmos. Sci., 65 (2008) 1049.
59. Y. Hu, Q. Fu, Atmos. Chem. Phys. 7 (2007) 5229.
60. N.C. Lau, A. Leetmaa, M.J. Nath, J. Clim. 19 (2006) 3607.
61. M. Hoerling, A. Kumar, Science 299 (2003) 691.
62. T.P. Barnett, D.W. Pierce, K.M. AchutaRao, P.J. Gleckler, B.D. Santer, J.M. Gregory, and W.M. Washington, Science, 309 (2005) 284.
63. C. Deser, A.S. Phillips, J.W. Hurrell, J. Clim. 17 (2004) 3109.
64. Y. Zhang, J.M. Wallace, D.S. Battisti, J. Clim. 10 (1997) 1004.
65. W.A. Robinson, in: T. Schneider, A. Sobel (Eds.), The Global Circulation of the A"mospheree, Princeton University Press, Pasadena, 2007.
66. R. Seager, N. Harnik, Y. Kushnir, W. Robinson, J. Miller, J. Clim. 16 (2003) 2960.
67. W.A. Robinson, Geophys. Res. Lett. 29 (2002) 1190.
68. I. Orlanski, J. Atmos. Sci. 62 (2005) 1367.
69. E. Yulaeva, J.M. Wallace, J. Clim. 7 (1994) 1719.
70. J.T. Abatzoglou, G. Magnusdottir, J. Clim. 19 (2006) 6139.
71. J. Lu, C. Deser, T. Reichler, Geophys. Res. Lett., 36, (2009) doi:10.1029/2008GL036076.
72. D.W.J. Thompson, J.M. Wallace, J. Clim. 13 (2000) 1000.
73. D.W.J. Thompson, J.M. Wallace, G.C. Hegerl, J. Clim. 13 (2000) 1018.
74. D.W.J. Thompson, S. Solomon, Science 296 (2002) 895.
75. G. Marshall, J. Clim. 16 (2003) 4134.
76. J. Marshall, et al., Int. J. Climatol. 21 (2001) 1863.
77. J. Cohen, M. Barlow, J. Clim. 18 (2005) 4498.
78. J.E. Overland, M. Wang, Geophys. Res. Lett. 32 (2005) doi:10.1029/ 2004GL021752.
79. J.W. Hurrell, Y. Kushnir, M. Visbeck, Science 291 (2001) 603.
80. J.H. Yin, Geophys. Res. Lett. 32 (2005) L18701.
81. J.M. Arblaster, G.A. Meehl, J. Clim. 19 (2006) 2896.
82. N.P. Gillett, D.W.J. Thompson, Science 302 (2003) 273.
83. N.P. Gillett, M.R. Allen, K.D. Williams, Quart. J. Roy. Meteorol. Soc. 129 (2003) 947.
84. M. Previdi, B.G. Liepert, Geophys. Res. Lett. 34 (2007) doi:10.1029/ 2007GL031243.
85. L. Bengtsson, K.I. Hodges, E. Roeckner, J. Clim. 19 (2006) 3518.
86. G. Chen, J. Lu, D.M.W. Frierson, J. Climate, 21 (2008) 5942.
87. M.P. Baldwin, T.J. Dunkerton, J. Geophys. Res. Atmos. 104 (1999) 30937.
88. L.M. Polvani, P.J. Kushner, Geophys. Res. Lett., 29 (2002) doi:10.129/2001GL014284.
89. J. Perlwitz, S. Pawson, R.L. Fogt, J.E. Nielsen, W.D. Neff, Geophys. Res. Lett. 35 (2008) doi:10.1029/2008GL033317.
90. P.W. Staten, T. Reichler, J. Geophys. Res. 113 (2008) doi:10.1029/ 2008JD009886.
91. R. Seager, M. Ting, I. Held, Y. Kushnir, J. Lu, G. Vecchi, H.-P. Huang, N. Harnik, A. Leetmaa, N.-C. Lau, C. Li, J. Velez, and N. Naik, Science, 316 (2007) 1181.
92. C.L. Archer, K. Caldeira, Geophys. Res. Lett. 35 (2008) doi:10.1029/ 2008GL033614.
93. D.E. Parker, M. Gordon, D. Cullum, D. Sexton, C. Folland, and N. Rayner, Geophys. Res. Lett., 24 (1997) 1499.

Chapter 8

Weather Pattern Changes in the Tropics and Mid-Latitudes as an Indicator of Global Changes

Ricardo M. Trigo
Centro de Geofisica da Universidade de Lisboa, IDL, Faculty of Sciences, University of Lisbon, Campo Grande, Ed C8, Piso 3, 1749-016 Lisbon, Portugal

Luis Gimeno
Departamento de Física Aplicada, Faculty of Sciences, University of Vigo, 32004 Ourense, Spain

1. **Introduction**
2. **Observed Changes in Extra-Tropical Patterns**
   2.1. North Atlantic Oscillation (NAO)
   2.2. Pacific North America (PNA)
3. **Changes in Tropical Patterns**
   3.1. El Niño Southern Oscillation (ENSO)
   3.2. Tropical Cyclones
   3.3. Monsoons
4. **Conclusion**
References

1. INTRODUCTION

It is now widely accepted in the earth-science scientific community that the emission of large amounts of greenhouse gases of anthropogenic origin (namely carbon dioxide and methane), into the atmosphere is partially responsible for recent trends in the climate of our planet at the global scale [1]. However, the separate of the roles of natural and human influences on climate change has only recently been elucidated [2]. This groundbreaking research has quantified the anthropogenic contribution to climate change through studies involving surface air temperature [3,4], precipitation [5] and sea level pressure (SLP) [6]. In spite of this, climate change at a regional level can be more difficult to understand than changes occurring at the global or hemispheric scales. Recent positive trends in temperature and sea level height have been amplified or partially offset, at the regional scale, by changes of atmospheric circulation. The same rationale seems to apply to climate change scenarios under a warmer planet [7]. It is therefore important to understand the climatic role of the most important large-scale patterns and to provide an assessment on their changes (variability and trends) over the recent historical period.

Two of the most important modes of atmospheric variability, namely the Southern Oscillation (SO) (later associated with El Niño and coined ENSO) and the North Atlantic Oscillation (NAO) were identified in the pioneering works of Gilbert Walker [8,9]. However, the majority of these large-scale circulation patterns (also known as teleconnections) were only identified unequivocally in the 1980s (e.g. [10,11]). These and subsequent studies confirm the climatic influence of both ENSO and NAO but also of the Pacific-North American Pattern (PNA). These teleconnections are known to have large impacts on the climate of entire continents due to their influence on the main physical mechanisms that rule near surface weather, namely controlling the main cyclone trajectories, enhancing heat advection, changing cloud cover and consequently the radiation balance [12–14]. It should be stressed that the relevance of these modes is seasonally dependent, that is, they only have a signature during part of the year [11]. Other modes, usually of a more regional nature, and only relevant during part of the year, may play a minor, albeit relevant role in modulating local climate.

None of the above mentioned teleconnections presents a distinctively symmetric behaviour over either hemisphere. However, two additional modes have been added in the last decade, the Northern Annular Mode (NAM) and the Southern Annular Mode (SAM) and these are characterised by a certain symmetry in their patterns [15]. The NAM is also known as the Artic Oscillation (AO) pattern and is closely related to the better established NAO pattern [16].

In this chapter, we will present a summary of the main results published in recent literature on changes in frequency and magnitude of the most important large-scale circulation patterns (NAO, PNA). Secondly, we will focus our attention on major trends in the occurrence of other important tropical patterns, such as ENSO, Tropical Cyclones (TCs) and Monsoons due to their relevance to tropical and sub-tropical climate regimes.

2. OBSERVED CHANGES IN EXTRA-TROPICAL PATTERNS

In the last two decades an increasing number of studies have gathered a wealth of information on changes of the most important circulation patterns that affect the climate conditions in the extra-tropical latitudes of both hemispheres. However, the imbalance between the northern and southern hemispheres in the extent of continental dry land, and affected population explains the bias towards Northern Hemisphere (NH) studies.

Different approaches have been developed to derive the main atmospheric circulation patterns that characterise the large-scale circulation over the entire

NH [10,11]. Here, the NAO and PNA teleconnection indices were obtained from the U.S. National Oceanographic and Atmospheric Administration (NOAA) Climate Prediction Center (CPC) (http://www.cpc.noaa.gov/data/-teledoc/nao.shtml). The methodology employed by CPC to identify the teleconnection patterns is based on rotated principal component analysis (RPCA) [11] applied to monthly mean standardized 50 kPa geopotential height anomalies. Spatial patterns of the NAO and PNA can be seen in Figs. 1 and 3, respectively, and represent the temporal correlation between the monthly standardized height anomalies at each point and the monthly teleconnection pattern time series from 1960 to 2000.

2.1. North Atlantic Oscillation (NAO)

The NAO was recognised more than 70 years ago as being one of the major patterns of atmospheric variability in the NH [8,9]. Historically, the NAO has been defined as a simple index that measures the difference in surface pressure between Ponta Delgada in the Azores and the Icelandic station of

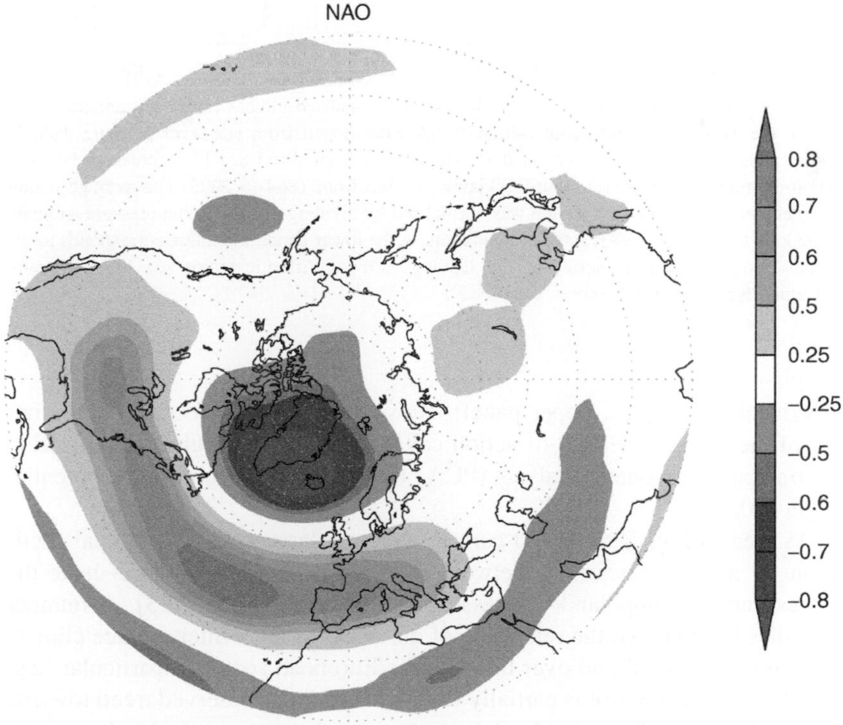

FIGURE 1 Spatial pattern of the NAO as given by the temporal correlation between the Winter (DJFM) monthly standardised 50 kPa geopotential height anomalies at each point and the monthly teleconnection pattern time series from 1960 to 2000. (See Color Plate 11).

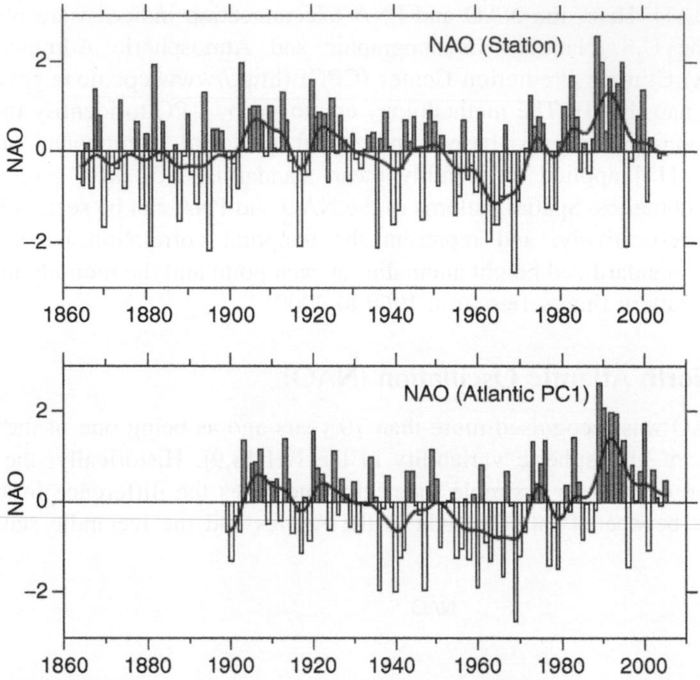

FIGURE 2 (Reprinted from Fig. 3.31 of Ref. [7]). Normalised indices (units of standard deviation) of the mean winter (December–March) NAO developed from sea level pressure data. In the top panel, the index is based on the difference of normalised sea level pressure between Lisbon, Portugal and Stykkisholmur/Reykjavik, Iceland from 1864 to 2005. The average winter sea level pressure data at each station was normalised by dividing each seasonal pressure anomaly by the long-term (1864–1983) standard deviation. In the lower panel, the index corresponds to the principal component time series of the leading EOF of Atlantic-sector sea level pressure (© Cambridge Press; IPCC report, Chapter 3).

Stykkisholmur (Fig. 2, upper panel). However, a more objective determination of the dipole's centres of action can be obtained through the application of principal component analysis (PCA) to SLP or tropospheric geopotential height [10,11].

As seen in Fig. 1, the NAO corresponds to a large-scale meridional oscillation of atmospheric mass between the subtropical anticyclone near the Azores and the subpolar low pressure system near Iceland [13]. A number of studies have shown the relevance of the NAO to the winter surface climate of the NH in general and over the Atlantic/European sector in particular (e.g. [13,14,17]). This control is partially responsible for the observed trend towards warmer Northern Eurasian land temperatures that occurred simultaneously with the trend towards a more positive phase of the NAO between the late 1960s and mid-1990s as observed in Fig. 2 [18,19]. Other works have clearly

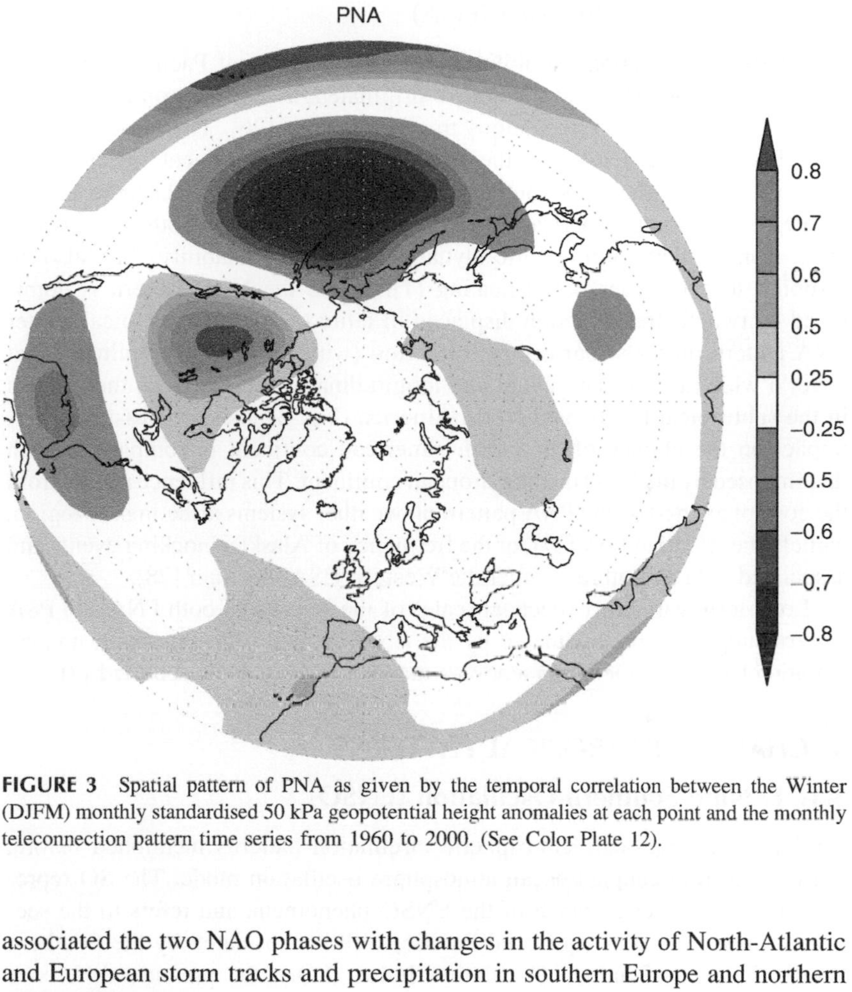

FIGURE 3 Spatial pattern of PNA as given by the temporal correlation between the Winter (DJFM) monthly standardised 50 kPa geopotential height anomalies at each point and the monthly teleconnection pattern time series from 1960 to 2000. (See Color Plate 12).

associated the two NAO phases with changes in the activity of North-Atlantic and European storm tracks and precipitation in southern Europe and northern Africa [20–23].

Analyses of SLP and 50-kPa geopotential height over the last five decades reveal negative trends over the Arctic, Antarctic and North Pacific, an increase over the subtropical North Atlantic, southern Europe and Northern Africa and a weakening of the Siberian High [6,7]. The increment in SLP gradients in the NH appears to significantly exceed simulated internal and anthropogenically forced variability [6]. Such changes in within the Euro-Atlantic sector are clearly associated with positive trends in the NAO index (Fig. 2). Moreover, Jung and Hilmer [24] pointed out that the NAO has undergone considerable changes in the location of the main centres, with the northern centre (the Icelandic low) being displaced towards Scandinavia. This shift has major implications for the NH climate, in general, but is particularly important for southern Europe and Northern Africa [22,25].

2.2. Pacific North America (PNA)

Atmospheric circulation variability over the extratropical Pacific is partially dominated by Rossby wave patterns originated in the subtropical western Pacific, associated with anomalous tropical heating [26]. The wave-like pattern that propagates towards the North American Continent is known as PNA pattern, similarly its southern hemisphere counterpart become known as the Pacific-South American (PSA) pattern (not shown). Both patterns can arise from natural atmospheric dynamic internal variability, but also in response to anomalous ocean heating [7]. While the NAO pattern is dominated by two centres of action displaced in latitude (Fig. 1) the typical winter PNA pattern presents four centres of action (with decreasing amplitude) that cover a wide range of latitudinal and longitudinal values between their origin in the subtropical Pacific and North America (Fig. 3). Nevertheless, the PNA impact on the climate of the North American continent is comparable with that imposed by the NAO on the European continent. This influence results from the control exerted by the PNA pattern on weather systems affecting the region, namely the Aleutian Low [27], or the frequency of Alaskan blocking events and associated cold air outbreaks over the Western USA in winter [28].

Long-term variability (decadal scale) of the activity of both PNA and PSA patterns appear to be modulated by the ENSO signal [27]. However, no systematic changes of their frequency or magnitude have been reported [7].

3. CHANGES IN TROPICAL PATTERNS

3.1. El Niño Southern Oscillation (ENSO)

Unlike other large-scale atmospheric circulation patterns mentioned before, ENSO is a truly coupled ocean-atmosphere oscillation mode. The SO represents the atmospheric branch of the ENSO phenomena and refers to the see-saw in pressure across equatorial Pacific, well encapsulated by the Southern Pacific Index or Southern Oscillation Index (SOI); the pressure difference between Tahiti in mid-Pacific and Darwin in northern Australia [8,9]. The El Niño is characterised by a strong warming of tropical waters in central and eastern Pacific following the decrease in strength of the trade winds. This pattern leads to an increase (decrease) of precipitation in central and eastern (western) tropical Pacific [29]. These changes occur intermittently (about once every three to seven years), alternating with the opposite phase – La Niña – that is characterised by below-average temperatures in central and eastern tropical waters. It is also worth noticing that, in contrast to the NAO and PNA patterns, the climatic impacts of ENSO are of a global scale and not restricted to the inter-tropical belt [29,30]. In fact, the signature of these events in SLP extends often into the extra-tropical latitudes (Fig. 4).

The frequency and strength of ENSO has varied over time at the decadal, centennial and millennia scales. A power spectrum analysis applied to the

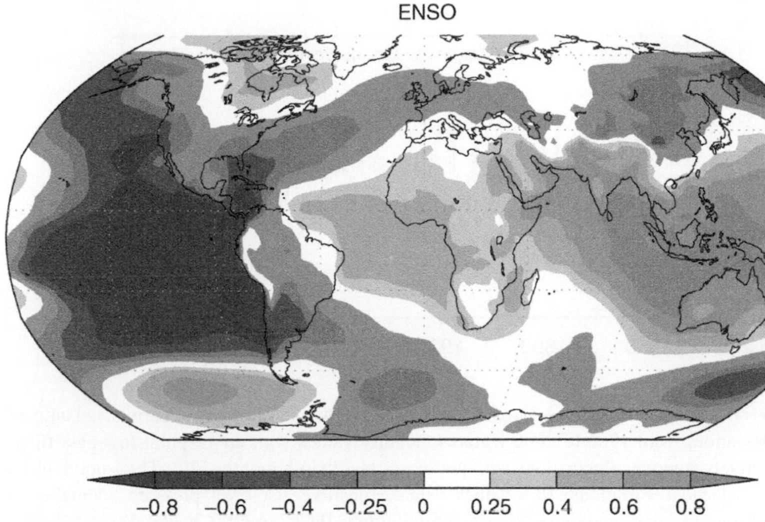

FIGURE 4 Spatial pattern of El Niño as given by the temporal correlation between the annual (May–April) standardised SLP anomalies at each point and the monthly teleconnection pattern time series from 1960 to 2000. (See Color Plate 13).

time series of El Niño events since 1500, obtained through the reconstruction of its impacts in Peru, presents a clear peak in power spectrum of about 80–100 a [31]. However, these reconstructions can be misleading because they are highly regionalised and depending on either limited documentary sources or specific natural proxies. Reliable time series of ENSO should take into account the multitude of impacts associated with this phenomenon and are therefore restricted to the last 130 a [7,30]. Based on these authoritative sources one can state that the period spanning between 1870 and 1920 was characterised by frequent El Niño events including several strong cases (Fig. 5). From the 1920s until the early 1970s the ENSO activity was relatively quieter (with the important exception of 1941–1942). However, over the last three decades there has been a resurgence of large (e.g. 1982–1983 and 1997–1998) and prolonged (1991–1995) El Niño episodes (Fig. 5), associated with a tendency towards positive SST anomalies in central and eastern equatorial Pacific. The large El Niño episode of 1997–1998 was the largest on record contributing significantly to the highest global average temperature recorded in 1998. Furthermore, hydrological cycle extremes associated with El Niño events (e.g. drought and floods) will probably be more frequent in a warmer world. The current generation of coupled ocean atmosphere climate models are capable of reproducing El Niño events and their impact relatively well. When forced with distinct climate change scenarios these very same models predict continued ENSO interannual variability [7].

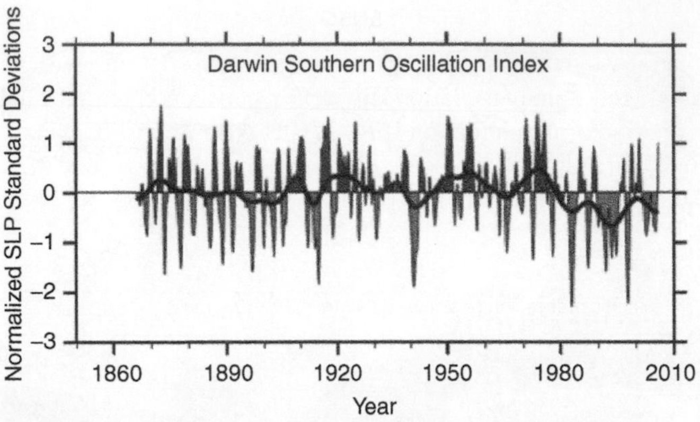

FIGURE 5 (Reprinted from Fig. 3.27 of Ref. [7]). The Darwin-based SOI, in normalised units of standard deviation, from 1866 to 2005 features monthly values with an 11-point low-pass filter, which effectively removes fluctuations with periods of less than 8 months [49]. The smooth black curve shows decadal variations. Red values indicate positive sea level pressure anomalies at Darwin and thus El Niño conditions. (© Cambridge Press; IPCC report, Chapter 3).

3.2. Tropical Cyclones

TCs are among the most destructive natural hazards in the world. Therefore, large fluctuations in tropical cyclone activity are of obvious importance to society, particularly in those coastal areas where populations are affected. In addition, the study of changes in tropical cyclone occurrence and activity has great scientific interest because of their large influence on regional and global climate.

The physical mechanisms responsible for TC development are complex and not fully understood, but it is known that they require high values of sea-surface temperatures (SST), a moderate Coriolis force, a pre-existent synoptic perturbation (usually a monsoon trough or easterly wave) and low wind shear [32]. These pre-conditions limit the development of TC to the five tropical oceanic bases represented in Fig. 6. The dependence of TC on high SST values has opened the debate on a possible increase in the frequency and intensity of TCs in a warmer climate. It has been proposed that a rise in SST induced by anthropogenic global warming has already led to a greater number of intense TCs in recent decades [33,34]. Whether this trend is real or an artefact of the short length and inhomogeneity of records is a matter of keen scientific argument [35]. Methods used for detecting and measuring the intensity of TC in different regions of the world have evolved, making it difficult to assess these trends. Furthermore, there is a considerable level of natural interannual and interdecadal variability, reducing the significance attributable to long-term trends. In particular, studies of TC variability in the North Atlantic basis (the most studied one) reveal large interannual and interdecadal swings in storm frequency that have been linked to different

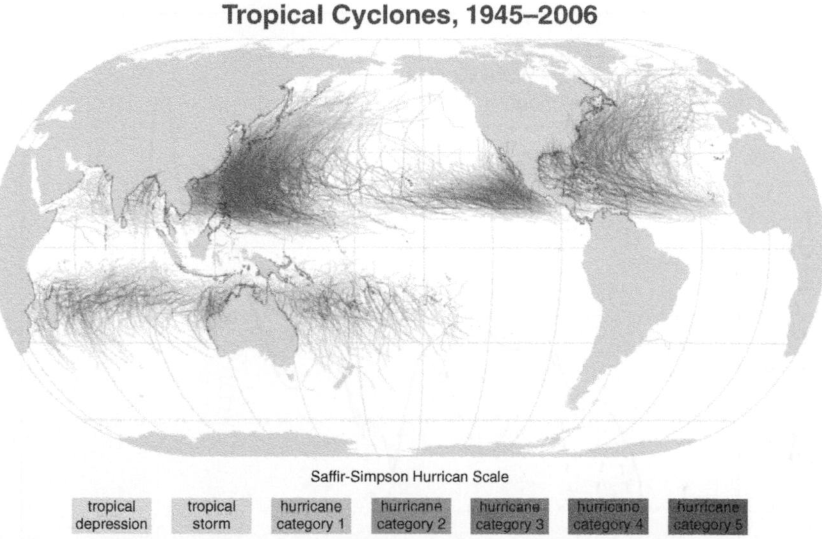

FIGURE 6 Tropical Cyclones, 1945–2006. Data from the Joint Typhoon Warning Center and the US National Oceanographic and Atmospheric Administration (NOAA). Permission is granted to copy, distribute and/or modify this document under the terms of the **GNU Free Documentation License**, Version 1.2 or any later version published by the Free Software Foundation with no Invariant Sections, no Front-Cover Texts and no Back-Cover Texts. (See Color Plate 14).

large-scale climate phenomena such as the El Niño/SO, the stratospheric quasi-biennial oscillation and multi-decadal oscillations in the North Atlantic region.

Traditionally, most studies on TC trends focus on the frequency of their occurrence (with weak or no trends detected); however, in recent years studies about trends in intensity have gained importance. In fact the latter is considered nowadays a more relevant index of TC activity (both in scientific and socio-economical terms) than the former, with a strong debate taking place between those authors that support for trends in intensity in the last decades and those that do not. We now summarise both arguments taking into account the most relevant works on this issue.

Two important papers published in 2005 [33,34] found a close relationship between increasing tropical SSTs and intense TCs. Emanuel [33] defined an index for the total power dissipation of a tropical cyclone that is proportional to the cube of wind speed (the Power Dissipation Index, PDI):

$$\mathrm{PDI} = \int_0^{\tau} V_{\max}^3 \, \mathrm{d}t \qquad (1)$$

where $V_{\max}$ is the maximum sustained wind speed at the conventional measurement altitude of 10 m. PDI measures the net power dissipation of a TC and, as such, a better indicator of the TC threat than storm frequency or

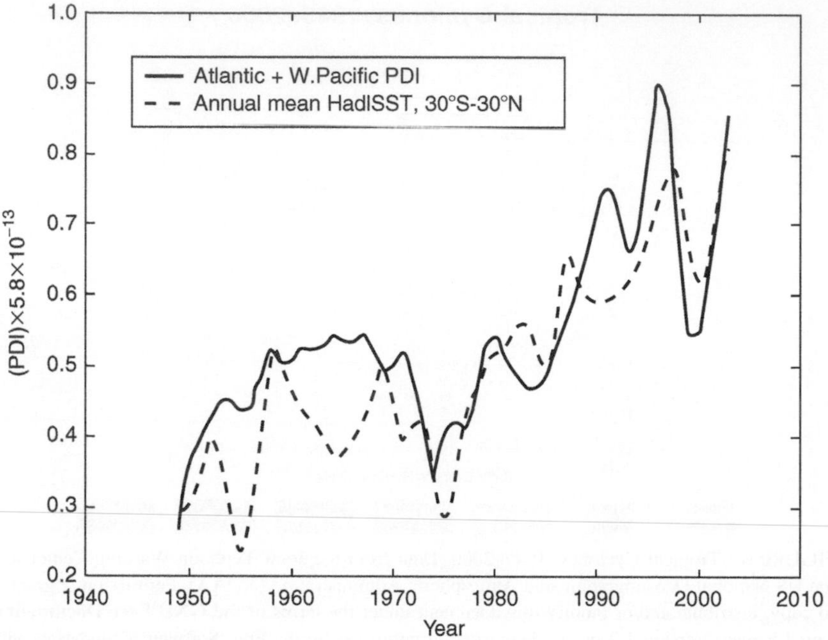

FIGURE 7 (Reprinted from Fig. 3 of Ref. [33]). Annually accumulated PDI for the western North Pacific and North Atlantic, compared to annually averaged SST. The HadISST is averaged between 30°S and 30°N (with a constant offset). Both quantities have been smoothed twice. This combined PDI has nearly doubled over the past 30 a (© Nature Publishing Group).

intensity alone. In a subsequent work [36] the author shows that the PDI has increased by ~50% for both the Atlantic and Northwestern Pacific basins since the mid-1970s (Fig. 7). Webster et al. [34,37] analysed trends and decadal variability of the most intense hurricanes, that is, category 4–5 in Saffir–Simpson classification (maximum sustained winds higher than 115 knots, where 1 knot = 0.5144 $m{\cdot}s^{-1}$) since 1970 for all TC basins. They have found a large increment in the number and proportion of the most intense hurricanes with their numbers nearly doubled between the two consecutive 15-year periods 1975–1989 and 1990–2004. These results have been questioned by other studies (e.g. [38,39]) mainly attending to the poor quality of data prior to 1986, the different intensity attributed to each storm by different research centres [7]. Furthermore, it has also been stressed that the strong association of hurricanes with El Niño events could result in artificial trends when this effect is not removed from the analysis.

Based on the Accumulated Cyclone Energy index (ACE), a wind energy index defined as the sum of the squares of the estimated 6 h maximum sustained wind, Klotzbach [39] found no statistically significant trend in any TC database since 1986 (Fig. 8). In spite of the differences in the results concerning trends, all the previous studies and indices are in agreement on

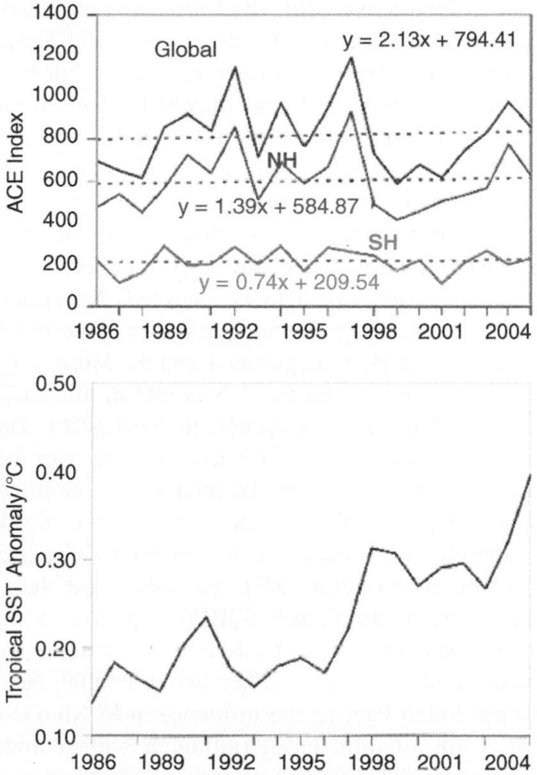

FIGURE 8 (Reprinted from Fig. 2 of Ref. [39]). The upper graph shows Accumulated Cyclone Energy (ACE) index values for 1986–2005 for the Northern Hemisphere (NH), the Southern Hemisphere (SH) and the Global. The dashed lines are linear trends that have been fitted to the three curves. The lower graph shows five-year running mean tropical NCEP reanalysis SST anomalies (23.52°N–23.5°S, all longitudes). The base period for tropical SSTs is 1951–1980 (© AGU).

the strong role played by the El Niño/SO in the activity and occurrence of TCs. Both TC activity indices (PDI and ACE) have attained their highest values during the major El Niño event that took place in 1997–1998. Again, it should be stressed that most of these TC-related index time series are relatively short, in particular the ACE index starts only in 1986 and therefore is not long enough to provide reliable trends [1].

Analysis of longer series and influences of major external forcings and modes of climate variability should be carefully considered due to strong inhomogenities of the series. Although there is evidence of important changes in the frequency of TC in pre-instrumental periods (e.g. [40,41]), we will limit this short review to the most homogeneous instrumental period. In any case this analysis should be performed for individual oceanic basins [7], in order to accommodate the different measuring methods of TC activity. In the North Atlantic basin, the one with longest register starting in 1851, the period from the 1930 to 1960 was very active while the following two decades (1970s and 1980s) were characterised with low TC occurrence and activity. Since the mid-1990s, the activity has risen significantly, with all but two seasons presenting activity above normal. The role played by both Atlantic SST and the occurrence of an El Niño appears to be two determinant factors for TC activity in the Atlantic. It is well known that the El Niño acts to reduce TC activity in the North Atlantic basin whereas la Niña acts to increase it. In the Western North Pacific basin there is a high degree of uncertainty in the significance of trends partially due to the high interdecadal variability and also taking into account the way data is analysed. Nevertheless, the most interesting result is the doubling of the PDI values since the 1950s and the trend towards more intense TC (categories 4 and 5). Moreover, the influence of changes in circulation associated to El Niño/SO in this basin seems to be much more important than changes related to local SSTs. During El Niño events TCs tend to be more intense and longer-lived than during La Niña years, a result that contrasts with that obtained for the North Atlantic basin. A similar influence of the El Niño event also occurs in the Eastern North Pacific basin, where above-normal tropical cyclone activity occurs during El Niño years. However, in this case, SST anomalies and the tropical lower stratospheric Quasibiennal Oscillation (QBO) play also a major role [7]. TCs have a higher intensity in the Eastern North Pacific basin when the QBO is in its westerly phase. For the other two remaining basins, the Indian Ocean, Australia and South Pacific, the influence of El Niño is similar to that observed for the North Atlantic basin (although with a minor amplitude) namely with more intense TCs occurring during la Niña years.

The occurrence of the only document TC in the South Atlantic basin close to Santa Catarina, Brazil in March 2004 should be taken carefully, particularly when it is presented as a signal of frequent phenomena in a warmer planet, since the structure of the cyclone did not agree completely the typical one in a TC.

3.3. Monsoons

According to Trenberth et al. [42] 'the global monsoon system is a global-scale persistent overturning of the atmosphere, throughout the tropics, that varies according to the time of year'. This means that over the tropics there is a region of intense heating where ascent is produced, which is balanced with adiabatic cooling. Such global atmospheric overturning is the common manifestation of the known regional monsoons over six zonal sectors, namely: Africa, Australia–Asia, North America, South America and the Pacific and Atlantic Oceans (Fig. 9). This is clearly associated with seasonal variation of the so-called monsoon precipitation, which plays a key role in driving monsoon circulations through latent heat release. Shaded regions in Fig. 9a delineate the six mentioned global monsoon domains and the main regional

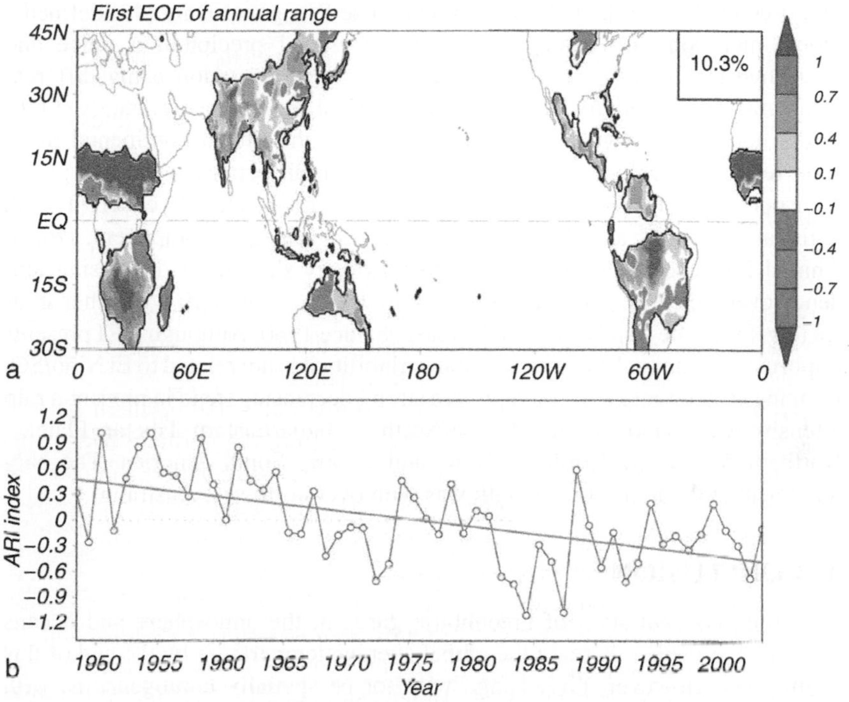

FIGURE 9 (Reprinted from Fig. 3 of Ref. [45]). (a) The spatial pattern of the leading Empirical Orthogonal Function (EOF) mode of the normalised annual range precipitation anomalies over the global continental monsoon regions. The bold contour indicates the boundaries of the monsoon domain; (b) the corresponding principal component or annual range index (ARI) (© Cambridge Press; IPCC report, Chapter 3 and © AGU). (See Color Plate 15).

monsoons. The three dimensional structure of monsoon circulation is complex involving three different planetary scale features, namely the Hadley circulation, the Walker circulation and the Inter-Tropical Convergence Zone (ITCZ).

Variability of regional monsoon activity depends on the different interacting circulations from other regions. Its quantification is dependant on the measures of the different monsoon strengths. Therefore, it is not straight forward to address the question of recent regional monsoon changes. Nevertheless, some of the works recently published do present trends in monsoon circulation activity, usually referring to a decrease in strength in the regional monsoonal systems. For instance, Stephenson et al. [43], using reanalysis data and simple shear indices, have shown that summer Asian monsoon strength has been decreasing at a rate of 1–3% per decade. Independently, Chase et al. [44] found that the monsoonal overturning circulations over the Australia-Maritime continent and African regions have diminished since 1950.

More conclusive evidence could be obtained by investigating the topic from a global perspective, that is, from the global monsoon system definition proposed by Trenberth et al. [42]. In this sense Wang and Ding [45] defined a global monsoon rain domain according to annual precipitation range and quantified the intensity of the global monsoon precipitation using different measures. These authors have applied empirical orthogonal functions (EOF) to the normalised annual range anomalies over the global continental monsoon regions, therefore identifying the leading EOF patterns with 'the global monsoon system' (Fig. 9a). The spatial pattern is clearly coherent with regional monsoons and the corresponding first principal component, coined Annual Range Index (ARI), shows a statistically significant decreasing tendency over the entire period examined (Fig. 9b). In accordance with other studies (e.g. [44]), the trend has been decreasing since 1980. Although ARI presents important interannual and interdecadal variability it is not related to El Niño/SO. Looking at different monsoon regions a strong decreasing trend in monsoon rain intensity was found for Bangladesh/Northern India/Eastern Tibetan Plateau, Northern Africa, the Northern China and Central South America. The only increment in the monsoon strength was seen over northwest Australia.

4. CONCLUSION

Increasing concentration of greenhouse gases in the atmosphere and oceans are bound to further increase the global average temperature by the end of this century [1]. However, this change will not be spatially homogeneous, with polar regions suffering a much higher increment than the tropical belt. Moreover trends in the large-scale driving patterns mentioned in this chapter are especially relevant since they may enhance or damp the warming at the regional scale. Relevant questions in this context are the possible changes of these tropical and extra-tropical modes under global climate change. The results to date seem to indicate that the so-called Annular Modes – the Arctic Oscillation and the Antarctic Oscillation – to which the NAO is linked will tend to become more intense in the future [46], although the signal to noise ratio may be not very large [47]. However, present climate models are still unable to replicate the observed amplitude of the interannual variability and of the multidecadal trends of some modes, for example, the NAO [48].

An enhanced hydrological cycle, with more evaporation and precipitation at the global scale, coupled with changes in frequency of El Niño and TCs and changes in the Monsoon systems may also raise the probability of extremes (floods and droughts). Nevertheless, there is no clear indication of a major shift in the frequency or magnitude of projected El Niño events [7]. Therefore, this ongoing effort should be continued with the aim of determining to what extent climate models yield a realistic picture of the variability in the present climate and evaluating the fraction of future regional climate change that can be attribute to future trends of both tropical and extra tropical circulation patterns, since these patterns will probably be partially responsible for regional differences in the future climate.

REFERENCES

1. S. Solomon, D. Qin, M. Manning, R.B. Alley, T. Berntsen, N.L. Bindoff, Z. Chen, A. Chidthaisong, J.M. Gregory, G.C. Hegerl, M. Heimann, B. Hewitson, B.J. Hoskins, F. Joos, J. Jouzel, V. Kattsov, U. Lohmann, T. Matsuno, M. Molina, N. Nicholls, J. Overpeck, G. Raga, V. Ramaswamy, J. Ren, M. Rusticucci, R. Somerville, T.F. Stocker, P. Whetton, R.A. Wood and D. Wratt, 2007: Technical Summary. In: Climate Change 2007: The Physical Science Basis. Contribution of Working Group I to the Fourth Assessment Report of the Intergovernmental Panel on Climate Change [S. Solomon, D. Qin, M. Manning, Z. Chen, M. Marquis, K.B. Averyt, M. Tignor and H.L. Miller (eds.) (2007) 19–91.
2. G.C. Hegerl, F.W. Zwiers, P. Braconnot, N.P. Gillett, Y. Luo, J.A. Marengo Orsini, N. Nicholls, J.E. Penner and P.A. Stott, 2007: Understanding and Attributing Climate Change. In: Climate Change 2007: The Physical Science Basis. Contribution of Working Group I to the Fourth Assessment Report of the Intergovernmental Panel on Climate Change (S. Solomon, D. Qin, M. Manning, Z. Chen, M. Marquis, K.B. Averyt, M. Tignor, H.L. Miller (eds.)). Cambridge University Press, Cambridge, United Kingdom and New York, NY, USA, 2007, 663–745.
3. P.A. Stott, Geophys. Res. Lett. 30 (2003) 1724 doi:10.1029/2003GL017324.
4. S.F.B. Tett, P.A. Stott, M.R. Allen, W. Ingram, J. Mitchell, Nature 339 (1999) 569–572.
5. X. Zhang, F.W. Zwiers, P.A. Stott, J. Clim. 19 (2006) 4294–4307.
6. N.P. Gillett, F.W. Zwiers, A.J. Weaver, P.A. Stott, Nature 422 (2003) 292–294.
7. K.E. Trenberth, P.D. Jones, P. Ambenje, R. Bojariu, D. Easterling, A. Klein Tank, D. Parker, F. Rahimzadeh, J.A. Renwick, M. Rusticucci, B. Soden, P. Zhai. 2007: Observations: Surface and Atmospheric Climate Change. In: Climate Change 2007: The Physical Science Basis. Contribution of Working Group I to the Fourth Assessment Report of the Intergovernmental Panel on Climate Change [S. Solomon, D. Qin, M. Manning, Z. Chen, M. Marquis, K.B. Averyt, M. Tignor, H.L. Miller (eds.)]. Cambridge University Press, Cambridge, United Kingdom and New York, NY, USA, 2007, 235–336.
8. G.T. Walker, Mem. Indian Meteorol. Dep. 24 (1924) 225–232.
9. G.T. Walker, E.W. Bliss, V. Mem. Roy. Meteorol. Soc. 4 (1932) 53–84.
10. J.M. Wallace, D.S. Gutzler, Mon. Weather Rev. 109 (1981) 784–812.
11. A.G. Barnston, R.E. Livezey, Mon. Weather Rev. 115 (1987) 1083–1126.
12. M. Glantz, R. Katz, N. Nicholls (Eds.), Teleconnections Linking Worldwide Climate Aomalies, Cambridge University Press, Cambridge, UK, 1991.
13. J.W. Hurrell, Science 269 (1995) 676–679.
14. R.M. Trigo, T.J. Osborn, J.M. Corte-Real, Clim. Res. 20 (2002) 9–17.
15. D.J.W. Thompson, J.M. Wallace, J. Clim. 13 (2000) 1000–1016.
16. D.J.W. Thompson, J.M. Wallace, Geophys. Res. Lett. 25 (1998) 1297–1300.
17. R. Bojariu, L. Gimeno, Geophys. Res. Lett. 30 (2003) doi:10.1029/2002GL015651.
18. J.W. Hurrell, Geophys. Res. Lett. 23 (1996) 665–668.
19. L. Gimeno, L. de la Torre, R. Nieto, R. García, E. Hernández, P. Ribera, Earth Planet. Sci. Lett. 206 (2003) 15–20.
20. T.J. Osborn, K.R. Briffa, S.F.B. Tett, P.D. Jones, R.M. Trigo, Clim. Dyn. 15 (1999) 685–702.
21. U. Ulbrich, M. Christoph, Clim. Dyn. 15 (1999) 551–559.
22. R.M. Trigo, I.F. Trigo, C.C. DaCamara, T.J. Osborn, Clim. Dyn. 23 (2004) 17–28.
23. I.F. Trigo, Clim. Dyn. 26 (2006) doi:10.1007/s00382-005-0065-9.
24. T. Jung, M. Hilmer, J. Clim. 14 (2001) 3932–3943.
25. J. Lu, R.J. Greatbatch, Geophys. Res. Lett. 29 (2002), doi:10.1029/2001GLO14052.
26. J.D. Horel, J.M. Wallace, Mon. Weather Rev. 109 (1981) 813–829.

27. D.M. Straus, J. Shukla, J. Clim. 15 (2002) 2340–2358.
28. G.P. Compo, P.D. Sardeshmukh, J. Clim. 17 (2004) 3701–3720.
29. S.G. Philander, El Niño, La Nina, and the Southern Oscillation, Elsevier, New York, 1989.
30. S. Brönnimann, Rev. Geophys. 45 (2007) (RG3003).
31. R. García-Herrera, H.F. Díaz, R.R. García, M.R. Prieto, D. Barriopedro, R. Moyano, E. Hernández, J. Clim. 21 (2008) 1948–1962.
32. A.H. Fink, P. Speth. Naturwissenschaften 85 (1998) 482–493.
33. K. Emanuel, Nature 436 (2005a) 686–688.
34. P.J. Webster, G.J. Holland, J.A. Curry, H.R. Chang, Science 309 (2005) 1844–1846.
35. G.J. Holland, P.J. Webster, Philos. T. Roy. Soc. A. 365 (2007) 2695–2716.
36. K. Emanuel, Nature 438 (2005b) E13 doi:10.1038/nature04427.
37. P.J. Webster, G.J. Holland, J.A. Curry, H.R. Chang, Science 311 (2006) 1713c.
38. C.W. Landsea, Nature 438 (2005) E11–E13, doi:10.1038/nature04477.
39. P.J. Klotzbach, Geophys. Res. Lett. 33 (2006) L10805 doi:10.1029/2006GL025881.
40. R. García-Herrera, L. Gimeno, P. Ribera, E. Hernández, J. Geophys. Res. 110 (2005) D03109 doi:10.1029/2004JD005272.
41. R. García-Herrera, P. Ribera, E. Hernández, L. Gimeno, J. Geophys. Res. 112 (2007) D06108 doi:10.1029/2006JD007370.
42. K.E.Trenberth, D.P. Stepaniak, J.M. Caron, J. Clim. 13 (2000) 3969–3993.
43. D.B. Stephenson, H. Douville, K. Rupa Kumar, Mausam 52 (2001) 213–220.
44. T.N.Chase, J.A. Knaff, R.A. Pielke, E. Kalnay, Nat. Hazards 29 (2003), 229–254.
45. B. Wang, Q. Ding, Geophys. Res. Lett. 33 (2006) L06711 doi:10.1029/2005GL025347.
46. N.P. Gillett, G.C. Hegerl, M.R. Allen, P.A. Scott, Geophys. Res. Lett. 27 (2000) 993–996.
47. E. Zorita, J.F. González-Rouco, Geophys. Res. Lett. 27 (2000) 1755–1758.
48. T.J. Osborn, Clim. Dyn. 22 (2004) 605–623.
49. K.E. Trenberth, Mon. Weather Rev. 112 (1984) 326–332.

Chapter 9

Bird Ecology as an Indicator of Climate and Global Change

Wolfgang Fiedler
Max Planck Institute for Ornithology, Department Vogelwarte Radolfzell, Germany

1. **Introduction**
2. **Indicators of Change**
   2.1. Range
   2.2. Migration
   2.3. Reproduction
3. **Conclusion**
References

1. INTRODUCTION

Birds are highly mobile and easy to observe. They are relatively easy to recognise and their occurrence and habits are noted by millions of passionate birdwatchers or just interested laymen. It is not surprising that changes in abundance or behaviour of birds are among the best documented changes known in the animal world. Changes in the arrival of migrating birds at their breeding grounds and their disappearance in autumn have been used as cues to forecast weather in many cultures for centuries. Modern biology understands bird behaviour not as a result of miraculous wisdom of individuals but as a result of the action of evolution through mutation, selection and reproduction. Since a central goal in evolution is adaptation to the environment, climate change, as well as global change in a wider sense will change selection pressures and reproductive success of various behavioural types. This is, indeed, what is presently being observed and birds show us that we are already in the middle of massive changes.

However, it is important to note that not all changes in bird behaviour, as they are currently observed, can be attributed to climate change. Other factors, such as changes in land use, can influence the migration behaviour of birds. Changes in agriculture, in industrial activities or in human behaviour may offer or destroy suitable wintering sites. Examples include a new food source for European Cranes *Grus grus* in fields of winter weed in northern France, ice-free waters for Coots *Fulica atra* due to power plant cooling in Lithuania or bird feeders for Blackcaps

Sylvia atricapilla wintering in Great Britain [1]. Effects can be accelerated or attenuated by climate change and in some cases it will not be possible to identify the primary source of change that affects a certain behavioural modification. Nevertheless, all of the environmental changes currently experienced, that top the list in terms of speed and extent, are very likely a result of human activity and thus share a common source. In this chapter, it will be shown that environmental changes affect all areas of a bird's life and that many indicators of this change can be found by observing birds and their ecology.

2. INDICATORS OF CHANGE

2.1. Range

2.1.1. Size and Position of Breeding Ranges

Changes in the distribution of birds, especially in their breeding range, were one of the earliest topics discussed among ornithologists in the context of climate change. Already in 1995, Burton [2] listed in his book 123 European bird species which extended their ranges in northern, western and northwestern directions and he attributed these changes to global warming. These observations meanwhile were supported by many other studies showing that range boundaries are moving poleward or upward in altitude as the climate gets warmer.

Sound data comes from comparing standardised breeding bird surveys. Such comparisons have been done on data from the United Kingdom and Ireland [5]. The comparison of the breeding bird atlas of 1968–1972 with the atlas of 1988–1991 showed that in 59 species with southerly distribution within the study area, there was a mean northward shift of their northern border of distribution of 18.9 km (see Table 1). This is equivalent to roughly a 1 kilometre northward shift of their northern range border per year. At the same time 42 northerly distributed species did not show any systematic movements of the southern border of their distribution area. In a comparable study with data from the time periods 1974–1979 and 1986–1989, it was found that the northern border of 119 southerly distributed species in Finland showed a mean northward movement of 18.8 km while the southern border of 34 northerly distributed species did not change [4]. Finally, data from the North American Breeding Bird Survey showed that 26 southerly distributed species moved their northern border of range on average 72.9 km northward between the periods 1968–1972 and 1988–1991 [3] while the southern border of northern species did not move.

Changes in breeding distribution registered, so far, are likely to be the first indications of rather severe ecological shifts and species rearrangements in some areas. Based on museum material of 1179 bird species and some mammal and butterfly species occurring in Mexico, ecological niche models have been developed with a genetic algorithm and were projected onto two predicted climate surfaces (conservative and liberal) for 2055. While extinctions

TABLE 1 Changes in borders of the range of bird species as revealed from breeding bird surveys

Region and source	Period 1	Period 2	Distribution within region	Number of species	Distance and type of shift	Shift/year
Great Britain and Ireland [5]	1968–1972	1988–1991	Southerly distribution	59	Northern border moved 18.9 km northward	ca. 1 km·a^{-1}
			Northerly distribution	42	Southern border did not show systematic movements	
Finland [4]	1974–1979	1986–1989	Southerly distribution	119	Northern border moved 18.8 km northward	ca. 1.7 km·a^{-1}
			Northerly distribution	34	Southern border did not show systematic movements	
North America [3]	1967–1971	1998–2002	Southerly distribution	26	Northern border moved 72.9 km northward	ca. 2.4 km·a^{-1}
			Northerly distribution	29	Southern border did not show systematic movements	

and dramatic range losses were expected to be few, the turnover in some regions was predicted to reach more than 40% of the species [6].

The studies mentioned above lead to the expectation that in the first instance species richness should increase in areas with incoming southern species and northern species which have not left. Indeed, the Lake Constance area in Central Europe could be an example for this. Based on 2 × 2 km grid cells breeding birds in an area of 1212 km^2 were counted in a semi-quantitative way in the periods 1980–1981, 1990–1992 and 2000–2002. During this time species numbers increased from 141 to 154, in the last decade with a significant increase of species with a southern centre of distribution [7].

A flagship species for a northward range extension of a southerly distributed bird species in Europe is the Bee-eater *Merops apiaster*. Distributed mainly in

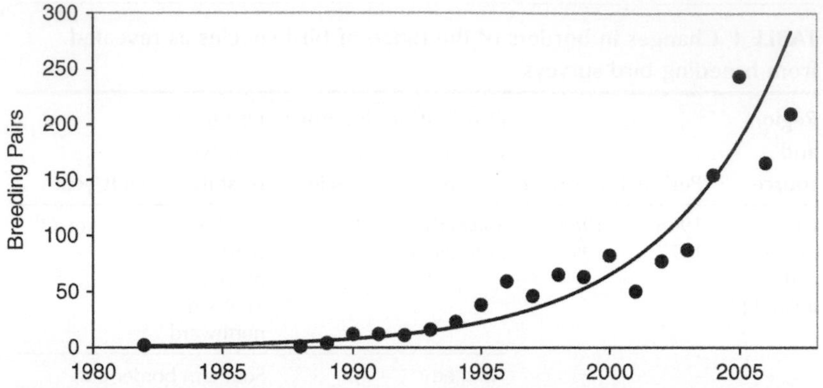

FIGURE 1 Number of breeding pairs of the Bee-eater *Merops apiaster* in the country of Baden-Württemberg (southwestern Germany) 1982–2007. Data from Boschert and Todte.

warmer areas such as the Mediterranean, the species starts breeding in higher European latitudes as soon as there are periods with warmer temperatures. This has been shown by a comparison between Bee-eater records in Central Europe and the size of growth rings in oak wood. The rings give evidence of warmer periods with higher annual growth rates in the oaks from the sixteenth Century onwards [8]. Currently, Bee-eaters are showing significant population increases (Fig. 1) and now breed as far north as Poland and Scandinavia.

On a European scale climatic variables in the actual breeding ranges of bird species have recently been used to forecast the future distributions based on the assumptions of first generation climate change models in the Climate Atlas of Breeding Birds in Europe [9]. This modelling leads to the prediction that between the two periods 1960–1990 and 2070–2100 breeding distribution areas of European birds shall move on average 550 km northwards and that many species shall suffer from area losses. This predicted value of a northward movement rate of 5.5 km$\cdot$a^{-1} is higher than has actually been found in the studies cited above.

One major problem with this approach is the lack of any account for habitat availability in Europe where many natural habitat types, due to human activity, will not be available, even when climate conditions would allow them to exist. This problem is not negligible as has been shown when bird distribution data from the breeding bird atlas of the United Kingdom and Ireland from 1968 to 1972 has been used to forecast the distribution in 1988–1991, using the same method as was used in the Climate Atlas of Breeding Birds [10]. The results were compared with real distributions in the second time period. For a series of species the forecasts did not match well and the real distributions were much smaller than the forecasted results. A remarkable example is the Red-backed Shrike *Lanius collurio* which in the 1970s was restricted to the South and Southwest of the United Kingdom. For the

1990s, models predicted a coverage of almost all of the United Kingdom, including the very north of Scotland. In reality, at that time, the species was almost extinct over the whole of the United Kingdom. However, two new breeding sites were established in Scotland, which was in accordance with the forecast.

2.1.2. Ranges during Nonbreeding Season

Besides breeding ranges, winter distributions are also changing. This is obviously the case where migrating birds can stay closer to their breeding grounds when closer areas become more suitable wintering areas for them, or when closer wintering areas become less suitable (e.g., dryer) and thus birds are forced to migrate longer distances [1].

Data from the Christmas Bird Count in North America between 1975 and 2004 showed a mean northward movement of the northern border of wintering ranges of migratory bird species of 1.5 $km{\cdot}a^{-1}$. At the same time, winter distributions of non-migrants also moved northward during that time period [11].

In Europe, the effect of the drying-up of the Sahel belt, (a dry savannah area south of the Sahara desert where many Palaearctic long-distance migrants have their wintering areas), has been considered as one of the main reasons for population declines in the Common Whitethroat *Sylvia communis*, the Sedge Warbler *Acrocephalus schoenobaenus* and in many other species [12]. This indicates that the potential for rather simple latitudinal shifts of wintering areas, corresponding to changes in climate, is limited and might not be an option for all species.

2.2. Migration

2.2.1. Timing of Migration to the Breeding Grounds

Despite its complexity and genetic component, bird migration behaviour appears to be highly flexible and changeable in many species. Changes in migration behaviour have been and still are the subject of numerous publications. For example, in an impressive dataset from Finland, the arrival dates of birds at their breeding grounds have been recorded between 1749 and end of twentieth century [13]. After being quite stable or only moderately shifting up to the 1960s, the arrival dates since then have advanced by about a month in the case of the Skylark *Alauda arvensis* and about half a month for the Wagtail *Motacilla alba* and the 'Swallows' (Barn Swallow *Hirundo rustica* and the House Martin *Delichon urbica*), while the Cuckoo *Cuculus canorus* and the Swift *Apus apus* showed little change.

A recent analysis of banding data of birds passing the island of Heligoland in the North Sea during pre-breeding migration, for the time period 1960–2007, showed that Blackbirds *Turdus merula* and Pied Flycatchers *Ficedula hypoleuca* were now arriving 11 days, Willow Warblers *Phylloscopus trochilus*

13 days and Blackcaps *Sylvia atricapilla* 17 days earlier than they had before. The mean advancement of 24 species was 8.6 d for the total period or 1.9 d every decade [14].

Lehikoinen et al. showed in a related analysis [13] of 21 long term studies of 10 European countries a consistent advancement of arrival times at the breeding grounds for Sand Martin *Riparia riparia*, Blackcap, Chiffchaff *Phylloscopus collybita*, Wagtail, Barn Swallow, Pied Flycatcher, Sedge Warbler, Tree Pipit *Anthus trivialis* and House Martin. In contrast, Whinchat *Saxicola rubetra*, Spotted Flycatcher *Muscicapa striata* and Cuckoo *Cuculus canorus* did not advance their arrival times in half of the reported studies.

In a very large compilation of different studies in Eurasia, Sparks et al. [15] calculated an advance in arrival times of 2.5–3.3 $d{\cdot}K^{-1}$ warmer mean temperature.

The evidence for earlier arrival of birds at their breeding grounds in concordance with the warming up of the climate is enormous and corresponds well with the finding of a consistent global advancement of phenological events in spring between 2 and 5 d per decade [16]. However, some species react stronger than others to the advancement of spring phenology in certain regions and a few species seem unable to follow the changes. It was generally found that among 56 species in Lithuania, those species that arrive early in spring, advance their arrival dates more than those species arriving later in the spring [16]. There is, however, a variation in response on the individual level, but in general the first birds to arrive at their breeding grounds advance their arrival by four days per decade, while the mean arrival date (average arrival date of a population) advances only by 1 d per decade [16].

From an evolutionary point of view this indicates that some birds might benefit from an earlier arrival at their breeding grounds and thus show a strong response to changed environmental or climate conditions, while others change their timing at a much slower rate. One reason for the variation in the rate of changes within populations can be explained by the proximate factors driving the advancement of arrival times. The earlier arrival of Pied Flycatchers in recent years in southern Finland correlates well with higher temperatures in the winter quarters and along the homeward migration routes. However, the last birds to arrive did not advance their arrival dates, and late spring temperatures did not change [17].

In Europe, many studies used the North Atlantic Oscillation (NAO) [18] as a measure of climatic conditions. Almost all bird species in those parts of Europe influenced by the NAO, can adjust their homeward migration timing to rising temperatures. This seems to be true for long distance migrants (migration routes from Europe to at least sub-Saharan Africa) as well as for short distance migrants (migration between Europe north of the Alps and the Mediterranean). However, it seems to be necessary that birds experience the warmer temperatures not only after arrival at the breeding grounds but along migration routes and also in their wintering quarters. In Europe and North

America, birds do not arrive early at their breeding grounds if temperatures in these breeding areas rise but do not rise along the migration routes [19]. However, correlations between arrival times and temperature in the breeding areas have been found. In the long term dataset from Finland mentioned above [13] spring arrival times were clearly earlier in years with higher mean temperature in the month before arrival. Also in passing migrants, over the Courish Spit (Southern Baltic), a strong negative correlation between April temperatures and passage times of 20 songbird species has been observed [20].

Positive NAO values in Europe can mean not only warmer temperatures but poor conditions in the Mediterranean and Sahel zone. For example, Barn Swallows in Italy arrive later in years when there are poor conditions in Africa [21]. In Spain, an increasing delay in the spring arrival of migrants in the 1970s and a current return to the level of the 1940s has been found [22] despite increasing local temperatures. It has been assumed that this is an effect of poor conditions in northern Africa (mainly due to low precipitation), resulting in a poor food supply which in turn means a delay of fat deposition and consequently a later takeoff to the breeding grounds [23].

2.2.2. Timing of Migration from the Breeding Grounds

In contrast to the fairly consistent patterns of more or less pronounced advancements of spring arrival at the breeding grounds, when mean temperatures rise, the post breeding migration timing shows a very different picture. From a 42-year dataset of 65 migrating bird species, passing the Swiss alpine pass Col de Bretolet, the autumn passage of migrants wintering south of the Sahara has advanced in recent years, while migrants wintering north of the Sahara have delayed their autumn passage [24]. This advancement of post breeding migration timing in long distance migrants might be seen under the light of a selection pressure to cross the Sahel before its seasonal dry period. Species with shorter migration routes might benefit from a less constrained time schedule for breeding and moulting during summer when autumns are warmer and the risk of bad weather during autumn is reduced. This assumption is supported by the additional finding that species with a variable rather than a fixed number of broods per year also delay their passage, possibly because they are free to attempt more broods [24]. Comparable results were also found in Oxfordshire on the British Isles [25].

This picture of advancements and delays in post breeding migration timing, being dependent on the species, seems to be consistent (at least) all over Europe, but the assumption of a rather simple division between advancing long distance migrants and delayed short distance migrants is not supported generally at other places [14]. While in most European Studies more species show a delay in post breeding migration timing [14,26], some studies like the one in southern Baltics clearly showed different trends at different time periods [27] and at the autumn passage on the Kola peninsula in Northern Russia the number of advances was much the same as the number of delays [28].

Despite the self-evident assumption that those birds advancing their autumn departure might benefit from an earlier arrival, an earlier onset of breeding and an earlier onset of post breeding moult [29], no marked relationship between timing in autumn and timing in the preceding spring has consistently been found [14].

2.2.3. Migration Routes and Wintering Areas

Results gained over more than a century of bird ringing enable us, at least in some regions with sufficient data, to detect possible changes in the migration routes and in the position of the wintering quarters. Birds marked with a small coded ring at the breeding grounds and recovered later outside the breeding season enable insights into the position of various areas used by the birds through the year as well as insights into the changes of the positions of these areas. Presumably wintering grounds and other areas used by birds during the non-breeding season like moulting areas or stopover sites during migration will change in the same way as changes of the breeding range have been described above. Generally, it can be expected that in regions with less severe winters migration routes will be shortened or that migration behaviour even will be reduced to zero. There is much evidence for a selection pressure towards earlier arrival at the breeding grounds for many bird species. Besides that, positions of wintering areas will also change when areas become unsuitable due to environmental changes. This may be true especially for birds wintering in areas endangered by desertification such as the Sahel Belt in Africa or parts of the Mediterranean Basin.

Studies available so far support these assumptions. Among 30 bird species investigated in Germany, 13 showed evidence of shorter migration routes, 11 showed evidence of a northward move of mean wintering latitude and 9 species showed increased numbers of winter recoveries within 100 km around the breeding place. Only a few species showed the opposite trend [30]. On a larger dataset of 66 species from the United Kingdom and Ireland it was found that 27 species showed increasingly northern wintering areas and 11 showed a northward move of the mean wintering latitude [31].

However, global warming might also lead to longer migration routes when breeding ranges are extended into higher latitudes and at the same time the wintering areas do not change much. For example, the European Bee-eaters showed a range expansion northwards and increased the intra-European part of their migration routes by up to 1000 km, but still winter south of the Sahara. Also the Black-Winged Stilt *Himantopus himantopus* expanded its breeding areas from the Mediterranean northward into France, Ucraine and Russia but still winters south of 40° latitude [1]. Evidence for increasing migratory activity also comes from White-rumped and Litte Swifts *Apus caffer* and *A. affinis* which colonise the Mediterranean area from the south, leaving these areas during non-breeding periods while they are resident in almost all of the rest of their African breeding ranges [32,33].

2.2.4. Partial Migration

Partial migration describes the widespread phenomenon of some birds of a population migrating, while others don't. This situation has been described as the turntable of migratory and sedentary behaviour which enables selection to favour either more migratory or more sedentary behaviour according to environmental conditions [34]. Increasing numbers of winter records of otherwise migratory bird species give evidence of the development of partial migratory populations in Europe and North America and presumably elsewhere [1]. The Central European Blackbird is a well known example of this phenomenon. It was once considered as a migrating thrush of European woodlands but in the early twentieth century it successfully started colonising human settlements and reduced migration to become the first entirely sedentary populations in recent decades [34,35].

2.2.5. Eruptions

The mass movements of parts of local populations, which may be directed but seldom are reversible, are commonly called eruptions or evasions. In less migratory species, with highly variable population sizes, living under highly variable food conditions such as tits in forest habitats and other boreal seedeaters, these eruptions occur repeatedly every few years. In a German Blue Tit population it has been shown that along with rising environmental temperatures, the numbers of eruptions have decreased remarkably [36]. While population size did not drop significantly, this observation (which might be a common phenomenon), may indicate a constant and improved food supply, making it unnecessary for parts of the population to emigrate.

2.3. Reproduction

2.3.1. Onset of Breeding Period

The reproduction of birds is influenced by weather and thus by climate change in many ways. It is known that temperature, precipitation and resulting food supply can trigger the start of breeding [37]. An analysis of the relationship between ambient temperature and time of the first egg laid showed that 45 out of 57 bird species advanced the time of the first egg, when temperatures were high. Therefore under current global warming it is not surprising that there are numerous studies indicating advancements in the onset of breeding in many species. With respect to migrating species the general advancement of arrival times in breeding areas has been mentioned above. Early breeding depends on fitness which stems from the availability of food insects which in turn depends on early leafing and flowering of plants under elevated spring temperatures [37,38].

Based on data from the British nest record scheme, for the period 1971–1995, Crick et al. [38] found significant trends towards earlier laying dates for 20 of 65 species analysed, with only one species having a delayed breeding date. The shift of the 20 species advancing their laying dates averaged 8.8 d. These species could neither be assigned to distinct migration strategies nor to ecological or taxonomic groups and comprise early and late breeders as well as long distance migrants and residents. Similarly, Tree Swallows *Tachycineta bicolor* throughout North America advanced their laying dates by up to nine days between 1959 and 1991 [39] and advancements of laying dates of six and nine days were also found in the German Great and Blue Tits *Parus major* and *P. caeruleus* between 1970 and 1995 [40]. Based on Danish bird ringing data of Arctic Terns *Sterna paradisaea*, A.P. Møller and colleagues reported an advancement of the ringing dates of chicks by 18 d during a 70 a period. This was explained by an increase in mean temperatures in April and May [41]. These are only few examples out of a long list of reports which in most cases indicated homologous trends.

2.3.2. Length of Breeding Period

As discussed briefly above, not only is the earlier onset of breeding beneficial, but it also may lead to an extension of the breeding period. In species with high nest predation rates, longer breeding periods can offer more time for replacement clutches or species might successfully raise more than one brood per season. Calculated durations of the stay of 20 migrating bird species at their breeding grounds, from passage data on the island of Heligoland, showed an average increase over a decade of 2.2 d [14]. A prolongation of the breeding period has also been shown in Reed Warblers *Acrocephalus scirpaceus* in Poland [42]. Between 1970 and 2006 the peak of egg laying advanced 18 d but the end of the breeding season did not change. Replacement clutches, in cases of nest failure, were produced in early years by 15% of breeding pairs while in recent years 35% of failing pairs started a second, third or up to a fifth laying attempt. For example, evidence for an increase in second broods (those are broods following a successful brood in the same season) comes from German Swifts: during the past few years Swifts have arrived at their breeding grounds earlier than before, have delayed post breeding migration [43,44] and have increased the number of second broods [45]. Also, correlations between weather, food availability and multiple broods per season have been shown in a series of studies on various bird species [46–49].

2.3.3. Breeding Success

Earlier arrival at breeding sites and earlier onset of egg laying in many bird species means also larger clutch sizes since there is a link between the length of daylight and the clutch size with clutches produced earlier often containing more eggs [50]. In a 30-year study of Reed Warblers breeding in Southern Germany the median of the date of the first egg advanced 15 d and the mean clutch size increased by about 0.5 eggs [51]. A similar relationship between onset of breeding, mean clutch size and breeding success can be found in Southern German Collared Flycatchers *Ficedula albicollis* (Fig. 2). However,

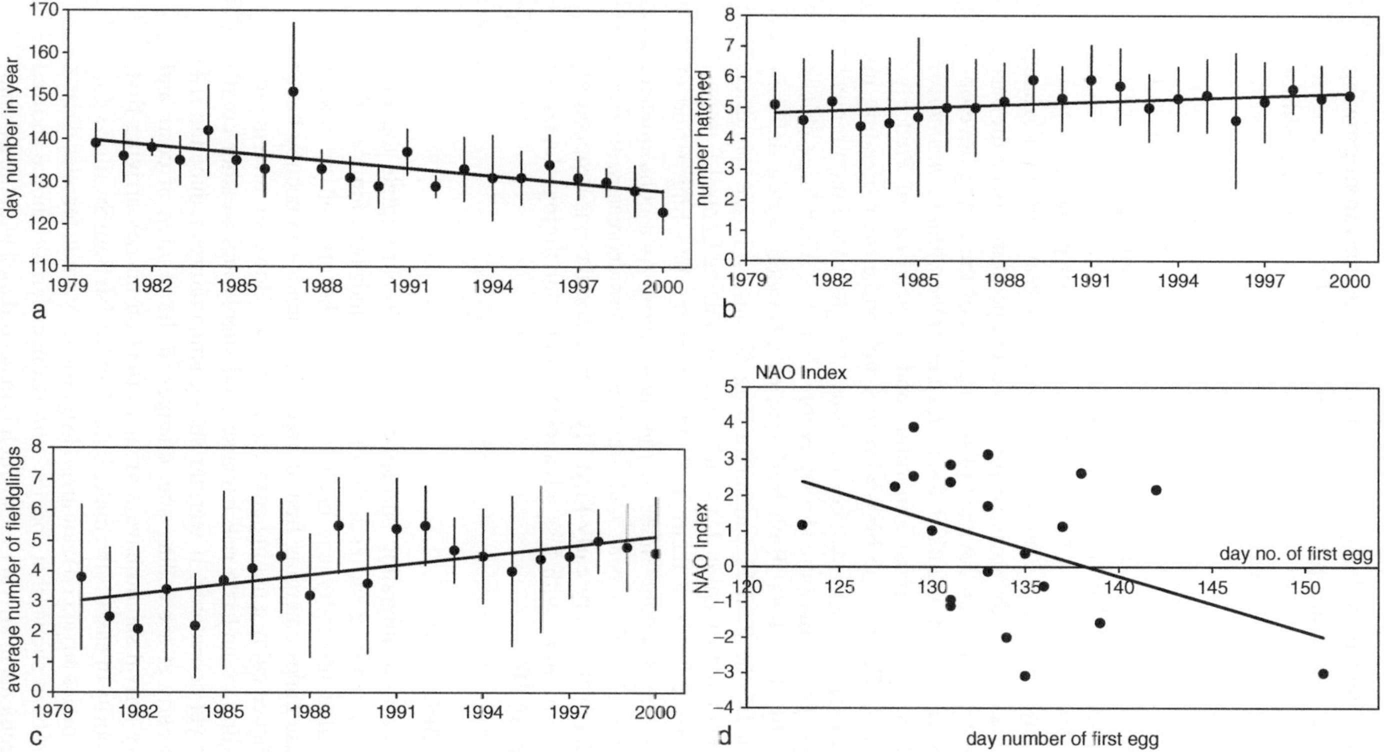

FIGURE 2 Date of first egg (a; day numbers counted from January 1), hatching success (b; average number of young hatched) and fledging success (c; average number of young fledged) of Collared Flycatchers *Ficedula albicollis* in a southwestern German study area. R^2 and ANOVA probabilities $>F$: a – 0.40, <0.003; b – 0.21, <0.04; c – 0.42, <0.002. Data from Renz, Dallmann and Braun, Analysis by Peintinger and Fiedler. (d) shows the correlation between the date of the first egg (day number in year) and the NAO Station based annual index (NAO Index Data provided by the Climate Analysis Section, NCAR, Boulder, USA, [18]); $R^2 = 0.19$, ANOVA probability $>F$ is 0.047.

reduced post fledging survival may prevent those populations from growing even when more young are produced. Capercaillies *Tetrao urogallus* in Scotland advanced the onset of breeding but suffer from a drop in breeding success, presumably due to seasonal changes in the insect supply for the chicks [52].

Optimal food supply of the young in the nest is crucial for reproductive success. Since timing of breeding as well as of moulting and migration is always a trade-off between multiple environmental and physiological requirements, phenological processes as induced by global warming may desynchronise. Marcel Visser, Christiaan Both and others presented a textbook example for this with Pied Flycatchers and Great Tits in Europe [53,54]. In nine Dutch study areas rising spring temperatures over the last 40 a, were connected with an advance of leafing and of the spring development of caterpillars of an abundant moth species (*Operophtera brumata*). These caterpillars form the most important food for nestlings of Pied Flycatchers and Great Tits and the birds aim to synchronize their breeding in a way that the caterpillar peak matches the time of highest food requirement for the nestlings. This is the time shortly before fledging, when large chicks have to be fed by the adults. Both bird species advanced laying dates in recent years but for the Pied Flycatcher (a long-distance migrant wintering south of the Sahara and spending 2/3 of it's lifetime outside Central Europe), other factors seem to prevent them from advancing the breeding period to match the advancing hatching times of the caterpillars. As a consequence, nestlings miss the caterpillar peak and breeding success decreases. In areas where caterpillars hatch very early Pied Flycatcher populations dropped by up to 90% while in areas with less advancing caterpillar timing, decreases only reached up to 10%.

2.3.4. Sexual Selection

In the large majority of migrating bird species, pairs do not migrate together and males arrive some time earlier at the breeding grounds than females. This phenomenon called protandry has been assumed to be affected by sexual selection because males emerging first at breeding grounds can occupy better territories and hence enjoy a mating advantage [55–57]. However, arriving too early at a breeding ground is a risk because food supply and weather conditions might not yet be suitable. If warmer spring temperatures reduce the risk of arriving too early at a breeding site, changes in the relation of costs and benefits of early arrival should have a greater effect on the sex arriving first, which in the majority of cases is the male. Indeed, in Danish Barn Swallows during 1971–2003, males advanced their arrival significantly while females did not [57]. It has also been shown that species with stronger female choice showed greatest advancements in arrival times which is in accordance with the assumption that early arrival of males is favoured by female choice [58,59].

In Blackcaps breeding in southwestern Germany and wintering either 1800 km southwest in Portugal and Spain or 1000 km northwest in the United Kingdom and Ireland, it has been shown that earlier arrival is not only related to a higher breeding success but also drives assortative mating among mates with comparable timing which drives evolution especially rapidly in one direction [60]. Birds wintering at higher latitudes not only face shorter distances to return to the breeding grounds but also experience a daylight-night-regime which triggers their circannual rhythms and accelerates pre-breeding migration, gonadal development and the onset of breeding [61,62].

3. CONCLUSION

The Ecology of birds can clearly serve as an indicator of climate and global change. Almost all aspects in the life cycle of birds, that have been regarded so far, show recent changes that can be linked to environmental changes. It is not surprising that birds show a high potential to adapt even complex behaviour such as breeding or migration to changing environments – either through evolutionary mechanisms acting on the genetic basis of behaviour or through available phenotypic plasticity. Ever since very early bird species evolved on earth 200 Ma ago, birds have had to cope with floating continents, rising and eroding mountains, ice ages and other massive environmental changes. A high degree of agility and mobility might have helped birds to adapt better to new conditions than other organisms might have done.

This is not to say that there is no conservation concern behind the reactions of birds to climate change. Some of the studies presented above clearly give evidence of problems that birds might face when they need to adapt their behaviour to rapid environmental and climatic changes. It is very likely that among bird species there will be winners and losers resulting from the current climate and global change and it might also be that the rate of losers will be high and extinctions of bird species will reach a level exceeding extinction rates seen in earlier times in bird's evolution. Since birds are easy to observe, are present in all parts of the world and are objects of interest to many people, they are ideal flagships to observe the consequences and the impacts of future environmental changes on organisms and on ecosystems.

REFERENCES

1. W. Fiedler, in: P. Berthold, E. Gwinner, E. Sonnenschein (Eds.), Avian Migration, Springer, Berlin, 2003, pp. 21–38.
2. J.F. Burton, Birds and Climate Change, Helm, London 1995.
3. A.T. Hitch, P.L. Leberg, Conservation Biology 21 (2007) 534–539.
4. M. Luoto, R. Virkkala, R.K. Heikkinen, Glob. Ecol. Biogeogr. 16 (2007) 34–42.
5. C.D. Thomas, J.J. Lennon, Nature 399 (1999) 213.
6. A. Townsend Peterson, M.A. Ortega-Huerta, J. Bartley, V. Sánchez-Cordero, J. Soberón, R.H. Buddemeier, D.R.B. Stockwell, Nature 416 (2002) 626–628.

7. N. Lemoine, H.-G. Bauer, M. Peintinger, K. Böhning-Gaese, Conserv. Biol. 21 (2007) 495–503.
8. R. Kinzelbach, B. Nicolai, R. Schlenker, J. Ornithol. 138 (1997) 297–308.
9. B. Huntley, R.E. Green, Y.C. Collingham, S.G. Willis, Climate Atlas of Breeding Birds in Europe, Lynx Editions, Barcelona, 2007.
10. M.B. Araúja, C. Rahbek, Science 313 (2006) 1396–1397.
11. F.A. La Sorte, F.R. Thompson, Ecology 88 (2007) 1803–1812.
12. P. Berthold, Naturwiss. Rundsch. 51 (1998) 337–346.
13. E. Lehikoinen, T.H. Sparks, M. Zalakevicius, in: A.P. Møller, W. Fiedler, P. Berthold (Eds.), Birds and Climate Change, Elsevier Science, London, 2004.
14. O. Hüppop, K. Hüppop, Proc. R. Soc. Lond. B 270 (2003) 233–240.
15. T.H. Sparks, F. Bairlein, J.G. Bojarinova, O. Hüppop, E.A. Lehikoinen, K. Rainio, L.V. Sokolov, D. Walker, Glob. Change Biol. 11 (2005) 22–30.
16. C. Parmesan, Glob. Change Biol. 13 (2007) 1860–1872.
17. M. Ahola, T. Laaksonen, K. Sippola, T. Eeva, K. Rainio, E. Lehikoinen, Glob. Change Biol. 10 (2004) 1610–1617.
18. J.W. Hurrell, Y. Kushnir, M. Visbeck, Science 291 (2001) 603–605.
19. P.P. Marra, C.M. Francis, R.S. Mulvihill, F.R. Moore, Oecologia 142 (2005) 307–315.
20. L.V. Sokolov, M.Y. Markovets, A.P. Shapoval, G.Y. Morozov, Zool. Zhurnal 78 (1999) 1102–1109.
21. N. Saino, T. Szep, M. Romano, D. Rubolini, F. Spina, A.P. Møller, Ecol. Lett. 7 (2004) 21–25.
22. O. Gordo, J.J. Sanz, Glob. Change Biol. 12 (2006) 1993–2004.
23. O. Gordo, L. Brotons, X. Ferrer, P. Comas, Glob. Change Biol. 11 (2005) 12–21.
24. L. Jenni, M. Kéry, Proc. R. Soc. Lond. Ser. B-Biol. Sci. 270 (2003) 1467–1471.
25. P.A. Cotton, Proc. Natl. Acad. Sci. 100 (2003) 12219–12222.
26. T.H. Sparks, C.F. Mason, Ibis 146 (2004) 57–60.
27. L.V. Sokolov, M.Y Markovets, Y.G Morozov, Avian Ecol. Behav. 2 (1999) 1–18.
28. A. Gilyazov, T.H. Sparks, Avian Ecol. Behav. 8 (2002) 35–47.
29. H.Q.P. Crick, T.H. Sparks, Nature 399 (1999) 423–424.
30. W. Fiedler, U. Köppen, F. Bairlein, in: A.P. Møller, W. Fiedler, P. Berthold (Eds.), Birds and Climate Change, Elsevier Science, London, 2004.
31. A. Soutullo, Dissertation, University of East Anglia, Norwich, 2003.
32. D.W. Snow, C.M. Perrins, Birds of the Western Palearctic, Oxford University Press, Oxford, 1998.
33. J.E. del Hoyo, A. Elliott, J. Sargatal (Eds.), Handbook of the Birds of the World, Lynx Editions, Barcelona, 1999.
34. P. Berthold, Bird Migration: A General Survey, Oxford University Press, Oxford, 2001.
35. H. Schwabl, J. Ornithol. 124 (1983) 101–116.
36. W. Winkel, M. Frantzen, J. Ornithol. 132 (1991) 81–96.
37. C.M. Perrins, Acta XIX Congr. Int. Ornithol. 1 (1988) 892–899.
38. H.Q.P. Crick, C. Dudley, D.E. Glue, D.L. Thomson, Nature 388 (1997) 526.
39. P.O. Dunn, D.W. Winkler, Proc. R. Soc. Lond. B 266 (1999) 2487–2490.
40. W. Winkel, H. Hudde, J. Avian Biol. 28 (1997) 187–190.
41. A.P. Møller, E. Flensted-Jensen, W. Mardal, J. Anim. Ecol. 75 (2006) 657–665.
42. L. Halupka, A. Dyrcz, M. Borowiec, J. Avian Biol. 39 (2008) 95–100.
43. M. Peintinger, S. Schuster, Vogelwarte 43 (2005) 161–169.
44. W. Gatter, Vogelzug und Vogelbestände in Mitteleuropa, Aula, Wiebelsheim, 2000.

45. E. Kaiser, Vogelwelt 125 (2004) 113–115.
46. M. Gucco, G. Malacarne, G. Orecchia, G. Boano, Ecography 15 (1992) 184–189.
47. J. Valencia, C. De la Cruz, J. Carranza, Etología 8 (2000) 25–28.
48. S. Verhulst, J.M. Tinbergen, S. Daan, Funct. Ecol. 11 (2003) 714–722.
49. S.F. Eden, A.G. Horn, M.L. Leonhard, Ibis 131 (2008) 429–432.
50. L. von Haartman, Proc. Int. Ornithol. Congr. 14 (1967) 155–164.
51. T. Schaefer, G. Ledebur, J. Beier, B. Leisler, J. Ornithol. 147 (2006) 47–56.
52. R. Moss, J. Oswald, D. Baines, J. Anim. Ecol. 70 (2001) 47–61.
53. M. Visser, A.J. van Noordwijk, J.M. Tinbergen, C.M. Lessells, Proc. R. Soc. Lond. B 265 (1998) 1867–1870.
54. C. Both, S. Bouwhuis, C.M. Lessells, M.E. Visser, Nature 441 (2006) 81–83.
55. R. Thornbill, J. Alcock, The Evolution of Insect Mating Systems, Harvard University Press, Cambridge, 1984.
56. M. Andersson, Sexual Selection, Princeton University Press, Princeton, 1994.
57. A.P. Møller, Glob. Change Biol. 10 (2004) 2028–2035.
58. D. Rubolini, F. Spina, N. Saino, Behav. Ecol. 15 (2004) 592–601.
59. C.N. Spottiswoode, A.P. Tøttrup, T. Coppack, Proc. R. Soc. Lond. B 273 (2006) 3023–3029.
60. S. Bearhop, W. Fiedler, R.W. Furness, S.C. Votier, S. Waldron, J. Newton, G.J. Bowen, P. Berthold, K. Farnsworth, Science 310 (2005) 502–504.
61. P. Berthold, S.B. Terill, Ringing Migration 9 (1988) 153–159.
62. T. Coppack, F. Pulido, M. Czisch, D.P. Auer, P. Berthold, Proc. R. Soc. Lond. B 270 (Suppl. 1) (2003) 43–46.

Chapter 10

Mammal Ecology as an Indicator of Climate Change

Murray M. Humphries
Department of Natural Resource Sciences, McGill University, Ste-Anne-de-Bellevue, Quebec, Canada H9X 3V9

1. **Introduction: A Primer on Mammal Thermoregulation and Climate Impacts**
2. **Demonstrated Impacts of Climate Change on Mammals**
   2.1. Temporal Approaches
   2.2. Spatial Approaches
3. **Linking Time and Space in Mammal Climate Responses**

Acknowledgements
References

On 12 January 2002, weather stations in New South Wales, Australia recorded air temperatures exceeding 42 °C, which is more than 16 °C hotter than the 30-year average daily maximum. On this day, more than 3500 flying foxes (large fruit bats in the genus Pteropus) from nine colonies in the region succumbed to hyperthermia. Mass die-offs of flying foxes associated with heat waves are known to have occurred 3 times in the century prior to 1990, 3 times in the decade between 1990 and 2000, and 13 times in 7 years between 2000 and 2007 [1].

1. INTRODUCTION: A PRIMER ON MAMMAL THERMOREGULATION AND CLIMATE IMPACTS

Fruit bats, like other mammals and birds, use a combination of physiological and behavioural mechanisms to regulate their body temperature [2]. This thermoregulatory capacity decouples their core body temperature from air temperature. Thus, despite exposure of the body surface to very cold or very hot air temperatures, appropriate physiological and behavioural thermoregulatory responses ensure that core body temperature never varies by more than a few degrees centigrade between birth and death [3]. Even birds and mammals that express torpor do not abandon thermoregulation, but rather lower their thermoregulatory setpoint [4]. For all endotherms, the abandonment of thermoregulation is fatal.

The capacity for mammals to thermoregulate might be expected to enable a degree of thermal independence that reduces their vulnerability to environmental conditions and their sensitivity to climate change. But 3500 dead flying foxes suggest any such expectation would be incorrect [1]. To understand why, we must broaden our consideration of how climate affects mammals, both directly and indirectly.

The defining feature of endotherms is their use of metabolic heat to regulate their body core at a constant set-point temperature that is independent of air temperature [2]. This means that under cool environmental conditions, where the body loses heat to the environment, maintenance of a constant body temperature requires that heat production, and thus metabolism, increases with declining air temperature along a slope that equals thermal conductance [2]. Under hot conditions, where the body gains heat from the environment, endotherms must begin actively dissipating heat through panting, perspiration, saliva spreading, and in the case of bats, wing fanning [2]. Because these responses increase heat production (i.e. contribute to the problem that it solves), the slope of the increase in metabolism at warm temperatures is always much steeper than the slope of the increase below the lower critical temperature. As a result of the inefficiencies of metabolic solutions to heat dissipation, endotherms are particularly vulnerable to heat stress and, whenever possible, occupy microenvironments that reduce heat stress [5]. Between the lower critical temperature (where thermoregulation begins to require heat production) and the upper critical temperature (where thermoregulation begins to require heat dissipation) is a region referred to as the thermal neutral zone where metabolic rate does not vary with air temperature because small, energetically insignificant adjustments in conductance (e.g. vasodilation, piloerection and postural changes) are sufficient to maintain a constant body temperature [2]. The metabolic rate (or energy expenditure) of an endotherm is minimised when they are at rest, in their thermoneutral zone, and not digesting food; metabolism measured under these circumstances is referred to as basal metabolic rate [2].

Thus, although endotherm thermoregulation permits maintenance of a constant body temperature that is independent of air temperature, air temperature has a direct and major effect on an endotherm's metabolic rate, which in turn determines their resource requirements. Endotherms exposed to environmental temperatures above or below their thermal neutral zone require more resources to stay alive than endotherms exposed to temperatures within their thermoneutral zone. Furthermore, the capacity for endotherms to produce and dissipate heat is not without limits. Exposure to extreme temperatures that cause thermoregulatory capacity to be exceeded, lead first to hypo- or hyperthermia, then, if exposure continues, to death.

Thus, air temperature has direct effects on the metabolism and resource requirements of endotherms and exposure to extreme air temperatures can have direct effects on survival.

Climate exerts additional, indirect effects on mammals through its effects on their resources, competitors and predators. Temperature has a fundamental effect on all biological processes [6], and thus climate variation should profoundly affect all organisms sharing the same environment. In fact, these indirect effects, acting via resources, competitors and predators, are likely to be so strong and pervasive that they will frequently supersede or mediate most direct effects of climate. The mass die off of flying foxes provides a potent example of a direct effect of climate operating independently of any indirect effects [1]. It was the heat that killed them, directly and outright. But even here, it is likely that more complex climate and biotic factors played a role. For example, although 1453 flying foxes from the Dallis Park colony succumbed to hyperthermia on 12 January 2002, more than 25 000 flying foxes present in the same colony and presumably exposed to the same thermal conditions survived [1]. Many factors are likely to dictate thermoregulatory capacity under such extreme situations, such as body size, age, social rank, reproductive condition, body composition and aerobic capacity [2], most of which will, in turn, be influenced by an individual's lifetime experience with resources, competitors and predators. More commonly, climate impacts on mammals are much more complex and multi-faceted, encompassing effects on thermoregulation and other forms of homeostasis, the distribution and abundance of resources, competitors and predators, as well as the interactions among all these biotic elements. Examples of climate impacts acting primarily on biotic interactions include the influence of snow cover on resource access [7] susceptibility to predation [8] and drought on the spatial overlap of competitors and predators [9]. Most climate impacts on mammals are perhaps best envisioned as climate setting the stage for a complex play involving competitors, resources and predators. Changing the stage changes the play, but often in indirect and nuanced ways.

2. DEMONSTRATED IMPACTS OF CLIMATE CHANGE ON MAMMALS

Demonstrated impacts of climate change on mammals is a broad topic, in evolutionary time, geographic scope and taxonomic diversity, which cannot be covered comprehensively in a short chapter. There is much research interest in this area, and many excellent reviews have appeared recently. For more detailed treatments, I refer the reader to the following reviews of climate change impacts on arctic marine mammals [10], Australian fauna [11], tropical ecosystems [12], fossil mammals [13], mammal morphology [14], mammal population dynamics [15] and mammal demographics [16].

2.1. Temporal Approaches

2.1.1. Geological Climate Variation, Mammal Assemblages and Body Size

Most paleo studies demonstrate that climate change has pronounced effects on the diversity and composition of mammal communities [17,18]. Evidence for long-period climate impacts on mammal diversity is provided by a 22 Ma (million years) record of fossil rodents from southern Spain that is characterised by pulses of species turnover at 1–2 Ma intervals coinciding with Milankovitch oscillations [19]. Gingerich [20] reviews faunal responses to the Paleocene–Eocene thermal maximum, a 20 000 a (year) interval of rapid, greenhouse warming marking the transition between the Paleocene and Eocene 55 Ma ago. The faunal responses to the resulting 5–7 °C warming of deep oceans and 4–5 °C warming of mid-latitude terrestrial regions were extreme, including the simultaneous disappearance of up to 50% of benthic foraminifera (associated with rising water temperature and associated reduction in dissolved oxygen) and the simultaneous appearance of several modern orders of mammals, including Artiodactyla, Perissodactyla and Primates. The small size of many of the mammals during this thermal maximum suggests a pattern of dwarfism, which Gingerich [20] attributes to the negative impacts of elevated CO_2 on plant growth and herbivore nutrition. Research using ancient DNA to reconstruct trends in the abundance of Beringian steppe bison (*Bison* spp.) during the Pleistocene has established that the onset of pre-extinction population declines coincided with periods of climate warming and forest expansion [21]. Related genetics research suggests a similar pattern of climate-driven population declines in Pleistocene bears, horses and mammoths [22], but there is also compelling evidence that humans contributed to several megafaunal extinctions in other regions of the world [23].

The strong associations between climate and faunal composition observed across paleo timescales suggest the maintenance of some degree of thermal niche conservatism during prolonged periods of climate change spanning evolutionary time. Direct evidence of this thermal niche conservatism is provided by Martinez-Meyer et al.'s [24] demonstration that 23 extant North American mammals with fossil records spanning back to the Last Glacial Maximum have spatially tracked consistent climate profiles for the last 18 000 a. Associations between paleoclimate reconstructions and the contemporary diversity and composition of fossil mammal assemblages are sufficiently strong that recent literature has advocated using fossil mammal assemblages to reconstruct paleo-climates (e.g. [25,26]).

2.1.2. Recent Climate Variation and Mammal Miscellany

Studies in this category examine mammal responses to annual-, decadal-, and century-scale climate variation. Responses examined are a miscellany of traits, including but not limited to morphology, phenology, life history traits, population abundance and species distribution. The consequences of anthropogenic climate change are best illustrated by long-term data extending back prior to the industrial revolution or, at least, the mid-twentieth century acceleration of greenhouse gas emissions and warming trends (e.g. [27–29]), but such long-term studies of mammals are largely lacking. Thus, most studies in this category examine mammal responses to climate variation occurring across much shorter timescales, ranging from a few years to several decades. Multi-decadal studies frequently encompass climatic variability induced by large scale climate drivers such as North Atlantic Oscillation (NAO), Pacific Decadal Oscillation (PDO), Arctic Oscillation or El Nino Southern Oscillation (ENSO). Interestingly, wildlife population responses to these climatic phenomena are often of greater magnitude and greater consistency than effects observed in meteorological records, suggesting that animal populations can be sensitive integrators and indicators of subtle and complex climatic events [30,31].

Several long-term studies have documented changes in body size that correlate with changing climate conditions. In many but not all cases, body size has been observed to decrease as climate warms [14], consistent with the biogeographical association between small body size and warm climates referred to as Bergmann's Rule [32,33]. For example, the body mass of woodrats (*Neotoma albigula*) in New Mexico decreased by 15% during a decade when summer temperatures warmed by 3 °C ([34]; Fig. 1a). Intriguingly, this short-term association between small body size and warm temperatures matches a much longer, evolutionary-timescale relationship between climate and woodrat body size as inferred from the size of faecal pellets preserved in paleomiddens ([35]; Fig. 1b.). However, other studies have documented the opposite trend, with body size increasing as climate warms, perhaps because of increased resources and enhanced growth rates [36,37]. These (relatively) short-term, observational studies of morphological and climate variation frequently have limited potential to discriminate phenotypic plasticity from evolutionary responses to climate, and to isolate responses to climate from other factors that may also vary over-time (e.g. habitat succession, density-dependence, competition, predation).

Changes in phenology, such as the annual timing of reproduction or dormancy, are among the best documented impacts of climate change on animals and plants in general [38,39]. Examples among mammals are few, but striking. The date when yellow-bellied marmots (*Marmota flaviventris*) emerge from hibernation in Colorado, USA advanced by 38 d (days) over 25 a with 60% of the observed variation in emergence date linked to variation in spring air temperature ([40]; Fig. 2). Although springs became warmer and earlier over the study period, increases in winter snow depth during the same time meant that snow cover persisted longer as marmots emerged earlier, creating a potential mismatch between energy demands and supply in early spring (see Ref. [41]). Réale et al. [42] showed that the parturition date of red squirrels (*Tamiasciurus hudsonicus*) in Yukon, Canada advanced by 18 d over 10 years. Applying

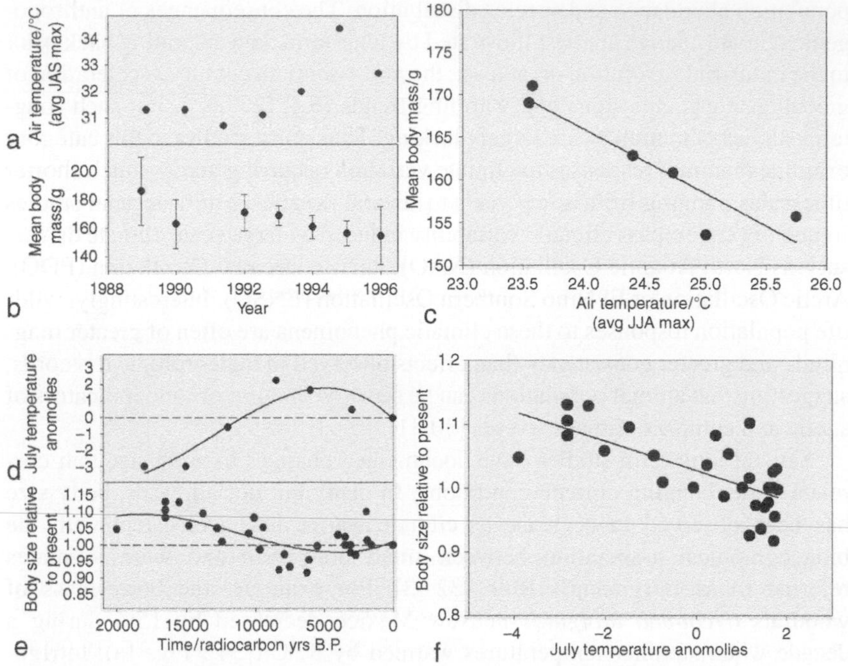

FIGURE 1 The negative influence of warm air temperatures, *t*, on the body size of *Neotoma* woodrats over decadal (a–c) and geological timescales (d–f) in southwestern USA. During a decade of warming air temperatures (a; with air temperature expressed as the average of monthly maximums in July, August and September; redrawn from Fig. 2b, pp. 143 of Smith et al. 1998) the average body mass of woodrats declined by more than 15% (b; redrawn from Fig. 5, pp. 147 of Smith et al. 1998), generating a negative correlation between summer air temperature and body mass (c; redrawn from Fig. 6b, pp. 147 of Smith et al. 1998). During 20 000 a of fluctuating summer air temperatures (d; redrawn from Fig. 3b, pp. 2013 of Smith et al. 1995), the body size of woodrats (as inferred from the size of fecal pellets in paleomiddens) decreased during warm intervals (e; redrawn from Fig. 3a, pp. 2013 of Smith et al. 1995), again generating a negative correlation between summer air temperature and body mass (f; redrawn from Fig. 3, pp. 2013 of Smith et al. 1995).

quantitative genetic approaches to this study population's known pedigree revealed that two-thirds of this advancement involved phenotypic plasticity whereas one-third involved genetically-based microevolution [42]. This study represents one of the few demonstrations that mammals have the evolutionary capacity to adapt to rapid, contemporary climate change, but the taxonomic generality and long-term sustainability of this evolutionary potential is unclear [43].

Many studies have documented climatic influences on mammal population dynamics, usually operating in combination with a complex array of biotic influences such as density dependence, competition, predation, and in the case of large mammals, human harvest [15]. Rigorous demonstration of population-level impacts of climate change impacts requires long-term monitoring of population abundance as well as potential climatic and biotic

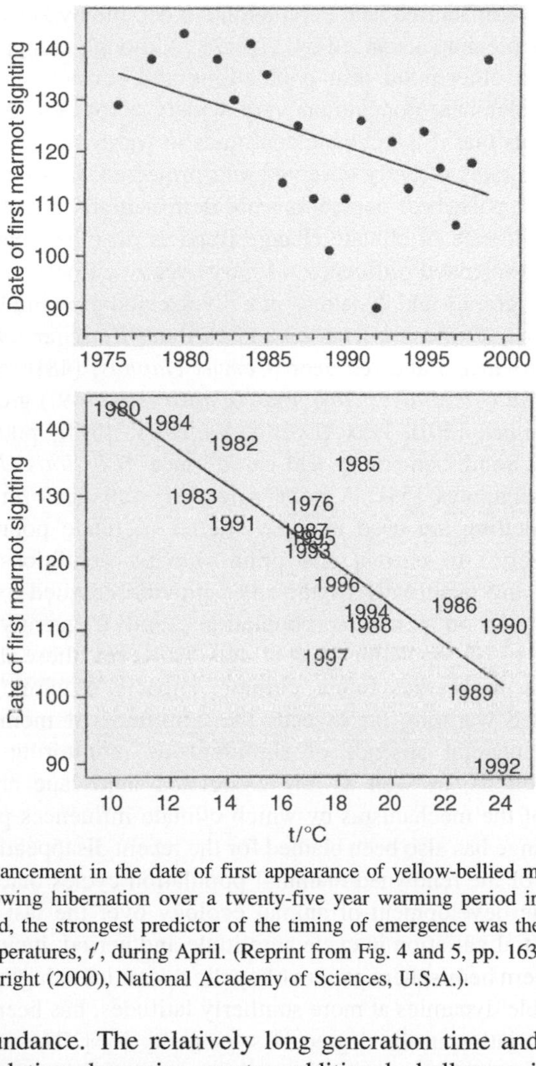

FIGURE 2 Advancement in the date of first appearance of yellow-bellied marmots (*Marmota flaviventris*) following hibernation over a twenty-five year warming period in Colorado, USA. During this period, the strongest predictor of the timing of emergence was the average of daily minimum air temperatures, t', during April. (Reprint from Fig. 4 and 5, pp. 1631–1632 of Inouye et al. 2000; Copyright (2000), National Academy of Sciences, U.S.A.).

drivers of abundance. The relatively long generation time and transience of mammal population dynamics creates additional challenges in establishing cause and effect relationships between climate and mammal population parameters. However, several case studies provide compelling evidence of localised impacts of climate change on mammals. Polar bears (*Ursus maritimus*) have played a prominent role in public and scientific discourse on the impacts of climate change on wildlife because they rely on sea ice that has been observed and projected to decline as a result of climate change [44]. The strongest evidence for a negative effect of climate change on polar bears comes from a 20 a time series on a population occurring at the southern edge of the species' distributional range [45]. This study establishes that reduced survival of juvenile, sub-adult and senescent-adult polar bears in years of early sea ice breakup caused bear populations to decline by 24% over a period when spring ice breakup advanced by 3 weeks. Although comparable data are lacking for most other polar bear populations and current trends in sea ice coverage and polar bear populations vary widely across the Arctic, there is general consensus that if the climate continues to warm and sea ice continues to decline at the rates recently observed and projected, it is only a matter of time before most polar bear populations are detrimentally affected by climate change [46]. Evidence of climate change impacts on other mammal species arise from demonstrated influence of large-scale climate oscillations on population demography and dynamics in a diverse and growing list of species, including muskox (*Ovibos moschatus*) and caribou (*Rangifer tarandus*; [47]), soay sheep (*Ovis aries*) and red deer (*Cervus elaphus*; [48]), wolves (*Canis lupus*) and moose (*Alces alces*; [8]), ibex (*Capra ibex*; [49]) grey-sided voles (*Myodes rufocanus*; [50]), lynx (*Lynx canadensis*; [51]), pikas (*Ochotona collaris*; [52]), South American leaf-eared mice (*Phyllotis darwini*; [53]) and Savanna ungulates [54]. A wide array of statistical time series and population modelling are used in these studies to relate population demographic time series to current and prior climatic conditions [15]. These long-term data and quantitative approaches provide detailed examination of climatic influences on particular populations, and frequently identify the demographic basis of population-level effects. However, these studies provide a weaker basis to predict future climate impacts requires extrapolation because expected warming far exceeds the amplitude of monitored climate variation. The general absence of simultaneous monitoring of prey and predator populations, as well as any associated landscape change, clouds interpretation of the mechanisms by which climate influences populations.

Climate change has also been blamed for the recent disappearance or dampening of many of the renowned mammal population cycles that have been so important in the development of animal ecology over the last century [55]. Large-scale spatial variation in cycle amplitude and period, including the well described northern hemisphere pattern of cyclic dynamics at high latitudes transitioning to stable dynamics at more southerly latitudes, has been attributed to geographical variation in the degree of seasonality [55]. The recent collapse of many of these population cycles from disparate high latitude localities coincides with a period of accelerated climate warming and reduced seasonality ([56;] Fig. 3.) Thus, the climate conditions responsible for the collapse of population cycles at low latitudes may be spreading north as the climate warms.

2.1.3. The Temporal Scale of Mammal Responses to Climate Change

The nature, extent, and significance of observed temporal responses to climate change depend on the timescale of the comparison [17]. Very short timescale comparisons offer insight into how mammals respond to and are affected by weather, as well as seasonal and annual climate differences. Frequently the

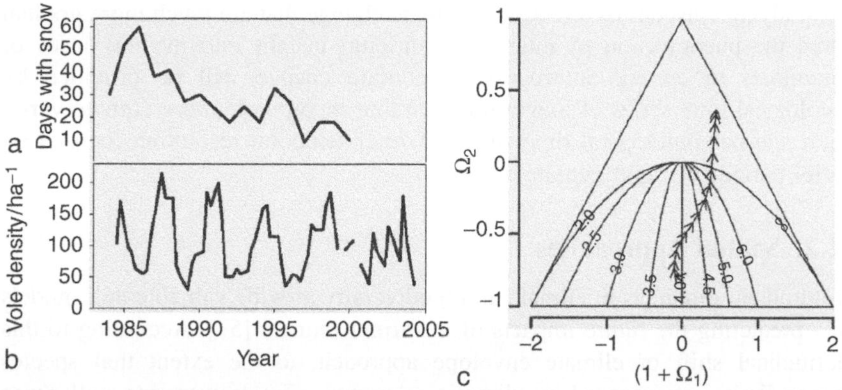

FIGURE 3 Twenty-year time series of declining snow cover duration (a) and disappearance of cyclic dynamics in a population of field voles (*Microtus agrestis*) from Kielder Forest, England (b). Changes over time in the form of direct $(1 + \Omega_1)$ and delayed density dependence (Ω_2) affecting vole population dynamics are shown in (c). Cyclic dynamics are predicted for combinations of the parameters that fall below the semi-circle, with the period of the cycle increasing from left to right. Stable dynamics are predicted for combinations lying above the semi-circle. The line with upward pointing arrows tracks the population from early in the time series, when dynamics are cyclic with a period of ~4 a, to later in the time series when the cycle period extends to >6 a then disappears. Population fluctuations apparent at the end of the time series in (b) are within-year seasonal variations, which appear to increase in magnitude following the transition to non-cyclic dynamics. Redrawn (a, b) and reprinted (c) from Ref. [56]. (Adapted from Am. Nat/University of Chicago Press).

responses detected by these comparisons involve behaviour and other forms of phenotypic plasticity [43]. But these studies reveal less about the likely impacts of longer-term directional climate change, particular when climate change will, by definition, involve conditions outside the contemporary range of variation. On the other hand, very long scales of comparison can obscure the significance of events occurring at shorter timescales. For example, Gingerich [20] argues that the failure of previous studies to detect a relationship between climate and diversity over the entire 65 Ma Cenozoic resulted from averaging of climate and diversity over 1 myr intervals when most significant climate fluctuations and evolutionary responses to them occurred on 1000 a time scales. The consequences of anthropogenic climate change might be similarly obscured if extinction rates during the twentieth and twenty-first century were averaged with extinction rates prevailing, for example, 10 000 a prior to and following these two centuries. Thus, there is a continuum of temporal scales available to study the effects of climate change on mammals, and studies situated at either end of this continuum contribute less to predicting the responses of mammals to current anthropogenic climate change. Short-term ecological studies are confined to studying climate fluctuations of much shorter period and lower amplitude than the phenomenon of interest. Long-term, paleontological studies are, for reasons of temporal resolution and the nature of past environmental change, confined to studying episodes of directional climate change that are much more gradual than the phenomenon of interest. Additional insight into the responses of mammals to current anthropogenic climate change will be provided by ecological time series of longer duration that incorporate more climate variation and paleontological time series of finer temporal resolution focused on brief periods of rapid climate change.

2.2. Spatial Approaches

Latitudinal gradients in climate and biodiversity provide valuable null models for predicting the future impacts of a warmer climate [57]. According to this latitudinal shift or climate envelope approach, to the extent that species range limits are imposed by climatic tolerance, climate warming will cause the plants and animals in a given region to more closely resemble the plants and animals currently found at lower latitudes [58]. The validity of these predictions depends critically on: (1) emission scenarios that correctly predict future greenhouse gas emissions based on current economic and demographic trends, (2) general circulation models (GCM) that correctly predict regional climates based on these emission scenarios and (3) species climate models that correctly predict species' responses to climate change based on a mechanistic understanding of how climate influences their distribution and abundance [59]. Although more research is needed on all fronts, there is enough consensus in emission scenarios and GCM's to render species climate models as far and away the weakest link in biotic climate change predictions. In fact, in many regions of the world, little is known about how much plants and animals actually vary across regional climate gradients (i.e. across hundreds to thousands of kilometres), much less how and why any observed variation across these scales is correlated to climate. In the following sections, I briefly review empirical relationships between spatial climate variation and mammal diversity as well as evidence for the mechanisms by which climate determines species range limits and spatial variation in abundance.

2.2.1. Spatial Climate Variation and Mammal Diversity

Latitudinal gradients in species diversity are among the strongest and most general patterns in ecology [60,61]. With the exception of extremely arid areas, warm regions of the globe host a much greater diversity of mammals (and other organisms) than cold regions. Within North America, measures of local mammal diversity (in this case, the number of terrestrial mammal species occurring in 58 275 km^2 quadrats) varies from 178 species in tropical regions of southern Nicaragua and northern Costa Rica to 20 species in arctic regions of north-central Canada ([62;] Fig. 4). Almost 90% of this variation can be accounted for by five environmental variables, representing seasonal extremes of temperature, annual energy and moisture and elevation [62]. Examples of latitudinal diversity gradients from regions outside of North

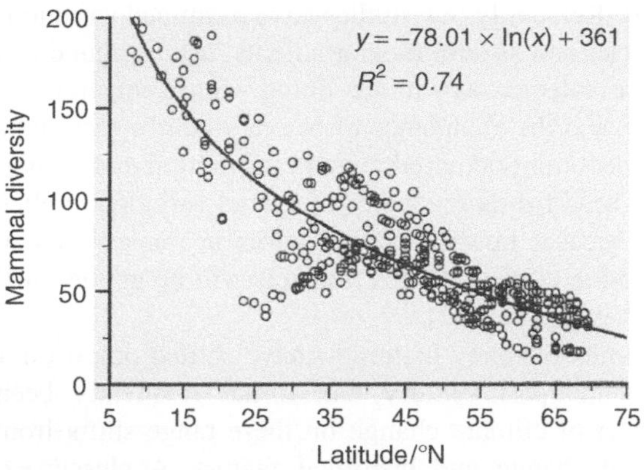

FIGURE 4 Latitudinal variation in the diversity of North American mammals. The diversity measure is the number of terrestrial mammal species present in 58 275 km^2 quadrats systematically distributed across the continent. Five environmental variables, representing seasonal extremes of temperature, annual energy and moisture, and elevation, predicted 88% of continental variation in this variable. From Ref. [62] (Adapted from Journal of Biogeography, Blackwell Science Ltd).

America are dominated by non-mammalian taxa, but the taxonomic and geographic consistency of latitudinal diversity gradients [61], suggests that the distribution of most of the world's mammal species will contribute to a pattern of more species in warmer regions. Climate variables usually outperform landcover variables as predictors of species diversity or distribution [63,64], particularly when scales of comparison are large [65], but it is unclear whether this is because climate is the more important mechanistic driver of diversity or because it is difficult to classify land cover appropriately for multiple species across a variety of landscapes. Regardless, the remarkable generality and strength of climate-diversity correlations across biogeographical space, together with their correspondence with patterns over geological time, has caused them to occupy a central role in approaches to examining and predicting the impacts of climate change on mammals and other animals.

2.2.2. Spatial Climate Variation and the Distribution and Abundance of Mammals

Climatic constraints are thought to impose biogeographic range boundaries on individual wildlife species. The breeding distribution of gray seals (*Halichoerus grypus*) may be limited by the effects of cold air temperatures on the fasting endurance of recently weaned pups [66]. The winter distribution of little brown bats (*Myotis lucifugus*) is limited to latitudes where hibernacula are warm enough to ensure that the energetic costs of hibernation do not exceed maximum autumn fat reserves [67]. The northward range limit of nine-banded armadillos (*Dasypus novemcinctus*) and virginia opossums (*Didelphis virginiana*) seems to be constrained by long bouts of cold winter weather [68,69]. Relatively few studies have examined variation in abundance of mammal species across latitudinal gradients, but the abundance of red foxes in northern Eurasia decreases with declining winter temperature and increased seasonality [70], and the abundance of beaver in north-eastern North America decreases with declining potential evapotranspiration and spring temperatures [71]. There is a need for more studies of spatial variation in abundance across species ranges, because how abundance varies as range limits are approached will dictate whether climate change impacts will be greater at the edge or in the core of species' ranges [71].

The range limits of many mammals have shifted poleward as the climate has warmed in the past century, but it has frequently been difficult to isolate the impact of climate change on these range shifts from other forms of environmental change and historical factors. A classic example is the southern range contraction of arctic foxes concomitant with the northern range expansion of red foxes, which is hypothesised to be mediated by climate driven changes in primary productivity and prey base [72]. This hypothesis is partially supported by recent work comparing ecosystem characteristics of sites abandoned and still occupied by arctic foxes, but there are many site-specific contingencies and alternative explanations [73]. In North America, many of the mammals characterised by prominent poleward range expansions are species that are also affiliated with human-modified habitats, including red foxes, Virginia opossums [73], raccoons [74], white-tailed deer [75] and coyotes [76], making it difficult to identify whether anthropogenic climate change or landscape modification are most responsible for the expansions. However, climate change is clearly contributing to the poleward range shifts of many mammals, as well as other animals and plants, and this will be an important impact on mammals over the next century [38,39,77]. Species of particular concern are those whose poleward range limit is imposed by physical barriers, such as coastlines or mountain ranges, leaving them nowhere to go as their non-poleward range limit moves towards the barrier.

2.2.3. Spatial Climate Variation and Mammal Morphology, Metabolism and Life Histories

All else being equal, mammals occupying warm environments are smaller, metabolically slower and have smaller litter sizes than mammals occupying cold environments. The tendency for animals to be larger in cooler environments, referred to as Bergmann's rule, is well supported by comparisons of populations or species of mammals distributed along latitudinal and climatic gradients [32,33,78]. There are many notable exceptions, but in general, more than two-thirds of mammals conform to this trend. The original explanation for Bergmann's rule, related to heat conservation in cold environments, is no longer widely accepted because the trend is supported by both large and small endotherms, as well as many ectotherms [32,78]. However, mammals occupying cold environments are characterised by increased heat production and reduced heat loss relative to their warm-climate counterparts [79,80]. Classic work in this area was conducted by Scholander comparing the metabolism and pelage insulation of arctic and tropical mammals [81,82], which has since been expanded using phylogenetically-informed analyses on a wider diversity of species [79,80]. An example of the influence of climate variability rather than average climate conditions is provided by Lovegrove's [83] analysis of mammal metabolism showing species inhabiting regions influenced by large scale climate fluctuations have lower resting metabolism than species located outside of these regions. The field metabolic rate of mammals tends to increase with declining air temperature, but also tends to be less variable in cold environments than in warm environments [85]. According to the metabolic niche hypothesis, this pattern suggests that cold environmental temperatures limit diversity by restricting the range of energetically feasible lifestyles [84,86]. Latitudinal and climatic variation in the life history traits of mammals has not been explored as thoroughly as in other vertebrates, but there are theoretical expectations and some empirical evidence for litter size being larger in cold, seasonal environments than warm environments [87–89].

2.2.4. The Spatial Scale of Mammal Responses to Climate Change

The nature, extent, and significance of correlations between mammal ecology and spatial variation in climate depend on the spatial scale of the comparison. At continental and hemispheric scales, climate frequently emerges as powerful predictor of ecological variation in phenomena ranging from diversity, abundance, body size and metabolism. Of course, the term climate captures a wide array of temperature, precipitation and related variables. Which climate variables have the highest predictive power varies according to the region and taxa under consideration, with, unsurprisingly, precipitation and temperature variables performing better in arid and cold regions, respectively. Furthermore, because most climate variables are highly correlated, it is difficult to proceed from correlative to causal models focused only on the best or the top few climate predictors without considering the potential influence of the many covarying climate variables. At a local and regional scale, habitat and other biotic and physical variables emerge as much more powerful predictors of ecological variation than climate. However, it would be erroneous to conclude that climate has not been a major driver of the form and extent of ecological variation observed at a local and regional scale (lack of observed correlation does not, necessarily, imply lack of causation). On the other hand, it would also be erroneous to conclude that climate is the major driver of the form and extent of ecological variation observed at continental and hemispheric scales (correlation does not imply causation). In both cases and at all spatial scales, it should be clear that ecological variation is driven by a complex interaction between climate, physical features, biotic interactions and historical contingencies. But our capacity to quantify and detect those influences varies with spatial scale. Returning to an earlier analogy of climate setting the stage for a complex play involving competitors, resources and predators, the perceived importance of the stage is diminished by watching different plays on the same stage and enhanced by watching the same play on different stages. Instead of sitting in the same theatre and waiting for the stage to change to see if it changes the play, why not visit some different stages where the same play is being performed. Instead of hopping from theatre to theatre and only counting the number of actors on the stage, why not sit down and watch some plays to understand what is happening on the stage. Opportunities to combine watching many different plays on the same stage and the same play on many different stages are rare, but are necessary to assess the relative importance of the stage (climate) and the play (biotic interactions).

3. LINKING TIME AND SPACE IN MAMMAL CLIMATE RESPONSES

Better documentation of the impacts of climate change on mammals awaits integration of temporal and spatial approaches, with careful attention paid to the scales of comparison. I end with a brief review of two approaches, one historical and one contemporary, with considerable promise in this regard.

Phylogeography, as its name implies, integrates spatial and temporal approaches by examining historical influences on geographical distribution. Much of phylogeography is based on genetic approaches, but palaeontology spread across geographical gradients provides similar insights [18]. For example, comparison of the geographical distribution of woolly mammoths (*Mammuthus primigenius*; based on dated fossil discoveries) and reconstructed climates at different time intervals during the Pleistocene provide insight into the role of climate change and human hunting in this species' range contraction and eventual extinction [90]. Genetic estimates of the timing of population and species divergence linked with paleoclimate and habitat reconstructions can provide critical insight into the influence of past environmental change on contemporary diversity. The divergence of Antarctic minke whales (*Balaenoptera bonaerensis*) and common minke whales (*Balaenoptera acutorostrata*) is estimated to have occurred during an extended warming period in the Pliocene, when elevated ocean temperatures would have disrupted the spatial homogeneity of oceanic upwelling and promoted allopatric speciation [91].

A contemporary analog to phylogeography is provided by spatially extensive, long-term monitoring of species diversity, population abundance and individual traits. Unfortunately, the best of these programs, such as breeding and winter bird surveys conducted for many decades over continental spatial scales [92], involve non-mammalian taxa. Nevertheless, these programs can serve as a model for the type and spatial extent of long-term data needed to document the effects of anthropogenic climate change on mammals. They also provide a warning of the difficulty in disentangling the effects of climate change from other forms of environmental change, even with fantastic data [93]. As anthropogenic climate change accelerates, we need to continue and expand the few global examples of spatially-extensive, long-term monitoring of mammal diversity (e.g. [94,95]) and for the many mammal taxa and regions currently excluded, we need to initiate rigorous monitoring programs before it is too late.

ACKNOWLEDGEMENTS

MMH acknowledges Yukon College for providing accommodation while writing the chapter, Christina Guillemette for assistance in compiling literature, Thomas Jung for comments on an earlier draft, and the Natural Sciences and Engineering Research Council of Canada (Northern Research Chairs Program) and the Canadian Networks of Centers of Excellence ArcticNet for financial support.

REFERENCES

1. J.A. Welbergen, S.M., Klose, N., Markus, P. Eby, Proc. R. Soc. B-Biol. Sci. 275 (2008) 419–425.
2. B.K. McNab, The Physiological Ecology of Vertebrates: A View from Energetics, Cornell University Press, Ithaca, New York, 2002, pp. 576.
3. B. Heinrich, Am. Nat. 111 (1977) 623–640.
4. F. Geiser, Ann. Rev. Physiol. 66 (2004) 239–274.
5. C.R. Vispo, G.S. Bakken, Ecology 74 (1993) 377–389.
6. J.F. Gillooly, J.H., Brown, G.B., West, V.M., Savage, E.L. Charnov, Science 293 (2001) 2248–2251.
7. K.S. Chan, A., Mysterud, N.A., Oritsland, T., Severinsen, N.C., Stenseth, Oecologia 145 (2005) 556–563.
8. E. Post, R.O., Peterson, N.C., Stenseth, B.E., McLaren, Nature 401 (1999) 905–907.
9. S.A.R. Mduma, A.R.E. Sinclair, R. Hilborn, J.Anim. Ecol. 68 (1999) 1101–1122.
10. H.P. Huntington, S.E. Moore, Ecol. Appl. 18 (2008) S1–S2.
11. L. Hughes, Austral Ecology 28 (2003) 423–443.
12. S.J. Wright, Trends Ecol. Evol. 20 (2005) 553–560.
13. A.D. Barnosky, B. P. Kraatz, Bioscience 57 (6) (2007) 523–532.
14. V. Millien, S.K., Lyons, L., Olson, F.A., Smith, A.B., Wilson, Yom-Tov, Y. Ecol. Lett. 9 (2006) 853–869.
15. M.C. Forchhammer, N.M. Schmidt, T.T., Hoye, T.B., Berg, D.K., Hendrichsen, E. Post, Population Dynamical Responses to Climate Change, in Advances in Ecological Research, vol. 40, Elsevier Academic Press Inc., San Diego, 2008, pp. 391–419.
16. V. Grosbois, O., Gimenez, J.M., Gaillard, R., Pradel, C., Barbraud, J., Clobert, A.P., Moller, H. Weimerskirch, Biol. Rev. 83 (2008) 357–399.
17. A.D. Barnosky, E.A. Hadly, C.J. Bell, J. Mammal. 84 (2003) 354–368.

18. G.M. MacDonald, K.D., Bennett, S.T., Jackson, L., Parducci, F.A., Smith, J.P., Smol, K.J., Willis, Prog. Phys. Geogr. 32 (2008) 139–172.
19. J.A. van Dam, H.A., Aziz, M.A.A., Sierra, F.J., Hilgen, L.J., Ostende, Lwvdh, Lourens, P., Mein, A.J., van der Meulen, P., Pelaez-Campomanes, Nature 443 (2006) 687–691.
20. P.D. Gingerich, Trends Ecol. Evol. 21 (2006) 246–253.
21. B. Shapiro, A.J., Drummond, A., Rambaut, M.C., Wilson, P.E., Matheus, A.V., Sher, O.G., Pybus, M.T.P., Gilbert, I., Barnes, J., Binladen, E., Willerslev, A.J., Hansen, G.F., Baryshnikov, J.A., Burns, S., Davydov, J.C., Driver, D.G., Froese, C.R., Harington, G., Keddie, P., Kosintsev, M.L., Kunz, L.D., Martin, R.O., Stephenson, J., Storer, R., Tedford, S., Zimov, A., Cooper, Science 306 (2004) 1561–1565.
22. Regis Debruyne, Genevieve Chu, Christine E. King, Kirsti Bos, Melanie Kuch, Carsten Schwarz, Paul Szpak, Darren R. Gröcke, Paul Matheus, Grant Zazula, Dale Guthrie, Duane Froese, Bernard Buigues, Christian de Marliave, Clare Flemming, Debi Poinar, Daniel Fisher, John Southon, Alexei N. Tikhonov, Ross D.E. MacPhee, Hendrik N. Poinar., Curr. Biol. 18 (2008) 1320–1326.
23. D.A. Burney, T.F. Flannery, Trends Ecol. Evol. 20 (2005) 395–401.
24. E. Martinez-Meyer, A. Townsend Peterson, W.W. Hargrove, Global Ecol. Biogeogr. 13 (2004) 305–314.
25. S. Legendre, S., Montuire, O., Maridet, G., Escarguel, Earth Planet. Sci. Lett. 235 (2005) 408–420.
26. M.H. Fernandez, P. Pelaez-Campomanes, Global Ecol. Biogeogr. 14 (2005) 39–56.
27. W. Buermann, B.R., Lintner, C.D., Koven, A., Angert, J.E., Pinzon, C.J., Tucker, I.Y., Fung, Proc. Natl. Acad. Sci. USA 104 (2007) 4249–4254.
28. J. Esper, U., Buntgen, D.C., Frank, D., Nievergelt, A., Liebhold, Proc. R. Soc. B – Biol. Sci. 2007 (2007) 671–679.
29. M.S.V. Douglas, J.P. Smol, W. Blake, Science 266 (1994) 416–419.
30. N.C. Stenseth, A. Mysterud, C. Ottersen, J.W. Hurrell, K.S. Chan, M. Lima, Science 297 (2002) 1292–1296.
31. T.B. Hallett, T., Coulson, J.G., Pilkington, T.H., Clutton-Brock, J.M., Pemberton, B.T., Grenfell, Nature 430 (2004) 71–75.
32. K.G. Ashton, M.C. Tracy, A. de Queiroz, Am. Nat. 156 (2000) 390–415.
33. T.M. Blackburn, B.A. Hawkins, Ecography 27 (2004) 715–724.
34. F.A. Smith, H. Browning, U.L. Shepherd, Ecography 21 (1998) 140–148.
35. F.A. Smith, J.L. Betancourt, J.H. Brown, Science 270 (1995) 2012–2014.
36. Y. Yom-Tov, S. Yom-Tov, Biol. J. Linn. Soc. 82 (2004) 263–267.
37. Y. Yom-Tov, J. Yom-Tov, J. Anim. Ecol. 74 (2005) 803–808.
38. G.R. Walther, E., Post, P., Convey, A., Menzel, C., Parmesan, T.J.C., Beebee, J.M., Fromentin, O., Hoegh-Guldberg, F., Bairlein, Nature 416 (2002) 389–395.
39. C. Parmesan, Ann. Rev. Ecol. Evol. Syst. 37 (2006) 637–669.
40. D.W. Inouye, B. Barr, K.B. Armitage, B.D. Inouye, Proc. Natl. Acad. Sci. USA 97 (2000) 1630–1633.
41. N. Pettorelli Pettorelli, N., Pelletier, F., von Hardenberg, A., Festa-Bianchet, M., Cote, S. D., Ecology 88 (2007) 381–390.
42. D. Reale, A.G., McAdam, S. Boutin, D., Berteaux, Proc. R. Soc. Lond. Ser. B – Biol. Sci. 270 (200) 591–596.
43. D. Berteaux, M.M., Humphries, C.J., Krebs, M., Lima, A.G., McAdam, N., Pettorelli, D., Reale, T., Saitoh, E., Tkadlec, R.B., Weladji, N.C., Stenseth, Int. Comp. Biol. 44 (2004) 140–151.

44. A.E. Derocher, N.J. Lunn, I. Stirling, Int. Comp. Biol. 44 (2004) 163–176.
45. E.V. Regehr, N.J., Lunn, S.C., Amstrup, L., Stirling, J. Wildlife Manage. 71 (2007) 2673–2683.
46. COSEWIC, COSEWIC Assessment and Update Status Report on the Polar Bear Ursus Maritimus in Canada, Committee on the Status of Endangered Wildlife in Canada: Ottawa, 2008, p. 75.
47. M.C. Forchhammer, E., Post, N.C., Stenseth, D.M., Boertmann, Popul. Ecol. 44 (2002) 113–120.
48. J. Hone, T.H. Clutton-Brock, J. Anim. Ecol. 76 (2007) 361–367.
49. V. Grotan, B.E., Saether, F., Filli, S., Engen, Global Change Biol. 14 (2008) 218–228.
50. T. Saitoh, B., Cazelles, J.O., Vik, H., Viljugrein, N.C., Stenseth, Clim. Res. 32 (2006) 109–118.
51. N.C. Stenseth, D., Ehrich, E.K., Rueness, O.C., Lingjaerde, K.S., Chan, S., Boutin, M., O'Donoghue, D.A., Robinson, H., Viljugrein, K.S., Jakobsen, Proc. Natl. Acad. Sci. USA 101 (2004) 6056–6061.
52. S.F. Morrison, D.S. Hik, J.Anim. Ecol. 76 (2007) 899–907.
53. M. Lima, N.C. Stenseth, F.M. Jaksic, Proc. R. Soc. Lond. Ser. B – Biol. Sci. 269 (2002) 2579–2586.
54. J.O. Ogutu, N. Owen-Smith, Ecol. Lett. 6 (2003) 412–419.
55. R.A. Ims, J.A. Henden, S.T. Killengreen, Trends Ecol. Evol. 23 (2008) 79–86.
56. S.M. Bierman, J.P., Fairbairn, S.J., Petty, D.A., Elston, D., Tidhar, X., Lambin, Am. Nat. 167 (2006) 583–590.
57. D.J. Currie, Ecosystems 4 (2001) 216–225.
58. M.B. Araujo, R.J., Whittaker, R.J., Ladle, M., Erhard, Global Ecol. Biogeogr. 14 (2005) 529–538.
59. R.K. Heikkinen, M., Luoto, M.B., Araujo, R., Virkkala, W., Thuiller, M.T., Sykes, Prog. Phys. Geogr. 30 (2006) 751–777.
60. D.J. Currie, Am. Nat. 137 (1991) 27–49.
61. K.J. Gaston, Nature 405 (2000) 220–227.
62. C. Badgley, D.L. Fox, J. Biogeogr. 27 (2000) 1437–1467.
63. W. Thuiller, M.B. Araujo, S. Lavorel, J. Biogeogr. 31 (2004) 353–361.
64. B. Huntley, R.E., Green, Y.C., Collingham, J.K., Hill, S.G., Willis, P. J., Bartlein, W., Cramer, W.J.M., Hagemeijer, C.J., Thomas, Ecol. Lett. 7 (2004) 417–426.
65. W. Thuiller, L., Brotons, M.B., Araujo, S., Lavorel, Ecography 27 (2004) 165–172.
66. S. Hansen, D.M. Lavigne, Physiol. Zool. 70 (1997) 436–443.
67. M.M. Humphries, D.W. Thomas, J.R. Speakman, Nature 418 (2002) 313–316.
68. L.L. Kanda, Ecography 28 (2005) 731–744.
69. J.F. Taulman, L.W. Robbins, J. Biogeogr. 23 (1996) 635–648.
70. K.A. Barton, A. Zalewski, Global Ecol. Biogeogr. 16 (2007) 281–289.
71. S.I. Jarema, J., Samson, B.J., McGill, M.M., Humphries, Global Change Biol. 15 (2009) 508–522.
72. P. Hersteinsson, D.W. Macdonald, Oikos 64 (1992) 505–515.
73. S.T. Killengreen, R.A., Ims, N.G., Yoccoz, K.A., Brathen, J.A., Henden, T., Schott, Biol. Conserv. 135 (2007) 459–472.
74. S. Lariviere, Wildlife Soc. Bull. 32 (2004) 955–963.
75. A.M. Veitch, Can. Field-Nat. 115 (2001) 172–175.
76. H.D. Cluff, Can. Field-Nat. 120 (2006) 67–70.

77. T.L. Root, J.T., Price, K.R., Hall, S.H., Schneider, C., Rosenzweig, J.A., Pounds, Nature 421 (2003) 57–60.
78. S. Meiri, T. Dayan, J. Biogeogr. 30 (2003) 331–351.
79. B.G. Lovegrove, J. Comp. Physiol. B – Biochem. Syst. Environ. Physiol. 173 (2003) 87–112.
80. B.G. Lovegrove, J. Comp. Physiol. B – Biochem. Syst. Environ. Physiol. 175 (2005) 231–247.
81. P.F. Scholander, V., Walters, R., Hock, L., Irving, Biol. Bull. 99 (1950) 225–236.
82. P.F. Scholander, R., Hock, V., Walters, L., Irving, Biol. Bull. 99 (1950) 259–271.
83. B.G. Lovegrove, Am. Nat. 156 (2000) 201–219.
84. K.J. Anderson, W. Jetz, Ecol. Lett. 8 (2005) 310–318.
85. M.M. Humphries, S. Boutin, D.W. Thomas, J.D. Ryan, C. Selman, A.G. McAdam, D. Berteaux, J.R. Speakman, Ecol. Lett. 8 (2005) 1326–1333.
86. A. Clarke, K.J. Gaston, Proc. R. Soc. B 273 (2006) 2257–2266.
87. R.D. Lord Jr., Am. Midland Nat. 64 (1960) 488–499.
88. M.S. Boyce, Am. Nat. 114 (1979) 569–583.
89. J.R. Whorley, G.J. Kenagy, J. Mammal. 88 (2007) 1404–1411.
90. D. Nogues-Bravo, J., Rodiguez, J., Hortal, P., Batra, M.B., Araujo, Plos Biol. 6 (2008) 685–692.
91. L.A. Pastene, M., Goto, N., Kanda, A.N., Zerbini, D., Kerem, K., Watanabe, Y., Bessho, M., Hasegawa, R., Nielsen, F., Larsen, P.J., Palsboll, Mol. Ecol. 16 (2007) 1481–1495.
92. A.H. Hurlbert, E.P. White, Ecol. Lett. 8 (2005) 319–327.
93. C. Rahbek, G.R. Graves, Proc. Natl. Acad. Sci. USA 98 (2001) 4534–4539.
94. J. Pellikka, et al., Annales Fennici Zoologici 42 (2005) 123–134.
95. S. Harris, D.W. Yalden, Mammal Rev. 34 (2004) 157–167.

Chapter 11

Climate Change and Temporal and Spatial Mismatches in Insect Communities

Shannon L. Pelini, Kirsten M. Prior, Derrick J. Parker, Jason D.K. Dzurisin, Richard L. Lindroth* and Jessica J. Hellmann

Department of Biological Sciences, University of Notre Dame, Notre Dame, Indiana 46556

**Department of Entomology, University of Wisconsin, Madison, Wisconsin*

1. **Introduction**
2. **Direct Effects of Climate Change on Insects**
   - 2.1. Temporal Changes
   - 2.2. Spatial Changes
   - 2.3. Genetic and Phenotypic Changes
3. **Host Plant-Mediated Effects on Insects**
   - 3.1. Temporal Mismatches Between Insects and Their Host Plants
   - 3.2. Spatial Mismatches Between Insects and Their Host Plants
4. **Predator-Mediated Effects on Insect Populations**
   - 4.1. Temporal Mismatches Between Insects and Their Predators
   - 4.2. Spatial Mismatches Between Insects and Their Predators
5. **Climate Change and Insect Pests**
6. **Conclusion**

Acknowledgements

References

1. INTRODUCTION

As ectotherms, insect performance is heavily dependent on climate. Warming from climate change will alter insect development time, voltinism, foraging behaviour, emergence time and survivorship [1]. These changes, which alter population size and distribution, will affect the temporal and spatial dynamics of insect communities. Many insects provide important ecosystem services (e.g. pollination, decomposition, etc.) or affect human activities (e.g. via pest activity) and the effects of climate change could alter these services or exacerbate these effects. To date, geographic range shifts and early emergence are the best documented responses of insects to climate change, but some species have experienced population increases, changes in cyclical population dynamics and local extinctions or genetic and phenotypic changes that may be examples of rapid evolution. Further, more research is needed to understand how species-level changes affect interacting species. Overall, the sensitivity of insects to climatic factors makes them ideal for tracking and understanding the effects of climate change on biodiversity.

This chapter provides an overview of the effects of climate change on insects and their interactions with host plants and predators. We provide examples from the literature using herbivorous insects and their host plants and predators because herbivores are a large, well-studied group of insects that affect human activities and provide important ecosystem services. While the effects of increasing carbon dioxide levels are also important, particularly the resulting changes in host plant chemistry [2,3], the focus of this review is on the effects of warming and increased variability in precipitation associated with climate change. We also offer a novel analysis of northern range limit comparisons between North American butterflies and their host plant species to delineate the potential for spatial disassociations (i.e. differential range shifts between herbivorous insects and their host plants under climate change). Finally, we discuss potential impacts on future communities of insects and altered community effects on ecosystem services.

Life history traits such as resource specificity, geographic location, trophic level and dispersal ability are potentially good predictors of the magnitude and direction of the response of insect species to climate change. For example, habitat and/or resource specificity may limit the tolerance of specialists to changing conditions, potentially leaving habitats dominated by generalists. Warren et al. [4] found that butterflies with strong habitat specificity and limited mobility have reduced distributions and have fared worse under changing climatic conditions than generalists that share the same geographic range. Another study on beetles identified climate change response groups based on host plant specificity and distribution size, and the authors project that cosmopolitan species may be the most resilient to climate change while specialists will be faced with extinction if they do not move with their host plants [5]. Furthermore, Deutsch et al. [6] concluded that tropical species that have narrow thermal limits are more likely to be negatively affected by climate change. Yet Bale et al. [1] also highlight other vulnerable groups: insects that are cold-adapted, restricted to montane areas and also those in polar areas, where proportionally larger temperature increases are expected. In addition, Voigt et al. [7] report that species in higher trophic levels are sensitive to climatic change due to the combined indirect effects of climate change on lower trophic groups. When possible, we relate our article back to life history traits to allow for generalisations because insects as a taxonomic group are too diverse to be studied exhaustively. Our only hope for understanding insect responses to climate change is through generalisations from well-studied species to others.

Climate change can decouple interactions between insects and their resources and/or predators. Variable responses of these interacting species to climate change can lead to differential changes in the geographic ranges of species (herein 'spatial mismatches') as well as differential changes in the phenology, or timing, of species ('temporal mismatches'). These mismatches could lead to places where and times when insect populations could greatly decline towards local extinction. In this review, we identify studies that demonstrate such mismatches, focusing, where possible, on those that have yielded negative effects on ecosystem services provided by insects.

2. DIRECT EFFECTS OF CLIMATE CHANGE ON INSECTS

The impact of climate change on insects is multifaceted. We begin by reviewing the direct effects of climate change on the temporal and spatial dynamics of insect populations. Thus far, enhanced individual and population growth in warmer temperatures have resulted in early emergence and changes in the location, through geographic range shifts, of insect species. A few studies also have linked climate change to genetic and phenotypic changes in insect populations, which we discuss here, but this response is not anticipated due to a number of constraints in evolution under human-caused, rapid climate change.

2.1. Temporal Changes

Insect life cycles depend on climatic variables such as degree days, minimum winter temperatures, average maximum summer temperatures, total precipitation and aridity. Generally, higher temperatures result in faster development, increased number of generations and increased overwinter survivorship [1]. As a result of enhanced and accelerated growth, climate change has been linked to phenological advances in many insects [8,9]. Gordo and Sanz [10] found earlier spring emergence in honey bee *Apis mellifera* (L.) and small white *Pieris rapae* (L.) populations in the Iberian Peninsula over the past 50 a. Several other studies of butterflies have documented early emergence [11–14], and it also has been observed in aphids [15,16] and in members of Heteroptera [17]. Decreased generation time due to warming has been observed in the mountain pine beetle (*Dendroctonus ponderosae*), leading to increases in the abundance of this pest species [18]. In some cases, the direct effect of increased temperature on insect development leads to simple population increases (e.g., *D. ponderosae*). In other cases, however, it can change synchrony of herbivores, host plants and predators, producing more complex community effects.

2.2. Spatial Changes

Increased overwinter survivorship, growth rates and generations have led to range shifts in many insect species. Many insects, including some pest species (e.g. the pine processionary moth (*Thaumetopoea pityocampa*)), have shifted their distributions poleward and to higher elevations to track recent climatic changes [19–27]. Still others have been unable to track the changing climate, and some have experienced range contractions or local extinctions [18,28–30]. These results and others have helped to establish a simple paradigm for the responses of species to climate change: populations will contract and go extinct in the equatorial portion of the range and poleward populations will expand and colonise new locations as the climate warms.

The ability of a species to shift its distribution is determined largely by its ability to disperse into newly suitable areas and the availability of suitable resources, that is breeding habitat and host plants, in those sites [4,31]. Thomas et al. [22] examined four insect species that recently expanded their ranges northward and found that two butterfly species increased the range of habitat types that they historically used while newly established populations of two bush cricket species had more longer-winged individuals than their source populations, suggesting that only the best dispersers were able to reach new habitat. Similarly, Warren et al. [4] have argued that climate change will leave habitats dominated by mobile generalists. Later, we discuss how limitations in resource use may lead to spatial mismatches between herbivorous insects and their host plants.

2.3. Genetic and Phenotypic Changes

Climate change has altered the selection pressures on insect populations, and some insects have responded via genetic and phenotypic change. Several groups have found that *Drosophila* populations are tracking climate change through genetic changes, for example, with genotypes characteristic of equatorial latitudes increasing in frequency with warming over the past few decades [32–34]. Rodríguez-Trelles and Rodríguez [35] found a decrease in diversity of a chromosome polymorphism in *Drosophila*, and they argue that this alteration correlates with climatic change. Rank and Dahlhoff [36] found allele frequency shifts in an enzyme related to heat stress in the leaf beetle (*Chrysomela aeneicollis*) that they linked to climatic changes occurring in the Sierra Nevada during the 1990s.

Phenotypic changes in insects also have been linked to recent climate change. For example, Bradshaw and Holzapfel [37] found that northern populations of the pitcher plant mosquito (*Wyeomyia smithii*) have shifted their critical diapause photoperiod towards that of their southern counterparts over the past 24 a. de Jong and Brakefield [38] found changes in melanism clines in the two-spot ladybird (*Adalia bipunctata*) that they also linked to recent climatic change. Such rapid evolutionary response under climate change, however, may be rare since high genetic diversity is required for adaptation to occur and co-variation among traits can slow the process [39,40]. Gienapp et al. [41] also caution that other studies claiming evidence of microevolution under climate change fail to separate genetic change from phenotypic plasticity.

Now that we have examined the direct effects of climate change on insect populations, we will explore the indirect effects of climate change. These indirect mechanisms of change can lead to spatial and temporal mismatches between insects and their food or predators. These mismatches ultimately affect the ecosystem services that we gain or lose due to insects.

3. HOST PLANT-MEDIATED EFFECTS ON INSECTS

The persistence of herbivorous insects depends highly on interactions with plants. Host plants impact herbivore populations directly through phenological and nutritional conditions [42–46]. For example, Nzekwu and Akingbohungre [47] showed that utilisation of different types of host plants has significant effects on insect development. Previous studies on butterflies also have shown that larvae need long-lasting host plants with the proper phenology to be able to withstand environmental changes [48]. Any variation in resources may influence the dynamics and abundances of herbivores [49]. Mismatches between herbivorous insects and their food plants are caused by differential responses in the two groups; for example, earlier emergence and range shifts in insects can occur at a faster rate than those of plants. These mismatches can lead to declines in insect populations. For those species that affect processes such as pollination, this could lead to a reduction in ecosystem services.

3.1. Temporal Mismatches Between Insects and Their Host Plants

The indirect effects of climate change on insects are largely occurring because of altered host plant phenologies and quality [50–53]. Reduced host plant quality results in increased mortality because insect larvae compensate for the decrease in nutritional value by increasing consumption and development time, which in turn increases their exposure to predators and other environmental stresses [2,54]. In addition, species that feed on ephemeral resources are more likely to be sensitive (e.g. increased population variability) to asynchronies with host plants [55]. For example, insects whose eggs hatch before bud burst in their host plants will likely starve while those that require young foliage that hatch late will be forced to eat leaves that are more heavily defended [56].

Experiments have shown that temperature increases are altering insect development proportionately more than that of their host plants [57,58]. In fact, asynchrony has already occurred between the winter moth (*Operophtera brumata*) and its host plant, pedunculate oak (*Quercus robur*) because *O. brumata* egg hatch has advanced more so than bud burst in *Q. robur* [59]. In this particular case, however, the authors speculate that high levels of genetic variation in *O. brumata* may allow rapid adjustment to phenology of *Q. robur* over time. Such information about the plasticity of species is lacking for most taxa; therefore, it is difficult to predict if this will be a general response (see Visser [60] for a discussion of adaptation to climate change).

Insects that provide important pollination services, for example, butterflies, are particularly vulnerable to temporal mismatches with their food resources. On average, butterfly species are advancing faster than herbs [61]. For example, in Britain, the migratory red admiral butterfly (*Vanessa atalanta*) has advanced its return flight over the past couple of decades, but its host plant, stinging nettle (*Urtica dioica*), has not advanced its flowering time, creating a mismatch in phenology [62]. In another study, McLaughlin et al. [28] linked increased variability in precipitation associated with climate change to extinctions in two populations of the Bay checkerspot butterfly (*Euphydryas editha bayensis*). This variability caused extinctions by accelerating plant senescence relative to larval development [28,48]. Simulations by Memmott et al. [63] concluded that 17–50% of pollinators, including insects, will suffer a disassociation with their food, and, as with other cases, small diet breadth was a greater risk factor for asynchrony with food. Later, we will discuss how other species that negatively affect human activities are benefiting from temporal mismatches with predators.

3.2. Spatial Mismatches Between Insects and Their Host Plants

In many insect systems, the dispersal ability of the insect is greater than its host plant. Such differential dispersal capacity of specialist herbivores, if they are to remain host plant-limited, could cause range contractions. Many insects have the ability to move long distances. For example, some butterflies in Europe have shifted their distributions northward 240 km in the past 30 a [19]. Recent shifts are not as well documented for plants, but post-glacial range expansions with past climate change were up to 100 km over 100 a with a median of only 20–40 km per century [64]. This is not surprising since plants, especially large trees, typically have long generation times and lower recruitment than insects [65]. These differences between plants and insects are further complicated by habitat fragmentation, which may be especially restrictive for plant migrations because of their limited dispersal ability [66–68].

To illustrate the potential impact of spatial mismatches under climate change, we analysed the relationship of specialist butterflies reaching a northern geographic limit in the United States with that of their primary host plants. A total of 74 butterfly species from 15 subfamilies have both a northern range boundary within the United States and utilise a single host species. Of these, 59 species have county-level distribution records for both butterfly and host plant.

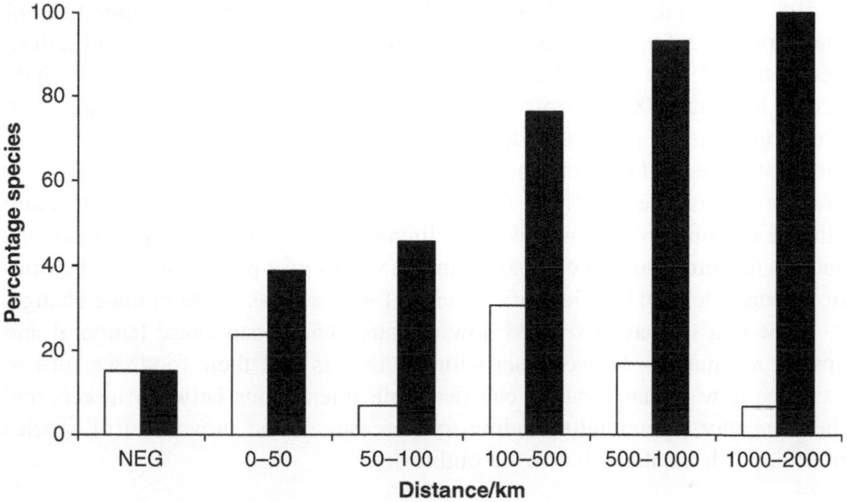

FIGURE 1 Categorisation of spatial distances between the northern-most geographic distributions of host-specific butterflies and their larval host plant in North America. Distances based on county-level centroid locations of observational butterfly and vegetation records (ET SpatialTechniques, ArcGIS, Version 9.0) [71]. White bars are totals within each category and black bars represent the additive totals across categories.

Geographic centroids were calculated (ET SpatialTechniques, ArcGIS, Version 9.0, ESRI) for the northern-most county within each species' range [69] and that of its host plant [70]. Spatial discontinuity distances were calculated by subtracting the centroid latitude of the host plant from that of the butterfly species (Fig. 1).

Of the 59 specialist species assessed, 46% show a northern-most geographic range distribution within 100 km, 76% within 500 km, 93% within 1000 km and 7% extend more than 1000 km beyond the range edge of their host (Fig. 1). Of these, ten species (17%) reach their current range limit in the same location as their host. Furthermore, nine species (15%) show distributions extending beyond their current food plant range margin (negative values in Fig. 1). This is likely a result of sampling error arising from the scale of the observational data used, broad longitudinal distributions leading to multiple northern range fronts, inaccurate assessments of range boundaries due to occasional migrants into sink habitats or undocumented range extensions of species.

Spatial differences between butterfly and host range margins will determine how far butterfly ranges can expand under climate change, assuming relatively static host plant margins and consistent host preference for specialist herbivores. Where the geographic distance between the boundary of the butterfly and its host is small, there may be a small distance available for butterfly range expansion. Given the extent of range change seen in European butterflies (upwards of 240 km over 30 a) [19], we would estimate that up to 59% of specialist butterflies in North America will reach the range limit of their host by 2050, assuming no change in host plant distribution and unlimited dispersal. For example, the Florida Leafwing, *Anaea floridalis*, which occurs less than 100 km from the edge of its host distribution, will approach host boundaries in just over a decade, as will other insect species. Up to 4% of the total butterfly fauna of North America could experience contracting geographic ranges where limiting food resource distributions interact with future climate warming to prevent adequate climate tracking. Simple risk assessments such as this procedure of comparing insect versus host plant poleward distribution boundaries can help identify species of greatest risk under climate change.

Now that we have explored how climate change can cause temporal and spatial mismatches between herbivorous insects and their food, we turn to examine how climate change can decouple interactions between insects and their predators, potentially leading to increases in and movement of species that diminish habitats, that is via outbreaks.

4. PREDATOR-MEDIATED EFFECTS ON INSECT POPULATIONS

Few studies have considered the influence of climate change on higher trophic levels (i.e. predators, parasitoids and pathogens) and its subsequent effects on insect herbivore populations. Climate change is expected to affect the relationship between predators and their prey by independently altering the dynamics of each trophic group [71]. In addition, higher trophic levels have been found to be more sensitive to climate change, likely due to the combined effects of climate change on lower trophic levels [7] and greater sensitivity to abiotic stress among higher trophic levels [72–74]. Changes in top trophic levels can cause trophic cascades. The loss of predators, therefore, can disrupt interactions between lower trophic levels including herbivores and primary producers [75,76].

Predators, particularly parasitoids, play an important role in controlling insect populations [77–82]. In some cases, they have been found to cause cyclical dynamics of herbivorous insects ([82,83]; but see Refs. [80,84] for exceptions). Recent studies of forest insects have linked population outbreaks to climate change as a result of phenological changes in insect life histories [85–87]. The role of higher trophic levels in causing insect outbreaks under climate change, however, has only started to be addressed [88]. We review the current observations and evidence to date of climate change causing temporal and spatial mismatches in predator–prey relationships and its subsequent effects on herbivore outbreaks.

4.1. Temporal Mismatches Between Insects and Their Predators

Herbivore development often changes rapidly and dramatically in response to changing climatic conditions, and this leads to large temporal variation in their populations. Parasitoids also may be affected by both climate variation and its effect on host dynamics. For example, a study by Stireman et al. [88] used data from multi-year inventories of caterpillars reared from natural ecosystems to relate parasitism frequency to interannual climate variability. They found that parasitism frequency was negatively related to variability in precipitation. This suggests that there will be increased outbreaks of herbivores occurring under climate change as a result of increased climate variability causing mismatches in parasitoid–herbivore interactions.

Temporal mismatches could occur if predators and their insect herbivores respond differentially to changes in climate. Hosts and parasitoids may have different thermal preferences or different capacities to survive extreme temperatures and differential responses to climate could disrupt synchronisation [89]. For example, there is evidence that parasitoids often have lower temperature tolerances than their hosts [90, and references therein]. Models predict that increasing phenological asynchrony between predators and prey due to differential responses to climate can cause destabilisation of their population dynamics and extinction of the system [91,92].

Few studies, however, have empirically shown differential responses of predators and prey to changes in climate. Van Nouhuys and Lei [93] found that the light-coloured parasitoid *Cotesia melitaearum*, which is restricted to shaded habitats, develops slowly, while the dark-coloured larvae of its butterfly host, *Melitaea cinxia*, seeks out warm microhabitats where the sun increases body temperature, allowing individuals to develop faster. In cool but sunny springs, therefore, parasitoids emerged after most of their larval hosts had already pupated. In warmer springs, the synchrony between host and parasitoid was maintained and parasitism rate was high. In this study, the authors detected no effect of synchrony on local host population size, but disruption of synchrony is likely important for overall host metapopulation dynamics.

There is evidence of increasing sensitivity to climatic variation and other forms of abiotic stress in higher trophic levels [7,73,74,94]. This could be due to combined effects of climate variation on lower trophic levels [7]. Menge and Sutherland's [72] theory of community regulation predicts that higher trophic levels are more sensitive to abiotic stress as organisms in high trophic levels are likely to be larger and more mobile, enabling them to disperse from suboptimal areas. Species in higher trophic levels have been found to have slower recovery rates after catastrophic events [95,96], for example. In addition, Spiller and Schoener [94] found reductions in predators after two hurricanes in 1999 and 2001 on ten small Caribbean islands that likely explained observed increases in herbivory. Preisser and Strong [74] also observed mass mortality of lupine, *Lupinus arboreus*, off of the coast of California after 2 years of below average precipitation that was associated with El Niño events. Lupine die-offs were due to an outbreak of a root herbivore, *Hepialus californicus*. *H. californicus* populations are regulated by the nematode *Heterorhabditis marelatus* [97], which is dependent on soil moisture for movement. In their study, Preisser and Strong experimentally increased soil moisture content to levels of wet El Niño years (above average precipitation levels) and found that increases in moisture levels directly affected the predatory nematode but not lupine or the ghost moth. Increases in the predator suppressed *H. californicus,* indirectly protecting lupines. Climate change is expected to increase such extreme events [98].

4.2. Spatial Mismatches Between Insects and Their Predators

Species are shifting their ranges in response to climate change, but as discussed above for the movement of herbivores and their plants, insect herbivores and their predators can shift at different rates. Predators that rely on close associations with their hosts such as parasites and parasitoids are expected to be especially affected by changes in the spatial distribution of their hosts. Colonising hosts could lack parasites or parasitoids and/or hosts that are infected could be less fit and therefore less successful in establishing in new locales [99]. In addition, native predators in newly colonised areas could switch from native hosts; however, some native predators may need to evolve phenological, behavioural or ecological specialisations before they can affect the colonising species. Predator species complexes, therefore, may be less rich with lower attack rates than those in the native range [100–102]. Reductions in parasitoids or other predators at the edges of expanding ranges of herbivorous insects could give expanding species an advantage (e.g. increases in population size) through release from predator control.

Only one study to date has investigated differential rates of spread of insect herbivores and their parasitoids under climate change. Menéndez et al. [103] found that the Brown Argus butterfly (*Aricia agestis*), a species that has shifted its range northward due to climate warming, experienced lower mortality from parasitoids in newly colonised areas. Other studies of invasive insects that have been introduced into new areas or of insects that have expanded their ranges as a result of their host plants being introduced also have found lower parasitoid rates in newly colonised areas [100,104]. Reductions in enemies of invasive species in their invaded ranges have been documented for a number of other plant and animal species as well (e.g. [105,106]).

Menéndez et al. [103] found no difference in parasitoid richness in the expanded and native ranges of *A. agestis*; they did, however, find a difference in parasitoid species composition. The majority of species attacking *A. agestis* in its newly colonised range were generalists that were already present in that area attacking *Polyommatus icarus,* a native butterfly. Other studies have found that specialist parasitoids were lost when insect hosts expanded their ranges and that the majority of species attacking hosts in their invaded range were generalists that switched from alternative native hosts [100,104,107].

Specialist parasitoids that follow hosts are likely to be more efficient predators than generalist parasitoids that have switched from native hosts. Although more generalist parasitoids attacked *A. agestis* in its expanded range, the most abundant parasitoid was one that is believed not to have been present historically and therefore is likely a specialist that expanded its range with *A. agestis* [103]. Other studies of invading hosts have found similar results- that the most abundant predator in the invaded range are specialist predators that followed from the native range [106].

Generalist species in the expanded range could be less effective predators on new hosts that have undergone a range expansion. For example, while parasitoid species richness was similar in the native and expanded range of *A. agestis*, attack rates were lower for *A. agestis* compared to the native host *P. icarus* [103]. Similar results were found for a recent invasion of the variegated leafhopper (*Erythroneura variabilis*) into California's San Joaquin Valley. *E. variabilis* experienced lower attack rates from a shared parasitoid than the native grape leafhopper, *E. elegantula* [108]. Therefore, there is evidence that native parasitoids are slow to shift to new hosts because they are locally adapted to native hosts.

Over time, assemblages of predators and hosts in the invaded range would be expected to become similar to assemblages in the native range as enemies catch up to expanded host distributions. Most studies have found an increase in predator richness over time in the hosts' invaded or expanded range, however, resulting in different species composition than in the hosts' native range. Schonrogge et al. [104] found that for several cynipid wasps that were expanding their ranges, parasitoid assemblages in the introduced range were determined more by the parasitoid assemblages attacking existing cynipid species than by the parasitoids in their previous ranges.

No study to date has linked reduced predation after a host range shift to increases in populations of species that have undergone an expansion. Experimental manipulation of predators of species undergoing range expansions in different portions of their ranges is needed to link reductions in parasitoids in expanding ranges to changes in population dynamics. The invasion biology literature provides frameworks to test a reduction in enemies to host release in expanded ranges [99,109,110]. However, no tests to date have been conducted for insects expanding their range in response to climate change. Since many insects are important pests, we now discuss how climate change could favour native and non-native pests that impact important human activities and ecosystem services, including loss of forests and reduction in carbon sequestration.

5. CLIMATE CHANGE AND INSECT PESTS

Traits that enable a species to respond favourably to climate change are those very traits that pest species (native and non-native) often possess. For example, pests often have wide diet breadth, are multi-voltine, have rapid growth rates, are highly mobile and/or are phenotypically plastic [111–113]. Since many pest species have these traits, they are likely to respond favourably to climate change, perhaps more so than those that are specialised and/or rare.

Chown et al. [114], for example, found that invasive springtails were more phenotypically plastic in their ability to cope with desiccation under warming conditions compared to indigenous springtails.

Insect pests affect many human activities and ecosystem services. Many insect species are already damaging, and some may become more so under climate change. Insect pests are the most important agents of disturbance in North American forests, affecting areas almost 50 times larger than fire [18,115,116] with an estimated average annual economic cost of 2.1 billion dollars in the United States [117]. Many insect pests are non-native; up to 40% of major insect pest species in North America are invasive [118] with 62% of 29 major insect forestry pests being non-native [117]. Insect pests, both native and non-native, not only have economic impacts, but also have significant ecological impacts. For example, the invasive balsam woolly adelgid (*Adelges piceae*) has caused extensive death in relict stands of Fraser fir (*Abies fraseri*) and threatens other native species that depend on *A. fraseri* [119].

Many pest species have expanded their ranges as a result of climate change. The pine processionary moth (*T. pityocampa*), a pest of *Pinus* spp. and other conifers in southern Europe, has shifted 87 km north and to higher altitudes as a result of increased winter survival due to increased winter temperatures in the region over the last 30 a [26]. The mountain pine beetle (*D. ponderosae*) has also moved northward and to higher elevations in western Canada due to increased summer and winter temperatures and reduced precipitation [120]. In addition to expanding their ranges, these and other pests also are outbreaking more frequently and for longer durations [85–87]. Although causes of insect outbreaks are complex, observed increases in outbreak severity have been linked to both the direct effect of climate change on insect physiology and indirect effects through changes in their host plants (e.g., if their host plants are more drought stressed) and predators (e.g. [85–88,121–123]).

Finally, the effect of climate change on forest insect pests can affect the current balance of forest budgets. Widespread tree mortality due to *D. ponderosae* causes forests to have low primary productivity and increased carbon emissions as a result of tree decay. Over the next 20 a, it is predicted that pine forests in British Columbia, Canada will become net carbon sources rather than net carbon sinks mediated by increases in the severity of *D. ponderosae* outbreaks from recent climate change [86,87,124].

6. CONCLUSION

In this review, we demonstrate that the direction and magnitude of the effects of climate change on insect species are multifaceted. Changes are occurring both spatially and temporally, and these changes result directly from changing climate and indirectly through interactions with species in lower (i.e. plants) and upper (i.e. predators) trophic levels. In addition, trophic cascades can occur such that changes in interactions between herbivores and their host plants affect higher trophic levels, and changes between predators and herbivorous prey can affect primary producers.

Further, species traits such as dispersal ability, trophic level and degree of specialisation are potentially good predictors of the effects of climate change. Our prediction is that future communities will be dominated by mobile generalists, species with fast generation times, those that have high dispersal capabilities or those that have been dispersed around the globe by humans. Many of these species are already pests, and new species could become pests on new host plants or in new locations under climate change. These opportunists may affect important human activities, such as forestry and agriculture, with extensions into the role of forests as carbon sinks. Insects that provide important ecosystems services such as pollination or biological control could not fare as well.

ACKNOWLEDGEMENTS

This paper benefited by funding from the Office of Science (Biological and Environmental Research) of the US Department of Energy, under award numbers DE-FG02-05ER64023 to JJH and DE-FG02-06ER64232 to RLL. SLP led sections 2 and 3; KMP led sections 4 and 5; DJP and JDKD conducted the analysis in Fig. 1 and JJH and RLL initiated the project and edited the manuscript.

REFERENCES

1. J.S. Bale, G.J. Masters, I.D. Hodkinson, C. Awmack, T.M. Bezemer, V.K. Brown, J. Butterfield, A. Buse, J.C. Coulson, J. Farrar, J.E.G. Good, R. Harrington, S. Hartley, T.H. Jones, R.L. Lindroth, M.C. Press, I. Symrnioudis, A.D. Watt, J.B. Whittaker, Glob. Change Biol. 8 (2002) 1–16.
2. T.M. Bezemer, T.H. Jones, Oikos 82 (1998) 212–222.
3. P. Stiling, T. Cornelissen, Glob. Change Biol. 13 (2007) 1823–1842.
4. M.S. Warren, J.K. Hill, J.A. Thomas, J. Asher, R. Fox, B. Huntley, D.B. Roy, M.G. Telfer, S. Jeffcoate, P. Harding, G. Jeffcoate, S.G. Willis, J.N. Greatorex-Davies, D. Moss, C.D. Thomas, Nature 41 (2001) 65–69.
5. N.R. Andrew, L. Hughes, Ecol. Entomol. 29 (2004) 527–542.
6. C.A. Deutsch, J.J. Tewksbury, R.B. Huey, K.S. Sheldon, C.K. Ghalambor, D.C. Haak, P.R. Martin, Proc. Natl. Acad. Sci. USA 105 (2008) 6668–6672.
7. W. Voigt, J. Perner, A.J. Davis, T. Eggers, J. Schumacher, R. Bahrman, B. Fabian, W. Heinrich, G. Kohler, D. Lichter, R. Marstaller, F.W. Sander, Ecology 84 (2003) 2444–2453.
8. C. Parmesan, G. Yohe, Nature 421 (2003) 37–42.
9. T.L. Root, J.T. Price, K.R. Hall, S.H. Schneider, C. Rosenzweig, J.A. Pounds, Nature 421 (2003) 57–60.
10. O. Gordo, J.J. Sanz, Ecol. Entomol. 31 (2006) 261–268.
11. W.N. Ellis, J.H. Donner, J.H. Kuchlein, Ent. Ber. Amst. 57 (1997) 66–72.

12. D.B. Roy, T.H. Sparks, Glob. Change Biol. 6 (2000) 407–416.
13. M.L. Forister, A.M. Shapiro, Glob. Change Biol. 9 (2003) 1130–1135.
14. C. Stefanescu, J. Peñuelas, I. Filella, Glob. Change Biol. 9 (2003) 1494–1506.
15. R.A. Fleming, G.M. Tatchell, in: R. Harrington, N. Stork (Eds.), Insects in a Changing Environment, Academic Press, London, 1995, pp. 505–508.
16. R. Harrington, S.J. Clark, S.J. Welham, P.J. Verrier, C.H. Denholm, M. Hullé, D. Maurice, M.D. Rounsevell, N. Cocu, Glob. Change Biol. 13 (2007) 1550–1564.
17. D.L. Musolin, Glob. Change Biol. 13 (2007) 1565–1585.
18. J.A. Logan, J. Regniere, J.A. Powell, Front. Ecol. Environ. 1 (2003) 130–137.
19. C. Parmesan, N. Ryrholm, C. Stefanescu, J.K. Hill, C.D. Thomas, H. Descimon, B. Huntley, L. Kaila, J. Kullberg, T. Tammaru, J. Tennent, J.A. Thomas, M. Warren, Nature 299 (1999) 579–583.
20. L. Crozier, Climate Change and Species Range Boundaries: A Case Study at The Northern Range Limits of *Atalopedes campestris* (lepidoptera: Hesperiidae), The Sachem Skipper, PhD thesis, University of Washington, Washington.
21. L. Crozier, Oecologia 135 (2003) 648–656.
22. J.K. Hill, Y.C. Collingham, C.D. Thomas, D.S. Blakeley, R. Fox, D. Moss, B. Huntley, Ecol. Lett. 4 (2001) 313–321.
23. C.D. Thomas, E.J. Bodsworth, R.J. Wilson, A.D. Simmons, Z.G. Davies, M. Musche, L. Conradt, Nature 411 (2001) 577–581.
24. R. Karban, S.Y. Strauss, Ecol. Entomol. 29 (2004) 251.
25. A.T. Peterson, E. Martinez-Meyer, C. Gonzalez-Salazaar, P.W. Hall, Can. J. Zool. 82 (2004) 851–858.
26. A. Battisti, M. Stastny, S. Netherer, C. Robinet, A. Schopf, A. Roques, S. Larsson, Ecol. Appl. 15 (2005) 2084–2096.
27. C. Parmesan, S. Gaines, L. Gonzalez, D.M. Kaufman, J. Kingsolver, A.T. Peterson, R. Sagarin, Oikos 108 (2005) 58–75.
28. J.F. McLaughlin, J.J. Hellmann, C.L. Boggs, P.R. Ehrlich, Proc. Natl. Acad. Sci. USA 99 (2002) 6070–6074.
29. J.F. McLaughlin, J.J. Hellmann, C.L. Boggs, P.R. Ehrlich, Oecologia 132 (2002) 538–548.
30. J.K. Hill, C.D. Thomas, R. Fox, M.G. Telfer, S.G. Willis, J. Asher, B. Huntley, Proc. R. Soc. Lond. Ser. B, 269 (2002) 2163–2171.
31. J.K. Hill, Y.C. Collingham, C.D. Thomas, D.S. Blakeley, R. Fox, D. Moss, B. Huntley, Ecol. Lett. 4 (2001) 313–321.
32. M. Levitan, Evol. Ecol. Res. 5 (2003) 597–604.
33. P.A. Umina, A.R. Weeks, M.R. Kearney, S.W. McKechnie, A.A. Hoffmann, Science 308 (2005) 691–693.
34. J. Balanyà, J.M. Oller, R.B. Huey, G.W. Gilchrist, L. Serra, Science 313 (2006) 1773.
35. F. Rodríguez-Trelles, M.A. Rodríguez, Evol. Ecol. 12 (1998) 829–838.
36. N.E. Rank, E.P. Dahlhoff, Evolution 56 (2002) 2278–2289.
37. W.E. Bradshaw, C.M. Holzapfel, Proc. Natl. Acad. Sci. USA 98 (2001) 14509–14511.
38. P.W. de Jong, P.M. Brakefield, Proc. R. Soc. Lond. Ser. B 265 (1998) 39–43.
39. J.R. Etterson, R.G. Shaw, Science 294 (2001) 151–154.
40. J.J. Hellmann, M. Pineda-Krch, Biol. Conserv. 137 (2007) 599–609.
41. P. Gienapp, C. Teplitsky, J.S. Alho, J.A. Mills, J. Merila, Mol. Ecol. 17 (2008) 167–178.
42. P. Feeny, Ecology 51 (1970) 565–581.
43. J.C. Schultz, P.J. Nothnagle, I.T. Baldwin, Am. J. Bot. 69 (1982) 753–759.

44. R.F. Chapman, in: R.F. Chapman, A. Joern (Eds.), Biology of Grasshoppers, Food Selection, Wiley, New York, 1990, pp. 36–72.
45. M.D. Hunter, P.W. Price, Ecology 73 (1992) 724–732.
46. P.W. Price, in: M.D. Hunter, T. Ohgushi, P.W. Price (Eds.), Effects of Resource Distribution on Animal–Plant Interactions, Academic Press, San Diego, CA, 1992, pp. 139–173.
47. A.N. Nzekwu, A.E. Akingbohungbe, J. Orthoptera Res. 11 (2002) 185–188.
48. J.J. Hellmann, J. Anim. Ecol. 71 (2002) 925–936.
49. N.J. Sanders, D.M. Gordon, Ecology 84 (2003) 1024–1031.
50. P.D. Coley, Clim. Change 39 (1998) 455–472.
51. G.J. Masters, V.K. Brown, I.P. Clarke, J.B. Whittaker, J.A. Hollier, Ecol. Entomol. 23 (1998) 45–52.
52. J.B. Whittaker, Eur. J. Entomol. 96 (1999) 149–156.
53. S.E. Hartley, T.H. Jones, Oikos 101 (2003) 6–17.
54. C.V. Johns, A. Hughes, Glob. Change Biol. 8 (2002) 142–152.
55. R.E. Forkner, R.J. Marquis, J.T. Lill, J.L.E. Corff, Ecol. Entomol. 33 (2008) 276–285.
56. R.D. Harrington, R.I. Woiwood, T. Sparks, Trends Ecol. Evol. 14 (1999) 147–149.
57. R.C. Dewar, A.D. Watt, Oecologia 89 (1992) 557–559.
58. J.K. Hill, I.D. Hodkinson, Ecol. Entomol. 20 (1992) 237–244.
59. M. van Asch, P.H. van Tienderen, L.J.M. Holleman, M. Visser, Glob. Change Biol. 13 (2007) 1596–1604.
60. M. Visser, Proc. R. Soc. Lond. Ser. B 275 (2008) 649–659.
61. C. Parmesan, Glob. Change Biol. 13 (2007) 1860–1872.
62. T.H. Sparks, D.B. Roy, R.L.H. Dennis, Glob. Change Biol. 11 (2005) 507–514.
63. J. Memmott, P.G. Craze, N.M. Waser, M.V. Price, Ecol. Lett. 10 (2007) 710–717.
64. M.B. Davis, R.G. Shaw, Science 292 (2001) 673–679.
65. S. Dullinger, T. Dirnböck, G. Grabherr, J. Ecol. 92 (2004) 241–252.
66. J.K. Hill, C.D. Thomas, B. Huntley, Proc. R. Soc. Lond. Ser. B 266 (1999) 1197–1206.
67. L.R. Iverson, A. Prasad, M.W. Schwartz, Ecol. Model. 115 (1999) 77–93.
68. O. Honnay, K. Verheyen, J. Butaye, H. Jacquemyn, B. Bossuyt, M. Hermy, Ecol. Lett. 5 (2002) 525–530.
69. P.A. Opler, H. Pavulaan, R.E. Stanford, M. Pogue, Butterflies and Moths of North America, Bozeman, MT, NBII Mountain Prairie Information Node, 2006, available from http://www.butterfliesandmoths.org/.
70. USDA, NRCS, The PLANTS Database, National Plant Data Center, Baton Rouge, LA, 2008, available from http://plants.usda.gov.
71. W.H. van der Putten, P.C. de Ruiterb, T.M. Harvey, M. Wassen, V. Wolters, Basic Appl. Ecol. 5 (2004) 487–494.
72. B.A. Menge, J.P. Sutherland, Am. Nat. 130 (1987) 730–757.
73. L. Cagnolo, S.I. Molina, G.R. Valladares, Biodivers. Conserv. 11 (2002) 407–420.
74. E.L. Preisser, D.R. Strong, Am. Nat. 163 (2004) 754–762.
75. M.L. Pace, J.J. Cole, S.R. Carpenter, J.F. Kitchell, Trends Ecol. Evol. 14 (1999) 483–488.
76. G.A. Polis, A.L.W. Sears, G.R. Huxel, D.R. Strong, J. Maron, Trends Ecol. Evol. 15 (2000) 473–475.
77. M.P. Hassell, J. Latto, R.M. May, J. Anim. Ecol. 58 (1989) 883–892.
78. H.V. Cornell, B.A. Hawkins, Am. Nat. 145 (1995) 563–593.
79. B.A. Hawkins, H.V. Cornell, M.E. Hochberg, Ecology 78 (1997) 2145–2152.
80. M.D. Hunter, G.C. Varley, G.R. Gradwell, Proc. Natl. Acad. Sci. USA 94 (1997) 9176–9181.
81. P. Turchin, A.D. Taylor, J.D. Reeve, Science 285 (1999) 1068–1071.

82. P. Turchin, S.N. Wood, S.P. Ellner, B.E. Kendall, W.W. Murdoch, A. Fischlin, J. Casas, E. McCauley, C.J. Briggs, Ecology 84 (2003) 1207–1214.
83. J.D. Reeve, Oecologia 112 (1997) 48–54.
84. J. Liebhold, J. Elkinton, D. Williams, R.M. Muzika, Popul. Ecol. 42 (2000) 257–266.
85. D.W. Williams, A.M. Liebhold, Environ. Entomol. 24 (1995) 1–9.
86. M.P. Ayers, M.J. Lombardo, Sci. Total Environ. 262 (2000) 263–286.
87. W.J.A. Volney, R.A. Flemming, Agric. Ecosyst. Environ. 82 (2000) 283–294.
88. J.O. Stireman III, L.A. Dyer, D.H. Janzen, M.S. Singer, J.T. Lill, R.J. Marquis, R.E. Ricklefs, G.L. Gentry, W. Hallwachs, P.D. Coley, J.A. Barone, H.F. Greeney, H. Connahs, P. Barbosa, H.C. Morais, R. Diniz, Proc. Natl. Acad. Sci. USA 102 (2005) 17384–17387.
89. T. Hance, J. van Baaren, P. Vernon, G. Bovin, Annu. Rev. Entomol. 52 (2007) 107–126.
90. H.K. Kaya, Y. Tanada, Ann. Entomol. Soc. Am. 62 (1969) 1303–1306.
91. F. Hérard, M.A. Keller, W.J. Lewis, J.H. Tumlinson, J. Chem. Ecol. 14 (1988) 1583–1596.
92. H.C.J. Godfray, M.P. Hassell, R.D. Holt, J. Anim. Ecol. 63 (1994) 1–10.
93. S. van Nouhuys, G. Lei, J. Anim. Ecol. 5 (2004) 526–535.
94. D.A. Spiller, T.W. Schoener, Ecology 88 (2007) 37–41.
95. R.D. Holt, in: G.A. Polis, K.O. Winemiller (Eds.), Food Webs: Integration of Pattern and Dynamics, Chapman and Hall, New York, 1996, pp. 313–326.
96. D.A. Spiller, J.B. Losos, T.W. Schoener, Science 281 (1998) 695–697.
97. D.R. Strong, A.V. Whipple, A.L. Child, B. Dennis, Ecology 80 (1999) 2750–2761.
98. IPCC, S. Solomon, D. Qin, M. Manning, Z. Chen, M. Marquis, K.B. Averyt, M. Tignor, H.L. Miller (Eds.), Climate Change 2007: The Physical Science Basis. Contribution of Working Group I to the Fourth Assessment Report of the Intergovernmental Panel on Climate Change, Cambridge University Press, Cambridge, UK and New York, 2007.
99. M.E. Torchin, C.E. Mitchell, Front. Ecol. Environ. 2 (2004) 183–190.
100. H.V. Cornell, B.A. Hawkins, Am. Nat. 141 (1993) 847–865.
101. J.B.F. Geervliet, M.S.W. Verdel, J. Schaub, H. Snellen, M. Dicke, L.E.M. Vet, Oecologia 124 (2000) 55–63.
102. M. Vos, L.E.M. Vet, Evol. Ecol. Res. 6 (2004) 1021–1035.
103. R. Menéndez, A. Gonzalez-Megias, O.T. Lewis, M.R. Shaw, C.D. Thomas, Ecol. Entomol. 33 (2008) 413–421.
104. K. Schönrogge, P. Walker, M.J. Crawley, Proc. R. Soc. Lond. B Biol. 265 (1998) 1643–1650.
105. C.E. Mitchell, A.G. Power, Nature 421 (2003) 625–627.
106. M.E. Torchin, K.D. Lafferty, A.P. Dobson, Nature 421 (2003) 628–630.
107. K. Schönrogge, G.N. Stone, M.J. Crawley, Oikos 77 (1996) 507–518.
108. W.H. Settle, L.T. Wilson, Ecology 71 (1990) 1461–1470.
109. K. Shea, P. Chesson, Trends Ecol. Evol. 17 (2003) 170–176.
110. R.I. Colautti, A. Ricciardi, L.A. Grigorovich, H.J. MacIsaac, Ecol. Lett. 7 (2004) 721–733.
111. D.B. Simberloff, in: J.A. Drake (Ed.), Biological Invasions: A Perspective, Wiley, New York, 2005, pp. 61–75.
112. J.H. Lawton, K.C. Brown, Proc. R. Soc. Lond. Ser. B 314 (1986) 607–617.
113. N.L. Ward, G.J. Masters, Glob. Change Biol. 13 (2007) 1605–1615.
114. S.L. Chown, S. Slabber, M. McGeoch, C. Janion, H.P. Leinaas, Proc. R. Soc. Lond. Ser. B 274 (2007) 2531–2537.
115. V.H. Dale, L.A. Joyce, S. McNulty, R.P. Neilson, M.P. Ayres, M.D. Flannigan, P.J. Hanson, L.C. Irland, A.E. Lugo, C.J. Peterson, D. Simberloff, F.J. Swanson, B.J. Stocks, B.M. Wotton, Bioscience 51 (2001) 723–733.

116. G.M. Lovett, C.D. Canham, M.A. Arthur, K.C. Weathers, R.D. Fizhugh, Bioscience 56 (2006) 395–405.
117. D. Pimental, in: D. Pimental (Ed.), Biological Invasions. Economic and Environmental Costs of Alien plant, Animal and Microbe Species, CRC Press, Boca Raton, FL, 2002, pp. 151–155.
118. P. Niemelä, M.J. Mattson, Bioscience 46 (1996) 741–753.
119. A.M. Liebhold, W.L. MacDonald, D. Bergdahl, V.C. Mastro, For. Sci. Monogr. 30 (1995) 1–49.
120. A.L. Carroll, S.W. Taylor, J. Regniere, L. Safranyik, in: T.L. Shore, J.E. Brooks, J.E. Stone (Eds.), Challenges and Solutions, Report number BC-X-399, Natural Resources Canada, Canadian Forests Service, Victoria, 2004, pp. 223–232.
121. E.E. Berg, J.D. Henry, C.L. Fastie, A.D. De Volder, S.M. Matsuoka, For. Ecol. Manage. 227 (2006) 219–232.
122. K.F. Raffa, B.H. Aukema, B.J. Bentz, A.L. Carroll, J.A. Hicke, M.G. Turner, W.H. Romme, Bioscience 58 (2008) 501–517.
123. L.C. Stige, K.S. Chan, Z. Zhang, D. Frank, N.C. Stenseth, Proc. Natl. Acad. Sci. USA 104 (2008) 16188–16193.
124. W.A. Kurz, C.C. Dymond, G. Stinson, G.J. Rampley, E.T. Neilson, A.L. Carroll, T. Ebata, L. Safranyik, Nature 452 (2008) 987–990.

Chapter 12

Sea Life (Pelagic and Planktonic Ecosystems) as an Indicator of Climate and Global Change

Martin Edwards

Sir Alister Hardy Foundation for Ocean Science, Citadel Hill, The Hoe, Plymouth PL1 2PB, United Kingdom

Marine Institute, University of Plymouth, Drake Circus, Plymouth PL4 8AA, United Kingdom

1. **Pelagic and Planktonic Ecosystems**
   - 1.1. Sensitivity of Pelagic and Planktonic Ecosystems to Climate and Global Change
   - 1.2. Marine and Terrestrial Biological Responses to Climate and Global Change
   - 1.3. Ocean Acidification and other Anthropogenic Influences on Pelagic and Planktonic Ecosystems
2. **Observed Impacts on Pelagic and Planktonic Ecosystems**
   - 2.1. Biogeographical Changes and Northward Shifts
   - 2.2. Life-Cycle Events and Pelagic Phenology
   - 2.3. Plankton Abundance and Pelagic Productivity
   - 2.4. Pelagic Biodiversity and Invasive Species
3. **Conclusion and Summary of Key Indicators**

References

1. PELAGIC AND PLANKTONIC ECOSYSTEMS

The marine pelagic realm is the largest ecological system on the planet occupying 71% of the planetary surface and a major part of the Earth's overall biosphere. As a consequence of this, pelagic ecosystems play a fundamental role in modulating the global environment via its regulatory effects on the Earth's climate and its role in biogeochemical cycling. Changes caused by increased warming on marine pelagic communities are likely to have important consequences on ecological structure and function thereby leading to significant feedbacks on the Earth's climate system.

This chapter will mainly concentrate on the epipelagic zone where biological production, biogeochemical cycles and marine food-webs are maintained by the inhabiting planktonic organisms. Apart from discussing the effects of climate on higher trophic organisms, particularly pelagic fish, the overall emphasis of this chapter is focused on the planktonic community. More specifically, the chapter will concentrate on observational evidence from contemporary plankton indicators over the past multidecadal period rather than palaeo planktonic indicators. The free floating photosynthesising life of the oceans (algal phytoplankton, bacteria and other photosynthesising protists), at the base of the marine food-web, provides food for the animal plankton (zooplankton) which, in turn, provide food for many other marine organisms ranging from the microscopic to whales. The carrying capacity of pelagic ecosystems in terms of the size of fish resources and recruitment to individual stocks as well as the abundance of marine wildlife (e.g., seabirds and marine mammals) is highly dependent on variations in the abundance, seasonal timing and composition of the plankton.

Phytoplankton also comprise approximately half of the total global primary production and play a crucial role in climate change through biogeochemical cycling and the export of the greenhouse gas to the deep ocean by carbon sequestration in what is known as the 'biological pump'. Phytoplankton have thus already helped to mitigate some of the climate effects of elevated CO_2 observed over the last 200 a with the oceans taking up $\sim$40% of anthropogenic CO_2 [1]. In terms of feedback mechanisms on Earth's climate, it is speculated that these biological pumps will be less efficient in a warmer world due to changes in the phytoplankton composition favouring small flagellates [2] and less overall nutrient mixing due to increased stratification (see Section 1.1). It is also predicted that warmer temperatures would shift the metabolic balance between production and respiration in the world's oceans towards an increase in respiration thus reducing the capacity of the oceans to capture CO_2 [3]. Apart from playing a fundamental role in the Earth's climate system and in marine food-webs, plankton are also highly sensitive contemporary and palaeo indicators of environmental change and provide rapid information on the 'ecological health' of our oceans. There is some evidence that suggest that plankton are more sensitive indicators than environmental variables themselves and can amplify weak environmental signals due to their nonlinear responses [4]. A plankton species, defined by its abiotic envelope, in affect has the capacity to simultaneously represent an integrated ecological, chemical and physical variable.

1.1. Sensitivity of Pelagic and Planktonic Ecosystems to Climate and Global Change

Temperature is a key driver of marine ecosystems and, in particular, its effects on pelagic populations are manifested very rapidly [5–7]. This is hardly surprising when more than 99% of pelagic and planktonic organisms are ectothermal making them highly sensitive to fluctuations in temperature [8]. The rapidity of the planktonic response is predominantly due to their short life-cycles and its passive response to advective changes. For example, phytoplankton fix as much CO_2 per year as all terrestrial plants but due to being unicellular they represent at any one time only 1% of the Earth's biomass. This means the rate of turnover in the world's oceans is huge and on average the global phytoplankton population is consumed in days to weeks [9]. This all makes plankton tightly coupled to fluctuations in the marine environment and highly sensitive indicators of environmental change such as nutrient availability, ocean current changes and climate variability.

In the marine environment the effect of short-term climate variability and inter-annual variability on populations of higher trophic levels such as seabirds and whales can to a degree be somewhat buffered due to their longer life-cycles. In the long-term, their ability to undergo large geographical migrations may also help them to mitigate some of the effects of global change; however, this hypothesis has not been investigated. In both cases, this is not applicable to planktonic organisms. Biologically speaking, changes in temperature have direct consequences on many physiological processes (e.g., oxygen metabolism, adult mortality, reproduction, respiration, reproductive development, etc.) and control virtually all life-processes from the molecular to the cellular to whole regional ecosystem level and biogeographical provinces. Ecologically speaking, temperature also modulates both directly and indirectly species interactions (e.g., competition, prey–predator interactions and food-web structures), ultimately, changes in temperatures can lead to impacts on the biodiversity, size structure and functioning of the whole pelagic ecosystem [10,11].

While, temperature has direct consequences on many biological and ecological traits, it also modifies the marine environment by influencing oceanic circulation and by enhancing the stability of the water-column and hence nutrient availability. The amount of nutrients available in surface waters directly dictates phytoplankton growth and is the key determinant of the plankton size, community and food-web structure. In terms of nutrient availability, warming of the surface layers increases water column stability, enhancing stratification and requiring more energy to mix deep, nutrient-rich waters into surface layers. Particularly warm winters will also limit the degree of deep convective mixing and thereby limit nutrient replenishment necessary for the following spring phytoplankton bloom. In summary, climatic warming of surface waters will increase the density contrast between the surface layer and the underlying nutrient-rich waters. The availability of one of the principle nutrients (nitrate) that limits phytoplankton growth has therefore been found to be negatively related to temperatures globally [12,13]. Similarly, a global analysis of satellite derived chlorophyll data shows a strong inverse relationship between Sea Surface Temperatures (SST) and chlorophyll concentration [9]. Furthermore, other abiotic variables like oxygen concentration (important to organism size and metabolism [14]), nitrate metabolism [15] and the viscosity of seawater (important for the maintenance of buoyancy for plankton) are also directly linked to temperature. So unlike terrestrial environments, where precipitation plays a key role, the chemical and upper-ocean temperature regime in open oceans and its consequent biological composition are inexorably entwined.

1.2. Marine and Terrestrial Biological Responses to Climate and Global Change

Many planktonic organisms live in narrow temperature ranges (stenothermal) and often undergo a much more rapidly observed change due to temperature, be it biogeographically or phenologically [10,11], in comparison to their terrestrial counterparts [16]. Apart from this and the fact that planktonic organisms having shorter life-cycles, already mentioned above, there are a number of distinct reasons why the speed of the response to climate and global change of pelagic organisms differs from those of terrestrial organisms. Some of the primary reasons are, firstly, due to the high specific heat of water in open ocean systems many planktonic organisms are largely buffered against extremes in daily and seasonal temperature fluctuations. Daily and seasonal variations in temperature are therefore less variable in comparison to the terrestrial domain allowing marine species to become firmly embedded in their optimum thermal envelope. Secondly, unlike terrestrial environments, many planktonic organisms can quickly track evolving bioclimatic envelopes by being largely free of geographical barriers hindering their dispersal range and do not need a large amount of energy expenditure to do so, being primarily passively advected. Ocean currents, therefore, provide an ideal mechanism for dispersal over large distances and this is seemingly why a vast many of marine organisms have evolved at least a portion of their life-cycles as planktonic entities. Thirdly, many terrestrial organisms are geographically and ecologically bound by their habitat type mainly dictated by the vegetative composition. In terrestrial systems, the development of these vegetative types can be particularly slow moving (e.g., forest ecosystems) and hence organisms that rely on this habitat will be restricted in terms of their geographical spread. This is not the case for phytoplankton that have extremely short life-cycles in comparison allowing rapid temporal and spatial spread of planktonic herbivores and associated communities. Furthermore, the presence of inimitably terrestrial anthropogenic pressures such as habitat fragmentation and habitat loss, which clearly limits the geographical spread of organisms in the terrestrial environment, is seemingly absent from open ocean systems [17].

1.3. Ocean Acidification and other Anthropogenic Influences on Pelagic and Planktonic Ecosystems

While temperature, light and nutrients are probably the most important physical variables structuring marine ecosystems, the pelagic realm will also have to contend with, apart from global climate change, with the impact of anthropogenic CO_2 directly influencing the pH of the oceans [18]. Evidence collected and modelled to date indicates that rising CO_2 has led to chemical changes in the ocean which has led to the oceans becoming more acidic. Ocean acidification has the potential to affect the process of calcification and therefore certain planktonic organisms (e.g., coccolithophores, foraminifera, pelagic molluscs) may be particularly vulnerable to future CO_2 emissions. Apart from climate warming, potential chemical changes to the oceans and its affect on the biology of the oceans could further reduce the ocean's ability to absorb additional CO_2 from the atmosphere which, in turn, could affect the rate and scale of global warming (see Chapter 21). Other anthropogenic driving forces of change that are operative in pelagic ecosystems are predominantly overfishing and its effect on modifying marine pelagic food-webs [19], (see Chapter 14) and in coastal regions nutrient input from terrestrial sources leading in some cases to enhanced biological production and Harmful Algal Blooms (HABs) and other general chemical and inorganic contaminants. The impacts of atmospheric derived anthropogenic nitrogen on the open ocean have only been recently investigated but may also play a significant role on annual new marine biological production [20].

2. OBSERVED IMPACTS ON PELAGIC AND PLANKTONIC ECOSYSTEMS

There is a large body of observed evidence to suggest that many pelagic ecosystems, both physically and biologically are responding to changes in regional climate caused predominately due to the warming of air and SST and to a lesser extent by the modification of precipitation regimes and wind patterns. The biological manifestations of rising SST have variously taken the form of biogeographical, phenological, biodiversity, physiological, species abundance changes and whole ecological regime shifts. Any observational change in the marine environment associated with climate change, however, should be considered against the background of natural variation on a variety of spatial and temporal scales. Recently, long-term decadal observational studies have focused on known natural modes of climatic oscillations at similar temporal scales such as the El Nino-Southern Oscillation (ENSO) in the

Pacific and the North Atlantic Oscillation (NAO) in the North Atlantic in relation to pelagic ecosystem changes (see reviews [21,22]). Many of the biological responses observed have been associated with rising temperatures. However, approximating the effects of climate change embedded in natural modes of variability, particularly multidecadal oscillations like the Atlantic Multidecadal Oscillation (AMO) [23], is extremely difficult and therefore observed evidence of planktonic changes directly attributable to anthropogenic climate and global change must be treated with a degree of scientific caution.

Evidence for observed pelagic changes is also biased towards regions, particularly seas around Europe and North America, which have had some form of biological monitoring in place over a consistently long period. Apart from a number of important long-term coastal research stations sampling plankton (e.g., Helgoland Roads time-series in the southern North Sea [24]) there are only a few long-term biological surveys that sample the open ocean. For this reason some of the strongest evidence detected for observed changes in open ocean ecosystems comes from the North Atlantic where an extensive spatial and long-term biological survey exists in the form of the Continuous Plankton Recorder (CPR) survey. The CPR survey has been in operation in the North Sea and North Atlantic since 1931 and has systematically sampled up to 500 planktonic taxa from the major regions of the North Atlantic at a monthly resolution [25]. Important multidecadal evidence from the Pacific is mainly derived from the Californian Cooperative Oceanic Fisheries Investigations (CalCOFI) survey operating off the coast of Californian since 1949.

2.1. Biogeographical Changes and Northward Shifts

Some of the strongest evidence of large-scale biogeographical changes observed in our oceans comes from the CPR survey. In a study geographically encompassing the whole NE Atlantic over a 50 year period, Beaugrand et al. [10] showed rapid northerly movements of the biodiversity of a key zooplankton group (calanoid copepods). During the last 50 years there has been a northerly movement of warmer water plankton by 10° latitude in the northeast Atlantic and a similar retreat of colder water plankton to the north (a mean poleward movement of between 200 and 250 km per decade) (Fig. 1). This geographical movement is much more pronounced than any documented terrestrial study, mainly due to advective processes and in particular the shelf-edge current running north along the northern European continental shelf. The rapid movement of plankton northward is only seen along the continental shelf, where deeper water is warming much more rapidly. Further along the shelf, plankton are upwelled from this deeper water to make an appearance in the surface plankton community. Hence the plankton have moved 10° latitude northward via mainly deep water advective processes not seen in the movement of surface isotherms. In other areas in the North

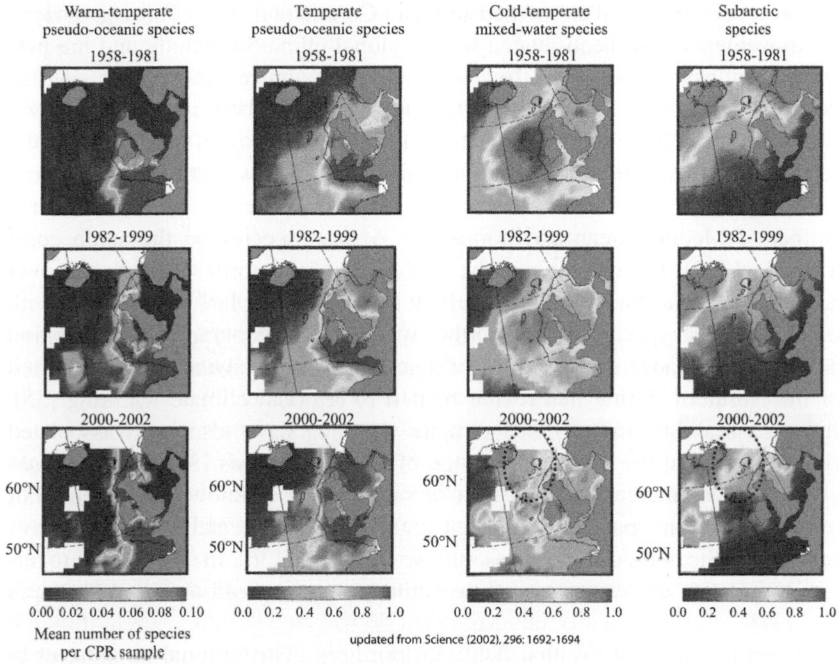

FIGURE 1 Changes in the geographical distribution of four different plankton assemblages over a multidecadal period. There has been a rapid northerly movement of warm-temperate species and a subsequent decline in sub-arctic species over 40 years. Particularly rapid movement is observed along the European Continental Shelf. Data derived from the Continuous Plankton Recorder survey. Updated from Ref. [76]. (See Color Plate 16).

East Atlantic the plankton shifts were more moderate and varied between 90 and 200 km per decade, still faster than any other documented terrestrial study which has a meta-analytic average of 6 km per decade [16]. Similar to the North Atlantic, in the north-east Pacific there has been a general increase in the frequency of southern species moving northward [26]. Interestingly, in the North West Atlantic, pelagic organisms have been moving southward [27]. This initially seems to contradict general thinking of homogenous global climate warming throughout the world's oceans. However, this movement has been linked to the strengthening of the Labrador Current which has spread colder water southward over the last decade carrying pelagic organisms with cold-water affinities as far south as Georges Bank.

These large-scale biogeographical shifts observed in the plankton have also seen paralleled latitudinal movements of fish species distribution [28–30]. Northerly geographical range extensions or changes in the geographical distribution of fish populations have been recently documented for European Continental shelf seas and along the European Continental shelf edge [31–33]. Similar to the plankton, the largest movements of fish species towards north have also been observed along the European Continental shelf. These geographical movements have been related with regional climate warming and are predominantly associated with the northerly geographical movement of fish species with more southern biogeographical affinities. These include the movement of pelagic fish species such as sardines and anchovies northward in the North Sea and red mullet and bass extending their ranges northward to western Norway [31,32]. New records were also observed over the last decade for a number of Mediterranean and north-west African species on the south coast of Portugal [32]. The cooling and the freshening of the north-west Atlantic over the last decade has had an opposite effect similar to the plankton patterns, with some groundfish species moving further south in their geographical distribution [34]. Northerly range extensions of pelagic fish species have also been reported for the Northern Bering Sea region related to regional climate warming [35]. Climate variability and regional climate warming have also been associated with variations in the geographic range of marine diseases [36]. New diseases typically have emerged through host or range shifts of known pathogens. For example, over the past few decades pathogens detrimental to oysters have spread from the mid-Atlantic states into New England [36]. In comparison to terrestrial systems, epidemics of marine pathogens can spread at extremely rapid rates [37].

Again it is noteworthy that fish with northern distributional boundaries in the North Sea have shifted northward at rates of up to three times faster than terrestrial species [16,30]. One of the largest biogeographical shifts ever observed for fish species is the dramatic increase and subsequent geographical spread northward of the Snake Pipefish (*Entelurus aequreus*). Once confined mainly to the south and west of the British Isles before 2003, it can now be found as far north as the Barents Sea and Spitzbergen [38]. While this present discussion has described surface geographical changes in epipelagic organisms it is worth remembering the three-dimensional nature of the pelagic environment. Recent research has observed not just changes in fish biogeography but also changes in fish species depth (towards deeper waters) in response to climate warming [39]. This change can be seen as analogous with the upward altitudinal movement of terrestrial organisms in alpine environments. All these studies highlight the consistency of pelagic organisms undergoing large-scale distributional changes in response to hydro-climatic variability.

2.2. Life-Cycle Events and Pelagic Phenology

Phenology, or repeated seasonal life-cycle events such as annual migrations or spawning, is highly a sensitive indicator of climate warming. This is because many terrestrial and marine organisms, apart from photoperiod, are dependant on temperature as a trigger for seasonal behaviour. In the terrestrial realm, phenology events such as bird migrations, egg-laying, butterfly emergence and flowering of certain plants are all getting earlier in response to milder spring weather [16]. In terms of the pelagic phenological response to climate warming, many plankton taxa have also been found moving forward in their seasonal cycles [11]. In some cases, a shift in seasonal cycles of over 6 weeks was detected, again a far larger shift than observed for terrestrial based observations. Summarising a terrestrial study of phenology using over 172 species of plants, birds, insects and amphibians, Parmesan & Yorke [16] calculated a mean phenological change of 2.3 d. It is thought that temperate pelagic environments are particularly vulnerable to phenological changes caused by climatic warming because the recruitment success of higher trophic levels is highly dependant on synchronisation with pulsed planktonic production [11]. Furthermore in the marine environment, and just as important, was the response to regional climate warming varied between different functional groups and trophic levels, leading to mismatch in timing between trophic levels (Fig. 2). For example, while the spring bloom has remained relatively stable in seasonal timing over five decades (mainly due to light limitation and photoperiod rather than temperature dictating seasonality [11,40]) many zooplankton organisms as well as fish larvae have moved rapidly forward in their seasonal cycles.

These changes, seen in the North Sea, have the potential to be of detriment to commercial fish stocks via trophic mismatch. For example, regional climate warming in the North Sea has affected cod recruitment via changes at the base of the food-web [41]. Cod, like many other fish species, are highly dependent on the availability of planktonic food during their pelagic larval stages. Key changes in the planktonic assemblage and phenology, significantly correlated with the warming of the North Sea over the last few decades, have resulted in a poor food environment for cod larvae and hence an eventual decline in overall recruitment success. The rapid changes in plankton communities observed over the last few decades in the North Atlantic and European regional seas, related to regional climate changes, have enormous consequences for other trophic levels and biogeochemical processes. Similarly, other pelagic phenology changes have been observed in the North Sea [24], the Mediterranean [42] and the Pacific [43,44].

2.3. Plankton Abundance and Pelagic Productivity

Contemporary observations of satellite-*in situ* blended ocean chlorophyll records indicate that global ocean net primary production has declined over the last decade [9]. Although this time-series is only 10 years in length it does show a strong negative relationship between primary production and SST and is evidence of the closely coupled relationship between ocean productivity and climate variability at a global scale. In the North Atlantic and over multidecadal periods, both changes in phytoplankton and zooplankton species and communities have been associated with Northern Hemisphere Temperature (NHT) trends and variations in the NAO index. These have included changes in species distributions and abundance, the occurrence of sub-tropical species

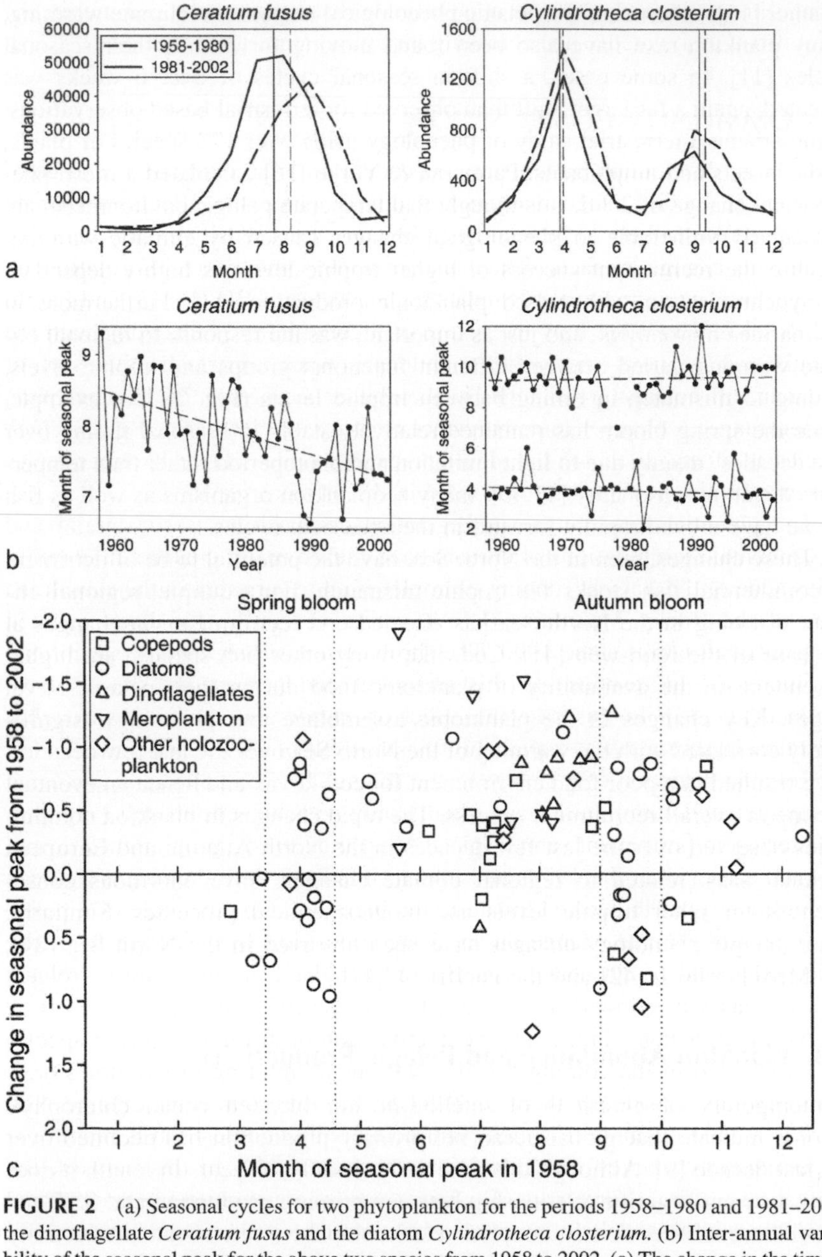

FIGURE 2 (a) Seasonal cycles for two phytoplankton for the periods 1958–1980 and 1981–2002: the dinoflagellate *Ceratium fusus* and the diatom *Cylindrotheca closterium.* (b) Inter-annual variability of the seasonal peak for the above two species from 1958 to 2002. (c) The change in the timing of the seasonal peaks (in months) for the 66 taxa over the 45 a (year) period from 1958 to 2002 plotted against the timing of their seasonal peak in 1958. For each taxon, the linear regression in (b) was used to estimate the difference between the seasonal peak in 1958 and 2002. A negative difference between 1958 and 2002 indicates seasonal cycles are becoming earlier. Standard linear regression was considered appropriate because there was minimal autocorrelation (determined by the Durbin–Watson statistic) in the phenology time series. From Ref. [11].

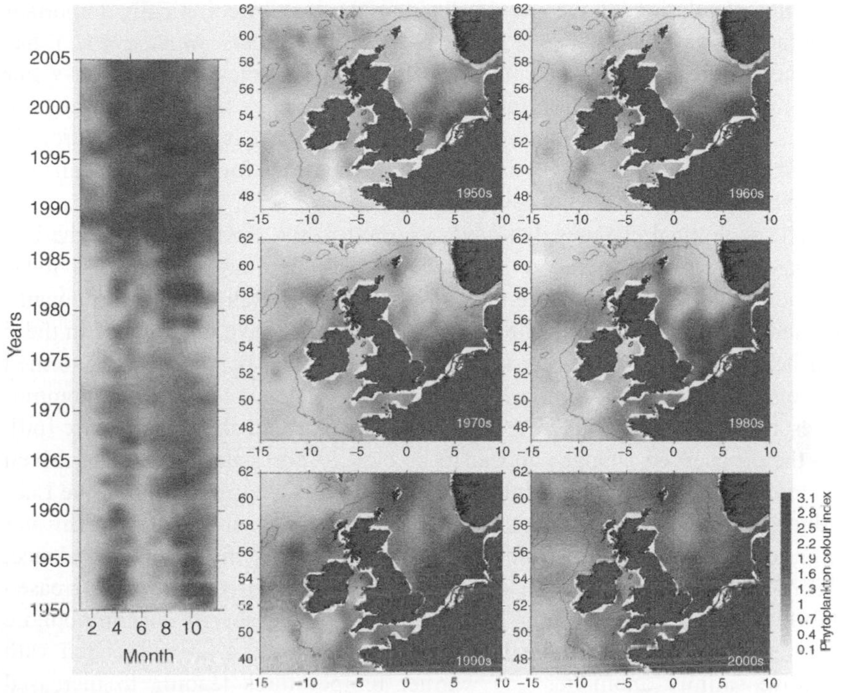

FIGURE 3 Spatial-temporal maps of the changes in the abundance of phytoplankton colour (an index of total phytoplankton biomass) for the NE Atlantic averaged per decade from the 1950s to the present. The contour plot shows monthly mean values from 1950 to 2005 of phytoplankton colour averaged for the North Sea. Large increases in phytoplankton colour are observed towards the end of the 1980s and have continued since. The increase in colour has been associated with a regime shift in the North Sea. Updated from Ref. [47]. (See Color Plate 17).

in temperate waters, changes in overall phytoplankton biomass and seasonal length (Fig. 3), changes in the ecosystem functioning and productivity of the North Atlantic [10,11,45–52]. The increase in overall phytoplankton biomass in the North Sea has been associated with an increase in smaller flagellates which favour more warmer and stratified conditions [46,47]. Over the whole NE Atlantic there has been an increase in phytoplankton biomass with increasing temperatures in cooler regions but a decrease in phytoplankton biomass in warmer regions [53]. Presumably, this is a trade-off between increased phytoplankton metabolic rates caused by temperature in cooler regions but a decrease in nutrient supply in warmer regions. Regional climate warming in the North Sea has also been associated with an increase in certain HABs in some areas of the North Sea [54]. A recent link has been established between the changes in the plankton in the North Sea to sandeels and eventual seabird breeding success (encompassing four trophic levels) [55]. In the North

Sea, the population of the previously dominant and ecologically important zooplankton species (the cold water species *Calanus finmarchicus*) has declined in biomass by 70% since the 1960s [56]. Species with warmer-water affinities are moving northward to replace this species but these species are not as numerically abundant or nutritionally beneficial to higher trophic levels. This has had inevitably important ramifications for the overall carrying capacity of the North Sea ecosystem.

The ecological changes that have occurred in the North Sea since the late 1980s (predominately driven by change in temperature regime and more warmer winters) have also been documented for the Baltic Sea for zooplankton and fish stocks [57,58]. The related changes that have taken place in these Northern European waters are sufficiently abrupt and persistent to be termed as 'regime shifts' [59]. Similarly in the Mediterranean, zooplankton communities have also been linked to regional warming and the NAO index [60]. All these observed changes appear to be closely correlated to climate-driven sea temperature fluctuations. Indirectly, the progressive freshening of the Labrador Sea region, attributed to climate warming and the increase in freshwater input to the ocean from melting ice, has resulted in the increasing abundance, blooms and shifts in seasonal cycles of dinoflagellates due to the increased stability of the water-column [61]. Similarly, increases in coccolithophore blooms in the Barents Sea and HABs in the North Sea are associated with negative salinity anomalies and warmer temperatures leading to increased stratification [54,62].

In the Benguela upwelling system in the South Atlantic, long-term trends in the abundance and community structure of coastal zooplankton have been related to large-scale climatic influences [63]. Similarly, changes in mesozooplankton abundance have also been related to large-scale climate influences in the Californian upwelling system [64]. The progressive warming in the Southern Ocean has been associated with a decline in krill [65] and an associated decline in the population sizes of many seabirds and seals monitored on several breeding sites [66,67]. In the Southern Ocean the long-term decline in krill stock has been linked to changes in winter ice extent which in turn has been related to warming temperatures [65]. Changes in the abundance of krill have profound implications for the Southern Ocean food-web. The progressive warming of the Southern Ocean has also been associated with the decline in the population sizes of many seabirds and seals monitored on several breeding sites [67]. Recent investigations of planktonic foraminifera from sediment cores encompassing the last 1400 a has revealed anomalous changes in the community structure over the last few decades. The study suggests that ocean warming as already exceeded the range of natural variability [68]. A recent major ecosystem shift in the northern Bering Sea has been attributed to regional climate warming and trends in the Arctic Oscillation [35]. Decadal changes in zooplankton related to climatic variability in the west sub arctic North Pacific have also been observed [69] and in the Japan/East Sea [70].

Many changes in abundance of marine commercial fish stocks have been observed over the last few decades in the Atlantic and Pacific Oceans but it is extremely difficult to separate, in terms of changes in population densities and recruitment, regional climate effects from direct anthropogenic influences like fishing. Geographical range extensions mentioned earlier or changes in the geographical distribution of fish populations, however, can be more confidently linked to hydro-climatic variation and regional climate warming. Similar to the observed changes in marine planktonic systems many long-term changes in pelagic fish populations have been associated with known natural modes of climatic oscillations such as ENSO and the Pacific Decadal Oscillation (PDO) in the Pacific and the NAO in the North Atlantic (see reviews: [5,21,22,26]). For example, variations in SST driven by NAO fluctuations have been linked to fluctuations in cod recruitment both off Labrador and Newfoundland and in the Barents Sea [71]. Populations of herring, sardine, salmon and tuna have also been related to fluctuations in the NAO index [5,45]. Warm events related to El Nino episodes and climate induced ecological regime shifts in the Pacific have been related to the disruption of many commercial fisheries [21,26,72]. These changes highlight the sensitivity of fish populations to environmental change. Direct evidence of biological impacts of anthropogenic climate change is, however, difficult to discern due to the background of natural variation on a variety of spatial and temporal scales and in particular natural oscillations in climate. A recent study based on a 50 a larval fish time-series from CalCOFI showed that exploited fish species were more vulnerable to the impacts of climate change than non-exploited species. The authors suggest that the enhanced response to environmental change of exploited species was due to a reduced spatial heterogeneity caused by fishery-induced age truncation and a restriction of geographic distribution that had accompanied fishing pressure [73].

2.4. Pelagic Biodiversity and Invasive Species

At the ocean basin scale studies on the pelagic biodiversity of zooplankton copepods are related to temperature and an increase in warming over the last few decades has been followed by an increase in diversity [74–76]. In particular, increases in diversity are seen when a previously low diversity system like Arctic and cold-boreal provinces undergo prolonged warming events. The overall diversity patterns of pelagic organisms, peaking between 20° and 30° north or south, follow temperature gradients in the world's oceans [77]. Similarly, phytoplankton show a relationship between temperature and diversity which is linked to the phytoplankton community having a higher diversity but an overall smaller size-fraction and a more complex food-web structure (i.e., microbial-based versus diatom based production) in warmer more stratified environments. Climate warming will therefore increase planktonic diversity throughout the cooler regions of the world's oceans as temperature isotherms shift poleward. However, the relationship between temperature and pelagic fish diversity is far more complex due to other anthropogenic pressures such as over fishing apparently playing a significant role in diversity patterns [19] (see Chapter 14).

Climate warming will open up new thermally defined habitats for previously denied non-indigenous species (e.g., sub-tropical species in the North Sea) and invasive species allowing them to establish viable populations in areas that were once environmentally unsuitable. Apart from these thermal boundaries limits moving progressively poleward and in some cases expanding, the rapid climate change observed the Arctic may have even larger consequences for the establishment of invasive species and the biodiversity of the North Atlantic. The thickness and areal coverage of summer ice in the Arctic have been melting at an increasingly rapid rate over the last two decades; to reach the lowest ever recorded extent in September 2007. In the spring following the unusually large ice free period in 1998 large numbers of a Pacific diatom *Neodenticula seminae* were found in samples taken by the CPR survey in the Labrador Sea in the North Atlantic. *N. seminae* is an abundant member of the phytoplankton in the subpolar North Pacific and has a well defined palaeo history based on deep sea cores. According to the palaeo evidence and modern surface sampling in the North Atlantic since 1948 this was the first record of this species in the North Atlantic for at least 800 000 a. The reappearance of *N. seminae* in the North Atlantic, and its subsequent spread southwards and eastwards to other areas in the North Atlantic, after such a long gap, could be an indicator of the scale and speed of changes that are taking place in the Arctic and North Atlantic oceans as a consequence of climate warming [78]. The diatom species may itself could be the first evidence of a trans-Arctic migration in modern times and be a harbinger of a potential inundation of new organisms into the North Atlantic. The consequences of such a change to the function, climatic feedbacks and biodiversity of Arctic systems are at present unknown.

3. CONCLUSION AND SUMMARY OF KEY INDICATORS

The case-studies highlighted in this review collectively indicate that there is substantial observational evidence that many pelagic ecosystems, both physically and biologically are responding to changes in regional climate caused predominately by the warming of SST, ocean current changes and to a lesser extent by the modification of precipitation regimes and wind patterns. The biological manifestations of climatic variability have rapidly taken the form of biogeographical, phenological, biodiversity, physiological, species abundance changes, community structural shifts and whole ecological regime shifts. Some of the most convincing evidence for the biological response to regional climate variability comes from the bottom of the marine pelagic food-web especially from phytoplankton and zooplankton communities. Many other responses associated with climate warming on higher trophic levels are also indirectly associated with changes in the plankton and imply bottom-up control of the marine pelagic environment. It is therefore assumed that one of the ways in which populations respond to climate is in part the determined by changes in the food-web structure where the population is embedded, with synchrony between predator and prey (match–mismatch) playing an important role.

At the species level, some of the first consequences of climate warming and global change are often seen in a species phenology (i.e., timing of annual occurring life-cycle events) and in species geographical distribution responses. This is mainly because temperature continually impacts the life cycle of the species and naturally the population will respond over time, providing it is not biotically restrained or spatially restricted, to its optimum position within its bioclimatic envelope. Whether this is within a temporal niche as in seasonal succession (observed as a phenological response) or in its overall biogeographical distribution (observed as a geographical movement in a population). These biological changes as well as those changes observed in biodiversity and planktonic abundance and productivity are perhaps the key indicators signifying the large scale changes occurring in our world's oceans as a consequence of climate and global change.

Summarising the observed case-studies, what particularly stands out in this review is the rapidity of the pelagic and planktonic response, be it biogeographically or phenologically, to climate warming and global change compared to their terrestrial counterparts. For example, plankton shifts of up to 200 km per decade [10] have been observed in the North East Atlantic compared with a meta-analytic terrestrial average of 6 km per decade [16]. Similarly, changes in phenology of up to 6 weeks have been observed in pelagic ecosystems [11] compared with a mean phenological change of 2.3 collectively observed for 172 species of plants, birds, insects and amphibians [16]. Of the myriad differences between the terrestrial and marine realm (see section 1.2 the rapidity of the planktonic response is predominantly due to their short life-cycles and in their mainly passive response to advective changes. These changes highlighted in this review are set to continue into the future following current climate warming projections. It is therefore thought that the currently observed and future warming have and are likely to continue altering the geographical distribution of primary and secondary planktonic production [53], affecting marine ecosystem services such as oxygen production, carbon sequestration and biogeochemical cycling and placing additional stress on already depleted fish and mammal populations.

In terms of feedback mechanisms on Earth's climate, it is thought that these biological pumps will be less efficient in a warmer world due to changes in the phytoplankton composition (floristic shifts) and less overall nutrient mixing (reduced bulk properties) due to increased stratification of the world's oceans. In particular, this will affect large areas of the tropical oceans that are permanently stratified [9]. There also exists a strong negative relationship between ocean productivity and SST (linked through nutrient availability) at a global scale [9,12,13]. Although climate change and its spatially heterogeneous effect on surface wind-patterns, wind strength, upwelling and deep-water mixing makes many regional predictions beset with uncertainty. It is also worth noting that potential habitat expansion for pelagic organisms in the Northern Hemisphere due to the melting of Arctic ice will be severely restricted by light limitations dictating seasonal phytoplankton production. However, many of these scenarios are still at their infancy stage and while it is relatively simpler to predict changing ocean physics under climate forcing, understanding the biological response due to the underlying complexity of biological communities and their quite often nonlinear responses to environmental change makes predicting floristic changes fraught with uncertainty. Investigating the importance of biological nitrogen fixation and the production of dimethylsulfide (DMS) by certain phytoplankton is currently needed to understand the biological consequences of increased stratification on nitrogen cycles [79,80] and biological feedbacks [81,82].

Ecologically speaking and on a planetary scale, plankton and pelagic ecosystems as a metaphorical collective entity, are perhaps some of the most sensitive organisms to environmental change and one of the most important biological communities on the planet. They are responsible for the overwhelming majority of marine biological production that fuel marine food-webs and nutrient cycling as well as contributing to approximately half of the world's oxygen production and carbon sequestration. Virtually, all the biological observations highlighted in this review result from financially fragile multidecadal monitoring programmes. Future biological monitoring of these ecosystems, through an integrated and sustained observational approach, will be essential in understanding the continuing impacts of climate and global change on our planetary system. This in turn may allow us through international collaboration to mitigate and adaptively manage some of their more detrimental impacts.

REFERENCES

1. R.E. Zeebe, J.C. Zachos, K. Caldeira, T. Tyrrell, Science 321 (2008) 51–52.
2. L. Bopp, O. Aumont, P. Cadule, S. Alvain, M. Gehlen, Geophys. Res. Lett. 32 (2005) Article No. L19606.
3. A. Lopez-Urrutia, E. San Martin, R.P. Harris, X. Irigoien, Proc. Natl. Acad. Sci. USA 103 (2006) 8739–8744.
4. A.H. Taylor, J.I. Allen, P.A. Clark, Nature 416 (2002) 629–632.
5. K.F. Drinkwater, A. Belgrano, A. Borja, C.A.M. Edwards, C.H. Greene, G. Ottersen, A.J. Pershing, H. Walker, Geophys. Monograph 134 (2003) 211–243.
6. G. Beaugrand, F. Ibañez, J.A. Lindley, Mar. Ecol. Prog. Ser. 219 (2001) 205–219.
7. G. Beaugrand, P.C. Reid, Glob. Change Biol. 9 (2003) 801–817.
8. D. Atkinson, R.M. Sibly, Trends Ecol. Evol. 12 (1997) 235–239.
9. M.J. Behrenfeld, R.T. O'Malley, D.A. Siegel, C.R. McClain, J.L. Sarmiento, G.C. Feldman, A.J. Milligan, P.G. Falkowski, R.M. Letelier, E.S. Boss, Nature 444 (2006) 752–755.

10. G. Beaugrand, P.C. Reid, F. Ibanez, J.A. Lindley, M. Edwards, Science 296 (2002) 1692–1694.
11. M. Edwards, A.J. Richardson, Nature 430 (2004) 881–884.
12. D. Kamykowski, S.J. Zentara, Deep Sea Res. (I Oceanogr. Res. Pap.) 33 (1986) 89–105.
13. D. Kamykowski, S.J. Zentara, Deep Sea Res. (I Oceanogr. Res. Pap.) 52 (2005) 1719–1744.
14. H.O. Portner, R. Knust, Science 315 (2007) 95–97.
15. J.A. Berges, D.E. Varela, P.J. Harrison, Mar. Ecol. Prog. Ser. 225 (2002) 139–146.
16. C. Parmesan, G. Yohe, Nature 421 (2003) 37–42.
17. L. Hannah, T.E. Lovejoy, S.H. Schneider, in: T.E. Lovejoy, L. Hannah (Eds.), Climate Change and Biodiversity, Yale University Press, New Haven, CT, 2005, pp. 3–14.
18. R.A. Feely, C.L. Sabine, K. Lee, W. Berelson, J. Kleypas, V.J. Fabry, F.J. Millero, Science 305 (2004) 362–366.
19. B. Worm, E.B. Barbier, N. Beaumont, J.E. Duffy, C. Folke, B.S. Halpern, J.B.C. Jackson, H.K. Lotze, F. Micheli, S.R. Palumbi, E. Sala, K.A. Selkoe, J.J. Stachowicz, R. Watson, Science 314 (2006) 787–790.
20. R.A. Duce, J. LaRoche, K. Altieri, K.R. Arrigo, A.R. Baker, D.G. Capone, S. Cornell, F. Dentener, J. Galloway, R.S. Ganeshram, R.J. Geider, T. Jickells, M.M. Kuypers, R. Langlois, P.S. Liss, S.M. Liu, J.J. Middelburg, C.M. Moore, S. Nickovic, A. Oschlies, T. Pedersen, J. Prospero, R. Schlitzer, S. Seitzinger, L.L. Sorensen, M. Uematsu, O. Ulloa, M. Voss, B. Ward, L. Zamora, Science 320 (2008) 893–897.
21. J. Overland, S. Rodionov, S. Minobe, N. Bond, Prog. Oceanogr. 77 (2008) 92–102.
22. N.C. Stenseth, G. Ottersen, J.W. Hurrell, A. Mysterud, M. Lima, K.S. Chan, N.G. Yoccoz, B. Aadlandsvik, Proc. R. Soc. Lond. Ser. B: Biol. Sci. 270 (2003) 2087–2096.
23. R.T. Sutton, D.L.R. Hodson, Science 309 (2005) 115–118.
24. W. Greve, S. Prinage, H. Zidowitz, J. Nast, F. Reiners, ICES J. Mar. Sci. 62 (2005) 1216–1223.
25. P.C. Reid, J.M. Colebrook, J.B.L. Matthews, J. Aiken, Prog. Oceanogr. 58 (2003) 117–173.
26. J.A. McGowan, D.R. Cayan, L.M. Dorman, Science 281 (1998) 210–217.
27. D.G. Johns, M. Edwards, S.D. Batten, Can. J. Fish. Aquat. Sci. 58 (2001) 2121–2124.
28. J.-C. Quero, M.-H. Du Buit, J.-J. Vayne, Oceanol. Acta 21 (1998) 345–351.
29. K. Brander, G. Blom, M.F. Borges, K. Erzini, G. Henderson, B.R. MacKensie, H. Mendes, J. Ribeiro, A.M.P. Santos, R. Toresen, ICES Mar. Sci. Symp. 219 (2003) 261–270.
30. A.L. Perry, P.J. Low, J.R. Ellis, J.D. Reynolds, Science 308 (2005) 1912–1915.
31. D. Beare, F. Burns, E. Jones, K. Peach, E. Portilla, T. Greig, E. McKenzie, D. Reid, Glob. Change Biol. 10 (2004) 1209–1213.
32. K. Brander, G. Blom, M.F. Borges, K. Erzini, G. Henderson, B.R. Mackenzie, H. Mendes, J. Ribeiro, A.M.P. Santos, R. Toresen, ICES Mar. Sci. Symp. 219 (2003) 261–270.
33. M.J. Genner, D.W. Sims, V.J. Wearmouth, E.J. Southall, A.J. Southward, P.A. Henderson, S.J. Hawkins, Proc. R. Soc. Lond. Ser. B: Biol. Sci. 271 (2004) 655–661.
34. G.C. Rose, R.L. O'Driscoll, ICES J. Mar. Sci. 59 (2002) 1018–1026.
35. J.M. Grebmeier, J.E. Overland, S.E. Moore, E.V. Farley, E.C. Carmack, L.W. Cooper, K.E. Frey, J.H. Helle, F.A. McLaughlin, S.L. McNutt, Science 311 (2006) 1461–1464.
36. C.D. Harvell, K. Kim, J.M. Burkholder, R.R. Colwell, P.R. Epstein, D.J. Grimes, E.E. Hofmann, E.K. Lipp, A. Osterhaus, R.M. Overstreet, J.W. Porter, G.W. Smith, G.R. Vasta, Science 285 (Suppl. 5433) (1999) 1505–1510.
37. H. McCallum, D. Harvell, A. Dobson, Ecol. Lett. 6 (Dec, 2003) 1062–1067.
38. M.P. Harris, D. Beare, R. Toresen, L. Nottestad, M. Kloppmann, H. Dorner, K. Peach, D.R.A. Rushton, J. Foster-Smith, S. Wanless, Mar. Biol. 151 (2007) 973–983.

39. N.K. Dulvy, S.I. Rogers, S. Jennings, V. Stelzenmuller, S.R. Dye, H.R. Skjodal, J. Appl. Ecol. (2008) doi:10.1111/j.1365–2664.2008.01488.x
40. U. Sommer, K. Lengfellner, Glob. Change Biol. 14 (2008) 1199–1208.
41. G. Beaugrand, K.M. Brander, J.A. Lindley, S. Souissi, P.C. Reid, Nature 426 (2003) 661–664.
42. J.C. Molinero, F. Ibanez, S. Souissi, M. Chifflet, P. Nival, Oecologia 145 (2005) 640–649.
43. D.L. Mackas, R. Goldblatt, A.G. Lewis, Can. J. Fish. Aquat. Sci. 55 (1998) 1878–1893.
44. S. Chiba, M.N. Aita, K. Tadokoro, T. Saino, H. Sugisaki, K. Nakata, Prog. Oceanogr. 77 (2008) 112–126.
45. G. Beaugrand, P.C. Reid, Glob. Change Biol. 9 (2003) 801–817.
46. M. Edwards, G. Beaugrand, P.C. Reid, A.A. Rowden, M.B. Jones, Mar. Ecol. Prog. Ser. 239 (2002) 1–10.
47. M. Edwards, P. Reid, B. Planque, ICES J. Mar. Sci. 58 (2001) 39–49.
48. J.M. Fromentin, B. Planque, Mar. Ecol. Prog. Ser. 134 (1996) 111–118.
49. R.R. Kirby, G. Beaugrand, J.A. Lindley, A.J. Richardson, M. Edwards, P.C. Reid, Mar. Ecol. Prog. Ser. 330 (2007) 31–38.
50. C.P. Lynam, S.J. Hay, A.S. Brierley, Limnol. Oceanogr. 49 (2004) 637–643.
51. P.C. Reid, M. Edwards, Senck. Marit. 31 (2001) 107–115.
52. P.C. Reid, M. Edwards, H.G. Hunt, A.J. Warner, Nature 391 (1998) 546.
53. A.J. Richardson, D.S. Schoeman, Science 305 (2004) 1609–1612.
54. M. Edwards, D.G. Johns, S.C. Leterme, E. Svendsen, A.J. Richardson, Limnol. Oceanogr. 51 (2006) 820–829.
55. M. Frederiksen, M. Edwards, A.J. Richardson, N.C. Halliday, S. Wanless, J. Anim. Ecol. 1259–1268.
56. M. Edwards, P. Licandro, D.G. Johns, A.W.G. John, D.P. Stevens, SAHFOS Tech. Rep. 3 (2006) 1–8.
57. F.W. Koster, C. Mollmann, H.H. Hinrichsen, K. Wieland, J. Tomkiewicz, G. Kraus, R. Voss, A. Makarchouk, B.R. MacKenzie, M.A. St. John, D. Schnack, N. Rohlf, T. Linkowski, J.E. Beyer, ICES J. Mar. Sci. 62 (2005) 1408–1425.
58. J. Alheit, C. Mollmann, J. Dutz, G. Kornilovs, P. Loewe, V. Mohrholz, N. Wasmund, ICES J. Mar. Sci. 62 (2005) 1205–1215.
59. G. Beaugrand, Prog. Oceanogr. 60 (2004) 245–262.
60. J.C. Molinero, F. Ibanez, P. Nival, E. Buecher, S. Souissi, Limnol. Oceanogr. 50 (2005) 1213–1220.
61. D.G. Johns, M. Edwards, A. Richardson, J.I. Spicer, Mar. Ecol. Prog. Ser. 265 (2003) 283–287.
62. T.J. Smyth, T. Tyrrell, B. Tarrant, Geophys. Res. Lett. 31 (2004) Article No. L11302.
63. H.M. Verheye, S. Afr. J. Mar. Sci./S.-Afr. Tydskr. Seewet. 19 (1998) 317–332.
64. B.E. Lavaniegos, M.D. Ohman, Deep Sea Res. (II Top. Stud. Oceanogr.) 50 (2003) 2473–2498.
65. A. Atkinson, V. Siegel, E. Pakhomov, P. Rothery, Nature 432 (2004) 100–103.
66. C. Barbraud, H. Weimerskirch, Nature 411 (2001) 183–186.
67. H. Weimerskirch, P. Inchausti, C. Guinet, C. Barbraud, Antarct. Sci. 15 (2003) 249–256.
68. D.B. Field, T.R. Baumgarter, C.D. Charles, V. Ferreira-Bartrina, M.D. Ohman, Science 311 (2006) 63–66.
69. S. Chiba, K. Tadokoro, H. Sugisaki, T. Saino, Glob. Change Biol. 12 (2006) 907–920.
70. S. Chiba, T. Saino, Prog. Oceanogr. 57 (2003) 317–339.
71. N.C. Stenseth, A. Mysterud, G. Ottersen, J.W. Hurrell, K.S. Chan, M. Lima, Science 297 (2002) 1292–1296.

72. F.P. Chavez, J. Ryan, S.E. Lluch-Cota, M. Niquen, Science 299 (2003) 217–221.
73. C.H. Hsieh, C.S. Reiss, R.P. Hewitt, G. Sugihara, Can. J. Fish. Aquat. Sci. 65 (2008) 947–961.
74. G. Beaugrand, F. Ibanez, Mar. Ecol. Prog. Ser. 232 (2002) 197–211.
75. G. Beaugrand, F. Ibanez, J.A. Lindley, Mar. Ecol. Prog. Ser. 219 (2001) 189–203.
76. G. Beaugrand, F. Ibanez, J.A. Lindley, P.C. Reid, Mar. Ecol. Prog. Ser. 232 (2002) 179–195.
77. S. Rutherford, S. D'Hondt, W. Prell, Nature 400 (1999) 749–753.
78. P.C. Reid, D.G. Johns, M. Edwards, M. Starr, M. Poulins, P. Snoeijs, Glob. Change Biol. 13 (2007) 1910–1921.
79. K.R. Arrigo, Nature 437 (2005) 349–355.
80. D.G. Capone, J.A. Burns, J.P. Montoya, A. Subramaniam, C. Mahaffey, T. Gunderson, A.F. Michaels, E.J. Carpenter, Glob. Biogeochem. Cycles 19 (2005) 1–17.
81. L. Bopp, O. Boucher, O. Aumont, S. Belviso, J.L. Dufresne, M. Pham, P. Monfray, Can. J. Fish. Aquat. Sci. 61 (2004) 826–835.
82. S.L. Strom, Science 320 (2008) 1043–1045.

Chapter 13

Changes in Coral Reef Ecosystems as an Indicator of Climate and Global Change

Martin J. Attrill
Marine Biology and Ecology Research Centre, Marine Institute, University of Plymouth, Drake Circus, Plymouth PL4 8AA, United Kingdom

1. Introduction
2. Tropical Coral Reef Ecosystems
3. The Associated Fauna of Coral Reefs
4. Conclusion
References

1. INTRODUCTION

In comparison with terrestrial systems and other components of the marine environment, relatively little is known about how climate and global change has been affecting the organisms of the seabed with one exception: tropical coral reefs. Recent reviews [1,2] and metanalyses [3,4] have pulled together all existing knowledge on how climate is impacting, for example, range shifts and phenology [4] of the world's species, but, corals aside, few examples within these overviews are from the marine seabed. In her excellent review, Parmesan [1] devotes a section to marine community shifts, but the majority of examples are from either the pelagic (Chapter 12) or intertidal (Chapter 15) zones; only two studies on fish [5,6] are associated with the subtidal seabed. Terrestrial and freshwater examples dominate these studies. A sizeable proportion of research assessing climate impact on marine systems has investigated the response to the major climate cycles, such as the El Niño Southern Oscillation (ENSO) and North Atlantic Oscillation (NAO). These provide information on how systems respond to cooling and warming trends across the extremes of these cycles, and thus provide a model of how potentially organisms and systems may respond to climate warming [7], particularly as the occurrence and severity of ENSO events is predicted to increase under warming scenarios [8]. This future relationship between atmospheric dynamics and temperature change is, however, uncertain, so there also needs to be some caution about how past responses of biological systems to these climatic cycles reflect ongoing and future climate change [1]. Nevertheless, such studies provide much of the information available on climate responses of marine systems.

In contrast to other subtidal benthic systems, the impact of global warming-related issues on coral reefs has had one of the highest profiles in recent years, particularly following the worldwide impact of the 1997–1998 extreme El Niño event [9], the extensive public concern about this ecosystem and recent reports predicting widespread losses of reef and extinction of species [10,11]. Climate change can potentially impact coral reefs through several key mechanisms, in particular increasing sea surface temperature (SST), ocean acidification, increasing storminess and sea level rise [12,13]. The latter three mechanisms are dealt with specifically in other chapters, so this chapter will focus on the impact of rising sea temperatures on coral reef ecosystems, particularly the effect of mass bleaching of the corals themselves.

2. TROPICAL CORAL REEF ECOSYSTEMS

Concern about the human impact on coral reefs has existed for decades and, until comparatively recently, the major threats to the integrity of reef systems have been considered to be overfishing and pollution [12,14]. Such impacts can be potentially managed at the local level, but any such management will be unsuccessful when put into the context of more recent recognised effects of global climate change [12]. Similarly, climate impacts may be exacerbated by the additional effect of these other local anthropogenic factors, which have made coral reefs systems in some areas of the world more susceptible to damage. The link between climate change and region-scale mass bleaching of corals is now incontrovertible [12,13], in particular the direct link between bleaching and SST anomalies [15]. There are no records of mass bleaching prior to the 1980s, though it is unclear how extensive such bleaching was earlier in the twentieth century before widespread reporting [16]; it is unlikely, however, that bleaching of the scale seen in recent years would have gone unnoticed. In the Great Barrier Reef, for example, bleaching events have become more widespread since the 1980s (Fig. 1a), coinciding with a decline in coral cover over this time [17]; globally, mass coral bleaching has become more frequent and intense in recent decades [11].

Bleaching is due to a whitening of the corals following the expulsion of the symbiotic zooxanthellae, the algae providing most of the coral's pigment. Loss of the zooxanthellae, therefore, leaves comparatively colourless coral tissue plus the white calcareous reef skeleton. The process is often considered as a response to increasing ambient temperatures above a threshold, ~0.8–1 °C above summer average temperatures for at least 4 weeks [18,19].

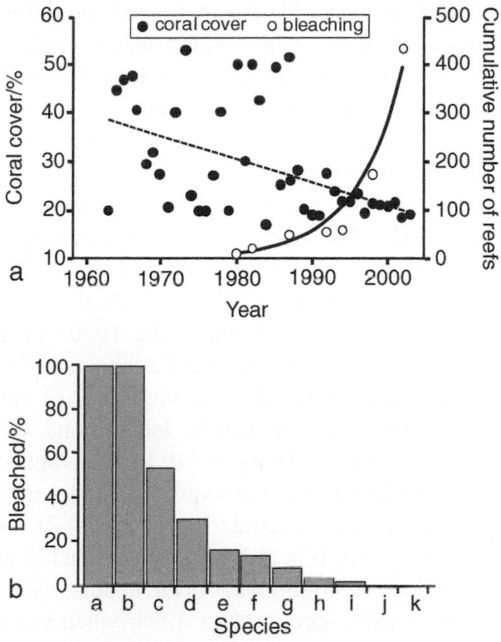

FIGURE 1 a, Trends in coral cover and number of reefs with mass bleaching on the Great Barrier Reef, Australia (Adapted by permission from Macmillan Publishers Ltd [Nature], Ref. [17]). b. Differential bleaching responses of nine species of corals in Raiatea, French Polynesia, during May 2002 (Redrawn from Ref. [12], reprinted with permission from AAAS). (a, *Acropora anthocercis*; b, *A. retusa*; c, *Montipora tuberculosa*; d, *Pocillopora verrucosa*; e, *M. caliculata*; f, *Leptastrea transversa*; g, *P. eydouxi*; h, *P. meandrina*; i, *L. bewickensis*; j, *Porites lobata*; k, *L. purpurea*).

Bleaching thresholds across coral species are likely to represent a broad spectrum of responses (Fig. 1b), however, and susceptibilities will change over time following phenotypic and genetic responses of the corals [12]. It is clear that many coral species exist over a wide biogeographical range of temperatures and individuals have subsequently different bleaching thresholds in terms of absolute temperature, indicating adaptive ability within species [20]. The key driver for bleaching, therefore, appears to be temperature increases above those generally experienced by corals in any given location. It has been hypothesised that the bleaching response is an adaptive process [21,22], the corals expelling susceptible symbionts and taking up more resistant ones; whilst there is some evidence for this [13], it does not appear to be supported by observations on the fate of bleached corals [12], perhaps being more accurately described as a stress response.

What is uncontested, however, is that major bleaching events can severely impact coral reefs in the long term: if bleaching is prolonged or exceeds 2 °C above seasonal maxima corals can die [13]. Major bleaching events were observed in 1982–1983, 1987–1988, 1994–1995, 1997–1998 [13], 2002

(GBR, [23]) and 2005 (Caribbean, [19]) and have often, but not always, been associated with intense El Niño events which enhance global sea temperatures. The 1997–1998 event was the most extreme El Niño on record [9] and resulted in extensive bleaching recorded across the world's coral reefs [24]. An estimated 16% of the world's coral was lost in this one event, in particular within the Indian Ocean/SE Asia [24] (Table 1), with only partial recovery evident. Overall, only approximately half of the reefs affected in 1998 have recovered [24]. The 2005 event, however, occurred without an El Niño and has provided evidence of the impact of the underlying increasing trend in global sea water temperatures (Chapter 19); this has been related to anthropogenic forcing in the Atlantic since the 1970s [19]. Anomalously warm temperatures were recorded across the Caribbean and tropical Atlantic [19], resulting in exceptional levels of bleaching: 90% of coral cover in the British Virgin Islands, 80% in US Virgin Islands and 66% in Trinidad and Tobago, for example [19]. Analysis of local temperature anomalies revealed that SSTs were higher than the expected annual maxima for longer than had been previously been recorded [19] (Fig. 2), resulting in the exceptional bleaching observed. It is also notable that such maxima have been exceeded every year since 1995 (Fig. 2); prior to this, such extremes were rare. The second highest value occurred in 1998, when extensive bleaching was also apparent in the Caribbean [25].

The only long-term data available on the impact of climate on coral species is from the geological record, as reviewed by Hughes et al. [12]. Many extant species of coral can be traced back in time to the Pliocene (1.8–5.3 Ma ago), so have experienced periods of extensive and rapid warming and cooling during the Pleistocene and Holocene prior to human impact [14]. In response to climatic changes, there is evidence species underwent large shifts in their distributional range [26]; for corals, this extended up to 500 km further south in Australia, for example [12]. Until very recently, there was no evidence of such shifts in response to modern climate change, but, following increasing sea temperatures, two species of *Acropora* have re-expanded their ranges 50 km northwards along the Florida Peninsula into areas where they have not been recorded for 6000 years [27].

Much evidence, therefore, exists that increasing SSTs are impacting coral reefs through extensive bleaching and subsequent mortality, particularly during exceptional years where average maximum temperatures are exceeded. Corals will also be impacted through changes in seawater pH, affecting their ability to produce calcareous reef skeletons as covered in Chapter 21 (see also Ref. [11]), and potentially further affected by severe storms [28] and sea level rise [29], other consequences of climate change [13]. Unlike past climate changes, however, coral reefs are now also markedly influenced by the synergistic effect of other anthropogenic activities, such as fishing and pollution, making them much more susceptible to changes associated with current climate warming [12].

TABLE 1 Summary of status of coral reefs in 17 regions of the world as of 2004 from Ref. [24], indicating proportion of coral reefs in each region that have been destroyed (i.e., 90% coral lost and unlikely to recover), plus proportion lost and recovered following the 1997–1998 El Niño event

Region	Coral reef area/km^2	Reefs destroyed/ (%)	Reefs destroyed in 1998/(%)	Reefs recovered/ (%)
The Gulfs	3800	65	15	2
South Asia	19 210	45	65	13
SE Asia	91 700	38	18	8
SW Indian Ocean	5270	22	41	20
US Caribbean	3040	16	NA	NA
S Tropical America	5120	15	NA	NA
E & N Asia	5400	14	10	3
East Africa	6800	12	31	22
East Antilles	1920	12	NA	NA
Central America	4630	10	NA	NA
Micronesian Islands	12 700	8	2	1
North Caribbean	9800	5	4	3
Red Sea	17 640	4	4	2
SW Pacific Islands	27 060	3	10	8
Australia and PNG	62 800	2	3	1
Polynesian Islands	6733	2	1	1
Hawaiian Islands	1180	1	NA	NA
TOTAL	284 803	20	16	6.4

NA = not applicable as no losses recorded in 1998.

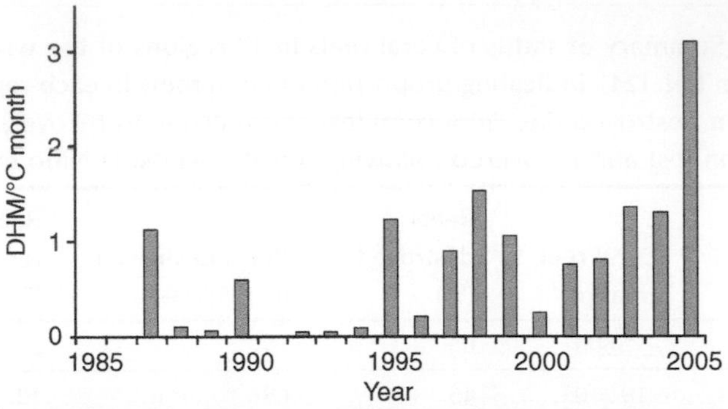

FIGURE 2 Maximum annual thermal stress on Eastern Caribbean coral reefs, presented as Degree Heating Months (DHM, °C month) since 1985. A DHM is equal to 1 month of SST that is 1°C greater than the monthly maximum climatology for that area (Adapted from Ref. [19], copyright (2007) National Academy of Sciences, U.S.A).

3. THE ASSOCIATED FAUNA OF CORAL REEFS

Unlike the corals themselves, comparatively little research has been focused on how climate and global change may be influencing the vast biodiversity associated with coral reefs (about a quarter of all known marine species [13,30]), perhaps due to the logistic difficulties and expense of constructing long-term data sets on these organisms. Additionally, studies on changes to fish populations and their immediate prey are greatly influenced by other stresses, particularly fishing pressure [13], making it more difficult to identify signals of climate change. However, although over 4000 species of fish are associated with coral reefs [31], remarkably few studies have addressed the impact of climate change on this group. There is little current evidence of direct impact of rising temperatures on coral reef fish, due mainly to lack of data on thermal tolerance [32], although one study in the tropical eastern Pacific [33] determined that the range of Critical Thermal Maxima for 15 fish species (34.7–40.8 °C) was higher than record temperatures (32 °C) recorded during the 1997–1998 El Niño event. Similarly, evidence of distributional shifts in coral fish is lacking, despite being predicted by climate impact reviews [31]. This may be due to a lack of habitat unless coral reefs also extend their range [27].

There is evidence, however, that coral reef fish can be impacted more severely by the indirect effects of mass coral bleaching [34–36]. In a study of Seychelles reefs following the 1997–1998 El Niño [37], intense bleaching resulted in decreased fish abundance and a shift within the assemblage from corallivores to species feeding on invertebrates. The size structure of fish also changed with an increase in large fish [38], a possible time-lag response due to a reduction in coral structural complexity affecting fish recruitment and thus the number of juveniles. However, a minimal effect of this 1997–1998 event was noted on the diversity and abundance of cryptobenthic fish species on the Great Barrier Reef [39], suggesting coral reef fishes may be comparatively resilient to short-term perturbations [39] as long as reef structure is sustained [36].

The most extensive study on the impact of a climate event on coral-associated invertebrates (>500 species) has been undertaken in Bahia, Brazil, by Kelmo and colleagues [40–42]. The 1997–1998 El Niño resulted in anomalous high temperatures and a reduction in the usual high turbidity of the area allowing more UV light to reach the reefs. All groups studied (except sponges) showed extensive reductions in diversity for several years following El Niño (Fig. 3), including the local extinction of one species of coral (*Porites astreoides*); this was most likely due to extensive neoplastic tumours on the corals following UV damage [43]. The density of the majority of species also decreased dramatically (Fig. 3), with only the urchin *Diadema antillarum* showing an opportunistic response to the changing conditions and disappearance of competitive taxa [44]. What is remarkable about this data set is the recovery of groups after 2001, with diversity returning to, or exceeding, levels prior to El Niño (Fig. 3). Reef assemblages clearly have the ability to recover from extreme climate events, but only if no further such events subsequently

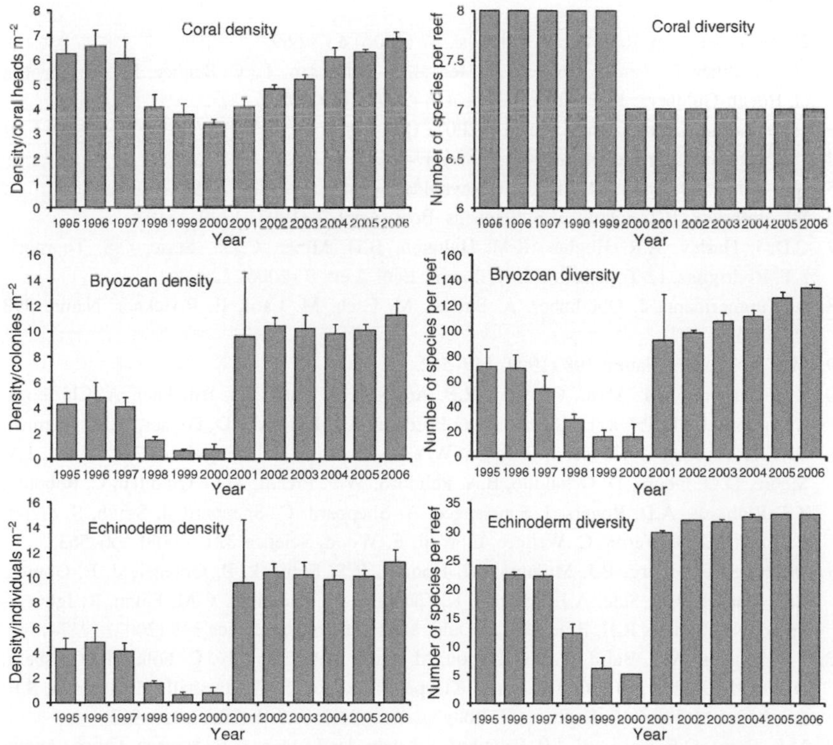

FIGURE 3 Changes in the density and diversity of three major groups of invertebrate (corals, bryozoans, echinoderms) on the patch reefs of Bahia, Brazil since 1995, indicating the impact of the 1997-8 El Niño on all measures, plus the marked recovery from three years after the event in all groups except coral diversity. Levels of echinoderm and bryozoan diversity are higher following recovery than during the pre-El Niño period.

occur; in Bahia, no major El Niño or bleaching event was evident after 1998. Models suggest ENSO events are likely to be more frequent and severe in the future [8], so this recovery ability of coral reef communities may be compromised by climate change.

4. CONCLUSION

There is clear, unequivocal evidence that climate change is affecting coral reef systems [12], with a particular concern about mass coral bleaching due to rising temperatures and the subsequent effects this will have on coral survival and thus the associated organisms. Whilst corals have an innate ability to acclimatise to such change [45], as evidenced in the past [12], severe and regular El Niño events coupled with the modern synergistic impact of other human activities such as fishing, pollution (including comparatively fast ocean acidification [11]) and tourism make coral reefs amongst the most vulnerable of the world's ecosystems under current scenarios of future climate change.

REFERENCES

1. C. Parmesan, Ann. Rev. Ecol. Evol. Syst. 37 (2006) 637–669.
2. G.R. Walther, E. Post, P. Convey, A. Menzel, C. Parmesan, T.J.C. Beebee, J.M. Fromentin, O. Hoegh-Guldberg, F. Bairlein, Nature 416 (2002) 389–395.
3. C. Parmesan, Glob. Change Biol. 13 (2007) 1860–1872.
4. C. Parmesan, G. Yohe, Nature 421 (2003) 37–42.
5. A.L. Perry, P.J. Low, J.R. Ellis, J.D. Reynolds, Science 308 (2005) 1912–1915.
6. S.J. Holbrook, R.J. Schmitt, J.S. Stephens, Ecol. Appl. 7 (1997) 1299–1310.
7. C.D.G. Harley, A.R. Hughes, K.M. Hultgren, B.G. Miner, C.J.B. Sorte, C.S. Thornber, L.F. Rodriguez, L. Tomanek, S.L. Williams, Ecol. Lett. 9 (2006) 228–241.
8. A. Timmermann, J. Oberhuber, A. Bacher, M. Esch, M. Latif, E. Roeckner, Nature 398 (1999) 694–697.
9. M.J. McPhaden, Nature 398 (1999) 559–562.
10. K.E. Carpenter, M. Abrar, G. Aeby, R.B. Aronson, S. Banks, A. Bruckner, A. Chiriboga, J. Cortes, J.C. Delbeek, L. DeVantier, G.J. Edgar, A.J. Edwards, D. Fenner, H.M. Guzman, B.W. Hoeksema, G. Hodgson, O. Johan, W.Y. Licuanan, S.R. Livingstone, E.R. Lovell, J.A. Moore, D.O. Obura, D. Ochavillo, B.A. Polidoro, W.F. Precht, M.C. Quibilan, C. Reboton, Z.T. Richards, A.D. Rogers, J. Sanciangco, A. Sheppard, C. Sheppard, J. Smith, S. Stuart, E. Turak, J.E.N. Veron, C. Wallace, E. Weil, E. Wood, Science 321 (2008) 560–563.
11. O. Hoegh-Guldberg, P.J. Mumby, A.J. Hooten, R.S. Steneck, P. Greenfield, E. Gomez, C.D. Harvell, P.F. Sale, A.J. Edwards, K. Caldeira, N. Knowlton, C.M. Eakin, R. Iglesias-Prieto, N. Muthiga, R.H. Bradbury, A. Dubi, M.E. Hatziolos, Science 318 (2007) 1737–1742.
12. T.P. Hughes, A.H. Baird, D.R. Bellwood, M. Card, S.R. Connolly, C. Folke, R. Grosberg, O. Hoegh-Guldberg, J.B.C. Jackson, J. Kleypas, J.M. Lough, P. Marshall, M. Nystrom, S.R. Palumbi, J.M. Pandolfi, B. Rosen, J. Roughgarden, Science 301 (2003) 929–933.
13. M.L. Parry, O.F. Canziani, J.P. Palutikof, P.J. van der Linden, C.E. Hanson, Cross-Chapter Case Study, Climate Change 2007: Impacts, Adaptation and Vulnerability. Contribution of Working Group II to the Fourth Assessment Report of the IPCC, Cambridge University Press, Cambridge, UK, 2007, pp. 843–868.

14. J.B.C. Jackson, M.X. Kirby, W.H. Berger, K.A. Bjorndal, L.W. Botsford, B.J. Bourque, R.H. Bradbury, R. Cooke, J. Erlandson, J.A. Estes, T.P. Hughes, S. Kidwell, C.B. Lange, H.S. Lenihan, J.M. Pandolfi, C.H. Peterson, R.S. Steneck, M.J. Tegner, R.R. Warner, Science 293 (2001) 629–638.
15. J.P. McWilliams, I.M. Cote, J.A. Gill, W.J. Sutherland, A.R. Watkinson, Ecology 86 (2005) 2055–2060.
16. A.D. Barton, K.S. Casey, Coral Reefs 24 (2005) 536–554.
17. D.R. Bellwood, T.P. Hughes, C. Folke, M. Nystrom, Nature 429 (2004) 827–833.
18. O. Hoegh-Guldberg, Mar. Freshw. Res. 50 (1999) 839–866.
19. S.D. Donner, T.R. Knutson, M. Oppenheimer, Proc. Natl. Acad. Sci. USA 104 (2007) 5483–5488.
20. S.D. Donner, W.J. Skirving, C.M. Little, M. Oppenheimer, O. Hoegh-Guldberg, Glob. Change Biol. 11 (2005) 2251–2265.
21. A.C. Baker, Nature 411 (2001) 765–766.
22. R.W. Buddemeier, D.G. Fautin, Bioscience 43 (1993) 320–326.
23. R. Berkelmans, G. De'ath, S. Kininmonth, W.J. Skirving, Coral Reefs 23 (2004) 74–83.
24. C. Wilkinson (Ed.), Status of Coral Reefs of the World: 2004, vol. 1, AIMS, Townsville, Australia, 2004, p. 316.
25. R.B. Aronson, W.F. Precht, M.A. Toscano, K.H. Koltes, Mar. Biol. 141 (2002) 435–447.
26. K. Roy, D. Jablonski, J.W. Valentine, Ecol. Lett. 4 (2001) 366–370.
27. W.F. Precht, R.B. Aronson, Front. Ecol. Environ. 2 (2004) 307–314.
28. T.A. Gardner, I.M. Cote, J.A. Gill, A. Grant, A.R. Watkinson, Ecology 86 (2005) 174–184.
29. N. Knowlton, Proc. Natl. Acad. Sci. USA 98 (2001) 5419–5425.
30. C.M. Roberts, C.J. McClean, J.E.N. Veron, J.P. Hawkins, G.R. Allen, D.E. McAllister, C.G. Mittermeier, F.W. Schueler, M. Spalding, F. Wells, C. Vynne, T.B. Werner, Science 295 (2002) 1280–1284.
31. P.L. Munday, G.P. Jones, M.S. Pratchett, A.J. Williams, Fish Fish. 9 (2008) 261–285.
32. J.M. Roessig, C.M. Woodley, J.J. Cech, L.J. Hansen, Rev. Fish Biol. Fish. 14 (2004) 251–275.
33. C. Mora, A.F. Ospina, Mar. Biol. 139 (2001) 765–769.
34. D.J. Booth, G.A. Beretta, Mar. Ecol. Prog. Ser. 245 (2002) 205–212.
35. K.C. Garpe, S.A.S. Yahya, U. Lindahl, M.C. Ohman, Mar. Ecol. Prog. Ser. 315 (2006) 237–247.
36. U. Lindahl, M.C. Ohman, C.K. Schelten, Mar. Pollut. Bull. 42 (2001) 127–131.
37. M.D. Spalding, G.E. Jarvis, Mar. Pollut. Bull. 44 (2002) 309–321.
38. N.A.J. Graham, S.K. Wilson, S. Jennings, N.V.C. Polunin, J. Robinson, J.P. Bijoux, T.M. Daw, Conserv. Biol. 21 (2007) 1291–1300.
39. D.R. Bellwood, A.S. Hoey, J.L. Ackerman, M. Depczynski, Glob. Change Biol. 12 (2006) 1587–1594.
40. M.J. Attrill, F. Kelmo, M.B. Jones, Clim. Res. 26 (2004) 151–158.
41. F. Kelmo, M.J. Attrill, R.C.T. Gomes, M.B. Jones, Biol. Conserv. 118 (2004) 609–617.
42. F. Kelmo, M.J. Attrill, M.B. Jones, Coral Reefs 22 (2003) 541–550.
43. F. Kelmo, Ecological Consequences of the 1997–98 El Niño Southern Oscillation on The Major Coral Reef Communities From Northern Bahia, Brazil, University of Plymouth, Plymouth, 2002, p. 245.
44. M.J. Attrill, F. Kelmo, Estuar. Coast. Shelf Sci. 73 (2007) 243–248.
45. P.J. Edmunds, R.D. Gates, Mar. Ecol. Prog. Ser. 361 (2008) 307–310.

Chapter 14

Changes in Marine Biodiversity as an Indicator of Climate Change

Boris Worm and Heike K. Lotze
Biology Department, Dalhousie University, Halifax, Nova Scotia, Canada B3H 4J1

1. Introduction
2. Climate Change and the Oceans
3. Effects of Climate Change on Biodiversity
3.1. Local Scale
3.2. Regional Scale
3.3. Global Scale
3.4. Other Factors Relating to Climate Change
4. Cumulative Impacts and Indirect Effects of Climate Change
5. Biodiversity as Insurance against Climate Change Impacts
6. Conclusions
Acknowledgements
References

1. INTRODUCTION

Our planet has a number of features that make it unique, namely the presence of large oceans, and the evolution of life forms therein. Biodiversity, commonly defined as the variability among living organisms from all sources [1], originated in the oceans and most of the larger taxonomic groups still reside there today. Over evolutionary time scales, there have been massive changes to the ocean's biodiversity, including several mass extinctions [2–4] that have shaped planetary diversity over millions of years [5]. Some, if not most of these events are thought to correlate with large-scale climate change that perturbed ocean temperature, chemistry, currents and productivity [6].

Today, we are living through another episode of rapid climate change, which is causing global changes in weather patterns and ocean temperature [7] that are beginning to change thermal stratification, currents and productivity [8–12]. Most studies on the ecological effects of climate change, whether on land or in the sea, have concentrated on individual species [13–16], as discussed elsewhere in this volume. Only quite recently have community metrics such as species diversity been studied in direct relation to climate change [17–19]. Here, we will build on this emerging literature while discussing how marine biodiversity may serve as an indicator of recent climate change. Biodiversity has three main components: diversity within species, between species and of ecosystems [1]. We will discuss changes in all three components, but note that studies to date have mostly focused on species composition and species richness, likely because these represent the most easily quantifiable aspects of biodiversity.

Despite its taxonomic prominence, marine biodiversity is sometimes overlooked in the climate change discussion undoubtedly because much of it is little known and less understood than its terrestrial counterpart. For example, Sala and others [20] projected 'global biodiversity scenarios' for the year 2100 but did not consider marine ecosystems at all. Yet, marine biodiversity needs to be accounted for, not just because of its geographic extent, but also as it provides important ecosystem goods and services such as fisheries yields, shoreline protection, carbon and nutrient cycling, detoxification of wastes and pharmaceuticals, to name a few [21,22]. The ocean's biodiversity should therefore be carefully studied in order to understand and project how it will change with climate change and what the consequences may be for human well-being [23,24].

Here we consider the role of marine biodiversity as a response variable and indicator of recent climate change. We first discuss observed changes in biodiversity at various scales: local, regional and global, and how they relate directly to warming and other climate-related factors. Then we outline some indirect effects of climate change that arise from complex interactions with biotic and abiotic factors, and the cumulative effects of climate and other global changes. Finally, we highlight the importance of biodiversity for maintaining ecosystem resilience and productivity in the face of climate change. We do not pretend to give a complete overview but instead discuss some prominent patterns by example, largely focusing on the effects of increasing temperature. Herein, we shall rely on documented changes from the published literature and highlight how these effects are projected to develop into the future. The primary question we are asking is whether diversity, here defined as the number of genotypes, species or habitats changes in some predictable way with climate change. A secondary question is how climate effects on marine biodiversity are modified by and interact with other, co-occurring aspects of global change such as overfishing or eutrophication.

2. CLIMATE CHANGE AND THE OCEANS

Climate change has a range of effects on the abiotic marine environment, which are documented in detail elsewhere in this volume. From a biodiversity perspective the prominent physical changes include ocean warming via greenhouse gas forcing [7,25], increased climatic variability leading to more frequent extreme events [26] and changes in sea level, thermal stratification and ocean currents [8,27,28]. These processes can act on biodiversity directly (e.g. where local temperatures exceed individual species' physiological tolerances [29,30]) or indirectly (e.g. by altering habitat availability, species interactions or productivity [8,11,27]). Furthermore, potentially complex interactions between climate change and other global change aspects, notably those due to fishing, eutrophication, ocean acidification, habitat destruction, invasions and disease may also be important [27,31–33] and are briefly highlighted in this review. This latter point suggests an important difference between the current episode of climate change and previous climate perturbations in Earth's history: recent changes in climate are superimposed on other stressors that have already compromised biodiversity in many places [22]. From a scientific standpoint this added complexity can make it more difficult to clearly attribute observed changes in diversity to a single factor.

3. EFFECTS OF CLIMATE CHANGE ON BIODIVERSITY

What are the recently observed changes in biodiversity, and how do they relate to climate?In the following Sections 3.1–3.3, we first review evidence for the effects of climate warming that are emerging at increasing scales, from local (0.1–10 km) to regional (10–1000 km) and global (1000–10 000 km), respectively. In Section 3.4, we discuss factors other than increases in temperature, that are related to climate. Observed effects are summarised in Table 1.

3.1. Local Scale

Changes in biodiversity at the local scale are often driven by the interplay of local and regional, abiotic and biotic factors. The effects of a regional change in sea surface temperature (SST), for example, may be mediated by local factors such as wave exposure, tidal mixing, upwelling and species composition. Nevertheless, some common patterns have been observed at local scales.

In temperate locations, slow changes in species composition have been observed that often lead to an overall net increase in diversity. Changes in species composition were first shown by Southward and colleagues in their classic long-term studies in the English Channel [34]. Both intertidal and pelagic communities changed predictably during periods of climate warming, with warm-adapted species increasing in abundance, and cold-adapted species decreasing, leading to overall increases in diversity. The reverse patterns were observed during periods of cooling [34]. Similar changes occurred in the northwest Pacific (Monterey Bay, California) where 8 out of 9 southern species of intertidal invertebrates increased between the 1930s and 1990s, while 5 out of 8 northern species decreased [35]. This change tracked observed increases in both mean and maximum temperature and led to an overall

TABLE 1 Summarising observed direct effects of climate change on marine biodiversity

Cause	Effect	Effect on diversity	References
Temperature increase (tropical)	Coral bleaching	↓	[38,39]
Temperature increase (temperate)	Warm-adapted species replace cold-adapted ones	↑	[19,35,45,47,52]
Temperature increase (polar)	Decline of polar endemics invasion of subpolar species	?	[41–44]
Increased climate variability (heat waves)	Increased rates of disturbance	↓	[67,68]
Increased upwelling intensity	Surface water hypoxia	↓	[69–71]
Increasing water column stratification	Lower nutrient supply and productivity	?	[8,11,12]
Sea level rise	Erosion of coastal habitat	↓	[74]
Changes in currents	Changes in larval transport	?	[96]

increase in invertebrate species richness by 7%, due to 3 species newly invading from the south [35]. A similar pattern of southern species invading and northern species declining was documented for a temperate reef fish community in southern California [36]. In this case, however, sudden warming in the 1970s also led to a decline in productivity, 80% loss of large zooplankton biomass and recruitment failure of many reef fish. This may explain why total biomass declined significantly, and total species richness also declined by 15–25% at the two study sites [36]. These two contrasting examples illustrate that predictions based on temperature alone can be misleading, at least on a local scale, if concomitant changes in productivity are involved. Moreover, it has been shown that local differences in tidal exposure render some northern sites more thermally stressful than southern sites, counteracting the poleward shift of southern species discussed above, and possibly causing localised extinctions [37].

In tropical locations warming can lead to species loss and a decline in diversity, as maximum temperature tolerances are exceeded. So far, this applies particularly to tropical coral reefs that are affected by warming-related bleaching events (reviewed, for example, by Refs. [33,38,39] and in Chapter 13 of this volume). Poised near their upper thermal limits, coral reefs have experienced mass bleaching where sea temperatures have exceeded long-term summer averages by more than 1 °C for several weeks [38]. The loss of coral species is likely to cause secondary losses of reef-associated fauna and flora through loss of critical habitat. This mirrors climate-related losses of tropical diversity on land [40]. Unfortunately, detailed estimates of how species richness and community structure have changed after bleaching events are scarce but such changes are suspected to be large [15].

Polar marine ecosystems are thought to be particularly sensitive to climate change because small temperature differences can have large effects on the extent and thickness of sea ice. Therefore, the rate of change in species abundances and composition has been very fast, much of it related to changes in sea ice cover. While sea-ice dependent species such as polar bears [41], krill [42] and some penguins [43,44] have sharply decreased in abundance at some locations, there are signs of increasing invasion of subpolar and ice-independent species in other places [43]. Little information on net changes in local species richness (increase or decrease in diversity) is available so far.

3.2. Regional Scale

A growing number of studies have examined changes in species composition and diversity at regional scales. Much of this work was done in relation to fisheries or plankton monitoring data. As on the local scale, a dominant observation is the replacement of cold-adapted by warm-adapted species. This appears to occur simultaneously at various levels in the food web, for example, in North Atlantic zooplankton [45,46], as well as fish communities [47]. These changes are not necessarily synchronised: Beaugrand and colleagues documented a growing mismatch between warming-related changes in zooplankton since the 1980s and the emergence of cod larvae and juveniles. Cod populations were directly affected by changes in temperature, but also indirectly by changes in their planktonic prey that compromised growth and survival of cod larvae. Perry et al. observed that larger species with slower life histories (such as cod) adapted their range much more slowly to changing conditions as compared to fast-growing species [47]. This finding has implications for fisheries, as species with slower life histories are already more vulnerable to overexploitation [48] and may also be less able to compensate for warming through rapid demographic responses. Constraints to range shifts, however, appear to be less important than on the land. In the North Sea, among species that shifted their range the average rate of northward change was 2.2 $km \cdot a^{-1}$, which is more than 3 times faster than observed range shifts in terrestrial environments, which reportedly average 0.6 $km \cdot a^{-1}$ [14]. This may not be surprising, given the lesser extent of physical boundaries in marine, and particularly pelagic environments.

The net effect of these compositional changes on species richness was surprisingly large: an almost 50% increase in the number of species recorded per year in North Sea bottom trawl surveys was documented between 1985 and 2006 [19]. This change correlated tightly with increasing water temperature during the same period [19]. The same trends have been found in the Bristol Channel, UK where fish species richness increased by 39% from 1982 to 1998 [49]. In both cases increases in richness were mainly driven by invasion of small-bodied southern species. It is noteworthy that similar regional changes have been observed on land, where species richness of British butterflies [18] and epiphytic lichen in the Netherlands [50] has increased with warming over time, mostly driven by southern species that were able to respond quickly to warming. The total magnitude of increase in species richness was quite variable, however: 10% increase in butterfly species, but a doubling in lichen richness over the last 2–3 decades.

These decadal changes in species richness and diversity are superimposed on significant year-to-year variation in temperature and diversity. In the NW Atlantic there is a well-documented latitudinal gradient in fish species richness that co-varies with temperature [51]. This latitudinal gradient in diversity has previously been treated as static. Recently it has been shown how temperature variability readjusts diversity gradients year-by-year [52]. Temperature variability is linked to large-scale pressure differences across the North Atlantic, known as the North Atlantic Oscillation (NAO) [53]. Positive NAO anomalies cause temperature gradients in the NW Atlantic to steepen, which leads to rapid adjustments in species diversity: northern areas decline, southern areas increase in diversity [52]. During NAO-negative years the gradient flattens: northern areas increase, southern areas decrease in diversity. Although the north–south trend of increasing diversity does not reverse, there are substantial differences in its slope. This dynamic pattern is mostly driven by expansions and contractions of species ranges at their northern or southern range limits [52]. Again, warming waters increase overall diversity in temperate regions; cooling waters have the opposite effect.

Similar mechanisms have been shown to affect pelagic fish diversity across the tropical to temperate Pacific Ocean. Here, pressure differences in the central Pacific lead to periodic warming and cooling of surface waters in the eastern tropical Pacific, the well-known El Niño Southern Oscillation (ENSO) that affects weather patterns around the planet [54]. Positive ENSO years are characterised by regional warming of the eastern tropical Pacific and an increase in species diversity in the following year [17]. Regional cooling leads to decrease in diversity [17]. Single species such as Blue Marlin [17] or skipjack tuna [55] are seen to readjust their distribution year-by-year in response to these temperature changes. These studies show how species diversity does not only serve as an indicator of long-term climate change, but accurately tracks short-term variability in climate as well. A caveat for exploited fish populations is of course that intense exploitation can override climate signals on diversity. In the Atlantic and Indian Oceans, for example, there has been a long term decline in tuna and billfish species richness, that is most likely explained by fishing [17]. In the Pacific, however, a similar decline is counteracted by increasing warming after 1977 [17].

In contrast to marine fish, plankton communities are not affected by exploitation, except maybe indirectly through trophic cascades [56]. For both phyto- and zooplankton phenological changes (e.g. the timing of the spring bloom), range shifts and changes in species composition have been shown to track changes in climate [9,57]. Recently, it has been suggested that plankton communities may in fact be more sensitive indicators of climate change than the environmental variables (like SST) themselves, because of nonlinear responses of biological communities that may amplify subtle environmental perturbations [58]. Thus, plankton communities are increasingly used as indicators of recent climate change [57].

3.3. Global Scale

There are few global scale studies of marine biodiversity and its response to climate variability and global change. The argument has been made on land, albeit controversially, that a large number of extinctions could be caused by climate change by compressing species thermal habitats, particularly for species of restricted ranges [59]. Whether to expect global marine extinctions due to climate change is yet unclear, although much concern is focusing on coral reefs worldwide that are simultaneously threatened by warming and acidifying waters [33]. Dulvy and co-workers [48] note the possible global extinction of two coral species due to bleaching (*Siderastrea glynni, Millepora boschmai*), both of which have limited geographic ranges in the Eastern Pacific. Moreover, some coral-associated fish have also disappeared over the course of recent bleaching events [48].

Although the question of projected extinctions due to climate change is contentious [60,61], there is little doubt that temperature is a major driver of marine diversity at the global scale. Global diversity patterns have so far been synthesised for single-celled (foraminiferan) zooplankton [62], tropical reef organisms [63], tuna and billfish [17], and most recently, marine mammals [64,65]. Global reef diversity peaks at tropical latitudes in the Philippine–Indonesian triangle [63], whereas fish, foraminifera and mammals all peak at intermediate latitudes, around 20–30° North or South [17,62,64,65]. These patterns are all most parsimoniously explained by variation in SST (Fig. 1a), which explains between 45 and 90% of the variation in species diversity for these groups [17,62,65]. As mentioned above, variation in SST well explains not just the broad spatial patterns but also much of the inter-annual variation in tuna and billfish richness in the Pacific [17] as well as seasonal variation in mammal diversity in the Atlantic [65]. Moreover, the global richness pattern of tuna and billfish could be independently reconstructed from individual species' temperature tolerances [30]. Therefore, it appears that temperature might indeed be a powerful and general determinant of species richness at

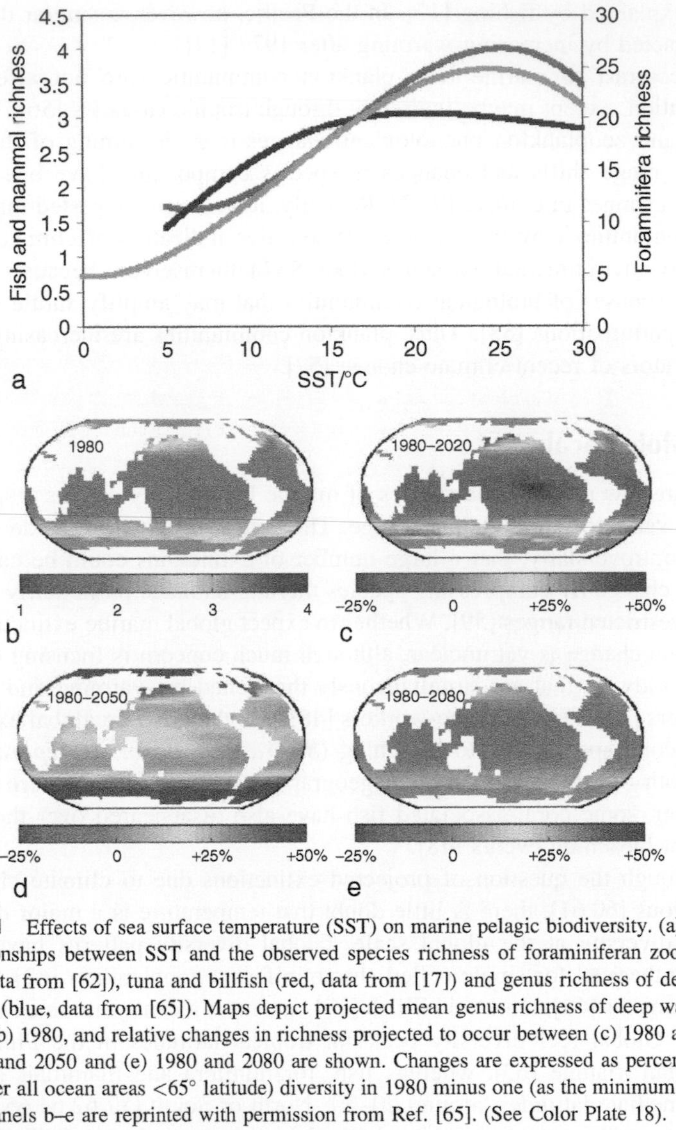

FIGURE 1 Effects of sea surface temperature (SST) on marine pelagic biodiversity. (a) Empirical relationships between SST and the observed species richness of foraminiferan zooplankton (green, data from [62]), tuna and billfish (red, data from [17]) and genus richness of deep-water cetaceans (blue, data from [65]). Maps depict projected mean genus richness of deep water cetaceans in (b) 1980, and relative changes in richness projected to occur between (c) 1980 and 2020, (d) 1980 and 2050 and (e) 1980 and 2080 are shown. Changes are expressed as percents of the mean (over all ocean areas <65° latitude) diversity in 1980 minus one (as the minimum diversity is 1.0). Panels b–e are reprinted with permission from Ref. [65]. (See Color Plate 18).

global scales. The empirically derived relationships between SST and species richness can be used to derive hypotheses about the potential effects of warming on large-scale patterns of species richness. An example is shown in Fig. 1, displaying the global pattern of deep-water cetacean genus richness (Fig. 1b) as derived from the empirical SST relationship (blue line in Fig. 1a), along with projected changes due to moderate warming (Fig. 1c–e, see Ref. [65] for more detail). Climate data were derived from the Intergovernmental Panel on Climate Change CGCM1 model using scenario A2a. Given the observed relationship with SST, diversity is projected to increase substantially at high latitudes, but to decrease in the tropical ocean. So far, the low availability of time series data does not allow testing this prediction for marine mammals, but this may change with improved tracking and monitoring capabilities.

3.4. Other Factors Relating to Climate Change

Despite the strong observed effects of temperature discussed above there are clearly other factors that are important in influencing diversity on local, regional and global scales. For tuna and billfish, for example, the availability of thermal fronts that act to concentrate food supply is of great importance, as is the availability of sufficient oxygen concentrations (>2 $ml{\cdot}l^{-1}$ at 100 m depth [17,66]). Many marine animals may also concentrate in areas of high productivity [64]. These factors are both directly and indirectly affected by climate change (Table 1). Increasing climate variability, for example, can affect biodiversity through extreme events, such as intense storms or heat waves, which can lead to large-scale die-offs, as recently seen in shallow-water corals or seagrass meadows [38,39,67,68]. Such events are likely leading to substantial losses in local diversity, at least on short to intermediate time scales. Similarly, increased variability in wind stress has been shown to affect the intensity of upwelling, leading to periodic hypoxia and death of marine organisms [69–71]. Furthermore, climate change is implicated in the observed shallowing of oxygen minimum zones in the tropical ocean [72], which is likely compromising local biodiversity at intermediate depths. Primary productivity is also affected by global warming, particularly through increased stratification and lower nutrient supply to the photic zone [8,9,11]. Because there are strong relationships between productivity, biomass and diversity in plankton [73], changes in stratification, nutrient supply and productivity are likely altering species diversity patterns.

Finally, climate change leads to sea level rise (Chapter 18, this volume) and changes in ocean currents (Chapter 20, this volume). Sea level rise in concert with increasing climate variability can lead to increasing coastal erosion and the loss of coastal habitats. This may compromise the diversity of species depending on wetlands, saltmarshes or mangroves [74]. Shorelines are increasingly fortified against rising water levels thereby preventing the adaptive inland movement of wetlands and upward movement of intertidal habitats, which decline or disappear over time together with their associated flora and fauna [74]. Ocean currents, fronts and upwelling zones are changing in response to alterations in temperature, precipitation, runoff, salinity and wind. These water movements strongly influence larval supply, species migrations and productivity [74]. So far, effects of changing currents on ocean diversity have not been studied, however, with the exception of upwelling studies mentioned above [69–71]. It can be concluded that both temperature as well as other climatic factors can modify patterns of diversity which may lead to interactive effects. Such complexities are discussed in more detail below.

4. CUMULATIVE IMPACTS AND INDIRECT EFFECTS OF CLIMATE CHANGE

A major challenge in ecological research is the disentanglement of multiple factors that are driving ecological change. Up to this point we have reviewed the direct effects of increasing temperature and climate variability, and resulting changes in upwelling, stratification, sea level and currents (Table 1). In reality, however, these processes are likely interacting with other impacts on biodiversity, such as exploitation, eutrophication, disease and physical disturbance, among others. Species composition and abundance are also influenced to a large degree by local species interactions, such as predation, competition and facilitation. Through changing species interactions, and by interacting with other drivers, climate change can have a number of indirect effects that are sometimes surprising and difficult to predict. Here we are highlighting such indirect effects, pointing towards some well-documented examples for illustration (Table 2).

TABLE 2 Examples of some indirect and interactive effects of climate change with other drivers of marine biodiversity

Primary cause	Secondary cause	Effect on species group	Effect on diversity	References
Increased upwelling intensity	Decline in keystone predator	Release of competitive dominant	↓	[76]
Warming	Disease	Increased pathogen development, disease transmission, and host susceptibility	↓	[32,77]
Increase climate variability	Fishing pressure	Fish more vulnerable to overexploitation	↓	[24]
Warming	Nutrient pollution	Increase in algal and jellyfish blooms	↓	[31,85–87]
Warming	Acidification and fishing	Coral reef loss due to bleaching, algal overgrowth and lower calcification	↓	[33,88]
Warming	Invasion	Faster establishment of invaders	?	[90]

Consider the classic example of a keystone predator, the starfish *Pisaster ochraceus*, which maintains intertidal diversity by feeding on competitively dominant mussels *Mytilus californianus* [75]. This interaction, however, is temperature-dependent: increases in upwelling lead to colder waters, lower predation rates and higher mussel cover [76]. Therefore, possible effects of climate change on diversity are mediated by a powerful interaction between a predator and a competitively dominant prey.

Another well-documented complexity concerns the interaction between warming temperatures and disease. There is good evidence that climate warming can increase pathogen development and survival, disease transmission, and host susceptibility (reviewed in Refs. [32,77]). This has become evident both in the sea and on land following large-scale warming events associated with ENSO, which are implicated with increases in several coral diseases, oyster pathogens, crop pathogens, rift valley fever and human cholera [32,77]. These effects occurred both in tropical and temperate location, with some documented range shifts of pathogens towards higher latitudes.

Climate change can also affect the interaction between humans and marine biodiversity. Over the past centuries human impacts have already had a marked impact on marine biodiversity, including a number of local, regional and global extinctions [48]. To date, exploitation and habitat destruction have probably had the most severe impacts [48,78]. The existing rate of habitat destruction will likely be accelerated by climate-driven habitat losses due to sea level rise, acidification and bleaching [33,74]. Similarly, the effects of exploitation are likely exacerbated by climate change. This is because most fisheries effectively truncate the age structure and size structure of target fish, by preferentially removing larger, older individuals. The fishery then becomes increasingly dependent on the recruitment of young (often immature) individuals to the fishery. Recruitment, however, is strongly affected by climate variability [79]. Removing the older age classes removes resilience to recruitment failure, and increases susceptibility both of the stock and the fishery to climatically induced fluctuations [24]. Another important factor is the removal of stock diversity by intense fisheries, which again increases vulnerability to climate by removing life-history variation and local adaptations [80]. Reducing fishing mortality in the majority of fisheries, which are currently fully exploited or overexploited, is the principal feasible means of managing fisheries for increased robustness to climate change [24,81].

Apart from fishing and habitat destruction, humans are affecting marine biodiversity through pollution, including nutrient pollution leading to eutrophication of coastal waters, algal blooms and hypoxic conditions [82]. These factors have documented negative effects on diversity, primarily by reducing susceptible species, but also by increasing dominance of fast-growing opportunists. The potential for complex indirect effects of climate change has been explored, for example, with respect to eutrophication and algal blooms [31]. Field and laboratory experiments have shown that increased nutrient availability (e.g. through sewage or fertilizer runoff) can trigger algal blooms, especially where herbivore populations are depressed [83,84]. Climate warming further accelerates algal growth but also feeding rates by grazers. The effect of climate warming on algal blooms depends therefore on the magnitudes of both nutrient input, and the composition and abundance of grazers [31]. Observation and experiments both suggest that as rates of nutrient input and climate warming grow, these could synergistically enhance bloom-forming species such as algae [31,85] and jellyfish [86,87].

Algal growth, particularly on tropical reefs can also be accelerated by the exploitation of herbivorous fishes, particularly parrotfish. This can synergistically enhance the effects of warming and acidification, which lead to bleaching and increase dissolution of calcareous exoskeletons, respectively [33]. Those disturbances open up new space for algae to colonise, which in the absence of herbivores can grow unchecked until they dominate reef structure and permanently alter the state of that community, as shown in recent field experiments [88].

Finally, human vectors are re-arranging marine biodiversity through the transport and release of non-indigenous organisms, both intentionally (as in aquaculture) and unintentionally (as in ship ballast water) [89]. Whether those species then become established or invasive in their new environment depends on a number of factors, such as temperature and salinity, habitat availability, predation and competition [89]. There is some evidence that ocean warming favours the establishment of invaders and hastens the displacement of native species [90]. Whether such invasions lead to a net loss of species, or even an increase in species richness as observed in some places [91], is not generally clear.

5. BIODIVERSITY AS INSURANCE AGAINST CLIMATE CHANGE IMPACTS

There is now good evidence that in addition to being a response variable to changes in temperature and climate, biodiversity may also provide resilience against climate change. This is because high genetic and species variation enhances the diversity of possible responses, and adaptive ability in the face of environmental variation [92,93]. For example, in a study on seagrass loss after the 2003 European heat wave, high genetic diversity (manipulated experimentally) led to faster recovery of damaged habitat [67]. This was driven both by selection of heat-adapted genotypes and by some form of facilitation that led to increased survival [67]. This observation was independently verified by laboratory experiments that manipulated temperature and genetic diversity in a controlled environment [68]. Another field study documented that high genetic diversity in seagrass also increased resilience to physical disturbance from overgrazing [94]. Theoretical studies have come to similar conclusions. For example, Yachi and Loreau [95] showed two major insurance effects of species richness on ecosystem productivity: (1) a reduction in the temporal variance of productivity and (2) an increase in the temporal mean of productivity despite stochastic disturbances.

From these studies follows the prediction that a loss in biodiversity should lead to a loss in productivity and resilience, which would enhance any effect of climate change (or other disturbances) on marine ecosystems. An increase in biodiversity should have the opposite effect. Evidence in support of this prediction comes from a series of meta-analyses examining local experiments, regional time series and global fisheries data [23]. The vulnerability to climate change in particular was examined by a regional study of Alaskan salmon fisheries that have been carefully managed to avoid loss of stock diversity [80]. These stock complexes show a remarkable resilience to climatic change due to a large number of local life-history adaptations that are preserved within the stock complex. As environmental conditions changed, overall productivity was maintained by different sub-stocks that were adapted to thrive under those conditions [80].

6. CONCLUSIONS

In this short (and necessarily incomplete) review, we examined whether marine biodiversity can serve as a useful indicator of climate and global change. It appears that indeed changes in diversity often indicate changes in climate, especially warming and increased climate variability. This is particularly true at large (regional and global) scales where diversity patterns are strongly linked to temperature. On local scales, this is less obvious because other factors may modify or override the underlying effects of climate change: (1) natural abiotic and biotic factors may alter the diversity response through changes in productivity, disturbance or species interactions and (2) other aspects of climate and global change may add complexity to the cumulative response of diversity. On a global ocean scale, it appears that, as on land, the tropics loose diversity, temperate regions show increased diversity, whereas polar environments so far mostly show declines in ice-dependent species as the climate warms. Underlying these dynamic patterns is a redistribution of species ranges, with range expansions of warm-adapted and range contractions of cold-adapted species towards the poles, as well as local extirpations and new invasions. On local scales, climate-change driven habitat losses, for example, through sea level rise, bleaching or acidification can accelerate the local loss of biodiversity. As a result, species communities and food webs on all scales reorganise. Sometimes this involves decoupling of predator populations from their prey or other mismatches in species interactions due to shifts in phenology and physiology. Little is known about how entire communities or food webs re-assemble with climate change; this should be a germane topic for further research.

From a biodiversity management perspective little can be done to change the shifting of species ranges and the reorganisation of ecosystems. It is important, however, to maintain as much as possible the response diversity both within and between species and habitats that is evidently so important for adaptation and resilience. This can be achieved by carefully adjusting the impacts of other factors that may reduce biodiversity and by minimising cumulative impacts. In an era of rapid climate change, complex and surprising effects are to be expected and any form of management must necessarily be highly adaptive and precautionary.

ACKNOWLEDGEMENTS

The authors wish to thank H. Whitehead and D. Tittensor for discussion and insight, as well as The Sloan Foundation (Census of Marine Life, FMAP Program) and NSERC for support.

REFERENCES

1. UN, Convention on biological diversity, United Nations, New York, URL: http://www.biodiv.org/, 1992.
2. D.M. Raup, J.J.J. Sepkoski, Science 215 (1982) 1501–1503.
3. A.I. Miller, Science 281 (1998) 1157–1160.
4. M.J. Benton, R.J. Twitchett, Trends Ecol. Evol. 18 (2003) 358–365.
5. J.W. Kirchner, A. Weil, Nature 404 (2000) 177–180.
6. G.J. Vermeij, Evol. Ecol. Res. 6 (2004) 315–337.
7. T.P. Barnett, D.W. Pierce, R. Schnur, Science 292 (2001) 270–273.
8. J.L. Sarmiento, R. Slater, R. Barber, L. Bopp, S.C. Doney, A.C. Hirst, J. Kleypas, R. Matear, U. Mikolajewicz, P. Monfray, V. Soldatov, S.A. Spall, R. Stouffer, Global Biogeochem. Cycles 18 (2004) doi:10.1029/2003GB002134.
9. A.J. Richardson, D.S. Schoeman, Science 305 (2004) 1609–1612.
10. H.L. Bryden, H.R. Longworth, S.A. Cunningham, Nature 438 (2005) 655–657.
11. M.J. Behrenfeld, R.T. O'Malley, D.A. Siegel, C.R. McClain, J.L. Sarmiento, G.C. Feldman, A.J. Milligan, P.G. Falkowski, R.M. Letelier, E.S. Boss, Nature 444 (2006) 752–755.
12. J.J. Polovina, E.A. Howell, M. Abecassis, Geophys. Res. Lett. 35 (2008) L03618.
13. T.L. Root, J.T. Price, K.R. Hall, S.H. Schneider, C. Rosenzweig, J.A. Pounds, Nature 421 (2003) 57–60.
14. C. Parmesan, G. Yohe, Nature 421 (2003) 37–42.
15. G.-R. Walther, E. Post, P. Convey, A. Menzel, C. Parmesan, T.J.C. Beebee, J.-M. Fromentin, O. Hoegh-Guldberg, F. Bairlein, Nature 416 (2002) 351–460.
16. C. Parmesan, Annu. Rev. Ecol. Evol. Syst. 37 (2006) 637–669.
17. B. Worm, M. Sandow, A. Oschlies, H.K. Lotze, R.A. Myers, Science 309 (2005) 1365–1369.
18. R. Menéndez, A.G. Megías, J.K. Hill, B. Braschler, S.G. Willis, Y. Collingham, R. Fox, D.B. Roy, C.D. Thomas, Proc. R. Soc. B 273 (2006) 1465–1470.
19. J.G. Hiddink, R. ter Hofstede, Global Change Biol. (2008) 453–460.
20. O.E. Sala, F.S. Chapin III, J.J. Armesto, E. Berlow, J. Bloomfield, R. Dirzo, E. Huber-Sanwald, L.F. Huenneke, R.B. Jackson, A. Kinzig, R. Leemans, D.M. Lodge, H.A. Mooney, M. Oesterheld, N.L. Poff, M.T. Sykes, B.H. Walker, M. Walker, D.H. Wall, Science 287 (2000) 1770–1774.

21. C.H. Peterson, J. Lubchenco, in: G.C. Daily (Ed.), Nature's Services: Societal Dependence on Natural Ecosystems, Marine Ecosystem Services, Island Press, Washington, DC, 1997, pp. 177–194.
22. MEA, Millenium Ecosytem Assessment. Ecosystems and Human Well-Being: Synthesis, Island Press, Washington, DC, USA, 2005.
23. B. Worm, E.B. Barbier, N. Beaumont, J.E. Duffy, C. Folke, B.S. Halpern, J.B.C. Jackson, H.K. Lotze, F. Micheli, S.R. Palumbi, E. Sala, K. Selkoe, J.J. Stachowicz, R. Watson, Science 314 (2006) 787–790.
24. K.M. Brander, Proc. Natl. Acad. Sci. USA 104 (2007) 19709–19714.
25. S. Levitus, J.I. Antonov, T.P. Boyer, C. Stephens, Science 287 (2000) 2225–2229.
26. C. Schär, P.L. Vidale, D. Lüthi, C. Frei, C. Häberli, M.A. Liniger, C. Appenzeller, Nature 427 (2004) 332–336.
27. C.D.G. Harley, A.R. Hughes, K.M. Hultgren, B.G. Miner, C.J.B. Sorte, C.S. Thornber, L.F. Rodriguez, L. Tomanek, S.L. Williams, Ecol. Lett. 9 (2006) 228–241.
28. A. Schmittner, Nature 434 (2005) 628–633.
29. H.O. Portner, R. Knust, Science 315 (2007) 95–97.
30. D. Boyce, D. Tittensor, B. Worm, Mar. Ecol. Prog. Ser. 355 (2008) 267–276.
31. H.K. Lotze, B. Worm, Limnol. Oceanogr. 47 (2002) 1734–1741.
32. C.D. Harvell, C.E. Mitchell, J.R. Ward, S. Altizer, A.P. Dobson, R.S. Ostfeld, M.D. Samuel, Science 296 (2002) 2158–2162.
33. O. Hoegh-Guldberg, P.J. Mumby, A.J. Hooten, R.S. Steneck, P. Greenfield, E. Gomez, C.D. Harvell, P.F. Sale, A.J. Edwards, K. Caldeira, N. Knowlton, C.M. Eakin, R. Iglesias-Prieto, N. Muthiga, R.H. Bradbury, A. Dubi, M.E. Hatziolos, Science 318 (2007) 1737–1742.
34. A.J. Southward, S.J. Hawkins, M.T. Burrows, J. Thermal Biol. 20 (1995) 127–155.
35. J.P. Barry, C.H. Baxter, R.D. Sagarin, S.E. Gilman, Science 267 (1995) 672–675.
36. S.J. Holbrook, R.J. Schmitt, J.S. Stephens Jr., Ecol. Appl. 7 (1997) 1299–310.
37. B. Helmuth, C.D.G. Harley, P.M. Halpin, M. O'Donnell, G.E. Hofmann, A. Blanchette, Science 298 (2002) 1015–1017.
38. O. Hoegh-Guldberg, Mar. Freshwat. Res. 50 (1999) 839–866.
39. T.P. Hughes, A.H. Baird, D.R. Bellwood, M. Card, S.R. Connolly, C. Folke, R. Grosberg, O. Hoegh-Guldberg, J.B.C. Jackson, J. Kleypas, J.M. Lough, P. Marshall, M. Nystrom, S.R. Palumbi, J.M. Pandolfi, B. Rosen, J. Roughgarden, Science 301 (2003) 929–933.
40. R.K. Colwell, G. Brehm, C.L. Cardelus, A.C. Gilman, J.T. Longino, Science 322 (2008) 258–261.
41. I. Stirling, N.J. Lunn, J.L. Iacozza, Arctic 52 (1999) 294–306.
42. A. Atkinson, V. Siegel, E. Pakhomov, P. Rothery, Nature 432 (2004) 100–103.
43. S.D. Emslie, W. Fraser, R.C. Smith, W. Walker, Antarct. Sci. 10 (1998) 257–268.
44. C. Barbraud, H. Weimerskirch, Nature 411 (2001) 183–186.
45. G. Beaugrand, P.C. Reid, F. Ibañez, J.A. Lindley, M. Edwards, Science 296 (2002) 1692–1694.
46. M. Edwards, A.J. Richardson, Nature 430 (2004) 881–884.
47. A.L. Perry, P.J. Low, J.R. Ellis, J.D. Reynolds, Science 308 (2005) 1912–1915.
48. N.K. Dulvy, Y. Sadovy, J.D. Reynolds, Fish Fish. 4 (2003) 25–64.
49. P.A. Henderson, J. Mar. Biol. Assoc. UK 87 (2007) 589–598.
50. C.M. van Herk, A. Aptroot, H.F. van Dobben, Lichenologist 34 (2002) 141–54.
51. K.T. Frank, B. Petrie, N.L. Shackell, Trends Ecol. Evol. 22 (2007) 236–242.
52. J.A.D. Fisher, K.T. Frank, B. Petrie, W.C. Leggett, N.L. Shackell, Ecol. Lett. 11 (2008) 883–897.
53. J.W. Hurrell, Science 269 (1995) 676–679.

54. M.J. McPhaden, S.E. Zebiak, M.H. Glantz, Science 314 (2006) 1740–1745.
55. P. Lehodey, M. Bertignac, J. Hampton, A. Lewis, J. Picaut, Nature 389 (1997) 715–718.
56. K.T. Frank, B. Petrie, J.S. Choi, W.C. Leggett, Science 308 (2005) 1621–1623.
57. G.C. Hays, A.J. Richardson, C. Robinson, Trends Ecol. Evol. 20 (2005) 337–344.
58. A.H. Taylor, J.I. Allen, P.A. Clark, Nature 416 (2002) 629–632.
59. C.D. Thomas, A. Cameron, R.E. Green, M. Bakkenes, L.J. Beaumont, Y.C. Collingham, B.F.N. Erasmus, M.F. de Siqueira, A. Grainger, L. Hannah, L. Hughes, B. Huntley, A.S. van Jaarsveld, G.F. Midgley, L. Miles, M.A. Ortega-Huerta, A.T. Peterson, O.L. Phillips, S.E. Williams, Nature 427 (2004) 145–148.
60. W. Thuiller, M.B. Araújo, R.G. Pearson, R.J. Whittaker, L. Brotons, S. Lavorel, Nature 430 (2004) doi:10.1038/nature02716.
61. D.B. Botkin, H. Saxe, M.B. Araujo, R. Betts, R.H.W. Bradshaw, T. Cedhagen, P. Chesson, T.P. Dawson, J.R. Etterson, D.P. Faith, S. Ferrier, A. Guisan, A.S. Hansen, D.W. Hilbert, C. Loehle, C. Margules, M. New, M.J. Sobel, D.R.B. Stockwell, BioScience 57 (2007) 227–236.
62. S. Rutherford, S. D'Hondt, W. Prell, Nature 400 (1999) 749–753.
63. C.M. Roberts, C.J. McClean, J.E.N. Veron, J.P. Hawkins, G.R. Allen, D.E. McAllister, C.G. Mittermeier, F.W. Schueler, M. Spalding, F. Wells, C. Vynne, T.B. Werner, Science 295 (2002) 1280–1284.
64. J. Schipper, J.S. Chanson, F. Chiozza, N.A. Cox, M. Hoffmann, V. Katariya, J. Lamoreux, A.S.L. Rodrigues, S.N. Stuart, H.J. Temple, J. Baillie, L. Boitani, T.E. Lacher Jr., R.A. Mittermeier, A.T. Smith, D. Absolon, J.M. Aguiar, G. Amori, N. Bakkour, R. Baldi, R.J. Berridge, J. Bielby, P.A. Black, J.J. Blanc, T.M. Brooks, J.A. Burton, T.M. Butynski, G. Catullo, R. Chapman, Z. Cokeliss, B. Collen, J. Conroy, J.G. Cooke, G.A.B. da Fonseca, A.E. Derocher, H.T. Dublin, J.W. Duckworth, L. Emmons, R.H. Emslie, M. Festa-Bianchet, M. Foster, S. Foster, D.L. Garshelis, C. Gates, M. Gimenez-Dixon, S. Gonzalez, J.F. Gonzalez-Maya, T.C. Good, G. Hammerson, P.S. Hammond, D. Happold, M. Happold, J. Hare, R.B. Harris, C.E. Hawkins, M. Haywood, L.R. Heaney, S. Hedges, K.M. Helgen, C. Hilton-Taylor, S.A. Hussain, N. Ishii, T.A. Jefferson, R.K.B. Jenkins, C.H. Johnston, M. Keith, J. Kingdon, D.H. Knox, K.M. Kovacs, P. Langhammer, K. Leus, R. Lewison, G. Lichtenstein, L.F. Lowry, Z. Macavoy, G.M. Mace, D.P. Mallon, M. Masi, M.W. McKnight, R.A. Medellin, P. Medici, G. Mills, P.D. Moehlman, S. Molur, A. Mora, K. Nowell, J.F. Oates, W. Olech, W.R.L. Oliver, M. Oprea, B.D. Patterson, W.F. Perrin, B.A. Polidoro, C. Pollock, A. Powel, Y. Protas, P. Racey, J. Ragle, P. Ramani, G. Rathbun, et al., Science 322 (2008) 225–230.
65. H. Whitehead, B. McGill, B. Worm, Ecol. Lett. 11 (2008) 1198–1207.
66. P.N. Sund, M. Blackburn, F. Williams, Oceanogr. Mar. Biol. Ann. Rev. 19 (1981) 443–512.
67. T.B.H. Reusch, A. Ehlers, A. Hämmerli, B. Worm, Proc. Natl. Acad. Sci. USA 102 (2005) 2826–2831.
68. A. Ehlers, B. Worm, T.B.H. Reusch, Mar. Ecol. Prog. Ser. 355 (2008) 1–7.
69. A. Bakun, S.J. Weeks, Ecol. Lett. 7 (2004) 1015–1023.
70. B.A. Grantham, F. Chan, K.J. Nielsen, D.S. Fox, J.A. Barth, A. Huyer, J. Lubchenco, B.A. Menge, Nature 429 (2004) 749–754.
71. F. Chan, J.A. Barth, J. Lubchenco, A. Kirincich, H. Weeks, W.T. Peterson, B.A. Menge, Science 319 (2008) 920.
72. L. Stramma, G.C. Johnson, J. Sprintall, V. Mohrholz, Science 320 (2008) 655–658.
73. X. Irigoien, J. Huisman, R.P. Harris, Nature 429 (2004) 863–867.
74. D. Scavia, J.C. Field, D.F. Boesch, R.W. Buddemeier, V. Burkett, D.R. Cayan, M.J. Fogarty, M.A. Harwell, R.W. Howarth, C. Mason, D.J. Reed, T.C. Royer, A.H. Sallenger, J.G. Titus, Estuaries 25 (2002) 149–164.

75. R.T. Paine, Am. Nat. 100 (1966) 65–76.
76. E. Sanford, Science 283 (1999) 2095–2097.
77. C.D. Harvell, K. Kim, J.M. Burkholder, R.R. Colwell, P.R. Epstein, D.J. Grimes, E.E. Hofmann, E.K. Lipp, A.D.M.E. Osterhaus, R.M. Overstreet, J.W. Porter, G.W. Smith, G.R. Vasta, Science 285 (1999) 1505–1510.
78. H.K. Lotze, H.S. Lenihan, B.J. Bourque, R. Bradbury, R.G. Cooke, M.C. Kay, S.M. Kidwell, M.X. Kirby, C.H. Peterson, J.B.C. Jackson, Science 312 (2006) 1806–1809.
79. N.C. Stenseth, A. Mysterud, G. Ottersen, J.W. Hurrell, K.-S. Chan, M. Lima, Science 297 (2002) 1292–1296.
80. R. Hilborn, T.P. Quinn, D.E. Schindler, D.E. Rogers, Proc. Natl. Acad. Sci. USA 100 (2003) 6564–6568.
81. B. Worm, R.A. Myers, Nature 429 (2004) 15.
82. J.E. Cloern, Mar. Ecol. Prog. Ser. 210 (2001) 223–253.
83. H.K. Lotze, B. Worm, U. Sommer, Limnol. Oceanogr. 46 (2001) 749–757.
84. B. Worm, H.K. Lotze, Limnol. Oceanogr. 51 (2006) 569–579.
85. J.C. Bintz, S.W. Nixon, B.A. Buckley, S.L. Granger, Estuaries 26 (2003).
86. J.E. Purcell, J. Mar. Biol. Assoc. UK 85 (2005) 461–476.
87. M.J. Attrill, J. Wright, M. Edwards, Limnol. Oceanogr. 52 (2007) 480–485.
88. T.P. Hughes, M.J. Rodrigues, D.R. Bellwood, D. Ceccarelli, O. Hoegh-Guldberg, L. McCook, N. Moltschaniwskyj, M.S. Pratchett, R.S. Steneck, B. Willis, Curr. Biol. 17 (2007) 360–365.
89. J.T. Carlton, J.B. Geller, Science 261 (1993) 78–82.
90. J.J. Stachowicz, J.R. Terwin, R.B. Whitlatch, R.W. Osman, Proc. Natl. Acad. Sci. USA 99 (2002) 15497–15500.
91. D.F. Sax, S.D. Gaines, Trends Ecol. Evol. 18 (2003) 561–566.
92. T. Elmqvist, C. Folke, M. Nyström, G. Peterson, J. Bengtsson, B. Walker, J. Norberg, Front. Ecol. Environ. 1 (2003) 488–494.
93. J.E. Duffy, Mar. Ecol. Prog. Ser. 311 (2006) 233–250.
94. A.R. Hughes, J.J. Stachowicz, Proc. Natl. Acad. Sci. USA 101 (2004) 8998–9002.
95. S. Yachi, M. Loreau, Proc. Natl. Acad. Sci. USA 96 (1999) 1463–1468.
96. C.J. Svensson, S.R. Jenkins, S.J. Hawkins, P. Aberg, Oecologia 142 (2005) 117–126.

Chapter 15

Intertidal Indicators of Climate and Global Change

Nova Mieszkowska
Marine Biological Association of the UK, The Laboratory, Citadel Hill, Plymouth PL1 2PB, United Kingdom

1. **Introduction**
2. **Climate Change and Biogeography**
   2.1. Using Long-Term Datasets to Detect Climate Change
   2.2. Responses of Intertidal Biota
   2.3. Extreme Weather Events
   2.4. Interactions
3. **Mechanisms**
   3.1. Thermotolerance
   3.2. Reproduction and Recruitment
4. **Additional impacts of Global Change**
   4.1. Ocean Acidification
   4.2. Coastal Zone Development
   4.3. Climate Change and Non-Native Species
5. **Conclusions**
Acknowledgements
References

1. INTRODUCTION

The rocky intertidal zone spans the region of the coastline from the highest vertical level reached at high water during spring tides (with associated wave splash) to the lowest level exposed to the air during low water springs. A wide variety of taxa inhabit the rocky intertidal zone, including algae, molluscs, echinoderms, cnidarians and crustaceans. Owing to the accessibility of rocky shores, intertidal species have been studied extensively throughout the nineteenth and twentieth centuries by amateur naturalists [1,2] and professional researchers as model systems for the development of ecological and biological theory [3–15].

Intertidal invertebrates and macroalgae are ectotherms of marine evolutionary origin, but due to the daily regime of emersion and immersion they must contend with both marine and terrestrial regimes. Therefore, they provide a unique insight into the impacts of changes in both aquatic and terrestrial climatic environments. Diurnal tidal cycles and seasonal fluctuations in both sea and air temperature mean that intertidal organisms are subject to extremes of temperature with resultant fluctuations in body temperature of over 30 °C frequently experienced [16]. Additional stressors such as desiccation [17], current and wave forces [18], rapid fluctuations in salinity [19–21], oxygen availability [22,23] and nutrient levels [24] mean that organisms are often living close to their physiological tolerance limits [25–32].

Marine ectothermic species often respond faster than terrestrial species to environmental change: the typically short lifespans [33] and sessile or sedentary nature of the adult and juvenile stages prevents escape from changing environmental regimes. The larval stage of most intertidal species is planktonic, and therefore also provides an indication of the impacts of environmental change in the pelagic zone. Changes in distribution and abundance are, therefore, likely to be driven by the direct response of organisms to changes in the environment. Intertidal invertebrates and marine macroalgae are from lower trophic levels, and thus would be expected to respond quicker to alterations in local conditions than species at higher trophic levels, often showing the first response in a cascade of effects up the food chain to tertiary and apex predators [34,35]. Variation in the abundance of keystone structural or functional species can alter the composition and dynamics of entire rocky communities [5,36,37] and these small changes in environmental conditions can lead to major alterations in community structure and functioning [38,39]. Taking all of the above factors into account, the rocky intertidal ecosystem is likely to be one of the most sensitive natural systems and show some of the earliest responses to climate change [40,41].

The geographical ranges of intertidal species are essentially one-dimensional, as they occupy a narrow strip of coastline between the low and high tide levels [42]. In combination with the highly stressful and fluctuating environment in which these species live, the rocky intertidal zone is thus ideal model system in which to study the effects of climate change. Rocky intertidal ecosystems occur all over the globe and thus facilitate spatial and temporal comparisons of the relative impacts of global environmental change. Responses to environmental change can be divided into two categories; proximate ecological responses which depend upon relationships between abiotic factors and organismal-level processes, population dynamics and community structure [43], and direct impacts on individual performance during various life stages through changes in physiology, morphology and behaviour. These impacts scale up to the population level response, which can be additionally affected by climate driven changes in hydrographic processes that affect dispersal of the pelagic larval life-stages and recruitment. All lead to alterations in distributions, biodiversity, productivity and micro-evolutionary processes.

2. CLIMATE CHANGE AND BIOGEOGRAPHY

Biogeographical studies were first introduced by Tournefort [44] in the 1700s, and work undertaken in the early 1900s [24,45–51] is used as the basis for ecological climate change research today. The major marine biogeographic provinces have been defined primarily on where clusters of biogeographic distributional limits occur for taxa of interest. Sea temperature has been assumed to ultimately set biogeographical ranges of marine species (see [24,52] for reviews). Low and high latitude biogeographic limits have been associated with August and February sea surface isotherms, respectively, for fauna and flora across a diverse range of taxonomic groups including marine algae [53–58], cirripedes [59] and molluscs [60,61]. However, the relationship between species' distributions and climate is not simple. Biogeographic studies are often complicated by covarying environmental parameters that prevent cause-and-effect relationships from being understood.

Suitable habitat exists beyond the distributional limits of many species of marine invertebrates [62–65] but the unsuitability of environmental conditions currently prevents their colonisation and therefore the ranges are assumed to be limited by climate. This principle is termed the 'climate envelope' of a species, and is the basis for many bioclimatic models in use today [66,67] but see Refs. [68,69]. Where environmental conditions alter to fall within the physiological tolerance range of a species, range extensions are predicted as organisms are able to colonise new sites of suitable habitat. In practice, however, the range edge may lie some distance inside this fundamental niche 'envelope'. Interactions between species and between organisms and environmental factors, and local influences such as a lack of suitable habitat, poor dispersal and connectivity of suitable habitat space act to set the realised niche for each species. The effects of climatic variability on the distributions of plants and animals and their interactions must, therefore, be measured in order to understand and ultimately forecast changes in marine ecosystems.

2.1. Using Long-Term Datasets to Detect Climate Change

Some of the most spatially and temporally extensive datasets in the world exist for the distribution and abundance of intertidal invertebrates and macroalgae along the coastline of the north east Atlantic. Intensive and wide ranging surveys were made in the 1930s, 1940s and 1950s by Fischer-Piette [70–73] along the Atlantic coastlines of France, Spain, Portugal and North Africa. Crisp and Southward made similar surveys around the coastlines of Britain and Ireland during the 1950s ([74,75], Southward and Crisp, unpublished data). These datasets are particularly valuable within the context of climate change monitoring as they provide extensive baselines from which to measure the rate and extent of changes in distribution and abundance of intertidal species during periods of warming and cooling over the past 70 years [76]. Time-series data for abundance and population structures for barnacles, trochids and limpets also exist for British shores dating back to the 1950s, 1970s and 1980s, respectively [74,75,77–79].

The Marine Biodiversity and Climate Change Project 'MarClim' was established by the Marine Biological Association of the United Kingdom in 2001 to assess and forecast the influence of climate change on rocky intertidal biodiversity in Britain and Ireland. It combined historical data with contemporary re-surveys at over 400 rocky shores (Fig. 1) to provide evidence of changes in abundance, population structure and geographical distribution of intertidal species in relation to recent climate change [64]. MarClim survey protocols were the same as those used in the original surveys made in the 1950s [74,75] to map the distribution and range limits of over 50 species of invertebrates and macroalgae of both cold and warm water origins. In addition, quantitative data on the abundance and population dynamics of key species of barnacles, limpets and trochids were collected. These surveys were carried out at locations spanning sites from the range edges to locations closer to the centers of distribution. These combined datasets have been used to track the changes in abundance and relative dominance of warm and cold water species on shores where they co-exist in response to fluctuating climatic conditions throughout the twentieth century.

2.2. Responses of Intertidal Biota

Contractions and expansions of geographic range edges due to global environmental change are resulting in species both being lost from and introduced to assemblages. Such changes are initially being recorded at the periphery of the geographic range of a species, where organisms are often already experiencing temperatures close to their thermal limits [26]. However, there can also be local or regional heterogeneity within the geographic range of a species as evidenced by environmental hotspots [31,40] or coldspots [79] occurring far from the distributional limits of sessile invertebrates. Such changes in turn influence the outcomes of species interactions for example competition, facilitation and predation, ultimately altering the structure of communities and marine ecosystem processes [41,43,80–82,85].

2.2.1. Europe

Alterations in distributional limits of a wide range of intertidal taxa have already occurred in Britain since rapid warming of the climate began in the mid-1980s. Northern and eastern range edges of warm water trochid gastropods such as *Osilinus lineatus* and *Gibbula umbilicalis*, barnacles including *Chthamalus montagui*, *C. stellatus* and *Perforatus (Balanus) perforatus* and the brown macroalga *Bifurcaria bifurcata* have extended between 85 and 180 km since previous records in the twentieth century [64,65,83] and at rates

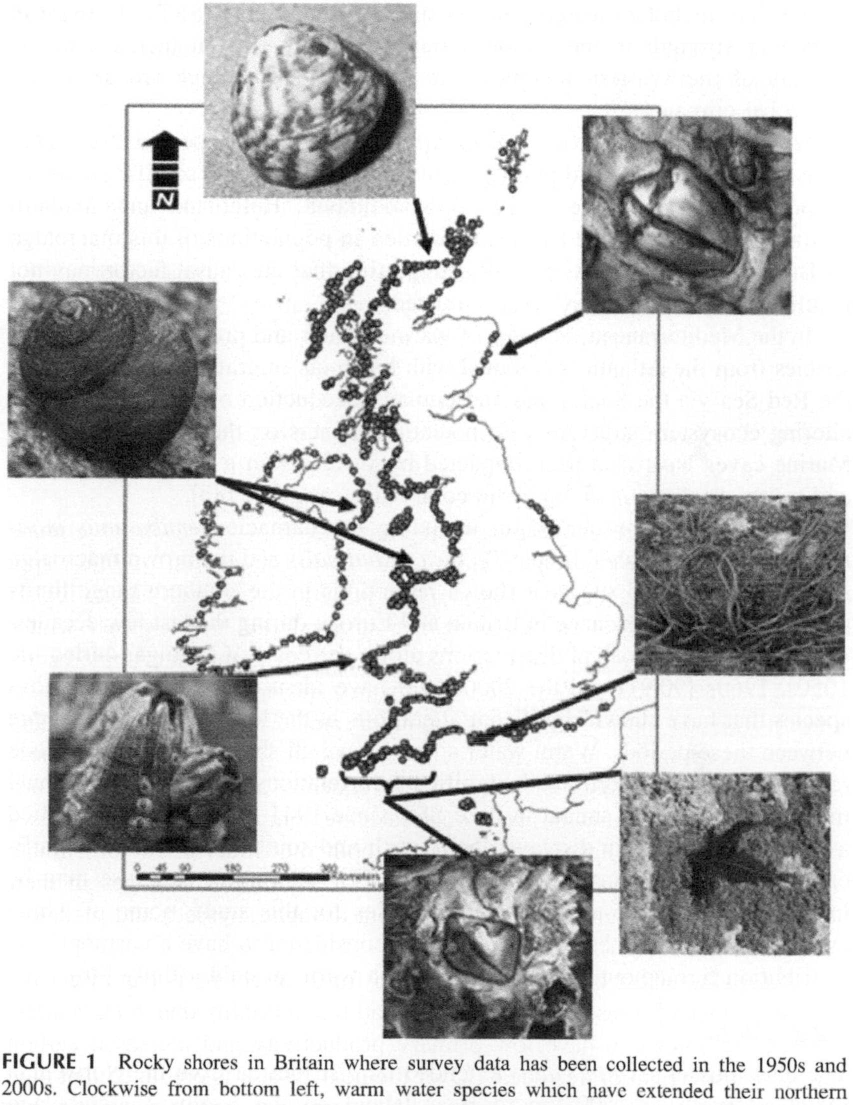

FIGURE 1 Rocky shores in Britain where survey data has been collected in the 1950s and 2000s. Clockwise from bottom left, warm water species which have extended their northern limits; *Perforatus perforatus*, *Osilinus lineatus*, *Gibbula umbilicalis*, *Chthamalus montagui*, *Bifurcaria bifurcata*. Cold water species which have shown contractions in their southern range edge: *Alaria esculenta*, *Semibalanus balanoides*. (See Color Plate 19).

of up to 50 km per decade. This rate is much faster than the average movement of 6.1 km per decade for terrestrial species' ranges [84] and is concordant with northward shifts in plankton recorded in British coastal waters (see Chapter 12 in this volume). The limpet *Patella rustica* has recently bridged a historical gap in its distribution in north Portugal during a period of weakened upwelling in coastal waters [79]. Several factors may be responsible, including increase in sea surface temperature (SST), decrease in upwelling strength in the southern Bay of Biscay and an increase in the strength of the Western Iberian Shelf Current, all of which are driven by the global climate.

Saccharina latissima (*Laminaria saccharina*) has disappeared from large areas across northern Europe during the last decade, with significant losses of populations along the coasts of Scandinavia, Helgoland and southern Brittany. No such decline has been recorded in populations of this macroalga on British coastlines, however [64], suggesting that the causal factor may not be climatic, but potentially disease-related.

In the Mediterranean, influxes of warmer water and propagules of tropical species from the Atlantic, combined with lessepian migration of species from the Red Sea via the Suez canal and human introduction of exotic species are altering ecosystem structure with potential impacts on the trophic web [86]. Marine caves have also been impacted by increases in a warm water mysid and severe declines in an endemic cold water congener [87].

Species of cold water origin including the barnacle *Semibalanus balanoides* [88], tortoiseshell limpet *Tectura testudinalis* and the brown macroalga *Alaria esculenta* [65,89] have shown retractions in the southern range limits and decreases in abundance in Britain and Europe during the last few decades.

Surveys of macroalgal distributions along the coast of Portugal during the 1950s, 1960s [90,91] and the 2000s [92] have identified ~120 conspicuous species that have shown significant alterations in the location of a range edge between these periods. Warm water species have all shifted their high latitude range limits further north, with significant correlations between distributional movement and mean annual inshore SST since 1941 [93]. Species classified as cold water in origin displayed both north and south shifts with no significant change when considered as a group. This alternative response may be in part due to the grouping of these species for this study, some of which are non-native, and others that are widely considered to have a cosmopolitan distribution throughout Europe rather than a warm or cold affinity [76].

2.2.2. Arctic

The blue mussel *Mytilus edulis* extended its distribution from the Norwegian mainland north, by 500 km to Bear Island on the Svalbard archipelago between 1977 and 1994 [94], and was found on the Arctic island of Svalbard itself for the first time in 1000 a after a period of anomalously warm Atlantic seawater influx between 2002 and 2004 [95,96]. This reappearance represented a huge poleward shift in distributional limits of ~1000 km, probably due to transport of pelagic larvae north in the warm water current. However, it is not known whether these populations are sustainable or if the prevailing climate is still too cold for this species to reproduce and survive at such high latitudes.

2.2.3. United States

Studies of rocky shores on both the Atlantic and Pacific coasts of the US date back to the early 1900s, but are mostly limited to recent decades, stemming from the growing awareness of the need for datasets of broad spatial and temporal coverage to track and predict impacts of global environmental change [97,98]. Both geography and oceanography have a large influence on intertidal community structure along the Pacific coast of the USA. In warm-regime years, some species from the Californian biogeographic province of the east Pacific have extended their ranges north into higher latitudes [99,100]. Additional biogeographic shifts have been inferred from increases in the abundance of warm versus decreases in the abundance of cold water gastropods, limpets and anthazoans between the early 1930s and the 1990s at a fixed site in Monterey Bay, California [38,102]. These alterations in the relative dominance of co-occurring species have changed the assemblage composition to a more typically warm-water community. The southern neogastropod *Kelletia kelletii* shifted its biogeographic range northwards in the California region of the northeast Pacific between the late 1970s/early 1980s and the 2000s [103,104]. Fossil records and surveys from the 1830s to the present illustrate that this is the first recorded extension beyond Point Conception, and coincided with strong warming of SST during the late 1900s [104]. *Codium fragile*, a warm water green alga has appeared in the Gulf of Maine on the east coast of the US since the 1970s as summer sea temperatures have increased sufficiently to facilitate successful reproduction [105].

2.2.4. Southern Hemisphere

Both Australia and New Zealand have a very high number of endemics due to their extensive history of geographic isolation from other temperature regions [106,107]. In such regions, reductions in abundance and geographic extent may lead to species becoming globally extinct. In Australia, new research programmes have been initiated to track the rate of biogeographic shifts of rocky intertidal species [108] but data is still sparse for this region. Tropical species of rockpool fishes are already being found at temperate latitude locations on the east coast as coastal water temperatures have increased [109]. Temperate species of kelp that form a dense zone from the low intertidal to shallow subtidal in Tasmania have been continually declining due to the direct impact of rising sea temperatures [110,111]. The decline has been exacerbated by intensive grazing from the spread of the warm water long-spined sea urchin *Centrostephanus rodgersii* from the mainland since the 1960s [112] and increases in abundance of the warm water urchin *Janus edwardsii* and the abalone *Haliotis rubra* [108]. The western rock lobster, *Parulirus cygnus* is the most important single species fishery in Australia [113]. Fisheries landings correlate strongly with the strength of the Leeuwin Current, which drives cross shelf transport of larvae and hence productivity of the stock [113]. The Leeuwin Current strength is highly correlated with ENSO events, with a weakening during El Niño years. Since the 1970s, El Niño events have become more common [114], resulting in more frequent depressions in the size of the lobster fishery.

New Zealand has little quantitative data for intertidal species covering the entire coastline, although extensive time-series exist for individual locations ([115] and Ballantine, unpublished data). Research began in 2008 to quantitatively survey the New Zealand rocky coastline to establish a baseline from which future climate-induced shifts in species distributions and ecosystem-level responses can be measured [116]. Experiments show that the mussel *Perna canaliculus* is less tolerant to warm air temperatures than the co-occurring *Mytilus* spp. on New Zealand shores [117] and exhibits reduced growth and reproductive effort when transplanted to higher intertidal elevations, indicating potentially deleterious effects of climate change. This species typically inhabits the lowshore and damp, shaded regions of the mid-shore and thus may not be subjected to as severe exposures as under experimentally manipulated conditions.

2.2.5. South Africa

Little climate related research has been published from the African continent. Range extensions and population-level changes were reported for warm water rocky intertidal species of limpets (*Patella longicosta*, *P. oculus*) and winkle (*Oxystele variegata*). Recruitment failure was observed in the cold water limpet *P. granatina* in South Africa in response to the unusually warm surface temperatures in the southern Benguela current (around South Africa) in 1982/1983 [118] probably connected to the strong 1982/1983 El Niño event in the Pacific [119]. Whilst there is plenty of evidence for species expanding their northern range limits there is less for contraction of southern limits. This issue stems in part from the lack of knowledge of both past and present locations of southern biological limits of many intertidal species, and the paucity of data collected from southern limit populations, which tend to lie in African or South American coastal waters.

2.3. Extreme Weather Events

The extreme cold winter of 1962/1963 in Britain substantially cut back northern range limits of many intertidal species as a direct result of exposure to sub-zero temperatures. Communities in north Wales were particularly severely impacted due to low water spring tides occurring in the early hours of the morning, when air temperatures are at their lowest. Populations were less affected further south in Wales and England due to low water occurring around mid-day, and northern limits were re-located in these regions [120]. The northern range edge of *O. lineatus* in north Wales did not show much recovery from the retraction caused by the extreme cold winter of 1962/1963 until the 1980s. In the intervening two decades, the range has re-extended by ~100 km around the coastline and multi-age, breeding populations have become established within 15 km of the previous limits [121].

Heatwave events are becoming more common during the current period of rapid global environmental change, and have also caused catastrophic mortalities of intertidal species including the Pacific seastar *Pisaster ochraceus* (Harley, personal communication), the mussel *M. edulis* and macroalgae (Mieszkowska, personal observation) due to exposure to high air temperatures. Interestingly, survival and foraging behavior of *P. ochraceus* does not appear to be negatively influenced by chronic, less severe increases in aerial temperature [122], suggesting that it will cope with global warming over long time periods as long as intense thermal shocks do not occur frequently. Although extreme weather events can cause temporary alterations in range limits of intertidal species, it is the longer-term climatic trends that appear to determine the biogeographic limits and large-scale distributional shifts.

2.4. Interactions

Increasing numbers of species from warm climatic regions are beginning to replace those with colder climate affinities in temporal regions, leading to alterations in the composition of local assemblages [76,101,102,123,124]. These local scale changes will also facilitate the pole-ward spread of species by altering the ratio of extinction to colonisation events within range edge populations. The northern cold water species of barnacle *S. balanoides* and limpet *Patella vulgata* have also declined in abundance relative to southern warm water barnacles (*Chthamalus* species) and limpets (*P. depressa*) in Britain and northern Europe [76,125,126]. Models built and tested using the long-term barnacle time-series for Britain show that this rate is increasing, and *S. balanoides* are predicted to have been completely replaced by *Chthamalus* spp. in south west England between 2060 and 2080. Climate change is also altering facilitative interactions. The cold water limpet, *P. vulgata*, preferentially locates its homescar under the shade of the macroalgae *Fucus vesiculosus*. When *F. vesiculosus* is experimentally removed from shores to simulate the impacts of climate warming, significant mortality is observed in *P. vulgata*, with surviving individuals relocating their homescars. In contrast, the warm water congener *P. depressa* does not suffer increased mortality and does not alter its location [72,127].

3. MECHANISMS

While correlational biogeographic studies can be used to obtain probabilistic maps of species occurrence and generate testable hypotheses, they cannot provide information on cause and effect [84,128]. In order to accurately predict the rate and extent of future biogeographic shifts in species distributions, the biological mechanisms driving these changes need to be better understood. Physical, ecological, evolutionary and physiological factors acting on the processes of reproduction, birth, dispersal, recruitment and mortality are all involved in shaping species' ranges [25,129–131] and must also be considered when studying the effects of a changing environment. Mechanistic responses to climate warming have been detected from the cellular and molecular to the organismal and population levels of biological organisation.

3.1. Thermotolerance

Most species of plants and animals are able to withstand a certain degree of thermal stress due to increased production of heat shock proteins (hsp), which reverse the unfolding of cellular proteins under elevated or reduced temperatures and remove badly damaged proteins from the cell [132]. Geographical trends in increased heat shock protein expression have been shown for intertidal species such as *Nucella canaliculata*, a temperate intertidal whelk occurring along rocky coastlines of the north east Pacific. The increase in hsp expression in southern populations indicates that individuals are more stressed at the southern range edge than in range centre [133]. The purple urchin, *Strongylocentrotus purpuratus* also displays geographic variation in genetic regulation of hsp induction in response to increasing temperatures, which reflects the level of thermal stress experienced at different locations along the biogeographic distribution [134]. hsps may therefore be a sensitive tool with which to monitor the extent of climate-induced stress in intertidal species.

3.2. Reproduction and Recruitment

Variations in sea and air temperatures have also long been known to influence the physiological performance and reproductive success of marine species [14,45,137–141]. Synchronous increases in abundance have been recorded in populations of southern trochids throughout Britain and northern France since the mid-1980s. These increases in abundance are linked to warming in the regional climate since the mid-1908s due to increased frequency of annual recruitment success [65]. The mechanisms behind these changes are earlier onset of annual reproductive cycles of southern trochid gastropods *Osilinus lineatus* and *Gibbula umbilicalis* in response to milder winters and warmer springs, coupled with increased survival of newly settled recruits (often the most sensitive lifestage to environmental stress) exposed to milder, shorter winters on the shore [121]. The annual reproductive cycles of southern limpets are also starting earlier and lasting longer in south-west Britain. In contrast, less than 20% of the population of northern limpets is reaching gonad development stages at which spawning can occur [135]. Recruitment can also be influenced by oceanographic features, which can control dispersal of the pelagic larval phase. Large-scale surveys of the Pacific coast of the USA have demonstrated a tight correlation between SST and recruitment, and large intraspecific differences in recruitment rate along the biogeographic range of keystone barnacle and mussel species [98] demonstrating the potential for alterations in recruitment success in a warming climate.

4. ADDITIONAL IMPACTS OF GLOBAL CHANGE

4.1. Ocean Acidification

Dealt with in Chapter 21, ocean acidification is predicted to impact upon rocky shore species such as barnacles, limpets and topshells during the second half of the twenty-first century. Potential synergistic effects of warming seas and decreases in oceanic pH are as yet unknown, and may also vary depending on the position of the organism within the latitudinal distribution of the species.

4.2. Coastal Zone Development

Other facets of global environmental change are rising sea levels and extreme weather events [142]. Many areas of low-lying coastline are facing greater risk of flooding around the world. This risk has led to proliferation of coastal defences to protect property, agricultural land and infrastructure such as roads and railways. Localised defences can scale up to whole coastlines when multiple structures are built along large sections of the land–sea interface. This has occurred in the Mediterranean [143,144] and on the coasts of the northern Irish Sea and eastern English Channel and North Sea. These developments can have severe impacts with loss of sedimentary habitats and their replacement with artificial rocky shores with an impoverished biota [144]. Such large-scale coastal modification can also influence biogeographic processes. Recent range extensions of rocky shore species in the eastern English Channel are probably the result of a combination of increased reproductive success and the provision of artificial habitat (sea defenses, marinas, seaside piers) as stepping stones for advance across patches of unfavourable habitat [145].

4.3. Climate Change and Non-Native Species

Introductions of non-native species are increasing globally due to increases in maritime shipping, aquaculture, mariculture, the aquarium trade and imported live bait. Ballast water, hull fouling and intensive culture practices have been identified as high risk vectors for the introduction of invasive non-native species into the marine environment. There is also evidence that such introductions are more likely in a warmer world [146]. The importation of shellfish species for aquaculture and mariculture businesses in the United Kingdom (e.g., the Pacific oyster *Crassostrea gigas*, the Mediterranean blue mussel

M. galloprovinicialis) have facilitated the spread of these species into natural ecosystems in Britain, Europe and the USA. Inshore sea temperatures are now warm enough to allow successful reproduction of these introduced species, resulting in aggressive colonisation of areas outside their site of cultivation, often to the detriment of native congeners which are outcompeted by the non-native species. Once non-natives become established in the natural environment there are few barriers to prevent further spread. The introduction of non-native species from different biogeographical realms can be considered as a facet of global environmental change with the potential for global homogenisation of biotas and hence decreased biodiversity.

5. CONCLUSIONS

Intertidal invertebrates and algae are already responding to global climate warming, with shifts in biogeographic distributions away from warmer low latitude regions towards the cooler poles. In general, the rate of recession of northern species is not as fast as the rate of advance of southern species. The rate and extent of change is also species specific, leading to alterations in community composition with knock-on effects for ecosystem structure and functioning and likely declines in biodiversity in temperate-tropical regions. These shifts are ultimately being driven by physiological responses to temperature, with additional indirect impacts from species interactions, oceanographic processes, coastal zone development and habitat availability. There is still a paucity of data with long temporal and wide spatial coverage, which is hampering the detection of environmentally driven changes in coastal ecosystems, especially in the southern hemisphere. It is of vital importance that research focuses on the combination of maintaining these valuable datasets with the development of experimental research to determine the mechanisms underpinning the observed responses, so that we may be better placed to predict the future impacts on intertidal systems from our rapidly changing environment.

ACKNOWLEDGEMENTS

The author wish to thank K. Richardson for comments and advice which helped to improve this chapter; the MarClim project www.mba.ac.uk/marclim and MarClim funding consortium; and B.T. Helmuth, B.R. Broitman and E.S. Poloczanska for assistance with publications.

REFERENCES

1. P.H. Gosse, Tenby: A Seaside Holiday, John van Voorst, Paternoster Row, London, 1856, pp. 1–397.
2. C. Kingsley, Glaucus; or, the Wonders of the Shore, Macmillan and Company, Cambridge, 1856, pp. 1–168.
3. J.H. Orton, Nature 123 (1929) 14–15.

4. M.S. Doty, Ecology 27 (1946) 315–328.
5. J.H. Connell, Ecology 42 (1961) 710–723.
6. R.T. Paine, Am. Nat. 100 (1966) 65–75.
7. R.T. Paine, Oecologia 15 (1974) 93–120.
8. T. Carefoot, Pacific Seashores, J.J. Douglas, Vancouver (1977) 1–208.
9. R.T. Paine, J.M. Levine, Ecol. Monogr. 51 (1981) 145–178.
10. J.R. Lewis, The Ecology of Rocky Shores, English Universities Press, London, 1964, pp. 1–323.
11. P.K Dayton, Ecol. Monogr. 41 (1971) 351–389.
12. B.A. Menge, Ecol. Monogr. 46 (1976) 355–393.
13. A.J. Southward, Nature 175 (1955) 1124–1125.
14. A.J. Southward, J. Mar. Biol. Assoc. UK 36 (1957) 323–334.
15. R.T. Paine, JAE 64 (1994) 425–427.
16. B.T. Helmuth, M.W. Denny, Limnol. Oceanogr. 48 (2003) 1338–1345.
17. B.A. Foster, Mar. Biol. 8 (1971) 12–29.
18. J.A. Kitzes, M.W. Denny, Biol. Bull. 8 (2005) 114–119.
19. J. Davenport, H. Macalister, J. Exp. Mar. Biol. Ecol. 76 (1996) 985–1002.
20. R. Li, S.H. Brawley, Mar. Biol. 144 (2004) 205–213.
21. L.E. Burnett, Am. Zool. 37 (1997) 633–640.
22. R.F. Service, Science 305 (2004) 1099.
23. E.P. Dahlhoff, B.A. Menge, Mar. Ecol. Prog. Ser. 144 (1996) 97–107.
24. L.W. Hutchins, Ecol. Monogr. 17 (3) (1947) 325–335.
25. R.M. MacArthur, Geographical Ecology: patterns in the distribution of species, Harper & Rowe, New York. pp. 269.
26. J.R. Lewis, Hydrobiologia 142 (1986) 1–13.
27. J.H. Brown, G.C. Stevens, D.M. Kaufman, Annu. Rev. Ecol. Syst. 27 (1996) 597–623.
28. J.H. Brown, Am. Nat. 124 (1984) 255–279.
29. J.H. Stillman, G.N. Somero, JEB. 199 (1996) 1845–55.
30. A.A. Hoffmann, P.A. Parsons, Extreme Environmental Change and Evolution, Cambridge University Press, Cambridge, 1997, p. 235.
31. B. Helmuth, C.D. Harley, P.M. Halpin, M. O'Donnell, G.E. Hofmann, C.A. Blanchette, Science 298 (2002) 1015–1017.
32. J.H. Stillman, Integr. Comp. Biol. 42 (2002) 790–796.
33. M.H. Carr, J.E. Neigel, J.A. Estes, S. Andelman, R.R. Warner, J.L. Largier, Ecol. Appl. 13, S90–S107.
34. P.E. Smith, Can. J. Fish. Aquat. Sci. 42 (1985) 69–82.
35. S. Jenouvrier, C. Barbraud, H. Weimerskirch, JAE 72 (2003) 576–587.
36. J.R. Lewis, Mar. Biol. Ann. Rev. 14 (1976) 371–390.
37. S.J. Hawkins, R.G. Hartnoll, J. Exp. Mar. Biol. Ecol. 62 (1982) 271–283.
38. J.P. Barry, C.H. Baxter, R.D. Sagarin, S.E. Gilman Science 267 (1995) 672–674.
39. J.R. Lewis, R.S. Bowman, M.A. Kendall, P. Williamson, Neth. J. Sea Res. 16 (1982) 18–28.
40. B. Helmuth, N. Mieszkowska, P. Moore, S.J. Hawkins, Annu. Rev. Ecol. Evol. Syst. 37 (2006) 373–404.
41. M.D. Bertness, G.H. Leonard, J.M. Levine, J.F. Bruno, Oecologia 120 (1999) 446–450.
42. R.D. Sagarin, S.D. Gaines, J. Biogeogr. 29 (2002) 985–997.
43. C.D.G. Harley, A.R. Hughes, K. Hultgren, B.G. Miner, C.J.B. Sorte, C.S. Thornber, L.F. Rodriguez, L. Tomanek, S.L. Williams, Ecol. Lett. 9 (2006) 228–241.
44. J.P. de Tournefort, Relation d'un voyage du Levant (1717). Paris. pp. 288.
45. J. Grinnell, Am. Nat. 51 (1917) 115–128.

46. J.H. Orton, J. Mar. Biol. Assoc. UK 2 (1920) 299–366.
47. W.A. Setchell, The temperature interval in the geographical distribution of marine algae (1920). Science 187–190.
48. S.A. Cain, Foundations of Plant Geography, Harper Brothers, New York, London, 1944.
49. L.W. Hutchins, Ecol. Monogr. 17 (1947) 325–335.
50. E.V. Wulff, An Introduction to Historical Plant Geography (translated from the Russian by E. Brissenden), Waltham, MA, 1950.
51. H.G. Andrewartha, L.C. Birch, The Distribution and Abundance of Animals, University of Chicago Press, Chicago, 1954.
52. E.C. Pielou, Biogeography, Wiley-Interscience, Chichester, 1979.
53. A.M. Breeman, Helgoland Marine Research 42 (1988) 199–241.
54. G. Michanek, Bot. Mar. 22 (1979) 375–391.
55. K. Luhning, Seaweeds their environment, biogeography and ecophysiology (1990) John Wiley & Sons, London, 1–61.
56. G.M. Voskoboinikov, A.M. Breeman, C. van den Hoek, V.N. Makarov, E.V. Shoshina, Bot. Mar. 39 (1996) 341–346.
57. F.J. Molenaar, A.M. Breeman, J. Phycol. 33 (1997) 330–343.
58. S. Orfanidis, A.M. Breeman, J. Phycol. 35 (1999) 919–930.
59. A.J. Southward, Nature 165 (1950) 408.
60. D.R. Franz, A.S. Merrill, Malacologia 19 (1980) 209–225.
61. G.J. Vermeij, Evolution 36 (1982) 561–580.
62. J.R. Lewis, The Ecology of Rocky Shores, English Universities Press, London, 1964.
63. M.A. Kendall, J.R. Lewis, Hydrobiologia 142 (1986) 15–22.
64. N. Mieszkowska, R. Leaper, P. Moore, M.A. Kendall, M.T. Burrows, D. Lear, E. Poloczanska, K. Hiscock, P.S. Moschella, R.C. Thompson, R.J. Herbert, D. Laffoley, J. Baxter, A.J. Southward, S.J. Hawkins, J. Mar. Biol. Assoc. UK Occasional Publ. 20 (2005) 1–55.
65. N. Mieszkowska, M.A. Kendall, S.J. Hawkins, R. Leaper, P. Williamson, N.J. Hardman-Mountford, A.J. Southward, Hydrobiologia 555 (2006) 241–251.
66. R.G. Pearson, T.P. Dawson, Glob. Ecol. Biogeogr. 12 (2003) 361–371.
67. G.R. Walther, E. Post, P. Convey, A. Menzel, C. Parmesan, T.J.C. Beebee, T.J.C. Fromentin, O.H. Guldberg, F. Bairlein, Nature 416 (2002) 389–395.
68. A.J. Davies, L.S. Jenkinson, J.H. Lawton, B. Shorrocks, S. Wood, Nature 391 (1998) 783–786.
69. R.W. Brooker, R.W. Travis, E.J. Clark, C. Dytham, J. Theor. Biol. 245 (2007) 59–65.
70. E. Fischer-Piette, J. Conch. Paris 79 (1935) 5–66.
71. E. Fischer-Piette, J. Linn. Soc. Zool. 40 (1936) 181–272.
72. E. Fischer-Piette, Ann. Insitut. Ocen. Monaco 31 (1955) 37–124.
73. D.J. Crisp, E. Fischer-Piette, Ann. Insitut. Ocen. Monaco 36 (1959) 276–381.
74. A.J. Southward, D.J. Crisp, Proc. R. Ir. Acad. 57 (1954) 1–29.
75. D.J. Crisp, A.J. Southward, J. Mar. Biol. Assoc. UK 37 (1958) 157–208.
76. A.J. Southward, S.J. Hawkins, M.T. Burrows, J. Therm. Biol. 20 (1995) 127–155.
77. M.A. Kendall, J.R. Lewis, Hydrobiologia 142 (1986) 15–22.
78. M.A. Kendall, J. Moll. Stud. 53 (1987) 213–222.
79. F.P. Lima, N. Queiroz, P.A. Ribeiro, S.J. Hawkins, A.M. Santos, J. Biogeogr. 33 (2006) 812–822.
80. B. Helmuth, J.G. Kingsolver, E. Carrington, Ann. Rev. Phys. 67 (2005) 177–201.
81. C. Parmesan, S. Gaines, L. Gonzales, D.M. Kaufman, J. Kingsolver, A.T. Peterson, R. Sagarin, Oikos 108 (2005) 58–75.
82. P. Moore, S.J. Hawkins, R.C. Thompson, Mar. Ecol. Prog. Ser. 334 (2007) 11–19.

83. R.J.H. Herbert, S.J. Hawkins, M. Sheader, A.J. Southward, J. Mar. Biol. Assoc. UK 83 (2003) 73–82.
84. C. Parmesan, G. Yohe, Nature 421 (2003) 37–42.
85. I. Bartsch, C. Wiencke, K. Bischof, C.M. Buchholz, B.H. Buck, A. Eggert, P. Feuerpfeil, D. Hanelt, S. Jacobsen, R. Karez, U. Karsten, M. Molis, M.Y. Roleda, H. Schubert, R. Schumann, K. Valentin, F. Weinberger, J. Wiese. The genus Laminaria sensu lato: recent insights and developments, European Journal of Phycology, 43(1) (2008) 1–86, doi:10.1080/09670260701711376.
86. C.N. Bianchi, Hydrobiologia 580 (2007) 7–21.
87. P. Chevaldonné, C. Lejeusne, Ecol. Lett. 6 (2003) 371–379.
88. D.S. Wethey, S.A. Woodin, Hydrobiologia 606 (2008) 139–151.
89. T. Vance, MRes Thesis, University of Plymouth, 2005.
90. F. Ardré, Port. Acta Biol. 10 (1970) 1–423.
91. F. Ardré, Bull. Centre d'Etudes Rech.Sci. Biarritz 8 (1971) 359–574.
92. F.P. Lima, P.A. Ribeiro, N. Queiroz, S.J. Hawkins, A.M. Santos, Glob. Change Biol. 13(12) (2007) 2592–2604.
93. R.T. Lemos, H.O. Pires, Intern. J. Clim. 24 (2004) 511–524.
94. J.M. Wesławski, M. Zajączkowski, J. Wiktor, M. Szymelfenig, Polar Biol. 18 (1997) 45–52.
95. O. Salvigsen, Nor. Geogr. Tidsskr. 56 (2002) 56–61.
96. J. Berge, G. Johnsen, F. Nilsen, B. Gulliksen, D. Slagstad, Mar. Ecol. Prog. Ser. 303 (2005) 167–175.
97. C.A. Blanchette, C.M. Miner, P.T. Raimondi, D. Lohse, K.E.K. Heady, B.R. Broitman, J. Biogeogr. (2008) in press.
98. B.R. Broitman, C.A. Blanchette, B.A. Menge, J. Lubchenco, P.A. Raimondi, C. Krenz, M. Foley, D. Lohse, S.D. Gaines, Ecol.Monogr. 78(3) (2008) 403–421.
99. R.S. Burton, Evolution 52 (1998) 734–745.
100. M.E. Hellberg, D.P. Balch, K. Roy, Science 292 (2001) 1707–1710.
101. J.P. Barry, C.H. Baxter, R.D. Sagarin, S.E. Gilman, Science 267 (1995) 672–674.
102. R.D. Sagarin, J.P. Barry, S.E. Gilman, C.H. Baxter, Ecol. Monogr. 69 (1999) 465–490.
103. T.J. Herrlinger, Veliger 24 (1981) 78.
104. D. Zacherl, S.D. Gaines, S.I. Lonhart, J. Biogeogr. 30 (2003) 913–924.
105. L.G. Harris, M.C. Tyrrell, Biol. Invasions 3 (2001) 9–21.
106. G.C.B. Poore, State of the Marine Environment Report for Australia: The Marine Environment - Technical Annex: 1 Compiled by Leon P. Zann Great Barrier Reef Marine Park Authority, Townsville Queensland, Ocean Rescue 2000 Program, Department of the Environment, Sport and Territories, Canberra, 1995 ISBN 0 642 17399 0.
107. R. Tsuchi, S. Nishimura, A.G. Beu, Tectonophysics 281 (1997) 83–97.
108. E.S. Poloczanska, R.C. Bobcock, A. Butler, A.J. Hobday, O. Hoegh-Guldberg, T.J. Kunz, R. Matear, D.A. Milton, T.A. Okey, A.J. Richardson, Oceanogr. Mar. Biol. Ann. Rev. 45 (2008) 407–478.
109. S.P. Griffiths, Est. Coast. Shelf Sci. 58 (2003) 173–186.
110. K.S. Edyvane, Final Report for Environment Australia, Department of Primary Industries, Water and Environment, Hobart, 2003.
111. G.J. Edgar, C.R. Samson, N.S. Barrett, Conserv. Biol. 19 (2005) 1294–1300.
112. C. Johnson, S. Ling, J. Ross, S. Shepherd, K. Miller, Tasmanian Aquaculture and Fisheries Institute, Australia, FRDC Project No. 2001/004, 2005.
113. N. Caputi, C. Chubb, A. Pearce, Mar. Freshw. Res. 52 (2001) 1167–1174.
114. K.E. Trenberth, Bull. Am. Met. Soc. 78 (12) (1997) 2771–2777.

115. J.E. Morton, V.J. Chapman, Rocky Shore Ecology of The Leigh Area, North Auckland, University of Auckland Press, Auckland, 1968, 44pp.
116. N. Mieszkowska, C. Lundquist, N. Z. J. Mar. Freshw. Res. in review.
117. L.E. Petes, B.A. Menge, G.D. Murphy, J. Exp. Mar. Biol. Ecol. 351 (2007) 83–91.
118. G.M. Branch, S.A. J. Sci. 80 (1984) 61–65.
119. R.T. Barber, F.P. Chavez, Science 222 (1983) 1203–2110.
120. D.J. Crisp, JAE 33 (1964) 165–210.
121. N. Mieszkowska, S.J. Hawkins, M.T. Burrows, M.A. Kendall, J. Mar. Biol. Assoc. UK 87 (2007) 537–545.
122. E. Sanford, J. Exp. Mar. Biol. Ecol. 273 (2002) 199–218.
123. J.A. McGowan, D.B. Chelton, A. Conversi, CalCOFI Report, 37 (1996).
124. S.J. Holbrook, R.J. Schmitt, J.S.J. Stevens, Ecol. Appl. 7 (1997). pp. 1299–1310.
125. A.J. Southward, J. Mar. Biol. Assoc. UK 71 (1991) 495–513.
126. S.J. Hawkins, P. Moore, M. Burrows, E. Poloczanska, N. Mieszkowska, S.R. Jenkins, R.C. Thompson, M. Genner, A.J. Southward, Hydrobiologia, Climate Research, 37 (2/3), (2008) pp. 123–133.
127. P. Moore, R.C. Thompson, S.J. Hawkins, J. Exp. Mar. Biol. Ecol. 344 (2007) 170–180.
128. M.J. Fortin, T.H. Keitt, B.A. Maurer, M.L. Taper, D.M. Kaufman, T.M. Blackburn, Oikos 108 (2005) 7–17.
129. R.N. Carter, S.D. Prince, Nature 293 (1981) 644–645.
130. D.M. Lodge, in: P.M. Karieva, J.G. Kingsolver, R.B. Huey (Eds.), Biotic Interactions and Global Change, Sinauer, Sunderland, MA, 1993, pp. 367–387.
131. J.J. Lennon, J.R.G. Turner, D. Connell, Oikos 78 (1997) 486–502.
132. S. Lindquist, Ann. Rev. Biochem. 55 (1986) 1151–1191.
133. C.J.B. Sorte, G.E. Hofmann, Mar. Ecol. Prog. Ser. 274 (2004) 263–268.
134. C.J. Osovitz, G.E. Hofmann, J. Exp. Mar. Biol. Ecol. 327 (2005) 134–143.
135. P. Moore, PhD Thesis, University of Plymouth, 2005.
136. V.L. Loosanoff, Science 102 (1945) 124–125.
137. R.A. Boolootian, in: R.A. Boolootian (Ed.), Physiology of Echinodermata, Wiley, New York, 1966, pp. 561–614.
138. B. Cocanour, K. Allen, Comp. Biochem. Physiol. 20 (1967) 327–331.
139. R.E. Stephens, Biol. Bull. 142 (1972) 132–134.
140. P.W. Frank, Mar. Biol. 31 (1975) 181–192.
141. B.T. Helmuth, G.E. Hofmann, Biol. Bull. 201 (2001) 371–381.
142. IPCC, Climate Change 2007: The Scientific Basis, IPCC 2007 1–996.
143. L. Airoldi, M. Abbiati, M.W. Beck, S.J. Hawkins, P.R. Jonsson, D. Martin, P.S. Moschella, A. Sundelöf, R.C. Thompson, P. Åberg, Coast. Eng. 52 (2005) 1073–1087.
144. P.M. Moschella, M. Abbiati, P. Åberg, L. Airoldi, J. Anderson, J.M. Bacchiocchi, F. Bulleri, G.E. Dinesen, M. Frost, E. Gacia, L. Granhag, P.R. Jonssonn, M.P. Satta, A. Sundelöf, R.C. Thompson, S.J. Hawkins, Coast. Eng. 52 (2005) 1053–1071.
145. S.J. Hawkins, N. Mieszkowska, P. Moschella, unpublished data.
146. J.J. Stachowicz, J.R. Terwin, R.B. Whitlatch, R.W. Osman, PNAS 99 (2002) 15497–15500.

Chapter 16

Plant Ecology as an Indicator of Climate and Global Change

Michael D. Morecroft
NERC Centre for Ecology and Hydrology, Crowmarsh Gifford, Wallingford OX10 8 BB, UK

Sally A. Keith
Centre for Conservation Ecology and Environmental Change, Bournemouth University, Fern Barrow, Poole BH12 5BB, UK

1. Introduction
2. Changes in Phenology
3. Changes in Distribution
4. Community Composition
5. Plant Growth
6. Conclusions
References

1. INTRODUCTION

The distribution of types of vegetation around the world is clearly related to climate. Different combinations of temperature, rainfall and seasonality produce the global variety of biomes, from rainforest to tundra, which we take for granted. At a finer scale we can see changes in vegetation with more localised changes in climate such as on a mountain as conditions become cooler with altitude [1,2]. Individual species also have distribution patterns, the boundaries of which are largely defined by climate at a global scale. These distribution patterns reflect the influence of climate on plant survival, physiology and growth, together with climatic effects on ecological interactions, such as competition, pollination and herbivory. Different types of plant are adapted to different climatic conditions, from cold-tolerant, but slow-growing alpine plants, to fast-growing trees in the wet tropics.

It is therefore reasonable to expect that changes in climate would lead to a change in species distributions and community composition. Evidence of such changes has been accumulating in recent decades [3–7]. However, before we come to evaluate this evidence, we should consider some general principles.

To identify the ecological impacts of climate change with confidence, it is necessary both to be able to detect a change in an ecosystem and to reliably attribute it to a change in climate [6,8]. *Detection* of any change in an environmental variable requires a reliable dataset with repeated measurements over a period of time. Good instrumental records of climate itself go back over 100 a in many countries, but very few biological datasets extend this far. In many cases climate change impacts must be inferred from re-surveys of early work carried out for quite different purposes. Attribution of impacts to climate change requires a relationship between climate and impact variables to be established and other potential causes of change ruled out. The effects of climate change on plants are complex (Fig. 1) and the presence or absence of a species from a particular location does not solely depend on its ability to tolerate physical conditions. In many cases climatic limits are determined by the influence of climate on a plant's ability to compete with other species [9]. Climate change may also disturb interactions between plants and

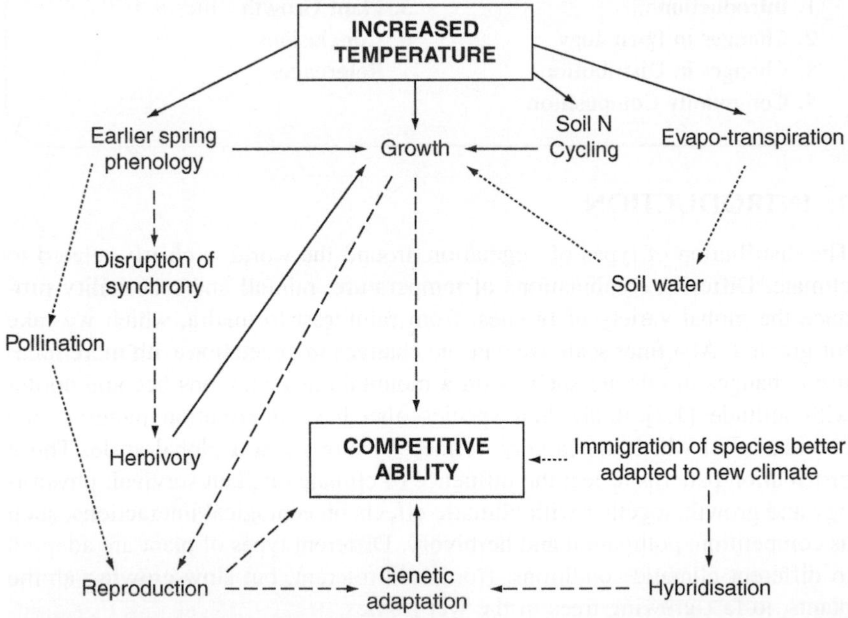

FIGURE 1 An example of complexity in plant responses to climate change. Factors influencing the effects of temperature on the competitive ability of a temperate, insect pollinated plant. Note that this is a simplified diagram and does not take account of all factors, interactions or the role of other climate change factors such as changes in precipitation or extreme events. Solid arrows indicate positive effects, dotted arrows indicate negative effects and dashed arrows indicate effects that could positive or negative.

their pollinators, mycorhizae, herbivores or pathogens. Rising temperatures are the best understood aspects of climate change but in the longer term changes in precipitation or one-off extreme events, which are harder to predict, may be more important. Changing atmospheric composition, including carbon dioxide concentration, can also have effects on plant performance and interactions [10].

A further issue is that many plant communities are composed of long-lived species and only change slowly in response to incremental changes in climate [11]. This contrasts with many invertebrate species for which clear signals of changing distributions have been found [12,13]; most of these species have short generation times and in many cases high mobility.

2. CHANGES IN PHENOLOGY

The recording of phenology – the seasonal timing of biological events such as leafing and flowering – provides several examples of unusually long-term data sets. A particularly good example is the Marsham family records for 'indications of spring' concerning over twenty plant and animal species for 200 a in Norfolk, UK [14]. Analyses of these data, particularly correlations with equally lengthy climate data, have provided important information on past effects of climate on phenological events, which in turn have been used to predict future responses of these species to projected climate change.

The Marsham data formed a component of a much larger, European wide meta-analysis of the relationship between phenology and temperature [15]. The meta-analysis included, *inter alia*, phenological trend data for 542 plant species from 21 countries. There was a clear correlation between warmer temperatures and the earlier onset of spring phenology (leaf-opening and flowering) in 78% of plants (31% significantly). In contrast, autumn onset indicators were more ambiguous, showing no overall pattern of correlation with temperature, although some individual events did correlate with temperature. The paper demonstrated a mean advance in spring and summer phenology of 2.5 days per decade in recent decades [15].

On a global scale, the most recent assessment report by the Intergovernmental Panel on Climate Change (IPCC) presented a synthesis of the current knowledge of climate change impacts on phenology. It concluded that the onset of spring has become earlier by 2.3–5.2 days per decade in the last 30 a, and that this is correlated with increasing temperatures [6]. However, of 16 studies cited, none are based in the southern hemisphere. This bias towards the northern hemisphere is a common theme throughout research into impacts of climate change on biodiversity. Satellite remote sensing has however allowed a different, more global approach to phenology. Indices based on the spectral composition of light reflected from the surface of the earth, such as the Normalised Difference Vegetation Index (NDVI), can quantify the 'greening up' of temperate zones in the spring. These techniques have broadly corroborated surface-based findings of an advancement of spring [16–18].

Phenology therefore provides a clear indicator of climate change impacts on plants. In itself, a change in phenology is arguably not a major issue if the species continues to persist in a current location. However, there is evidence that the changing phenology of a species can have important ecological consequences for pollinators [19], herbivores [20] and competitors [21].

3. CHANGES IN DISTRIBUTION

After phenology, the most frequently reported changes in plant ecology in response to climate change are changes in species' geographical distributions. The mapping of distributions of species and vegetation types, whether local, national or international in scale, pre-dates contemporary interest in climate change by several decades or more. Re-surveys of distributions provide an opportunity to test whether changes consistent with the impacts of climate change are taking place [22,23]. Studies of this sort have been an important component of the impacts reported in the IPCC's third and fourth assessment reports [6].

Good examples of changes in distribution can be seen in altitudinal studies. Temperatures typically fall with altitude by ~6.5 °C·km^{-1} [24], although this varies with other factors, such as humidity. Plant communities consequently change markedly with altitude. The clearest example of this is the presence of tree lines, beyond which trees do not grow. Many explanations for the occurrence of tree lines have been offered, but plants are thought to respond to combinations of temperature change, atmospheric CO_2 concentration, nutrient availability and solar radiation [25]. Regardless of the exact mechanism, which may vary between situations, natural tree lines (those not changed by forest management) are determined primarily by climate, particularly temperature. A warming of climate would therefore be expected to lead to tree lines shifting to higher altitudes. Evidence of this has been found with tree lines shifting at rates of 0.01–7.5 m·a^{-1}, depending on the species of tree involved and the type of climatic forcing [3]. The length of data collection is also likely to affect the mean shift each year (in this and other variables) because longer datasets will be subject to a smoothing of the trend through natural variation and sign switching. Latitudinal tree line shifts have also been observed, correlated with warmer summer temperatures [25].

Tree line shifts are subject to time lags in their response to environmental change because of trees' long generation time, therefore, changes in non-woody plants and dwarf shrubs might be expected to be more sensitive [3]. Evidence of changes in altitudinal distribution have been found for alpine plants [22]. In a re-survey of vascular plants in the Alps of northern Italy, 52 of the 93 monitored vascular plants were found at a higher altitude than in the 1950s, moving upwards at a rate of 23.9 m per decade [26]. The largest change in species richness was at an altitude that had experienced melting of permafrost, associated with increasing air temperature [26].

4. COMMUNITY COMPOSITION

Changes in distribution patterns are dependent on local extinctions and colonisations at species range margins. As this is where the effects of climate change are most likely to be seen first, they provide a sensitive early indicator of climate change impacts. They also provide some basic information on changes in plant communities. Studies such as those of Walther [3] and Parolo and Rossi [26] indicate how the nature of a community is changing with the colonisation of new, more thermophilous species. However, this sort of research will not capture changes in the abundance of species in other parts of their range. A change from abundance to rarity, or vice versa, for any given species is of major ecological significance, but undetectable if only species presence or absence has been recorded in the original survey.

The potential for changes in vegetation composition is substantial and experimental manipulations of climate have caused major changes in communities. One of the longest-running examples is an experiment in the sub-alpine zone of the Rocky Mountains (USA), where vegetation has been warmed using infrared lamps since 1990. The shrub, *Artemisia tridentata* (sagebrush) has increased in response to this treatment and herbaceous species have declined [27–30]. In this case the effect of warming is mediated through a reduction in summer water availability as a result of earlier snow melt.

Reliable detection of a change in the balance between different species in un-manipulated communities can usually only be achieved through long-term monitoring in permanent sample plots. Most monitoring programmes do not go back earlier than the 1970s and to date it is hard to find changes that can be confidently attributed to climate in the literature. One example of a possible impact of climate change on species composition was reported by Kirby et al. [31], who found changes in British woodland ground flora correlated with increases in growing season length between 1971 and 2001.

As major changes in the relative composition of different types of species are anticipated in the coming decades, various monitoring programmes have been developed to detect them. In the UK, the Environmental Change Network is a good example in which plant community composition is monitored in permanently marked quadrats [32] (www.ecn.ac.uk). In this network the vegetation quadrats form part of a larger ecosystem monitoring programme in which animal populations are also monitored, together with climate, soil nutrients and water content and other potential causes of change such as air pollution. This demonstrated a change in species composition of grasslands, specifically an increase in ruderal species in response to drought [33]. Ruderal plants are those which grow and reproduce quickly and they colonised gaps which opened up in the grassland in response to drought, before being excluded by competitors as wetter conditions returned.

Another network is the Amazon Forest-Inventory Network (RAINFOR) comprised of long-term forest monitoring plots throughout the Amazon rainforest [34]. The network plots have provided evidence for a change in community composition of old-growth Amazonian forest, whereby slow growing tree genera are decreasing and fast growing tree genera are increasing in dominance or density [34,35]. There has also been an increase in density and dominance of lianas within these forests. These changes have been attributed with relative confidence to an increase in atmospheric CO_2 concentration [34].

Individualistic species responses and changes in the nature of interactions will lead to changes in the nature of plant communities. It is possible that assemblages of species may sometimes change from one currently recognised community to another. It is, however, likely that in many cases, novel combinations of species will develop as species respond to changing climate at different rates. Palaeoecology provides evidence of this happening during previous climate change events, indicating the formation of non-analogous communities, that were of a different composition from anything currently recognised [36]. This will have important implications for the functioning of communities and ecosystems and present challenges where current conservation policy is based on defined, historical communities.

5. PLANT GROWTH

Any change in species distribution or community composition is likely to be preceded by a change in plant growth. Plant growth may therefore be a sensitive indicator of climate change impacts. It is also of interest in its own right as it drives the production of food, materials and fuel and is responsible for the sequestration of carbon. The two main categories of plants whose growth is measured are crops and trees. Crops are dealt with in Chapter 17, but we will consider tree growth here.

The growth response of trees, as well as other plants, to climate differs between species, depending on their ecophysiology and life history characteristics. For example, Morecroft et al. [37] showed that the growth of sycamore (*Acer pseudoplatanus*) was adversely affected by drought to a greater extent than pedunculate oak (*Quercus robur*) and ash (*Fraxinus excelsior*) in a British woodland. This was associated with reduced photosynthetic rates in dry soil conditions and may reflect relatively shallow rooting. Broadmeadow et al. [38] modelled broadleaved tree species' growth responses to future climate using a model based on empirical data for species specific growth rates and their correlations with aspects of climate. They found that water limitation in southern England was likely to lead to reductions in growth and increased mortality, with beech (*Fagus sylvatica*) the worst affected.

One of the areas in which tree growth rates have been a particular subject of research interest has been the Amazon rain forest, with the RAINFOR network of old-growth forest plots again providing long-term observational evidence of changes [34]. The plots have shown evidence of an increase in growth rates and biomass in recent decades. More importantly when considering the carbon sink function of the Amazon forest, there is also an increase in turnover of tropical forest trees that is thought to be a function of increased mortality following more rapid growth. These responses are, like the associated changes in community composition, most parsimoniously explained as a response to higher CO_2 concentrations, possibly combined with nutrient enrichment resulting from ash deposition from an increasing number of forest fires [34]. Under certain recruitment/mortality rate ratios, an increase in forest turnover could decrease the carbon sink potential of the Amazon [34].

The trunks of most temperate and some tropical trees have annual rings, reflecting seasonal differences in growth rates. These provide a particularly valuable historical record of growth rates and are often used as proxies for the estimation of past climates. Tree ring data are useful indicators of climate change because they provide a 'self-kept' record of climate response over the lifetime of an individual tree, thereby circumventing the challenge of obtaining long-term monitoring data. Width of tree rings can be correlated with environmental data. In addition to assessing general trends in tree growth to trends in climate, tree rings are very useful for examining the response of trees to extreme climatic events. A reduction in productivity demonstrated by reduced tree ring width of old beech forests in Italy, has been attributed to recent drought during the growing season [39].

A further strength of the use of tree ring data as an indicator of climate change is that changes can be explored in the context of a longer time frame, potentially increasing our understanding of current trends. Touchan et al. [40] analysed tree ring records from North West Africa for approximately the last 600 a to ascertain the influence of drought and found that the most recent drought (1999–2002) was probably the most severe since the fifteenth century and consistent with projections from global circulation models.

6. CONCLUSIONS

There is clear evidence that plants are responding to climate change through changing phenology and distribution patterns, with species tending to disperse towards cooler areas. More far reaching changes in community composition are starting to be recognised and are likely to become increasingly obvious in the coming decades. Responses to temperature have been clearest to date at a global scale, but in the long term, local changes in precipitation or extreme events may be more important than the global trend in temperatures. There are also likely to be complex interactions within ecosystems and with other pressures, which we need to understand and model if attempts to mitigate climate change and adapt to it are to be successful.

REFERENCES

1. A. Hemp, Plant Ecol. 184 (2006) 27–42.
2. B. Sieg, F.J.A. Daniels, Phytocoenologia 35 (2005) 887–908.
3. G.R. Walther, Perspect. Plant Ecol. 6 (2004) 169–185.
4. C. Parmesan, G. Yohe, Nature 421 (2003) 37–42.
5. T.L. Root, J.T. Price, K.R. Hall, S.H. Schneider, C. Rosenzweig, J.A. Pounds, Nature 421 (2003) 57–60.
6. C. Rosenzweig, G. Casassa, D.J. Karoly, A. Imeson, C. Liu, A. Menzel, S. Rawlins, T.L. Root, B. Seguin, P. Tryjanowski, in: M.L. Parry, O.F. Canziani, J.P. Palutikof, P.J. van der Linden, C.E. Hanson (Eds.), Assessment of Observed Changes and Responses in Natural and Managed Systems. Climate change 2007: Impacts, Adaptation and Vulnerability. Contribution of Working Group II to the Fourth Assessment Report of the Intergovernmental Panel on Climate Change, Cambridge University Press, Cambridge, 2007, pp. 79–131.
7. G.R. Walther, E. Post, P. Convey, A. Menzel, C. Parmesan, T.J.C. Beebee, J.M. Fromentin, O. Hoegh-Guldberg, F. Bairlein, Nature 416 (2002) 389–395.
8. G.F. Midgley, S.L. Chown, B.S. Kgope, S. Afr. J. Sci. 103 (2007) 282–286.
9. M. Morecroft, J. Paterson, in: J. Morison, M. Morecroft, (Eds.), Effects of Temperature and Precipitation Changes on Plant Communities, Blackwell Publishing, Oxford, 2006, pp. 146–164.
10. E.A. Ainsworth, S.P. Long, New Phytol. 165 (2005) 351–371.
11. J.P. Grime, J.D. Fridley, A.P. Askew, K. Thompson, J.G. Hodgson, C.R. Bennett, Proc. Natl. Acad. Sci. USA 105 (2008) 10028–10032.
12. R. Hickling, D.B. Roy, J.K. Hill, C.D. Thomas, Global Change Biol. 11 (2005) 502–506.
13. C. Parmesan, N. Ryrholm, C. Stefanescu, J.K. Hill, C.D. Thomas, H. Descimon, B. Huntley, L. Kaila, J. Kullberg, T. Tammaru, W.J. Tennent, J.A. Thomas, M. Warren, Nature 399 (1999) 579–583.
14. T.H. Sparks, P.D. Carey, J. Ecol. 83 (1995) 321–329.
15. A. Menzel, T.H. Sparks, N. Estrella, E. Koch, A. Aasa, R. Ahas, K. Alm-Kubler, P. Bissolli, O. Braslavska, A. Briede, F.M. Chmielewski, Z. Crepinsek, Y. Curnel, A. Dahl, C. Defila, A. Donnelly, Y. Filella, K. Jatcza, F. Mage, A. Mestre, O. Nordli, J. Penuelas, P. Pirinen, V. Remisova, H. Scheifinger, M. Striz, A. Susnik, A.J.H. Van Vliet, F.E. Wielgolaski, S. Zach, A. Zust, Global Change Biol. 12 (2006) 1969–1976.
16. W. Lucht, I.C. Prentice, R.B. Myneni, S. Sitch, P. Friedlingstein, W. Cramer, P. Bousquet, W. Buermann, B. Smith, Science 296 (2002) 1687–1689.
17. R.B. Myneni, C.D. Keeling, C.J. Tucker, G. Asrar, R.R. Nemani, Nature 386 (1997) 698–702.
18. L.M. Zhou, C.J. Tucker, R.K. Kaufmann, D. Slayback, N.V. Shabanov, R.B. Myneni, J. Geophys. Res.-Atmos. 106 (2001) 20069–20083.
19. J. Memmott, P.G. Craze, N.M. Waser, M.V. Price, Ecol. Lett. 10 (2007) 710–717.
20. E. Post M.C. Forchhammer, Philos. Trans. R. Soc. B 363 (2008) 2369–2375.
21. E.E. Cleland, N.R. Chiariello, S.R. Loarie, H.A. Mooney, C.B. Field, Proc. Natl. Acad. Sci. USA 103 (2006) 13740–13744.
22. G.R. Walther, S. Beissner, C.A. Burga, J. Veg. Sci. 16 (2005) 541–548.
23. A.E. Kelly, M.L. Goulden, Proc. Natl. Acad. Sci. USA 105 (2008) 11823–11826.
24. C.D. Whiteman, Mountain Meteorology Fundamentals and Applications, Oxford University Press, Oxford, 2000.
25. J. Grace, F. Berninger, L. Nagy, Ann. Bot-Lond. 90 (2002) 537–544.
26. G. Parolo, G. Rossi, Basic Appl. Ecol. 9 (2008) 100–107.

27. J. Harte, Bioscience 51 (2001) 332–333.
28. J. Harte, R. Shaw, Science 267 (1995) 876–880.
29. T. Perfors, J. Harte, S.E. Alter, Global Change Biol. 9 (2003) 736–742.
30. F. Saavedra, D.W. Inouye, M.V. Price, J. Harte, Global Change Biol. 9 (2003) 885–894.
31. K.J. Kirby, S.M. Smart, H.I.J. Black, R.G.H. Bunce, P.M. Corney, R.J. Smithers, Long-term ecological change in British woodland (1971–2001). A re-survey and analysis of change based on the 103 sites in the nature conservancy 'Bunce 1971' woodland survey, (English Nature Research Report 653) English Nature, Peterborough, 2005, p. 137.
32. J.M. Sykes, A.M.J. Lane, The United Kingdom Environmental Change Network: Protocols for Standard Measurements at Terrestrial Sites. Stationary Office, London, 1996.
33. M.D. Morecroft, C.E. Bealey, E. Howells, S. Rennie, I.P. Woiwod, Global Ecol. Biogeogr. 11 (2002) 7–22.
34. O.L. Phillips, S.L. Lewis, T.R. Baker, K.J. Chao, N. Higuchi, Philos. Trans. R. Soc. B 363 (2008) 1819–1827.
35. W.F. Laurance, A.A. Oliveira, S.G. Laurance, R. Condit, C.W. Dick, A. Andrade, H.E.M. Nascimento, T.E. Lovejoy, J. Ribeiro, Biotropica 37 (2005) 160–162.
36. J.W. Williams, S.T. Jackson, Front. Ecol. Environ. 5 (2007) 475–482.
37. M.D. Morecroft, V.J. Stokes, M.E. Taylor, J.I.L. Morison, Forestry 81 (2008) 59–74.
38. M.S.J. Broadmeadow, D. Ray, C.J.A. Samuel, Forestry 78 (2005) 145–161.
39. G. Piovesan, F. Biondi, A. Di Filippo, A. Alessandrini, M. Maugeri, Global Change Biol. 14 (2008) 1265–1281.
40. R. Touchan, K.J. Anchukaitis, D.M. Meko, S. Attalah, C. Baisan, A. Aloui, Geophys. Res. Lett. 35 (2008) L13705.

Chapter 17

The Impact of Climate and Global Change on Crop Production

Geoffrey R. Dixon

Centre for Horticulture and Landscape, School of Biological Sciences, Whiteknights, The University of Reading, Reading, Berkshire RG6 6AS, United Kingdom

1. Introduction
2. Impact on Plant Growth and Reproduction
3. Scale of the Problems
4. Climate Change Models
5. Winners and Losers
6. Adaptation
References

1. INTRODUCTION

Changing climate adds a very significant dimension to the complex problem of ensuring that agriculture worldwide can feed the burgeoning human population. Ensuring food security must reduce environmental damage, not add to it. Population growth, the loss of fertile land through degradation and its use for housing and industry, reduced water supplies and aspirations for an increasingly protein-based diet are integral parts of this problem. Supplying adequate and appropriate food against a background of changing climate is the paramount problem that scientists of all disciplines and politicians must solve collectively. Without a solution that is equitable to the environment and mankind, the spectres of famine and war stalk our Planet.

Historical analyses such as that of Therrell et al. [1] for maize yield over the period 1474–2001 demonstrate the close link between food supply and climatic change. The implications of changing climate have been recognised scientifically for well over a century [2–5]. Change may be beneficial, at least in the short term, as demonstrated by Magrin et al. [6] who showed that recently Argentinian yields of wheat, maize, sunflower and soybean have benefited from increased precipitation, decreased maximum and increased minimum temperatures.

Climate Change: Observed Impacts on Planet Earth

The first formal scientifically validated link between observed global changes in physical and biological systems and human-induced climate change predominantly from increasing concentrations of greenhouse gases was demonstrated by Rosenzweig et al. [7]. They surveyed 29 500 data series of which 90% ($P \ll 0.001$) demonstrated that changes at the global scale are in the direction that would be expected as responses to global warming. In biological systems, 90% of the data sets showed that plants and animals are responding consistently to temperature change. This is mostly illustrated by phenological change with earlier blooming, leaf unfolding and spring arrivals. Events on the current scale have not visited the Earth in the past three quarters of a million years [8]. Previously, however, no one single species (man) has gained full control of the Planet's entire resources and reproduced itself in unprecedented numbers at a rapid rate. The Earth's resources are in imminent danger of exhaustion and its environment is changing in a manner that enhances the process.

Stern [9] identified that 'if no action is taken to reduce emissions, the concentration of greenhouse gases in the atmosphere could reach double its pre-industrial levels as early as 2035, virtually committing us to a global average temperature rise of over 2 °C. In the longer term there would be more than a 50% chance that temperature rise would exceed 5 °C'.

2. IMPACT ON PLANT GROWTH AND REPRODUCTION

Blackman's Principle of Limiting Factors [10] – 'when a process is conditioned as to its rapidity by a number of separate factors, the rate of the process is limited by the pace of the "slowest" factor' – applies equally now as it did a century ago. The basic principles of plant physiology likely to govern responses to climate change are broadly understood [11,12]. While some elements in the changing environment may promote plant growth and reproduction, others will be in short supply and cause physiological stresses. What differs now is that the magnitude of stress is more substantial. Growth of C_3 plants[1] (temporal and boreal) increases with rising carbon dioxide levels more than with C_4 plants[2] (warm tropical). The relatively small group of plants using the Crassulacean acid metabolisms (CAM) pathway (such as members of the Cactaceae) may be favoured by increased carbon dioxide concentrations and temperatures [13]. Where the C_3 plants are in association with benign nitrogen fixing microbes (e.g., legumes) there appears to be added benefit. Benefits of additional carbon dioxide concentration are greater for annuals as compared with perennial plants. Leaf area increases as a result of raised photosynthesis with earlier and more complete light interception and

---

[1] C_3 plants form the three carbon compound 3-phosphoglyceric acid as a first stage in photosynthesis.

[2] C_4 plants form the compound 4-carbon oxaloacetate as a first stage in photosynthesis.

resultant greater biomass production. But maintenance costs increase with higher demands for energy and rising respiration. Leaf turnover rises partly due to shading effects consequently photosynthesis per leaf falls. Stomatal opening is reduced with increased carbon dioxide. This is beneficial in limiting the impact of aerial pollutants like nitrogen oxides (NO_X), sulphur dioxide (SO_2) and ozone (O_3) but does inhibit water uptake. Stomatal conductance and transpiration rates drop as carbon dioxide concentrations rise. This effect is less marked when measured on a ground area (canopy evapotranspiration) basis versus consumption measured against leaf-area. There is an increase in water use-efficiency in terms of dry matter formed relative to unit of water transpired. Consequently, leaf temperature increases raising the rate of plant development especially in early growth stages.

Ultimately, however, reduced transpiration and resultant higher temperatures in the leaves leads to accelerated tissue senescence. Whether effects are beneficial or not depends on the extent to which temperatures rise and exceed the optimum for efficient photosynthesis. Overall, the data suggest that elevated carbon dioxide may have positive benefits for C_3 plants including yield stimulation, improved resource-use efficiency, more successful competition with C_4 weeds, less damage from ozone toxicity and in some cases better pest and pathogen resistance [14].

Benefits from increased atmospheric carbon dioxide may be counterbalanced by adverse effects of rising temperatures. Although warming accelerates plant development it reduces grain filling, limits nutrient-use efficiency, increases water consumption and favours C_4 weeds over C_3 crops plants. Changes in the water balance and amount of water available in the soil are crucial for crop growth. In grasslands, 90% of the variance in primary production can be accounted for by annual precipitation [15]. Calculations using the Penman–Monteith equation predict that potential evaporation increases by about 2–3% for each 1 °C rise in temperature [16]. While biomass and yield increase with rising carbon dioxide concentrations dry matter allocation patterns to roots, shoots and leaves also change. Root to shoot ratios increase with elevated carbon dioxide favouring root and tuber crops. Conversely, rising temperature and reduced transpiration limit biomass and seed production drops. Non-structural carbohydrate levels increase but protein and mineral nutrient content fall hence food quality declines both for herbivores and for humans [17].

Currently, ~25% of crop production is lost to the ravages of pests and pathogens between the field and consumer's plate. Climate change will alter phasing of life cycle stages and their rates of development for pests and pathogens and associated antagonistic organisms. It may modify mechanisms of host resistance and host–pathogen relationships. The geographical distribution of hosts and pathogens will alter. The level of crop losses will increase while the efficacy of control measures [18] could fall when faced with greater populations of pests and pathogens. Increased fecundity of fungi results from elevated carbon dioxide. The rate of insect development accelerates as temperatures rise. In warmer conditions they grow and reproduce more quickly and there are more generations per season. For example, the common house fly (*Musca domestica*), although not a direct crop pest is a disease vector and nuisance, populations are predicted to rise by 244% by 2080 [19] as a result of rising temperatures. More aggressive pest and pathogen strains are postulated to develop under elevated carbon dioxide. Increased rainfall events would reduce weather-windows for spray application and allow greater likelihood of contact sprays being washed off. Raised carbon dioxide could increase the thickness of epicuticular waxes resulting in slower penetration of pesticides. Raised aerial carbon dioxide concentrations are unlikely to have much impact within the soil since they are already 10–15 times higher than in air. Rising temperatures could increase the range of pathogens as suggested for *Phytophthora cinnamomi* by Brasier [20]. Similarly, increased spread is likely for rice blast (*Magnaporthe grisea*), wheat scab (*Fusarium* spp.), stripe rust (*Puccinia striiformis*) and powdery mildew (*Blumeria graminis*). Boag et al. [21] estimated that each 1 °C rise in temperature would allow soil-borne nematodes to migrate northwards by 160–200 km. A similar rise would allow leaf rusts of wheat and barley and powdery mildew infection to rise by 2–5-fold [22]. The effects of climate change on pest and pathogen outbreaks are already being seen in the United Kingdom and Western Europe for example, insect pests such as Diamondback moth (*Plutella xylostella*), pathogens like bacterial black rot (*Xanthomonas campestris* pv. *campestris*) and various *Phytophthora* spp. have become well established causing damage respectively, to field brassicas and a wide range ornamentals.

One of the most dramatic examples of the interaction between climate change and husbandry change that exacerbates disease problems is that of the soil-borne microbe *Plasmodiophora brassicae* which causes clubroot disease of brassicas. Previously, this pathogen was held in check in the British winter oil seed rape (*B. napus*) plants because it was a predominantly winter crop. This meant that it was drilled in late August to early September into cooling soil. The seed germinated and produced rosette plants by November which formed the components of yield before growth recommenced in mid to late February. The pathogen was inactive in the cold winter soils. Consequently, the crop could grow and yield in summer with little damage from *P. brassicae* in contradistinction to the spring drilled crops of Continental Europe which succumbed to clubroot as both developed as the soils were warming. Now the British crop is being sown in late July to early August and the soils retain heat through the winter as a consequence *P. brassicae* remains actively causing damage throughout the year. Greater soil moisture content in the autumn and winter because of increased rainfall has only served to offer the pathogen improved opportunities for spread and multiplication [23].

Significant increases in mammalian vermin such as rats (*Rattus* spp.) are noted. Warmer conditions for extended periods enable them to retain activity without any forms of hibernation and in consequence more litters are produced. These are becoming major problems for field vegetables especially in late autumn. Means of control are limited especially in crowded sub-urban areas which exacerbates the problem. Similarly, avian vermin like wood pigeons (*Columba* spp.) are increasingly despoiling food crops. Both these animals contaminate produce with urine and excrement that frequently is infected with bacterial pathogens capable of causing human diseases. Overall, while temperature increases would have significantly increase the severity and spread of plant diseases, precipitation will act as a regulator [24]. Climate change models are not yet sufficiently sensitive or detailed to incorporate estimations of the impact of change on microbial activity. Extreme weather events such as excessive rainfall and consequent flooding are most likely to worsen the incidence of crop pathogens. A major effect of climate warming in temperate zones could be increased winter survival of pests and pathogens.

In more northerly latitudes there will be shifts in patterns for growth and reproduction especially woody perennial plants. There is a substantial body of information dating back to the early 1700s in Great Britain on which predictions of the effects of climate change may be based [25]. Climate change disturbs the synchrony between temperature and photoperiod and because insects and pathogens show individual patterns of response to temperature, carbon dioxide and photoperiod there will be a loss of evolved phasing which damages the relationships between plants and the environment. This adversely affects the temporal and spatial associations between species interacting within specific ecosystems and at different trophic levels. Rosenzweig et al. [7] identify shifts in blossoming, leaf unfolding, migrations and time of reproduction, species distributions and community structure. Both in nature and in crop production there will be a shortage of 'chilling events' in the autumn which encourage perennial plants to acclimate and ultimately enter a dormant state. Dormancy is likely to be much less profound and more easily broken [26].

Phenology studies (the study of times of recurring natural phenomena especially in relation to climatic conditions) already show clearly that flowering times of bulbous and deciduous woody species have advanced by anything up to 1 month in the last 30 years. Freezing events will become sporadic, unpredictable and frequently severe. The result for woody plants that have developed early, season growth will be the loss of flowering and fruiting tissue. Most of these plants are incapable of replacing these organs until the following year. As a consequence an entire season's growth and reproduction fails. Fruit and seed production is lost. If this happens over successive seasons then ultimately the plants will die. This will be a substantial problem for commercial fruit crops and for amenity and natural plantings. The likelihood is that the 'forest giant' trees (oak, *Quercus* spp., beech, *Fagus* spp., elm, *Ulmus* spp.) will suffer most. It also means that top fruit such as apples (*Malus sylestris*) and pears (*Pyrus communis*) and stone fruit such as apricot (*P. armeniaca*), cherry (*P. avium*) and plum (*P. domestica*) will be forced into earlier-flowering and will have entire crops destroyed.

Nutrient acquisition is closely associated with overall plant biomass and is strongly influenced by the available root surface area. When climate change alters root exploration in the soil a restriction of nutrient acquisition follows leading to stress and reduced growth. Nutrient replacement management will be required where crop spectra change following the effects of temperature and carbon dioxide availability [27].

Climate change has both direct and indirect effects on soil erosion. Devastating soil erosion results from even modest rainfall falling onto bare soil. Increased soil erosion accelerates the loss of crop productive land. An avoidance of erosion prone crops, that is, those which are either slow or fail completely to provide full canopy closure is one strategy. An increase from 24% to 46% of the total land area of England and Wales which has a moderate to high risk of erosion is predicted as a result of climate change [28]. Heavy rainfall events in the Great Brittain such as the extensive flooding in 2007 wiped out crops and opportunities for autumn planting because of soil degradation at a direct cost of £3 $\times$ 10^9 [29].

Increasingly, severe wind events are thought likely [30]. Because of the technical complexity of analysing wind effects there is little data that identifies the consequences of this prediction. In northern Europe, winter and early spring winds are frequently associated with periods of intense cold. These are disastrous events for all types of plant but especially young emerging seedlings. Wind damage to young seedlings is underestimated in its impact on yield and quality. Even relatively low speed winds pick up soil particles that then abrade the leaves and stem tissues of emerging seedlings. Abrasion of this type causes cryptic stress in crops which is manifested later at harvest. Winds of greater intensity rock and twist seedlings leading in severe cases to breakage of the stems at ground level. Seedlings that remain in the soil are frequently badly damaged and the disrupted stem tissue permits invasion by collar rotting pathogens.

The odours emitted by damaged tissue are powerful attractants for pests such as root flies (e.g., *Delia* spp.). Physiological disorders that may become manifested later in the plant's life can be initiated by stress in the seedling stage. Where wind gusts are very powerful then both soil and seedlings are collected and transported many hundreds of metres or even further. Spring winds do substantial damage to woody perennials especially fruit trees. The damage may not be apparent in the year of the event. With large trees root damage may take a least one season to cause an effect and then lead to foliar chlorosis and die back. This is frequently followed by pest and pathogen invasion which compounds the damage. Wind in summer is damaging because trees are in full foliage. This makes the aerial parts much heavier and limbs are more easily removed. Wind will also cause significant damage to horticultural structures. Glasshouses, polyethylene tunnels and low level field covers are susceptible to wind damage. Swedish research indicates that climate change will increase the damage to forest trees [31]. Increased intensity of wind, changed direction and frequency of wind events each contributes to these effects.

3. SCALE OF THE PROBLEMS

Some 1.5×10^9 ha of land is used worldwide for crop production and of this 960×10^6 are in developing countries [32]. In the last 30 years, the world's cropped area has expanded by $\sim 5 \times 10^6$ ha annually with Latin American countries accounting for 35% of this increase by deforestation. Land is the basic resource that cannot be created. There is, therefore, a finite point beyond which the cropped area cannot rise. About 40% of the world's arable land is now degraded to some degree, most of this land is in the poorer nations in densely populated, rain-fed farming areas where overgrazing, deforestation and inappropriate land-use compound other problems. About 3×10^9 ha (one fifth of the world's land surface) is under forest ecosystems. Russia, Brazil, Canada, USA, China, Australia, Congo and Indonesia account for 60% of the world's forest land. In the decade of 1990s, 127×10^6 ha of forests were cleared and 36×10^6 ha replanted. Africa lost 53×10^6 ha of forest mainly converted into cropped land. Two-thirds of the world population live in areas receiving 25% of the annual rainfall. About 70% of the world's fresh water goes to agriculture and that figure rises to 90% in nations relying on extensive irrigation. Currently, 30 developing nations face water shortages and by 2050 this will reach 50 nations mostly in the 'developing country' grade. Water scarcity and the degradation of arable crop land are the most serious obstacles that inhibit increases in food production.

Against this background, Smith and Almarez [33] have summarised the dangers of climate change to crop production. Extremes in temperature are dangerous to crop production especially where growth has accelerated due to added carbon dioxide. More northerly zones become wetter and warmer which could benefit crop production in the short term but the tropics and subtropics become hotter and drier. Calculations based on three out of four Climate Change Models show consistent increases in areas of arid land in developing countries. Africa is thought to be the region most vulnerable to negative impacts of climate change on crop production [34].

Currently, 1.080×10^9 ha of land in Africa has a growing period of less than 120 days. With climate change by 2080s this expands by 5–8% (equal to 58–92 $\times 10^6$ ha). This change is accompanied by a loss of 31–51 $\times 10^6$ ha of land in favourable growing zones with growing period lengths of 120–270 days per year. About 1×10^9 people worldwide and of that 180×10^6 in Africa live in vulnerable zones currently relying on agriculture for their living. By 2080s land areas with increasingly severe constraints for crop production in the world zones amounts to: Central America and Caribbean (1.2–2.9% of 271×10^6 ha); Oceania and Polynesia (0.3–4.3% of 848×10^6 ha); Northern Africa (1.9–3.4% of

547 × 10^6 ha) and West Asia (0.1–1.0% of 433 × 10^6 ha). In Southern Africa, an extra 11% (of 266 × 10^6 ha) could suffer severe constraints to cropping. By 2080s decreases in potentially good agricultural land are: Northern Europe 1.5–1.9% (with Great Britain and Ireland particularly affected); Southern Europe 0.2–5.9% (especially Spain); Northern Africa 0.5–1.3% (especially Algeria, Morocco and Tunisia); Southern Africa 0.1–1.5% (especially South Africa) and in East Asia and Japan 0.9–2.5% (especially China and Japan).

Venezuela, New Zealand, Mozambique, Sudan and Uganda are individually nations with good agricultural land that is especially vulnerable. Some economists make the assumption that by 2080 consumers will be much richer than today and separated even more from agricultural production processes earning their income in non-agricultural industries. Hence, they postulate, that changes in consumption will depend more on food prices and on income differences than on local agricultural production. They suggest further that the share of undernourished in the world total population falls below 20% when an arbitrary index of 130 is reached whereby aggregate food supply exceeds aggregate food requirements by 30%. Hunger is completely eliminated where this index reaches 170. Fischer et al. [32] postulate that 'the trade system will (*only*) mitigate local climate-change impacts when consumers can afford to buy food on the international market... (*but*) food prices rising due to climate change may put an extra burden on those consumers who depend on imports, even without a region experiencing direct local climate-change impacts on production conditions' (*my parenthesis*). The economic and climate change models give starkly different prospective outcomes for 2080. Either 'climate change impacts on agriculture will increase the number of people at risk of hunger' or 'with rapid economic growth *and a transition to stable population levels*, poverty, and with it hunger – though negatively affected by climate change – would become a much less prevalent phenomenon than it is today' (*my italics*).

4. CLIMATE CHANGE MODELS

Estimating the effects of climate change depends on the climate change model used and postulates applied for the response or adaptation of the farming community and the new husbandry practices developed from scientific and technological advances. Evidence suggests, as might be expected, that higher resolution land surveying models provide more realistic postulated responses in terms of the effects of climatic change and crop response compared with coarser scale models.

This is especially the case for regions with complex geomorphology such as areas with high relief, the mountainous areas, complex coastlines or complex patterns of land use [35]. Currently much prediction is 'clairvoyance'. Considerable changes to agricultural practice will be needed not least in the characteristics of cultivars bred to withstand the impact of climate change

[36]. Where refinement was increased and the scale of study decreased from hundreds of kilometres to more regional levels then it became apparent that for a wide range of crops in the USA (corn, *Zea mays*, cotton, *Gossypium* spp., soybean, *Glycine max.*, hard red spring wheat, hard red winter wheat, soft white wheat, durum wheat (*Triticum* spp and sorghum, *Sorghum vulgare*) climate change correlated with increasing yield reduction. This proved correct for regions such as the Lakes States, Corn Belt, Northern Plains, Delta States and Southern Plains. Considering this aspect for soybean and sorghum crops in detail [37], fine scale (50 km Regional Climate Model, RCM) compared with coarse definition (300 km Commonwealth Scientific and Industrial Research Organisation, CSIRO Model) considerably raised the level of yield loss irrespective of adaptive husbandry effects for these two crops. With other crops such as cotton the use of irrigation could mitigate the effects of climate change [38]. But this does not allow for decreased availability of water which may accompany climate change compounded by other factors such as population growth and migration. Determinants of variability differ across crops such that for winter wheat the key effect comes from temperature applied during the vernalisation growth stage while for corn (maize) it is the availability of water during grain filling [39]. Recognition of such environmental effects at specific stages in the growth and reproduction of crops has been achieved by agronomists and plant breeders long before climate change emerged as an issue.

5. WINNERS AND LOSERS

While climate change is a global problem, at least initially the biggest losers are likely to be in under developed and developing regions, particularly Africa. Although African farmers are already adapted to local conditions, net revenues would fall with more warming or drying [40]. Dryland crop and livestock farmers are especially vulnerable, with temperature elasticities of –1.9 and –5.4, respectively. Irrigated cropland tends to benefit from marginal warming because irrigation mutes climatic impacts. But these farms are currently located in relatively cool regions of Africa. With precipitation elasticities of 0.4 for dryland crops and 0.8 for livestock across Africa, net revenues for dryland crops and livestock will increase if precipitation increases with climatic change and fall where precipitation decreases. Net revenues for irrigated land follow in the same direction but to a lesser extent (elasticity of 0.1). Increases in precipitation have unambiguously beneficial effects on African farms. As temperatures warm the effects on African farms becomes steadily more harmful. Farms located in currently hotter and drier areas are at greater risk because they are already in a precarious state for agriculture. Dryland farming throughout Sub-Saharan Africa is vulnerable to warming. In the East, West and Sahel regions dryland farming is especially risky.

By contrast irrigated crops in parts that are relatively cool now such as the Nile Delta and the Highlands of Kenya enjoy marginal gains from warming. Because Sub-Saharan African economies depend more heavily on agriculture, total gross domestic product (GDP) and *per capita* income are also vulnerable. By contrast, non-agricultural GDP in Northern Africa is more diversified and so the economies of these countries are less vulnerable to climate change. Adaptation through scientific and technological advance has moved too slowly in Africa compared with the rest of the World. As a consequence the risks from climate change are far greater there than elsewhere.

Specific crop studies of maize and sorghum production in Botswana by Chipanshi et al. [41] using the African core climate change scenario showed that simulated yields declined by 363% for maize and 31% for sorghum in the sand veldt region. Yield reductions in the hard veldt were 10% for maize and sorghum. Growing season became shorter, reduction in the sand veldt being 5 and 8 days for maize and sorghum, respectively, and correspondingly 3 and 4 days in the hard veldt region. Currently, lack of water is the main crop yield constraint. Both maize and sorghum are C_4 with optimal photosynthesis at higher temperatures (30–35 °C) and insolation than C_3 plants. But elevated carbon dioxide concentrations may well negate these benefits. Instead C_3 plants outperform C_4 plants with elevated carbon dioxide [42] and most weeds of maize and sorghum are C_3 types.

Weed competition will, therefore increase. Also the problems of the sandy environment such as degraded fertility and erosion will increase. Arenosol soils that cover more than half of Botswana are most liable to wind erosion and a drier warmer climate can only exacerbate erosion and nutrient loss. Since 1990 satellite evidence shows that soil exposure around settlements and boreholes and the encroachment of woody weeds on bare soil areas have been taking place [43] and are likely to result from a combination of climate warming and over grazing. Similar conclusions come from a study of Kenyan agriculture by Kabubo-Mariara and Karanja [44], showing that climate change produced adverse effects with substantial negative impact on net crop revenue. Temperature rises were more important than changes to precipitation there is a nonlinear relationship between temperature and revenue on the one hand and precipitation and revenue on the other.

The key food crop for at least half of the world's population is rice (*Oryza sativa*). Reliance is greatest in under developed and developing nations. Studies of the rice cultivar IR36 simulating yield changes with increasing carbon dioxide levels and temperature have been made using the INFOCROP model for the Tamil Nadu region of India. Crop duration, days to anthesis, leaf area index and dry matter percentage (DMP) all fell resulting in with lower grain yield per square metre. The authors conclude that crop husbandry will need to improve substantially [45] in order to offer any chance of sustaining the food supply. Bangladesh is also a region highly vulnerable to the impact of climate change and requires adaptation strategies to reduce this risk. Here suggestions are made that greater use could be made of locally adapted plants such as *Jatropa curca* and *Simmondsia chinensis* as supplies of biofuel extracted these oilseeds [46].

Broad level analyses of Chinese agriculture [47] used country – level cross sectional data on agricultural net revenue, climate and other economic and geographical data for 1275 agriculturally dominated counties. Under most climate change models higher temperature and more precipitation would have an overall positive impact on China's agricultural output. But impacts vary seasonally and regionally. The autumnal effects are most significant and the spring time ones most negative. Applying the model to five climatic scenarios in the year 2050 shows that the East, the Central part, the South, the northwest part of North East and the Plateau would benefit from climate change. The South West, North West and southern part of the North East may be negatively affected. The authors reach the general conclusion that overall China benefits from climate change. But this neglects the impact on many millions of people living in those parts where the effects are deleterious.

A realistic study comes from Russia where it is suggested that the shortage of water for irrigation may override any advantages accrued from temperature increases and the availability of high grade soils for grain and other crop production. As a consequence Dronin and Kirilenko [48] analysed strategies for food security based on previous systems in Russian agricultural history, viz Free Market, Big Commune/War Communism, Developed Socialism and Fortress Market employed to provide interregional food exchange. They deliberately omitted the strategy of compensating for short falls in food by substituting imports. The Free market model outperformed the others but the Fortress Market also succeeded as no regions were threatened by grain shortage. Several adaptation measures are identified such as moving meat production northwards and the exploitation of genetically modified cultivars. The authors note that increased irrigation could mitigate some effects of climate change especially in Southern Russia. But they admit that water supply will come under severe restrictions and hence this should not be seen as a route for adaptation.

One of the prime sources of food exports is the USA. Hence studies of the impact of climate change there have ramifications for the world's population collectively. A broad scale review of major crops effects by Chen et al. [49] identified as might be expected that climate change effects varied for different crops. For corn (maize) precipitation and temperature have opposing effects on yield levels and variability, increased rainfall raises yield and decreases variance. Temperature has the reverse effect. For sorghum higher temperatures reduced yields and yield variability. Increased rainfall raised sorghum yields and its variability. The authors used the Hadley and Canadian climate change models, and these indicated that future variability decreased for corn and cotton but increased for soybeans while effects for wheat and sorghum were mixed. Increased variability equates with unreliability in harvest volume which is an unwelcome outcome for all sections of the food chain from field to plate.

Reviewing climate change effects in more detail Changnon and Hollinger [50] studied the production of corn (maize) in the Mid-Western USA. There appears to be a potential for up to 40% increases in rainfall since there has been steadily increasing rainfall over past 50 years in the Midwest. But this translated into little effect on yield unless the rainfall coincided with the drought stressed summer period. The impact of increased soil moisture depends on timing and season. Using two climate change models, the United Kingdom Hadley Centre for Climate Prediction and Research model and the Canadian Centre for Climate Modelling and Analysis model for studies of wheat production in the Great Plains region of the USA, Weiss et al. [51] concluded that yield and percent kernel nitrogen could not be sustained at current levels especially in the arid part of Nebraska. This translates into a loss of quality in the flour required for bread making [52]. The authors identify needs for new cultivars to increase nitrogen uptake and translocation, simply adding extra quantities of nitrogen fertiliser is not the agronomic, economic or environmental answer.

Perennial crops are affected by climate change, not only during the growing season but also while they are dormant. Winter chill hours and chill degree hours are diminishing across the fruit and nut producing regions of California, losses range from 50 to 260 h per decade [53]. By the end of the twenty-first century, Californian orchards are expected to receive less than 500 chill hours per winter which will have a significant deleterious effect on the fruit and nut industries of that State.

Further north there are evaluations of spring wheat, maize, soybean and potato crops in seven agricultural regions of Southern Quebec. These were made in relation to increased carbon dioxide and temperature with resultant acceleration in crop maturation caused by reduced soil moisture availability. Adaptive moves would be needed to cancel out negative effects caused by climate change [54]. A similar conclusion comes from studies of wheat production in parts of South Australia which will cease to be economically viable [55] based on critical yield thresholds. Farmers' adaptation options and adaptive capacity, market fluctuations and agricultural technology levels including genetic alteration and the products of plant breeding will affect future levels of critical yield threshold.

The impact of climate change on European agriculture has received considerable attention. As in the wider world there are some initial beneficiaries particularly in more northerly areas. In northern Europe yields increase as new crops and cultivars emerge. For example, analyses indicate that in the short term German farmers may benefit from climate change, with maximum gains where the temperature increase is +0.6 °C. In the longer term there may be losses [56]. This work is based on theoretical modelling which is unable to take all effects into account. Similarly, in North Eastern Austria studies using the Global Circulation Model (GCM) predict a rise in temperature of between 0.9 and 4.8 °C between 2020s and 2080s. Warming decreases crop growing period which reduces yields, but increased precipitation linked to higher temperatures and carbon dioxide raises yield for crops such as winter wheat and soybean [57]. Spring barley (*Hordeum vulgare*) is the most important cereal crop in Central and Western Europe [58] because of its use for animal feed. In the Czech Republic soil water content increases. This is a key factor in determining yield. Yields increase by 54–101 $kg{\cdot}ha^{-1}$ per 1% increase in available soil water content on sowing day. Doubling carbon dioxide increased yield by 13–52% and opportunities for earlier sowing further enhances yield. Adverse effects can be expected in southern Europe, water shortages reduce yield but farmers could adapt their husbandry to prevailing conditions aided by technological progress. Adaptation needs to be quantified and built into simulation models determining the impact of climate change [59]. The variability of sugar beet (*Beta vulgaris*) yield (measured as coefficient of variation) [60] will increase by half from 10% to 15% compared with 1961–1990 with serious implications for commercial planning in the sugar industry. Climate change is expected to bring yield increases of around 1 $t{\cdot}ha^{-1}$ of sugar in northern Europe and comparable losses in yield in France, Belgium and west/central Poland over 2021–2050. The figures mask significant increases in yield potential due to earlier springs and accelerated growth offset by losses due to drought stress. The effects of carbon dioxide concentration on biomass production are approximately linear from 360 to 700 ppm CO_2 [61]. Areas with existing drought problems will suffer from a doubling of losses and they will become a serious new problem in North Eastern France and Belgium. Overall west and central Europe will potentially see losses from drought rise from 7% (1961–1990) to 18% (2021–2050).

In Spain High Resolution Climate Models (HRCMs) were used to study potential yields and showed crop failures of winter wheat in the south but yield increases for spring wheat in northern and high altitude areas [62]. While in Turkey a study by Umetsu et al. [63] considered the Lower Seyhan Irrigation Project in Turkey using an expected value–variance (E–V) model. Under water constraints farmers chose to grow high value added crops such as watermelon (*Citrullus vulgaris*), citrus (*Citrus* spp.), cotton, fruits and vegetables. But because of rising cost of water gross revenue fell even with this business model. Adaptation by increasing irrigation and nitrogen use are advocated by Haim et al. [64] as means of mitigating the adverse implications of climate change by 2070–2100 for Israeli wheat and cotton production. Since water supply for this entire region will be at a premium by then such strategies for adaptation may not be feasible.

Wine production is a good example of a worldwide product where clear differences in advantage or disadvantage emerge from climate modelling. Changes in cool climate areas such as the Mosel Valley, Alsace, Champagne and the Rhine Valley could lead to more consistent vintage quality and potentially the ripening of warmer climate cultivars [65]. But those regions currently growing close to the climatic optimum for grape (*Vitis vinifera*) cultivars, for example, Southern California, southern Portugal, the Barossa Valley and the Hunter Valley may become too hot for quality wine production. Winter temperature changes would also affect viticulture by making regions that experience hard winter freezes (e.g., Mosel Valley, Alsace and Washington) less prone to vine damage, while other regions (e.g., California and Australia) would have such mild winters that latent bud hardening may not be achieved and cold-limited pests and pathogens may increase in both number and severity.

As a general conclusion, climate gets warmer and where temperature rise is extreme then this is dangerous both directly to humans and indirectly through the effects on food supply. The spring, summer and autumn seasons get longer and this effect becomes more dramatic in higher latitudes. In these areas the climate becomes drier, for example, in parts of Canada this reduces the area available to produce hard red spring wheat, in Quebec fruit trees are moved northwards and reduced snow cover makes it difficult for forage legumes to survive in winter. Glaciers have retreated round the world by up to 30% in the twentieth century. The result is less water flowing through the rivers hence reduced amounts available for irrigation. Extreme weather events become more common, increasing droughts and tropical storms make crop production more difficult. Potentially there are changes to soil organic matter. Higher temperatures accelerate the breakdown of soil organic matter. Less organic matter means lower yields because of a lack of nutrients and water. As a counterweight increased carbon dioxide concentration raises soil organic matter content resulting in greater microbial activity. Increased carbon dioxide means more photosynthate for nitrogen fixation encouraging the growth of C_3 plants. Soil erosion increases as a result of more severe wind events. Rising sea levels also mean that adjacent land becomes more saline. Pests and pathogens move and propagate more quickly resulting in greater losses to crops.

6. ADAPTATION

The assessment of winners and losers is solely a snap shot of potential effects and implications. It becomes abundantly evident that as the twenty-first century progresses food production worldwide is threatened. But adaptation and consequent mitigation can be achieved through science and technology. As in the 1960s their capacities to provide means by which agriculture can adapt to its changing environment are crucial for the security of the food chain [66]. In general intensive systems such as horticulture have greater potential to adapt, or be adapted, to changing climates than extensive and low-input systems. As an example, the assessment by Weatherhead et al.

[67], showed that in East Anglia, Great Britain's area of intensive field crop agriculture, water availability for crops will decrease by the 2020s, but farmers could still produce high value irrigated crops such as fresh vegetables and potatoes by reducing irrigation to other crops, installing more reservoirs to hold water from expectedly increased winter rainfall, using winter abstraction into the reservoirs and using more efficient irrigation systems such as low level drip fertigation[3]. But farming will be more financially vulnerable because of reduced net margins. The availability of water and nutrient resources and ability of plants to make efficient and effective use of them become crucial factors. This contention is supported by Hopkins et al. [68], who identify the agricultural responses needed for adaptation to climate change as: new crops, increased irrigation and changes in land use patterns for crops and livestock.

The biodiversity changes that will affect the availability of benign organisms that aid crop growth will be affected by: the timing of seasonal events and hence loss of synchrony between species and the food and other resources that they require; changes in species abundance and range, habitat alterations in chemical, physical and biological terms; altered water regimes and increased decomposition. Farming can respond to climate change with modifications to current husbandry systems. This only allows compensation for the immediate short term effects of climate change over the next decade or so. More radical change demands substantial programmes of research and development on a co-ordinated worldwide basis. Only genotype change can provide the level of mitigation needed, without this as illustrated by Challinor et al. [34], with studies of ground nut (*Arachnis hypogaea*) yields can drop by 70% as a result of increased temperatures as the century progresses. The importance of genetics and breeding linked with environmental and husbandry measures cannot be over emphasised [69]. Fortunately, science through current studies of the molecular processes of inheritance has considerable knowledge in store. Deploying this basic knowledge into applied science and technology demands united political decisions at intergovernmental level. As part of such decisions there must be recognition that coherent provision for knowledge transfer from the scientists' laboratories in to farming practice is a paramount necessity [70]. For the past generation at least governments worldwide have expected that free market forces would facilitate knowledge transfer. This has not happened as the free market system is not capable of providing knowledge transfer to individual farming enterprises in an effective manner. Providing adequate numbers of properly educated and experienced crop specialists capable of translating science and technology into farming practice is an essential component of coping with climate change and feeding mankind.

---

[3] Fertigation = combined irrigation and liquid nutrient supply.

REFERENCES

1. M. Therrell, D.W. Stahle, J.V. Diaz, E.H.C. Oviedo, M.K. Cleaveland, Clim. Change 74 (2006) 493–504.
2. C.G. Johnson, L.P. Smith (Eds.), The Biological Significance of Climate Change in Britain. Symposium of the Institute of Biology No. 14, The Institute of Biology and Academic Press, London, New York, 1965.
3. L.P. Smith, In: C.G. Johnson, L.P. Smith (Eds.), The Biological Significance of Climate Change in Britain. Symposium of the Institute of Biology No. 14, The Institute of Biology and Academic Press, London, New York, 1965.
4. R.J. Froud-Williams, R. Harrington, T.J. Hocking, H.G. Smith, T.H. Thomas, Implications of "Global Environmental Change" for crops in Europe. Aspects of Applied Biology No. 45, The Association of Applied Biologists, Wellesbourne, UK, 1996.
5. P. Simons, Weather Eye: How Monsoon Floods Fit into the Global Picture, The Times, Friday, August 10, 2007, p. 467.
6. G.O. Magrin, M.I. Travasso, G.R. Rodríguez, Clim. Change 72 (2006) 229–249.
7. C. Rosenzweig, D. Karoly, M. Vicarelli, P. Neofotis, G. Casassa, A. Menzel, T.L. Root, N. Estrella, B. Seguin, P. Tryjanowski, C. Liu, S. Rawlins, A. Imeson, Nature 453 (2008) 353–357.
8. D. King, J. Appl. Ecol. 42 (2005) 779–783.
9. N. Stern, The Economics of Climate Change: The Stern Review, Cambridge University Press, Cambridge, 2008.
10. F.F. Blackman, Ann. Bot. 19 (1905) 281–295.
11. J.I.L Morison, M.D. Morecroft, Plant Growth and Climate Change, Blackwell Publishing, Oxford, UK, 2006.
12. J. Grace, R. Zhang, In: J.I.L. Morison, M.D. Morecroft (Eds.), Plant Growth and Climate Change, Blackwell Publishing, Oxford, UK, 2006 (Chapter 9), pp. 187–208.
13. P.S. Nobel, J. Arid Environ. 34 (1996) 187–196.
14. J. Fuhrer, Agric. Ecosyst. Environ. 97 (2003) 1–20.
15. B.D. Campbell, M.D. Stafford Smith, G.M. MecKeon, Glob. Change Biol. 3 (1997) 177–187.
16. J.G. Lockwood, Clim. Change 41 (1999) 193–212.
17. G.R. Dixon, Vegetable Brassicas and Related Crucifers. CAB International, Oxford, UK, 2007.
18. S.M. Coakley, H. Scherm, S. Chakraborty, Ann. Rev. Phytopathol. 37 (1999) 399–426.
19. D. Goulson, L. Derwent, M.E. Hanley, D.W. Dunn. S.R. Abolins, J. Appl. Ecol. 42 (2005) 795–804.
20. C.M. Brasier, Ann. Sci. For. 53 (1996) 347–358.
21. B. Boag, J.W. Crawford, R. Neilson, Nematologica 37 (1991) 312–323.
22. M. Jahn, E. Kluge, S. Enzian, Asp. Appl. Biol. 45 (1996) 247–252.
23. G.R. Dixon, Acta Hortic. 706 (2006) 271–282.
24. L.V. Madden, G. Hughes, F. van den Bosch, The Study of Plant Disease Epidemics, The American Phytopathological Society, St. Paul, MN, 2007.
25. F. Last, A. Roberts, & D. Patterson, (2003). Climate Change? A statistical account of flowering in East Lothian 1978–2001, *In*: S. Baker (editor) East Lothian 1945–2000: Fourth Statistical Account. Volume one: The County, published by East Lothian Library Service for The East Lothian Fourth Statistical Account Society, pp. 22–29.
26. G.R. Dixon, M.P. Biggs, Asp. Appl. Biol. 45 (1996) 93–100.
27. S.M. Brouder, J.J. Volenec, Proceedings of the International Fertiliser Society Annual Conference, vol. 609, International Fertiliser Society, York, UK, 2007, pp. 1–31.

28. J. Boardman, D.T. Favis-Mortlock, Proceedings of the International Symposium, Honolulu, Hawaii, USA, 3–5 January 2001, American Society of Agricultural Engineers, St. Joseph, USA, 2001, pp. 498–501.
29. Anon., Together – Make a Difference to Your Climate With The Met Office, The Meteorological Office, Exeter, UK, 2007.
30. P.E. Hulme, J. Appl. Ecol. 42 (2005) 784–794.
31. K. Blennow, E. Olofsson, Clim. Change 87 (2008) 347–360.
32. G. Fischer, M. Shah, H. van Velthuizen, A special report prepared by the International Institute for Applied Systems Analysis under United Nations Institutional Contract Agreement No. 1113 on "Climate change and agricultural vulnerability" as a contribution to the World Summit on Sustainable Development, Johannesburg, South Africa, 2002.
33. D.L. Smith, J.J. Almaraz, Can. J. Plant Pathol. 26 (2004) 253–266.
34. A.J. Challinor, T.R. Wheeler, P.Q. Craufurd, C.A.T. Ferro, D.B. Stephenson, Agric. Ecosyst. Environ. 119 (2007) 190–204.
35. F. Giorgi, in: J.T. Houghton, G.J. Jenkins & J.J. Ephraums (Eds.), IPCC Third Assessment Report. The Science of Climate Change, Cambridge University Press, Cambridge, UK, 2001, pp. 583–638.
36. R.M. Adams, B.A. McCarl, L.O. Mearns, Clim. Change 60 (2003) 131–148.
37. G.J. Carbone, W. Kiechle, C. Locke, L.O. Mearns, L. McDaniel, M.W. Downton, Clim. Change 60 (2003) 73–98.
38. R.M. Doherty, L.O. Mearns, K.R. Reddy, M.W. Downton, L. McDaniel, Clim. Change 60 (2003) 99–129.
39. E.A. Tsvetsinskaya, L.O. Mearns, T. Mavromatis, W. Gao, L. McDaniel, M.W. Downton, Clim. Change 60 (2003) 37–71.
40. P. Kurukulasuriya, R. Mendelson, R. Hassan, J. Benhin, T. Deressa, M. Diop, M. Eid, K.Y. Fosu, G. Gbetibouno, S. Jain, A. Mahamadou, R. Mano, J. Kabubo-Mariara, S. El-Marsafawy, E. Molua, S. Ouda, M. Ouedraogog, I. Séne, D. Maddison, S. N. Seo, A. Dinar, World Bank Econ. Rev. 20 (2006) 367–388.
41. A.C. Chipanshi, R. Chanda, O. Totolo, Clim. Change 61 (2003) 339–360.
42. L. Ringius, T.E. Downing, M. Hukme, R. Selrod, Climate Change in Africa: Issues and Regional Strategies, Cicero Report No. 1996 (2), Cicero, Oslo, Norway, 1996.
43. S. Ringrose, W. Matheson, Int. J. Remote Sens. 12 (1990) 1023–1051.
44. J. Kabubo-Mariara, F.K. Karanja, Glob. Planet. Change 57 (2007) 319–330.
45. O. Srivani, V. Geethalaakshmi, R. Jagannathan, K. Bhuvaneswari, L. Guruswamy, Asian J. Agric. Res. 1 (2007) 119–124.
46. C. Bowe, Tropical Agriculture Association Newsletter, vol. 27, Tropical Agriculture Association, Penicuik, Edinburgh, UK, 2007, pp. 9–13.
47. H. Liu, X. Li, G. Fischer, L. Sun, Clim. Change 65 (2004) 125–148.
48. N. Dronin, A. Kirilenko, Clim. Change 86 (2008) 123–150.
49. C-C. Chen, B.A. McCar, D.R. Schimmelpfennig, Clim. Change 66 (2004) 239–261.
50. S.A. Changno, S.E. Hollinger, Clim. Change 58 (2003) 109–118.
51. A. Weiss, C.J. Hayes, J. Won, Clim. Change 58 (2003) 119–147.
52. C. Blumenthal, H.M. Rawson, E. McKenzie, P.W. Gras, E.W.R. Barlow, C.W. Wrigley, Cereal Chem. 73 (1996) 762–766.
53. D. Baldocchi, S. Wong, Clim. Change 87 (Suppl. 1) (2008) S153–S166.
54. J.P. Brassard, B. Singh, Mitigation and Adaptation Strategies for Glob. Change 13 (2008) 241–265.
55. Q. Luo, W. Bellotti, M. Williams, I. Coper, B. Bryan, Clim. Change 85 (2007) 89–110.
56. G. Lang, Clim. Change 84 (2007) 423–439.

57. V. Alexandrov, J. Eaitzinger,V. Cajic, M. Oberforster, Glob. Change Biol. 8 (2002) 372–389.
58. M. Trnka, M. Dubrovsky, Z. Žalud, Clim. Change 64 (2004) 227–255.
59. P. Reidsma, Adaptation to Climate Change: European Agriculture, Wageningen University, The Netherlands, 2007.
60. P.D. Jones, D.H. Lister, K.W. Jaggard, J.D. Pidgeon, Clim. Change 58 (2003) 93–108.
61. H. Demmers-Derks, R.A.C. Mitchell, V.J. Mitchell, D.W. Lawlor, Plant Cell Environ. 21 (1998) 829–836.
62. M.I. Minguez, M. Ruiz-Ramos, C.H. Diaz-Ambrona, M. Quemada, F. Sau, Clim. Change 8 (Suppl. 1) (2007) 343–355.
63. C. Umetsu, K. Palanisami, Z. Coskun, S. Donma,T. Nagano, Y. Fujihara, K. Tanaka, J. Rural Econ. 79 (2007) 567–574.
64. D. Haim, M. Shechter, P. Berliner, Clim. Change 86 (2008) 425–440.
65. G.V. Jones, M.A. White, O.R. Cooper, K. Storchmann, Clim. Change 73 (2005) 319–343.
66. L.S. Hardin, Nature 455 (7212) (2008) 470–471.
67. E.K. Weatherhead, J.W. Knox, T.T. Vries, S. Ramsden, J. Gibons, N.W. Arnell, N. Odoni, K. Hiscock, C. Sandhu, A. Saich, D. Conway, C. Warwick, S. Bharwani, J. Hossell, B. Clemence, Sustainable Water Resources: A Framework for Assessing Options in the Rural Sector. Technical Report – The Tyndall Centre for Climate Change Research, Norwich, UK, 2005.
68. J.J. Hopkins, H.M. Allison, C.A. Walmsley, M. Gaywood, G. Thurgate, Conserving Biodiversity in a Changing Climate: Guidance on Building Capacity to Adapt, The Department for Environment, Food & Rural Affairs, London, UK, 2007.
69. B. Kobiljski, S. Denic, J. Genet. Breed. 55 (2001) 83–90.
70. G.R. Dixon, L.V. Hardiman, C.C. Payne, S.M. Swan, T. Harwood, J. Deen, D.N. Whalley, J. Wood, Acta Hortic. 641 (2003) 125–130.

Chapter 18

Rising Sea Levels as an Indicator of Global Change

Roland Gehrels
School of Geography, University of Plymouth, Plymouth PL4 8AA, United Kingdom

1. Introduction
2. Is Sea Level Rising?
3. Why Is Sea Level Rising?
4. Are Contemporary Rates of Sea-Level Rise Unusual?
5. Conclusion
Acknowledgement
References

1. INTRODUCTION

The release of the fourth assessment report of the Intergovernmental Panel on Climate Change (IPCC) in 2007 was followed by much debate, both in the media and in the scientific community, on the sea-level rise predictions it contained [1]. This was partly because sea-level rise is an effect of global climate change that will have far-reaching consequences for a majority of the world's population. However, attention also focussed on the predictions themselves – how accurate were they? The 2007 predictions contained similar uncertainties as the previous IPCC report from 2001 – so what was new? Despite perceived shortcomings, however, the last IPCC volume contains a wealth of new data on recent sea-level changes, in particular from the last 50 years, a period when oceanographic data collection underwent a true revolution. Collectively, the data summarised by the IPCC reflect our improved understanding of the causes of sea-level rise and they identify rising sea levels as one of the major indicators of ongoing global change.

2. IS SEA LEVEL RISING?

Although this question appears almost rhetorical and can at first glance be answered with a resounding 'yes', the direction of sea-level changes, positive or negative, depends on the time scale of observations and the spatial scale under consideration. This is obvious on very short timescales, for example in the case of tidal fluctuations. However, sea-level changes are also highly variable in time and space on decadal timescales. Satellite observations since the early 1990s have revealed the complex regional patterns of sea-level changes (Fig. 1a). Linear trends over the decade 1993–2003 show that some parts of the world's oceans have experienced high rates of sea-level rise (>10 $mm{\cdot}a^{-1}$ in places), while in other places

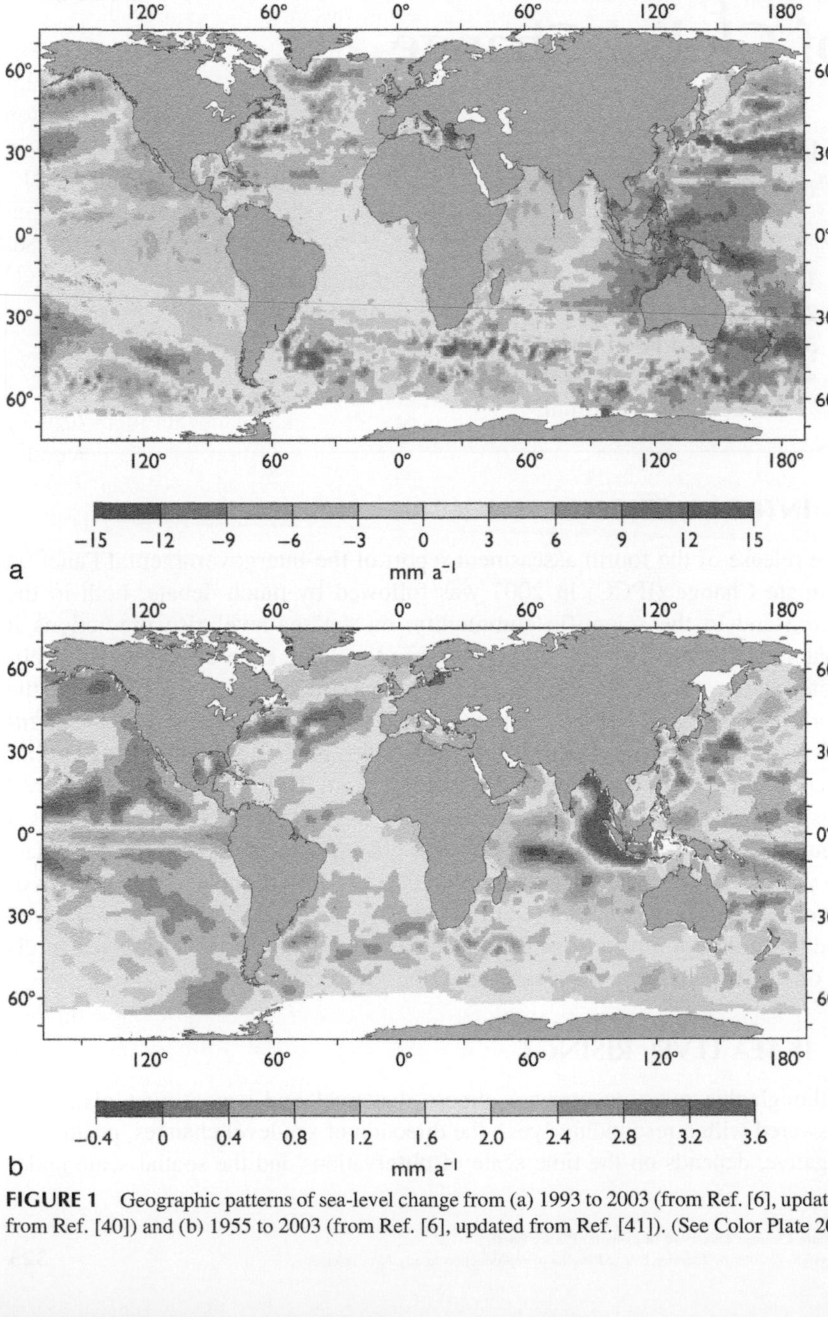

FIGURE 1 Geographic patterns of sea-level change from (a) 1993 to 2003 (from Ref. [6], updated from Ref. [40]) and (b) 1955 to 2003 (from Ref. [6], updated from Ref. [41]). (See Color Plate 20).

sea level has fallen by similar amounts. When averaged over the past 50 years (Fig. 1b), linear rates are an order of magnitude smaller, but the pattern, which is derived from tide-gauge and satellite measurements, is still complex. Most areas have been subjected to sea-level rise, but, again, there are places, most notably in the Indian Ocean and the tropical Pacific, where sea level has fallen, albeit by a small amount. This picture does not take account of land-level movements which in many coastal locations need to be added to, or subtracted from, the mean sea-level change to derive a figure that represents the relative change at a coastline. This is, after all, the number that is of most practical value for coastal management.

To answer the question adequately it is clear that sea-level observations need to be averaged in some way, and, indeed, since the first attempt by Gutenberg [2] many scientists have derived a global value of average sea-level rise by using a range of statistical techniques and various datasets of measurements, many of which have been corrected for vertical movements of the coastlines to which tide gauges are attached. The IPCC consensus is that global sea level has risen during the twentieth century by about 17 cm [1]. An updated estimate is 1.6 $\pm$ 0.2 mm·a^{-1} for the period 1961–2003 [3]. Since the early 1990s, the rate of sea-level rise has been about 3 mm·a^{-1} [1], but it is too early to conclude that this change represents a true deviation of the twentieth century global trend. This becomes evident when the global rate is analysed at decadal timescales (Fig. 2). Rates of sea-level rise, similar to those of the past few years, have occurred in previous decades. However, the maximum rates have increased from ~2.5 mm·a^{-1} in the decade centred on 1970, to ~6 mm·a^{-1} in the

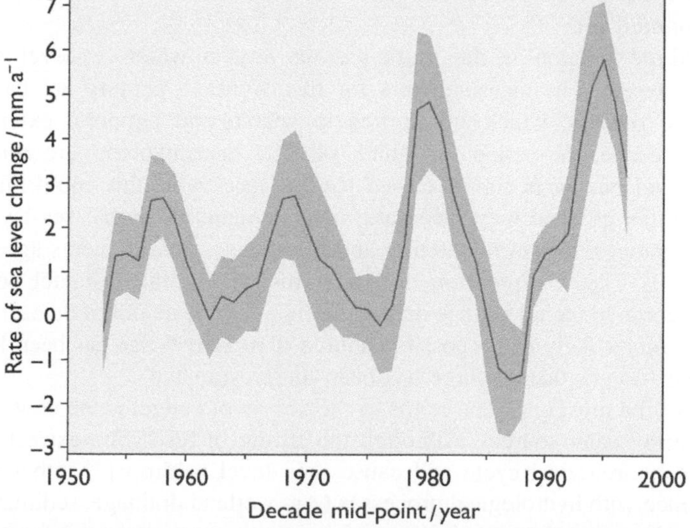

FIGURE 2 Global rates of sea-level change since 1950s, averaged over a decade, based on 177 tide-gauge stations [42,43].

late 1990s. The question is, of course, whether the lengths of the records limit us in the conclusions we can draw. Are measurements during three decadal cycles sufficient to conclude that the rates of sea-level rise are on the increase (i.e. that sea-level rise is accelerating)? Longer timescales of observations are necessary to provide the appropriate context to identify possible accelerations in the rate of sea-level rise. Long tide-gauge records and reconstructions based on proxy data are therefore crucial to inform the climate change debate (see Section 4).

3. WHY IS SEA LEVEL RISING?

Many factors contribute to the changes in sea level that the globe is experiencing today. Some cause sea levels to rise, others make a negative contribution. Not all factors are well constrained and herein lies one of the bigger challenges of sea-level science: can we explain the sea-level rise we are observing?

According to the IPCC, the main contributors to sea-level rise since 1961 are thermal expansion ($0.4\ mm{\cdot}a^{-1}$) and melting ice from small glaciers and ice caps ($0.5\ mm{\cdot}a^{-1}$), with small amounts (that carry relatively large uncertainties) from the Greenland Ice Sheet ($0.1\ mm{\cdot}a^{-1}$) and the Antarctic Ice Sheet ($0.1\ mm{\cdot}a^{-1}$). In the IPCC assessment for the latter half of the twentieth century, these known sources of sea-level rise do not match the measurements and only add up to a little over half of the observed rise. This issue has been called the 'sea-level enigma' [4] or the 'attribution problem' [5]. The discrepancy implies one of three things: either the measurements produce a global value that is too high, or the contributions are underestimated, or there are sources of sea-level rise that are not accounted for.

Could the 'enigma' be due to the various ways in which sea-level rise has been measured? The measurements for the twentieth century are based on tide-gauge records, which are limited in spatial and temporal extent. For the past decade, the period for which satellite measurements are available, the sea-level budget is almost closed [6]. At face value this could highlight that satellites produce more accurate measurements of global sea-level rise than tide gauges. However, satellite and tide-gauge measurements agree well in the 1990s. They diverge from 1999 onwards [3], but this mismatch could be due to records in the tide-gauge dataset being a few years out of date. It seems therefore more likely that a possible source of sea-level rise has been left out of the equation or that a source has been underestimated.

One of the most uncertain terms in the sea-level budget is the contribution of terrestrial water sources. Although the filling of reservoirs extracts water from the hydrological cycle and causes sea level to drop [7], other human interference with hydrological processes (e.g. wetland drainage, sedimentation in reservoirs, groundwater mining, surface water consumption, deforestation) contribute positively to sea-level rise. In fact, it has been argued that the transfer of terrestrial water sources to the ocean could represent the 'missing' term in the sea-level budget of the twentieth century [8].

Others have argued that the ice-melt term has been underestimated [5]. The contribution of small glaciers and ice caps is reasonably well constrained, but monitoring of mass-balance changes in land-based ice sheets is a relatively new activity so that the volumes of discharge of the polar ice sheets are not well known before the 1970s. It is now clear that dynamical changes in the outlet glaciers of the Greenland and Antarctic Ice Sheets can lead to large sea-level contributions. Outlet glaciers in Greenland, including the Jakobshavn Glacier, have contributed about 0.1 $mm{\cdot}a^{-1}$ of sea-level rise since the mid 1990s [9]. The Amundsen Sea glaciers in West Antarctica, including the Pine Island Glacier, produced sea-level rise contributions of 0.15 $mm{\cdot}a^{-1}$ in the 1990s [10], and possibly as much as 0.24 $mm{\cdot}a^{-1}$ between 2002 and 2005 [11]. Is it possible that a hitherto unknown ice shelf disappeared, for example somewhere in West Antarctica, which led to a rapid discharge of parts of the interior of the West Antarctic Ice Sheet? Seemingly, one can only speculate, but there are indirect methods by which this question may be addressed.

One method by which the ice-mass term in the sea-level budget can be tackled is by determining the rates at which the ocean has freshened as a result of melting ice [5,12]. However, this method is, again, limited by the record length of hydrographic data. An arguably more intriguing route of investigation is to map the 'sea-level fingerprint' that would have been left by a melting ice mass. A shrinking ice mass produces a diminishing gravitational pull on the ocean surface and perturbs the sea surface as far as thousands of kilometres away, so that sea-level rise near the melt source is less than the sea-level rise in the far field [13,14]. In theory, therefore, it should be possible to determine from the patterns of sea-level change measured by tide gauges the ice-mass contribution to sea-level rise. For example, this method has been used to estimate a 1.0 ± 0.6 $mm{\cdot}a^{-1}$ contribution of melting of the Greenland Ice Sheet to global sea-level rise since 1960 [15]. Although other attempts to find a systematic pattern in tide-gauge measurements have been less successful [16], possibly due to steric (density) and isostatic overprints, it is clear that sea-level fingerprinting has wide ranging applications in sea-level research [17,18].

In a potentially highly significant study, a new suggestion has recently been made which can resolve the enigma [3]. This study concludes that the contribution of thermal expansion in past assessments has been underestimated, because of biases in the way the expansion was calculated from observational data. This appears to be a convincing explanation and it is satisfying to see that the sea-level budget since 1961 now appears to be closed (Fig. 3). With the revised steric estimates, the sum of the contributions is 1.5 ± 0.4 $mm{\cdot}a^{-1}$ for the period 1961–2003, very close to what has been measured. It is probably too early to tell whether this is the final word on the enigma debate, but the

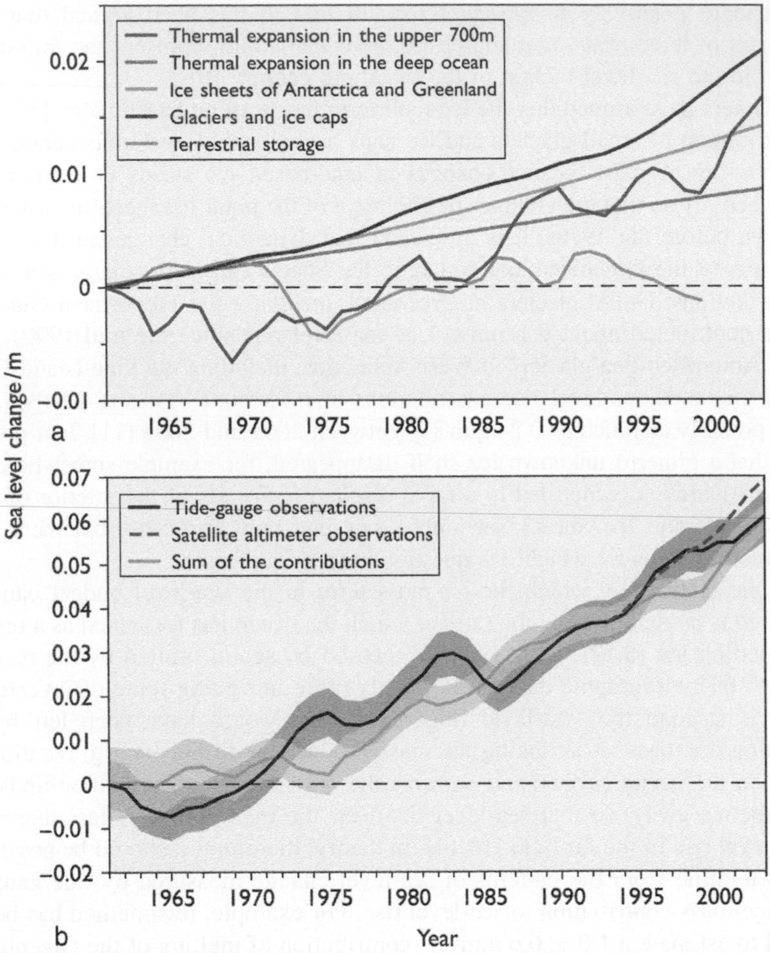

FIGURE 3 (a) Contributions to global sea-level change since 1960, including thermal expansion in the upper 700 m of the oceans, thermal expansion in the deep ocean, polar ice sheets, glaciers and ice caps, and terrestrial water storage. (b) Sea-level change estimated from global measurements and the sum of the contributions in (a). One standard deviation errors are also shown. From Ref. [3]. (See Color Plate 21).

advances made are of some significance. A better understanding of the past contributions to sea-level rise will enable modellers to improve their predictions of future sea-level rise.

Humans have had a measurable impact on sea-level rise since about 1900 (Fig. 4). Model experiments demonstrate the influence of greenhouse gas emissions, producing increased thermal expansion and greater glacier melt [19]. Volcanic eruptions have slowed down sea-level rise, and some of the twentieth century rise in sea level was delayed by the eruptions of

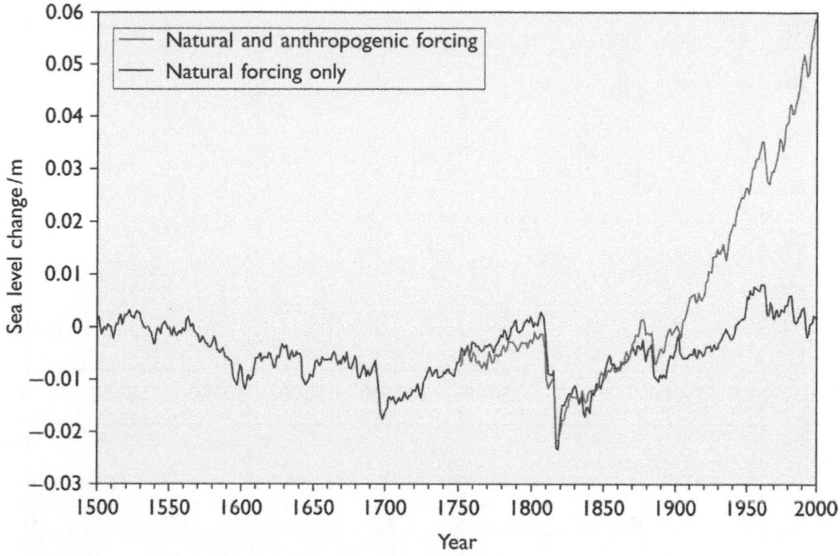

FIGURE 4 A model simulation for the past 500 years of natural and anthropogenically forced sea-level change (red) and sea-level changed forced by natural factors only (blue). Although the model simulations fail to reproduce the magnitude of both the observed long-term sea-level trend and interannual and decadal variability, the onset of twentieth century sea-level rise appears to be controlled by anthropogenic forcing. Sea-level rise in these model runs started after the eruption of Tambora in 1815, but was driven by natural factors during the nineteenth century. From Ref. [19].

Krakatoa in 1886 and Pinatubo in 1991 [20]. This is important, because these eruptions temporarily masked the impact of anthropogenic effects on sea-level rise.

The relationship between sea-level change and greenhouse gas concentrations is well known on geological timescales. For example, when CO_2 concentrations were higher than 1000 ppm around 70 Ma, ice was absent from the planet and sea level was 73 m higher than today [9]. Figure 5 shows the Red Sea sea-level record during the past 400 000 years [21] and a comparison with CO_2 concentrations measured in the Vostok ice core [22]. Although chronologies of both records have inherent uncertainties, and CO_2 fluctuations may in fact lead temperature change by several centuries [23], the correlation between CO_2 and sea-level change is obvious: higher CO_2 levels correspond with increased sea levels. If the almost linear trend, shown in Fig. 5c, persists for CO_2 concentrations in excess of 300 ppm, it implies that sea levels will continue to rise significantly in the future. Indeed, climate models predict that with stabilisation of CO_2 levels by the year 2100 at 550 ppm [9] or 700 ppm [1], sea level will continue to rise for another 1000 years. If CO_2 levels reach 1000 ppm, the Greenland Ice Sheet will disappear in 3000 years, raising sea level by about 7 m [9].

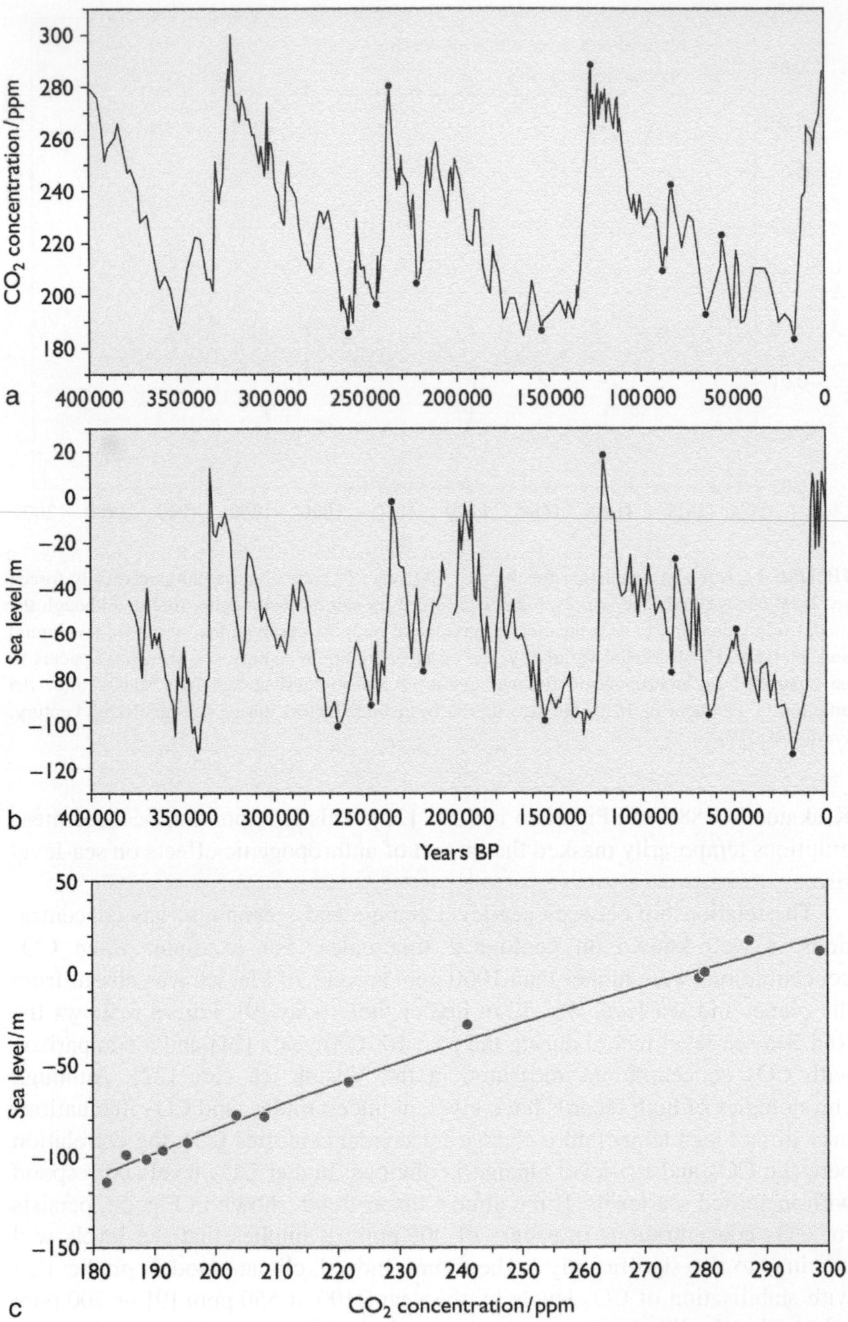

FIGURE 5 (a) CO_2 concentrations during the past 400 ka as measured in the Vostok ice core from Antarctica [22]. (b) Sea-level changes in the Red Sea during the past 400 ka [21]. (c) Relationship between CO_2 concentrations and sea level, assuming minima and maxima in (a) and (b) are of similar age. Points in (c) correspond to dots on the curves in (a) and (b).

4. ARE CONTEMPORARY RATES OF SEA-LEVEL RISE UNUSUAL?

It is a well known fact that rates of sea-level rise in the past have been much higher than the ones we are experiencing today. For example, during the last deglaciation around 14 000 years ago, rapid melting of ice sheets during Meltwater Pulse 1A produced rates of sea-level rise in excess of 40 mm·a^{-1} [24]. However, the world was then emerging from an ice age, and many ice sheets contained unstable marine components which have now largely disappeared. The only marine-based ice sheet left is in West Antarctica, and this ice sheet is situated in the coldest region of our planet. A comparison with the late glacial sea-level history, therefore, does not provide a suitable analogue for modern (or future) conditions. Instead, it is more instructive to examine periods in the Earth's history when the cryosphere contained roughly the same volume of ice as today (or slightly less) and temperatures were similar (or slightly higher) than today's. Periods often cited as useful analogues include the Last Interglacial, the middle Holocene and the Medieval Climatic Optimum.

The position of sea level during the Last Interglacial is generally estimated at 4–6 m above present [25], but the exact height is difficult to determine due to uncertainties about land movements that have occurred since the Last Interglacial. Most evidence points at sea levels close to the present level, or slightly higher, for the time interval 128–116 ka [26]. Only one study on sediment cores from the Red Sea provides a detailed assessment of the rates of sea-level rise during the Last Interglacial [25]. It is estimated that the 'full potential range' of rates of sea-level rise was between 0.6 and 2.5 m per century. It is interesting to note that this estimate is within the same ballpark as some predictions made for the twenty-first century [27], although it is higher than those provided by the IPCC. Most of the sea-level rise during the Last Interglacial is thought to have come from Greenland, with possibly a small contribution from Antarctica [28].

Evidence from many parts around the world suggests that temperatures in the current interglacial reached their maximum in the middle Holocene, although the Holocene 'thermal optimum' is spatially variable and not globally synchronous. There are many published sea-level studies that argue for sea-level fluctuations, some up to several meters in amplitude, during the Holocene, but these are almost always based on data with large vertical and age uncertainties and often use a 'connect-the dot' approach that ignores these uncertainties [29]. Many Holocene sea-level histories are only resolved on millennial time scales. The more robust sea-level reconstructions that contain evidence for century-scale sea-level oscillations are arguably from microatolls in Australia and have recorded rates of 0.1–0.2 m per century in the middle Holocene [30]. Even during the 8.2 ka event, which was caused by the final draining of a huge glacial lake (Lake Agassiz-Ojibway) and thus is not truly representative as a modern analogue, sea-level rise may not have been more than 0.4 m [31].

Sea-level changes during the Medieval Climatic Optimum were small, and have not been clearly resolved in palaeo-records. Salt marshes in eastern North America provide evidence that rates did not exceed 0.2 m per century during the past millennium before the twentieth century [32]. In the North Atlantic Ocean [33] and in the Southwest Pacific [34], the recent acceleration of sea-level rise started about 100 years ago, although there are also signs of an earlier sea-level acceleration in the beginning of the nineteenth century [35]. These findings are based on microfossil evidence and high-precision dating of salt-marsh sediments and are supported by some analyses of long tide-gauge records [36]. However, tide-gauge measurements that extend back into the eighteenth century are only available for a few stations in western Europe (Amsterdam since 1700, Stockholm since 1774, Liverpool since 1768 [37]). The earliest global acceleration of sea-level rise that can be clearly demonstrated in instrumental sea-level data [38] and in global reconstructions based on tide-gauge data [39] occurred in the 1930s.

5. CONCLUSION

Sea-level rise is a major indicator of ongoing global change. Sea level has been rising at rates of up to 0.06 m per decade in the twentieth century. Since the 1950s, every subsequent decade has experienced increased rates of sea-level rise. Model experiments show that twentieth century sea-level rise cannot be explained by natural processes alone. Anthropogenic forcing by greenhouse gasses has become a dominant cause for recent sea-level change. The geological record of the past three glacial-interglacial cycles shows a strong positive relationship between atmospheric CO_2 concentrations and sea level. Modern rates of sea-level rise started about 100 years ago and the rate of twentieth century sea-level rise appears to be faster than rates reconstructed for the warm intervals of the Medieval Climatic Optimum and the middle Holocene. However, during the Last Interglacial rates of sea-level rise were possibly higher and were similar to those predicted in some future climate-change scenarios.

ACKNOWLEDGEMENT

I am grateful to Phil Woodworth for helpful comments which improved this chapter.

REFERENCES

1. G.A. Meehl, T.F. Stocker, W.D. Collins, P. Friedlingstein, A.T. Gaye, J.M. Gregory, A. Kitoh, R. Knutti, J.M. Murphy, A. Noda, S.C.B. Raper, I.G. Watterson, A.J. Weaver, Z.-C. Zhao, in: S. Solomon, D. Qin, M. Manning, Z. Chen, M. Marquis, K.B. Averyt, M. Tignor, H.L. Miller (Eds.), Climate Change 2007: The Physical Science Basis. Contribution of Working Group I to the Fourth Assessment Report of the Intergovernmental Panel on Climate Change, Cambridge University Press, Cambridge, UK and New York, NY, USA, 2007, pp. 747–845.

2. B. Gutenberg, Geol. Soc. Am. Bull. 52 (1941) 721–772.
3. C.M. Domingues, J.A. Church, N.J. White, P.J. Gleckler, S.E. Wijffels, P.M. Barker, J.R. Dunn, Nature 453 (2008), 1090–1094, doi:10.1038/nature07080.
4. W. Munk, Proc. Natl. Acad. Sci. 99 (2002) 6550–6555.
5. L. Miller, B.C. Douglas, Nature 428 (2004) 406–409.
6. N.L. Bindoff, J. Willebrand, V. Artale, A. Cazenave, J. Gregory, S. Gulev, K. Hanawa, C. Le Quéré, S. Levitus, Y. Nojiri, C.K. Shum, L.D. Talley, A. Unnikrishnan, in: S. Solomon, D. Qin, M. Manning, Z. Chen, M. Marquis, K.B. Averyt, M. Tignor H.L. Miller (Eds.), Climate Change 2007: The Physical Science Basis. Contribution of Working Group I to the Fourth Assessment Report of the Intergovernmental Panel on Climate Change, Cambridge University Press, Cambridge, UK and New York, NY, USA, 2007, pp. 385–432.
7. B.F. Chao, Y.H. Wu, Y.S. Li, Science 320 (2008) 212–214, doi:10.1126/science.1154580.
8. T.G. Huntingdon, Hydr. Proc. 22 (2008) 717–723.
9. R.B. Alley, P.U. Clark, P. Huybrechts, I. Joughin, Science 310 (2005) 456–460.
10. A. Shepherd, D. Wingham, Science 315 (2007) 1529–1532.
11. I. Velicogna, J. Wahr, Science 311 (2006) 1754–1756.
12. J.I. Antonov, S. Levitus, T.P. Boyer, J. Geophys. Res. 107 (C12) (2002), doi:10.1029/2001JC000964.
13. J.A. Clark, J.A. Primus, in: M.J. Tooley, I. Shennan (Eds.), Institute of British Geographers, London, 1987, pp. 356–370.
14. J.X. Mitrovica, M.E. Tamisica, J.L. Davis, G.A. Milne, Nature 409 (2001) 1026–1029.
15. M. Marcos, M.N. Tsimplis, Geophys. Res. Lett. 34 (2008) doi:10.1029/2007GL030641.
16. B.C. Douglas, J. Coastal. Res. 24 (2008) 218–227.
17. M.E. Tamisiea, J.X. Mitrovica, J.L. Davis, Earth Planet. Sci. Lett. 213 (2003) 447–485.
18. H.-P. Plag, Phil. Trans. R. Soc. A 364 (2006) 821–844.
19. J.M. Gregory, J.A. Lowe, S.F.B. Tett, J. Clim. 19 (2006) 4576–4591.
20. J.A. Church, N.J. White, J.M. Arblaster, Nature 438 (2005) 74–77.
21. M. Siddall, E.J. Rohling, A. Almogi-Labin, Ch. Hemleben, D. Meischner, I. Schmelzer, D.A. Smeed, Nature 423 (2003) 853–858.
22. J.R. Petit, J. Jouzel, D. Raynaud, N.I. Barkov, J.M. Barnola, I. Basile, M. Bender, J. Chappellaz, J. Davis, G. Delaygue, M. Delmotte, V.M. Kotlyakov, M. Legrand, V. Lipenkov, C. Lorius, L. Pépin, C. Ritz, E. Saltzman, M. Stievenard, Nature 399 (1999) 429–436.
23. J. Ahn, E.J. Brook, Geoph. Res. Lett. 34, L10703, doi:10.1029/2007GL029551.
24. J.D. Stanford, E.J. Rohling, S.E. Hunter, A.P. Roberts, S.O. Rasmussen, E. Bard, J. McManus, R.G. Fairbanks, Paleoceanography 21 (2006), doi:10.1029/2006PA001340.
25. E.J. Rohling, K. Grant, Ch. Hemleben, M. Siddall, B.A.A. Hoogakker, M. Bolshaw and M. Kucera, Nat. Geosci. 1 (2008), 38–42, doi:10.1038/ngeo.2007.8.
26. D.R. Muhs, Quat. Res. 58 (2002) 36–40.
27. S. Rahmstorf, Science 315 (2007), 368–370, doi:10.1126/science1135456.
28. E.J. Jansen, J. Overpeck, K.R. Briffa, J.-C. Duplessy, F. Joos, V. Masson-Delmotte, D. Olago, B. Otto-Bliesner, W.R. Peltier, S. Rahmstorf, R. Ramesh, D. Raynaud, D. Rind, O. Solomina, R. Villalba, D. Zhang, in: S. Solomon, D. Qin, M. Manning, Z. Chen, M. Marquis, K.B. Averyt, M. Tignor, H.L. Miller (Eds.), Climate Change 2007: The Physical Science Basis. Contribution of Working Group I to the Fourth Assessment Report of the Intergovernmental Panel on Climate Change, Cambridge University Press, Cambridge, UK and New York, NY, USA, 2007, pp. 433–497.
29. W.R. Gehrels, J. Coast. Res. 17 (2000) 244–245.
30. S.E. Lewis, R.A.J. Wüst, J.M. Webster, G.A. Shields, Terra Nova 20 (2008), 74–81, doi: 10.1111/j.1365-3121.2007.00789.x.

31. R.A. Kendall, J.X. Mitrovica, G.A. Milne, T. Törnqvist, Y. Li, Geology 36 (2008), 423–426, doi:10.1130/G24550A.1.
32. O. van de Plassche, K. van der Borg, A.F.M. de Jong, Geology 26 (1998) 319–322.
33. W.R. Gehrels, J.R. Kirby, A. Prokoph, R.M. Newnham, E.P. Achterberg, E.H. Evans, S. Black, D.B. Scott, Quat. Sci. Rev. 24 (2005) 2083–2100.
34. W.R. Gehrels, B.W. Hayward, R.M. Newnham, K.E. Southall, Geophys. Res. Lett. 35 (2008), doi:10.1029/2007GL032632.
35. W.R. Gehrels, W.A. Marshall, M.J. Gehrels, G. Larsen, J.R. Kirby, J. Eiriksson, J. Heinemeier, T. Shimmield, The Holocene 16 (2006) 948–964.
36. S. Jevrejeva, J.C. Moore, A. Grinsted, P.L. Woodworth, Geophys. Res. Lett. 35 (2008), doi: 10.1029/2008GL033611.
37. P.L. Woodworth, Geoph. Res. Lett. 26 (1999) 1589–1592.
38. P.L. Woodworth, N.J. White, S. Jevrejeva, S.J. Holgate, J.A. Church and W.R. Gehrels (2008), Int. J. Clim., doi:10.1002/JOC.1771.
39. J.A. Church, N.J. White, Geoph. Res. Lett. 33, L01602, doi:10.1029/2005GL024826.
40. A. Cazenave, R.S. Nerem, Rev. Geophys. 42 (2004) RG3001.
41. J.A. Church, N.J. White, R. Coleman, K.Lambeck, J.X. Mitrovica, J. Clim. 17 (2004) 2609–2625.
42. S.J. Holgate, P.L. Woodworth, Geophys. Res. Lett. 31 (2004), doi:10.1029/2004GL019626.
43. S.J. Holgate, Geophys. Res. Lett. 34 (2007), doi:10.1029/2006GL028492.

Chapter 19

Sea Temperature Change as an Indicator of Global Change

Martin J Attrill
Marine Biology and Ecology Research Centre, Marine Institute, University of Plymouth, Drake Circus, Plymouth PL4 8AA, United Kingdom

1. Introduction: Role of Ocean, Mechanisms and Correction of Bias
2. Long-term Trends in Sea Temperature: The Historical Context
3. Global and Regional Patterns of Sea Temperature over the Last 100–150 Years
4. Conclusion: Anthropogenic Influence
References

1. INTRODUCTION: ROLE OF OCEAN, MECHANISMS AND CORRECTION OF BIAS

The oceans play the pivotal role in Earth's climate variability and as early as 1959 it was suggested that, due to their physical properties and volume, the heat content of oceans may dominate changes in the Earth's heat balance [1]. Data collected over the last 40 a suggest that 84% of the total heating of the Earth's systems has been due to warming of the oceans [2], their heat capacity being ~1000 times larger than the atmosphere [3]. Therefore, as Barnett et al. [2] stated, 'if one wished to understand and explain this warming, the oceans are clearly the place to look'. Understanding the variability, and long-term changes, in the Earth's climate therefore requires an estimation of the relative contribution of different parts of the Earth system to absorbing heat over the last 50 years [1]. Over this time period, the energy content of the oceans has increased by $\sim 14.2 \times 10^{22}$ J (Fig. 1, [3]) compared with $<1 \times 10^{22}$ J for the atmosphere and land mass, with ~57% of this change occurring since 1993 [3]. Assimilation of heat into the oceans will, therefore, effectively be stemming the potential build-up of heat in the atmosphere.

Two main measures of temperature of the oceans have been employed to assess changes over time: sea surface temperature (SST), taken from the

Climate Change: Observed Impacts on Planet Earth

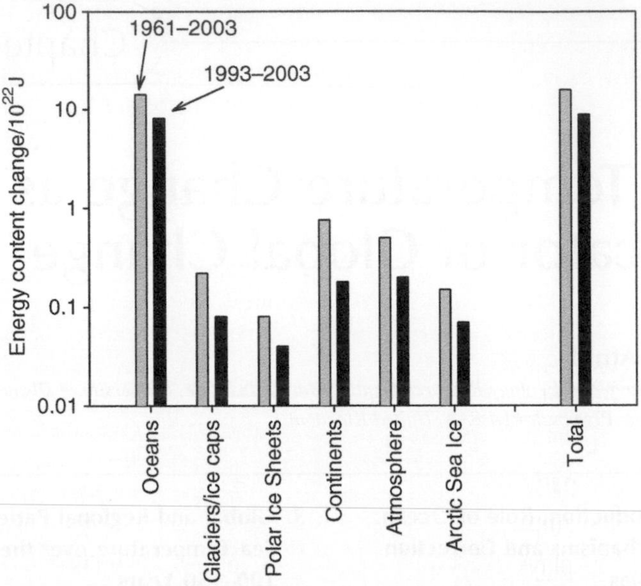

FIGURE 1 Energy content change across different parts of the Earth system. Light shaded bars are 1961–2003, dark bars 1993–2003. Data from Ref. [3] and papers therein. Values for the ocean are 14.2×10^{22} J for the 1961–2003 period and 8.11×10^{22} J for the period 1993–2003.

top few metres, and heat content, which integrates measurements from a larger depth of the water column (up to 3000 m). Traditionally, as it has been measured for decades, SST has been used as the main indicator of global ocean temperatures and thus has fed into overall trends in global surface warming. However, unlike the situation for temperature records on land, which have been relatively consistent and reliable due to a fixed network of measuring stations [4], the methodology utilised to record SST has varied over time and space [5]. Up until the 1970s, sea water temperature readings were made entirely from ships; after 1970 measurements were also taken using drifting buoys and, from the 1980s, satellites [4]. Primarily, therefore, the historical record of SST change has relied on ship-based measurement, but methods have varied over the years which affect the temperature recorded. For example, earlier in the SST time series (mainly pre-1940), temperatures were recorded from uninsulated buckets on the decks of vessels which tend to produce slightly colder temperatures due to the evaporative effect of a moving ship and standing in air [6]. A more subtle bias was introduced over time as ships generally got taller and faster and the cooling effect more enhanced [5]. After 1940, a greater proportion of temperature records were made using the ship's intake water; these records are more likely to be biased towards warmer temperatures [4]. Generally, global and regional SST values have been calculated by averaging all raw data records on the database (e.g., International Comprehensive Ocean–Atmosphere Data Set, ICOADS [7]), so major bias problems in the record can arise when there have been temporal shifts in the main methodology, or certain practices have become dominant for a period of time. Over recent years, much effort has been targeted at correcting these biases [5,6,8–10], and thus constructing a more realistic picture of how SST have varied over the last 150 years. The result of these revisions has been to alter the original trends in raw SST data and thus alter our perception of how global ocean temperatures, and therefore overall global temperature patterns, have changed over the twentieth century.

Figure 2 displays the trends in the raw SST data (ICOADS), highlighting a cool period early in the twentieth century, followed by warming to a peak during the 1940s. Following this peak, temperatures tended to cool again, before rising from the late 1960s – the 'familiar' pattern of climate change during the twentieth century. The top trace in Fig. 2 displays SST values [9] corrected for the bias associated with the uninsulated buckets prior to 1941, the correction allowing parity with the mixed methods used after WW2. This has the effect of raising temperatures prior to 1941, although the warming trend up to 1945 is still apparent. Thompson et al. [9] have, however, noted a major

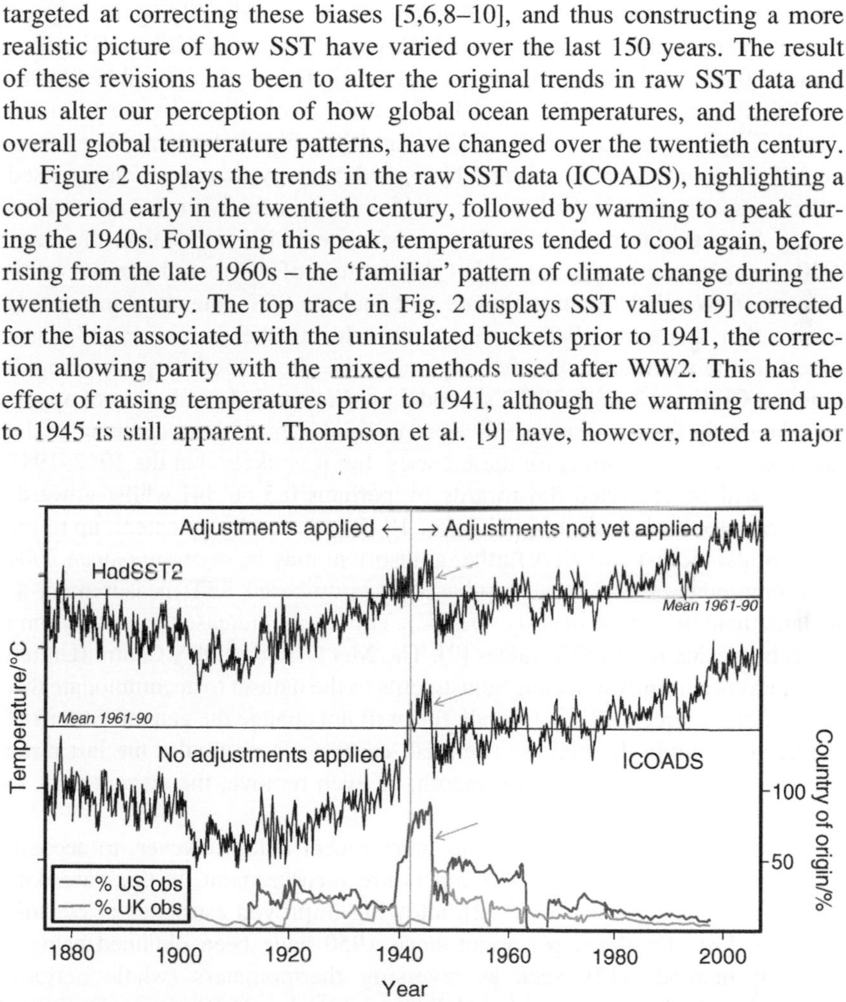

FIGURE 2 Detailed SST values since 1870s. Top, the global-mean SST time series corrected for ENSO fluctuations and pre-1941 methodological artefact (use of on-deck buckets). Middle, As in the top time series but for uncorrected data from ICOADS highlighting apparent cold period pre-1940 due to change in sampling method. Bottom, the percentage of observations which can be positively identified as coming from US (dark line) and UK (light line) ships. The vertical line denotes December 1941. All data sets show the clear discontinuity in 1945 (sharp drop in temperature) due to shift in sampling from US (engine intake) to UK (bucket) programmes. Left vertical axis shows temperature anomalies; tickmarks indicate steps of 0.5 °C. Right vertical axis shows percentage of observations. Figure redrawn with permission from Macmillan Publishers Ltd [Nature] from Thompson et al. [9].

discontinuity in the data in 1945, where temperatures cooled dramatically, resulting in peak temperatures during the early 1940s (Fig. 2, top). As would be expected, the number of SST measurements achieved plummeted during both world wars (see Fig. 1 in Ref. [4]); during WW2, around 80% of measurements were from ships of US origin, these vessels relying mainly on engine room intake measurements. Following 1945, the United Kingdom restarted their monitoring programme, but continued to use uninsulated buckets at this time; between 1945 and 1949 ~50% of observations are from the United Kingdom and only 30% of US origin [9] (Fig. 2, bottom). Therefore, the WW2 records were dominated by a methodology that was warm biased and the sudden drop in SST during 1945 is consistent with an uncorrected change from engine room to bucket measurements [9] rather than the early 1940s being exceptionally warm, but this feature of the record persists in all patterns of twentieth century climate that include SST data. It is interesting to note that this early 1940s warm period, due to the dominance of warm biased engine room data, was the only one to lie above the Intergovernmental Panel on Climate Change (IPCC's) model predictions [11] and was not apparent when only land measurements were utilised. Current reassessment of the data is underway to correct for these biases, but it is likely that the 1942–1945 records will be corrected downwards by perhaps 0.3 °C [4] whilst upwards adjustment to the data immediately after 1945 and, to a lesser extent, up to the 1960s is also necessary [9]. A further adjustment may be necessary since 2001 to accommodate a shift from ship-based to buoy-based SST measurement as the latter tend to be cool-biased (~0.1 °C). This could increase the century long trends by raising recent SST values [9]. The Met Office Hadley Centre (United Kingdom) is currently assessing adjustments to the dataset to accommodate this range of bias corrections [9]. Overall, this will not change the general pattern of increased warming through the twentieth century, in particular the last three decades, but it is more likely to smooth, or even remove, the current peak in 1940s temperatures.

Corrections have been applied to more recent data, however, to account for biases due to the method of temperature measurement, in this case data on heat content of the ocean [8,12], allowing improved estimates of oceanic warming. Data for the upper-ocean since 1950 have been obtained using a range of methods [12], such as reversing thermometers (whole period), expendable bathythermographs (XBTs since the 1960s), conductivity-temperature-depth probes from ships (since 1980s) and, since 2001, Argo floats. The biggest differential between these methods is between XBTs and CTDs [8], with XBTs having a warm bias of 0.2–0.4 °C; XBTs comprise the largest proportion of the dataset. Rates for the 1990s in particular have a positive bias due to instrumental errors [12], so adjusted temperatures to account for the range of these recent biases has resulted in a heat-content trend showing a continual upwards progression since the 1950s (Fig. 3). Domingues et al. [12] suggest that actual ocean warming

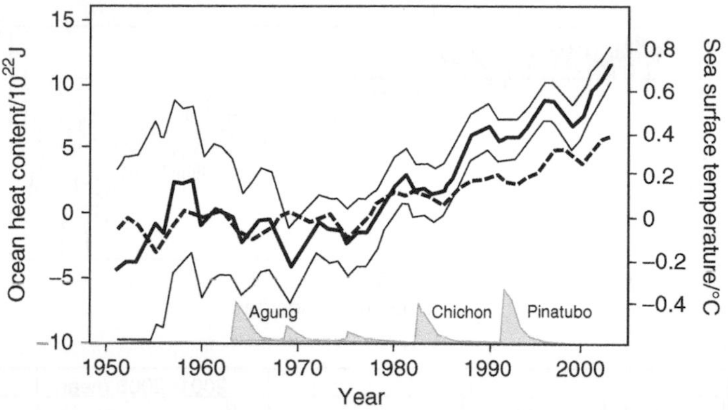

FIGURE 3 Improved estimates of upper-ocean warming since 1950 as presented by Domingues et al. [12] and redrawn from that source with permission from Macmillan Publishers Ltd [Nature]. Thick black line is upper-ocean heat content (thin lines 1SD) following application of recent methodological corrections (see text). Broken line is sea surface temperature. All time series were smoothed with a three-year running average and are relative to 1961.

trends from 1950 to 2003 are 50% larger than earlier estimates, but the 1993–2003 trend is about 40% smaller (so will impact Fig. 1). It is notable (Fig. 3) that the ocean heat content warming trend for the upper 700 m is increasing faster than the equivalent for SST.

2. LONG-TERM TRENDS IN SEA TEMPERATURE: THE HISTORICAL CONTEXT

A variety of proxies enables the reconstruction of ocean temperatures through geological time and thus assessment of global climate trends. In particular, the oxygen isotope ratio ($\delta^{18}O$) of calcite depends on the ambient water temperature from which it has been precipitated [13], so analysing the shells of fossil calcareous planktonic organisms (such as Foraminifera and coccolithophores) allows estimation of past surface ocean temperatures. For much of earth's long-term history, the oceans (and the global climate) have been warmer than today [14] and have been gradually declining at the millions-of-years scale since the Cretaceous [15]. This is particularly marked for the deep-ocean which has seen a general near-linear drop in temperature of at least 12 °C over the last 70 Ma [16]. Over the last 5 Ma, a general downwards trend has also been apparent ([17], Fig. 4, top) from a warm Pliocene [18], although there has been increased variability with time due to the Milankovitch cycles and onset of ice ages ~2.75 Ma ago; recent inter-glacial temperature peaks almost match the warm temperatures evident >3 Ma ago (Fig. 4, top).

The current interglacial, however, shows little sign of receding (Fig. 4, middle): despite an apparent original peak 8200 a ago [19], temperatures have recently increased again with the average SSTs for 2001–2005 being amongst the highest during the last 1.4 Ma (Fig. 4, middle; [20]).

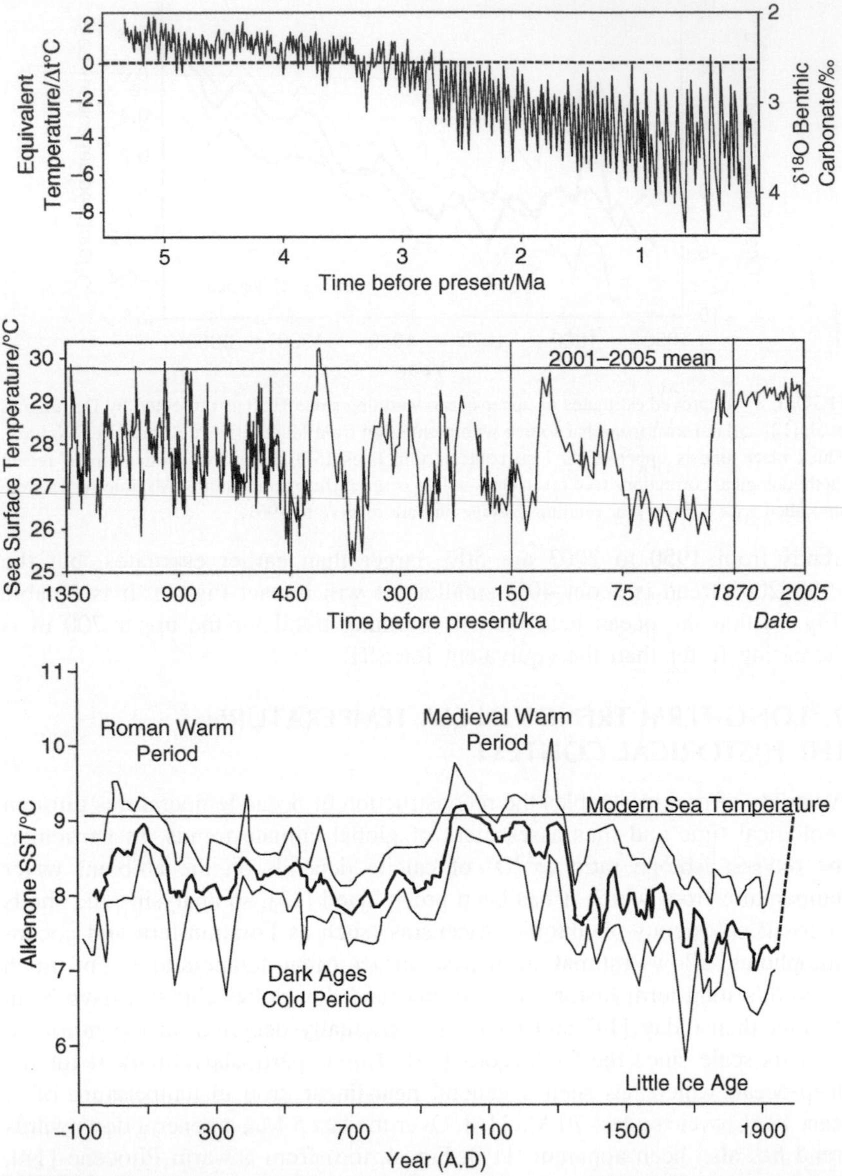

FIGURE 4 Top, climate record of Lisiecki and Raymo [17] constructed by combining measurements from 57 globally distributed deep sea sediment cores. Original figure by Robert A. Rohde of Global Warming Art from published data (http://www.globalwarmingart.com/wiki/Image:Five_Myr_Climate_Change_Rev_png). Middle, modern sea surface temperatures in the Western Equatorial Pacific compared with paleoclimate proxy data. Modern data are the 5-a running mean, while the paleoclimate data have a resolution of the order of 1000 a. Figure redrawn from Hansen et al. [20], copyright (2006) National Academy of Sciences, U.S.A. Bottom. Alkenone data from sediment cores off Iceland reconstructing temperatures over last 2000 a. Solid line is 10 point running mean, thin lines indicate range of temperature data. Redrawn from Sicre et al. [23] with permission from Elsevier.

Variations away from this overall pattern are evident, however, with the western tropical Pacific appearing to have generally cooled by ~0.5 °C over the last 10 000 a [21]. Ocean temperature trends have also been reconstructed for the last 2000 a using techniques such as Mg/Ca ratios in sediment cores [22] and alkenone biomarkers from the coccolithophore *Emiliania huxleyi* [23], providing more detail on the historical context of recent trends. Most data sets demonstrate a trend of a medieval warm period around 900–1300 and a general decrease in ocean temperatures from this point in time [24] which has been markedly reversed over the last century. Chesapeake Bay records suggest anomalous recent behaviour of the climate system over a 2000 a record [22], this modern increase in temperature being much more recent (but equally marked) in records off Iceland [23] where previously ocean temperatures had been steadily falling since 1300 (Fig. 4, bottom); modern records are an equivalent temperature to the medieval warm period.

3. GLOBAL AND REGIONAL PATTERNS OF SEA TEMPERATURE OVER THE LAST 100–150 YEARS

Global trends in sea temperature since the late nineteenth century can be split into several clear periods (Fig. 2), although the magnitude and clarity of some of the decadal trends have been partly due to the sampling artefacts discussed in Section 1 [5,6,8,9,12]. The early twentieth century generally witnessed a trend of cooling SSTs to around 1910 that has now primarily been attributed to the lasting impact of aerosols from major volcanic eruptions such as Krakatoa (1883) and Santa Maria (1902) ([4,9,25,26]; see Chapter 4), with the volcanic cooling signature clearly visible in subsurface ocean temperatures into the middle part of the twentieth century [27]. Individual volcanic eruptions have resulted in several discontinuities within the ocean temperature record over the twentieth century [4,9], particularly the eruption of Mt. Pinatubo in 1991 ([25]; Fig. 3); in simulations, recovery of temperatures from this eruption was not complete by 2000, depressing the underlying warming trend [25].

Two distinct warming periods have been evident during the twentieth century [28]: the recovery of depressed temperatures from the 1920s to the 1940s and the pronounced warming from 1978 till the present. The first phase has been exaggerated over time due to the problems with the warm-biased SSTs obtained during the 1940s and cold-biased records prior to 1941 [9], although there is also evidence of increasing human-induced radiative forcing due to greenhouse gases and a particularly large realisation of the decadal ocean-climatic variability during this time [28].These records, plus the major discontinuity in SSTs in 1945 discussed earlier [9], will also dampen the cooling trend apparent in uncorrected data from 1945 to 1970 (Fig. 2). It is, therefore, most likely that finally corrected global SST records will demonstrate an overall gradual warming from 1920s to 1970s, followed by the modern period of accelerated warming. This last 30 a period has also seen some variability in ocean temperatures, particularly a levelling off of the warming trend since 1998 [29; Fig. 2]. In addition to required methodological adjustments for the 1990s detailed earlier [12], this trend is most likely a function of the behaviour of the El Niño Southern Oscillation (ENSO) cycle, which has a major controlling influence over the world's climate [30]; the 1997–1998 'super' El Niño part of the cycle was the most extreme on record [30] and lifted temperatures 0.2°C above the trend line [20]. The Pacific ENSO cycle more recently (late 2000s) has moved into the cooling La Niña phase, but during 2005 near-record temperatures were also recorded without the boost from El Niño [20]. Throughout the ocean temperature record, ENSO has resulted in fluctuations around any warming trend and so can be corrected for in simulations to understand the underlying trend [9]. The most modern improved estimates of ocean warming smooth out the effect of ENSO and adjust for methodological biases [12], resulting in a clear, continued upwards trend in global temperatures since 1950 (Fig. 3), with no underlying evidence of long-term cooling since 1998 [29].

Global trends in ocean temperatures have not been consistent across all seas, however. Whilst most records do demonstrate upwards trajectories in temperature comparable with the global trend [5,26], northern hemisphere seas have warmed more since 1850 than those in the southern hemisphere [5]. Decadally filtered differences in SST for the Northern Hemisphere are 0.71°C ± 0.06; for the Southern Hemisphere warming has been on average 0.64°C ± 0.07 [5]. Such a warming trend differential is even more marked for the Arctic, where the ice–ocean system has been warming faster than the global average since 1966 [31]. A clear global anomaly in terms of SST (together with part of the North Pacific and south of Greenland [32]) is the east equatorial Pacific region where ENSO events originate and are most marked. Here long-term trends have only shown modest upwards trends in SST [5], if any [10], since 1870, due primarily to increased trade-winds and upwelling [32], but this has resulted in another trend in ocean temperatures with major global consequences. Over the course of the twentieth century, there has been an increase in the temperature gradient across the equatorial Pacific [32], the build up of such a temperature gradient being generally a precursor of El Niño events [30]. Hansen et al. [20] suggested this trend will increase the likelihood of strong El Niños. There is evidence that El Niño events are becoming more frequent and severe over recent decades [33,34], resulting in increased variability (and thus more extreme peaks) in SST in the east Pacific region [34] and thus affecting the world's climate and ocean temperatures. Such Pacific temperature distributions may have been apparent during the warm Pliocene which had a permanent El Niño-like climate: paleoceanographic data suggest Pacific SST distribution pre-Ice Ages most resembled that of the 1997–1998 El Niño [35].

4. CONCLUSION: ANTHROPOGENIC INFLUENCE

In summary, the oceans have been warming over the last century, with the latest most accurate adjusted data [5,8,9,12] that has accounted for methodological artefacts (e.g., the 1940s) suggesting this trend has been more consistent and continuous than previously thought, with a particularly marked increase in sea temperatures since the 1970s (Figs. 2 and 3). Debate about the causes of global warming has been discussed in earlier chapters (e.g., Chapters 1–6), but for the oceans there is clear evidence of an anthropogenic signal in the pattern of warming over the last 40 years [2,36]. The penetration of this human-induced warming is evident across the top 700 m and apparent in all oceans, but the signal is complex and varies widely by ocean [2]. Figure 5 displays examples of the change in sea temperature at depth since 1960 for northern parts of the three major oceans (see [2] for full set of data) and illustrates how warming at depth has varied. The North Atlantic demonstrates a strong warming pattern down to 700 m, with an increase in the rate of change from depth to the surface. However, warming in the Pacific and Indian oceans is more confined to the upper 100 m, with the North Pacific in particular actually demonstrating cooling at depth (Fig. 5, right panel). Deep convection is characteristic of the Atlantic, whereas in the Pacific the shallow meridional overturning circulation isolates the surface layer and thus confines the signal to the upper ocean [2]. In order to assess cause of the warming trend, Barnett et al. [2] have modelled the warming effect of all natural internal variability; the grey polygons in Fig. 5 display the 90% confidence limits of this natural signal strength. As can be seen, observed warming patterns bear little resemblance to what would be expected from warming due to internal variability. Observed warming also bears no resemblance to a signal forced by solar and volcanic variability (Fig. 5, open circles), but does fit closely to modelled anthropogenic forcing signal strength [2]. Evidence compiled over recent years [2,36,37], therefore, strongly demonstrates a human-induced warming signal in the ocean temperature record.

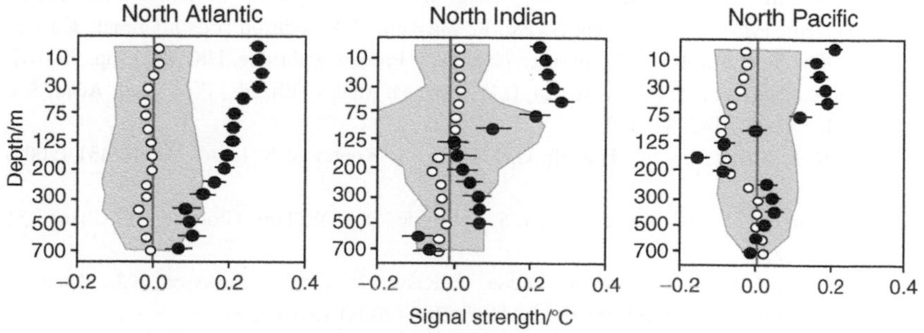

FIGURE 5 Warming signal strength since 1960 by ocean and depth (black circles, ± 2SD). Grey polygons reflect the 90% probability distribution of warming signals associated with internal variability. Open circles are the warming signal forced by solar and volcanic variability. Figure redrawn from Barnett et al. [2], reprinted with permission from AAAS.

REFERENCES

1. S. Levitus, J. Antonov, T. Boyer, Geophys. Res. Lett. 32 (2005) L02604 doi:10.1029/2004GL021592.
2. T.P. Barnett, D.W. Pierce, K.M. AchutaRao, P.J. Gleckler, B.D. Santer, J.M. Gregory, W.M. Washington, Science 309 (2005) 284–287.
3. N.L. Bindoff, V. Willebrand, V. Artale, A. Cazenave, J. Gregory, S. Gulev, K. Hanawa, C. Le Quere, S. Levitus, Y. Nojiri, C.K. Shum, L.D. Talley, A. Unnikrishnan, in: S. Solomon, et al. (Eds.), Climate Change 2007: The Physical Science Basis.Contribution of Working Group I to The Fourth Assessment Report of The IPCC, Cambridge University Press, Cambridge, UK, 2007, pp. 385–2428.
4. C.E. Forest, R.W. Reynolds, Nature 453 (2008) 601–602.
5. N.A. Rayner, P. Brohan, D.E. Parker, C.K. Folland, J.J. Kennedy, M. Vanicek, T.J. Ansell, S.F.B. Tett, J. Clim. 19 (2006) 446–469.
6. C.K. Folland, D.E. Parker, Q. J. R. Meteorol. Soc. 121 (1995) 319–367.
7. S.J. Worley, S.D. Woodruff, R.W. Reynolds, S.J. Lubker, N. Lott, Int. J. Climatol. 25 (2005) 823–842.
8. V. Gouretski, K.P. Koltermann, Geophys. Res. Lett. 34 (2007). Article Number L01610.
9. D.W.J. Thompson, J.J. Kennedy, J.M. Wallace, P.D. Jones, Nature 453 (2008) 646–649.
10. T.M. Smith, R.W. Reynolds, J. Clim. 17 (2004) 2466–2477.
11. IPCC, Summary for Policymakers, in: S. Solomon, et al. (Eds.), Climate Change 2007: The Physical Science Basis. Contribution of Working Group I to the fourth assessment report of the IPCC, Cambridge University Press, Cambridge, UK, 2007, pp. 1–18.
12. C.M. Domingues, J.A. Church, N.J. White, P.J. Gleckler, S.E. Wijffels, P.M. Barker, J.R. Dunn, Nature 453 (2008) 1090–1093.
13. P.N. Pearson, P.W. Ditchfield, J. Singano, K.G. Harcourt-Brown, C.J. Nicholas, R.K. Olsson, N.J. Shackleton, M.A. Hall, Nature 413 (2001) 481–487.
14. J. Veizer, Y. Godderis, L.M. Francois, Nature 408 (2000) 698–701.
15. J. Zachos, M. Pagani, L. Sloan, E. Thomas, K. Billups, Science 292 (2001) 686–693.
16. S.M. Savin, Ann. Rev. Earth Planet. Sci. 5 (1977) 319–355.
17. L.E. Lisiecki, M.E. Raymo, Paleoceanography 20 (2005) 1–17.
18. A.C. Ravelo, D.H. Andreasen, M. Lyle, A. Olivarez Lyle, M.W. Wara, Nature 429 (2004) 263–267.
19. C.K. Folland, T.R. Karl, J.R. Christy, R.A. Clarke, G.V. Gruza, J. Jouzel, M.E. Mann, J. Oerlemans, M.J. Salinger, S.-W. Wang, in: J.T. Houghton, Y. Ding, D.J. Griggs, M. Nogeur, P.J. van der Linden, X. Pai, K. Maskell, C.A. Johnson (Eds.), Climate Change 2001: The Scientific Basis, Cambridge University Press, Cambridge, UK, 2001, pp. 99–182.
20. J. Hansen, M. Sato, R. Ruedy, K. Lo, D.W. Lea, M. Medina-Elizade, Proc. Natl. Acad. Sci. USA 103 (2006) 14288–14293.
21. L. Stott, K. Cannariato, R. Thunell, G.H. Haug, A. Koutavas, S. Lund, Nature 431 (2004) 56–59.
22. T.M. Cronin, G.S. Dwyer, T. Kamiya, S. Schwede, D.A. Willard, Glob. Planet. Change 36 (2003) 17–29.
23. M.A. Sicre, J. Jacob, U. Ezat, S. Rousse, C. Kissel, P. Yiou, J. Eiriksson, K.L. Knudsen, E. Jansen, J.L. Turon, Earth Planet. Sci. Lett. 268 (2008) 137–142.
24. P.D. Jones, K.R. Briffa, T.P. Barnett, S.F.B. Tett, The Holocene 8 (1998) 455–471.
25. J.A. Church, N.J. White, J.M. Arblaster, Nature 438 (2005) 74–77.

26. T.R. Knutson, T.L. Delworth, K.W. Dixon, I.M. Held, J. Lu, V. Ramaswamy, M.D. Schwarzkopf, G. Stenchikov, R.J. Stouffer, J. Clim. 19 (2006) 1624–1651.
27. T.L. Delworth, V. Ramaswamy, G.L. Stenchikov, Geophys. Res. Lett. 32 (2005). Article No. L24709.
28. T.L. Delworth, T.R. Knutson, Science 287 (2000) 2246–2250.
29. R. Fawcett, Bull. Aust. Meteorol. Oceanogr. Soc. 20 (2007) 141–148.
30. M.J. McPhaden, Science 283 (1999) 950–954.
31. J.L. Zhang, Geophys. Res. Lett. 32 (2005). Article No. L19602.
32. M.A. Cane, A.C. Clement, A. Kaplan, Y. Kushnir, D. Pozdnyakov, R. Seager, S.E. Zebiak, R. Murtugudde, Science 275 (1997) 957–960.
33. M.J. McPhaden, S.E. Zebiak, M.H. Glantz, Science 314 (2006) 1740–1745.
34. A. Timmermann, J. Oberhuber, A. Bacher, M. Esch, M. Latif, E. Roeckner, Nature 398 (1999) 694–697.
35. P. Molnar, M.A. Cane, Geosphere 3 (2007) 337–365.
36. T.P. Barnett, D.W. Pierce, R. Schnur, Science 292 (2001) 270–274.
37. K.M. AchutaRao, B.D. Santer, P.J. Gleckler, K.E. Taylor, D.W. Pierce, T.P. Barnett, T.M.L. Wigley, J. Geophys. Res. 111 (2006) C05019.1–C05019.

Chapter 20

Ocean Current Changes as an Indicator of Global Change

T. Kanzow
National Oceanographic Centre, Southampton, United Kingdom

M. Visbeck
Leibniz Institute of Marine Sciences, Kiel, Germany

1. **Introduction**
2. **The Variable Ocean**
3. **Oceanographers' Tools**
4. **The Atlantic Meridional Overturning Circulation**
   4.1. Motivation
   4.2. Circulation, Driving Mechanisms
5. **The AMOC's Role in Heat Transport, Oceanic Uptake of Carbon and Ventilation of the Deep Ocean**
   5.1. Simultaneous Changes of the AMOC and Atlantic Climate in the Past
   5.2. Why should the AMOC Change as Part of Ongoing Climate Change?
6. **Can We Detect Changes in the AMOC? Is the AMOC Changing Already?**
7. **Conclusion**
References

1. INTRODUCTION

The high heat capacity of seawater and the relatively slow ocean circulation allow the oceans to provide significant 'memory' for the climate system. Bodies of water that descend from the sea surface may reside in the ocean interior for decades and centuries, while preserving their temperature and salinity signature, before they surface again to interact with the overlying atmosphere. In contrast to that, the residence time of water in the atmosphere is about 10 days and the persistence of dynamical states of the atmospheric circulation may last up to a few weeks. Thus, on long time scales ocean dynamics becomes important for climate, which implies that climate variations and climate change can only partially be understood without consideration of ocean dynamics and the

Climate Change: Observed Impacts on Planet Earth

intricate ocean–atmosphere interaction. The El Nino/Southern Oscillation phenomenon in the tropical Pacific is a prominent example of tightly coupled ocean–atmosphere dynamics on interannual time scales, other more weakly coupled interactions exist throughout the system.

The oceans' role in climate and climate change is manifold. Ocean circulation transports large amounts of heat and freshwater on hemispheric space scales which have significant impacts on regional climate in the ocean itself but also noticeable consequences via atmospheric teleconnections on land. What is well known for the seasonal cycle with only moderate temperature changes between summer and winter in marine climates compared with much larger swings within the continents, is also true on decadal time scales. Since 1960 the heat uptake of the oceans has been 20 times larger than that of the atmosphere. Thus the oceans have been able to reduce the otherwise much more pronounced temperature rise in the atmospheric climate. Also, over the last 200 a, the oceans have absorbed about half of the CO_2 release into the atmosphere by human activities (fossil fuel combustion, de-forestation, cement production), thereby reducing the direct effect of greenhouse gases on atmospheric temperatures.

2. THE VARIABLE OCEAN

Bodies of water circulate throughout the oceans – both horizontally and vertically – as a consequence of physical forces exerted on them according to Newton's Law. The oceanic circulation is not steady in time. Rather motions of water bodies in the ocean are known to vary on a broad range of spatial and temporal scales. The following four examples serve to highlight natural variations of large-scale circulation patterns:

(i) Seasonal variations of the strength of the North Atlantic subtropical gyre at 26°N have the amplitudes of 25 Sv[1] (peak to peak). This range is comparable to the time mean strength of the wind-driven, anti-cyclonic basin-scale gyre at this latitude [1].

(ii) The Pacific subtropical cells (STC) – a meridional, upper-ocean pattern of circulation that links the subtropical subduction regions north and south of the equator to the equatorial thermocline – has seen a decline of 11 Sv or 30% since the 1950s, however, displaying decadal variations of the same order of magnitude [2]. The observed strong decadal and multi-decadal variations in sea surface temperature (SST) in the equatorial Pacific have been shown to be related to changes in STC strength.

(iii) The cyclonic circulation of the North Atlantic subpolar gyre has possibly weakened by 25% and shrunk in size since the mid-1990s [3,4]. This

---

[1] 1 Sv = $1 \times 10^6\ m^3 \cdot s^{-1}$ (unit for volumetric transport, named after Harald Ulrik Sverdrup). For comparison, the Amazon River discharge in the Atlantic is about 0.2 Sv.

has been attributed to the transition of the North-Atlantic Oscillation[2] (NAO, [5]) from a comparably strong phase between 1960 and 1995 (manifesting itself in stronger than average westerly winds at mid-latitudes) to a significantly weaker one after 1995. The gyre's weakening and westward retreat has allowed large quantities saline subtropical upper-ocean waters to flow northward past its eastern flank, as a consequence of which a drastic increase in salinities in the Nordic Seas[3] has been observed [6]. This is thought to have an impact on the sinking of waters as part of the Atlantic Meridional Overturning Circulation (AMOC), the latter being the primary focus of this chapter.

(iv) Although not having been observed directly, it is commonly thought that temporal changes in the strength of the AMOC – a basin wide meridional circulation pattern that links upper-ocean net northward flow of warm, saline waters with cold southward return flow below roughly 1000 m throughout the Atlantic – explain large parts of the observed multi-decadal North-Atlantic SST changes [7]. A recent summary and discussion of climate variability and its predictability in the Atlantic sector [8] provides a perspective on the difficulties one has, to distinguish decadal variability from long term, possibly anthropogenic induced trends.

The reason why the different components of the ocean circulation have the potential to change substantially over time is a consequence of the complex forcing at the sea surface (exchange of momentum, heat and freshwater between ocean and atmosphere) on the one hand and internal ocean dynamics on the other. Examples of internal ocean dynamics include advection of water of anomalous density by the mean large-scale ocean circulation, westward energy transfer by off-equatorial planetary waves, the equatorial wave guide, horizontal mixing by meso-scale eddies, deep-water formation due to convection or small-scale vertical mixing, acting to push the cold waters of the oceans' abyss upwards. The large variations that basin-scale circulation patterns may exhibit have the potential to delay the detectability of climate change related shifts in the flow field.

3. OCEANOGRAPHERS' TOOLS

Oceanographers have developed direct and indirect techniques for the observation of ocean currents in order to document and analyse the strength of the interior ocean circulation and its changes in space and time. Direct current measurements can be divided in two classes, Eulerian and Lagrangian ones.

---

[2] The NAO is the dominant mode of (winter) climate variability in the North Atlantic region ranging from central North America to Europe and into Northern Asia.

[3] Nordic Seas is used as collective term for Greenland Sea, Norwegian Sea and Iceland Sea.

The Eulerian approach measures current velocity and direction at a fixed location. Current meters are typically used in this context that can be mounted on a stationary platform or 'mooring' [9], installed in a vessel's hull [10] or lowered from a vessel [11]. The scope of Lagrangian observations aims at deriving the streamlines of regional or ocean basin-scale flow fields. For this drifters are used, that move passively with the flow at a specified depth horizon and whose displacements are monitored over time [12].

The equation of motion for a fluid (i.e. Newton's Law applied to a fluid) requires the various different physical forces acting on a unit mass of water to balance one another. The two probably most-widely used indirect methods to study the flow field in the open ocean – (i) the Ekman balance and (ii) the geostrophic balance – rely on well-founded simplifications of the equation of motion, that are valid on time scales longer than 1 or 2 days and spatial scales in excess of a several tens of kilometres.

(i) Within roughly the top 50 m of the water column (near-surface Ekman layer), the flow field results from a balance between the stress the atmospheric wind field exerts onto the sea surface and the Coriolis force [13]. To first order, the horizontal flow in the Ekman layer depends on strength of the wind speed but moves at right angles to the direction of the wind as a consequence of the Earth's rotation.[4] The availability of daily wind fields over the global ocean from space borne measurements makes the Ekman balance a very powerful diagnostic tool of large-scale near-surface flows.

(ii) In the vast ocean interior below the Ekman layer, the horizontal movement results in long surfaces of constant pressure as result of the near geostrophic balance between the horizontal pressures gradient force and the Coriolis force.[5] This is analogous to atmospheric conditions as depicted in weather charts, where (to first order) wind flows *around* cells of high or low pressure (and not from high to low pressure), again as a consequence of the Earth's rotation. As a result, the strength of the flow across a section between two points is approximately proportional to the difference in pressure at the section's two end points (both in the ocean and in the atmosphere). The geostrophic balance is therefore an effective tool to diagnose the net strength of basin-scale ocean circulation patterns. All that is required are measurements of the pressure field at the section end points, while the actual horizontal structure of the flow in between does not need to be resolved. Practically, the pressure field of the ocean cannot be measured directly in the water column to derive reliable estimates of the strength of the flow.

---

4 The net flow in the Ekman layer is to the right of the direction of the wind stress in the Northern Hemisphere and to the left in the Southern Hemisphere.

5 The geostrophic flow is directed such that the pressure increases to the right in the Northern Hemisphere and to the left in the Southern Hemisphere.

However, profiles of water density allow the computation of the ocean's pressure and velocity field relative to reference pressure level. This has served oceanographers for many decades to study strength and vertical structure the ocean circulation [14].

Besides observing ocean currents the numerical simulation of the circulation using so-called ocean general circulation models (OGCM) has emerged as an important discipline in physical oceanography. To this end the equation of motion (if applicable also the heat and salt conservation equations) are solved numerically on a pre-defined spatial model grid sequentially for each time step. This is commonly referred to as model 'integration'. For each time step, the so-called boundary conditions have to be prescribed, that drive the model ocean. The boundary conditions mainly include observed fluxes of momentum, heat and freshwater between the ocean and the atmosphere. Typical horizontal resolutions of currently used OGCMs range between 10 and 100 km. The finer the resolution (and the larger the model region) is chosen, the more computationally expensive it becomes to run the model, such that the integration periods that can be reached, become shorter.

Typical integration periods are up to several decades. Besides OGCMs the class of climate models has found widespread use in oceanography. Here, the equations of motion for the ocean and the atmosphere (and for ice sheets, if applicable) are solved simultaneously, and both model components are coupled at the air sea interface. In climate models, the ocean is not passively forced, but instead can feed back on the atmosphere. Climate models are driven by orbital forcing (i.e. insolation at the top of the atmosphere). As climate models are typically integrated over climate-relevant time scales (centuries and longer) their horizontal resolution is often sparse compared to OGCMs, typically between 100 and 500 km.

The main advantage of numerical model simulations is that self-consistent estimates of the ocean circulation can be obtained for the entire spatial and temporal domain of interest. The degree of their shortcomings, however, is difficult to evaluate. One problem common to all numerical models is their finite resolution. That means that physical processes that take place on spatial scales smaller than the grid size in the real ocean are not included in the model physics and have to be parameterised. Other uncertainties derive from errors in the boundary conditions and the numerical integration itself.

4. THE ATLANTIC MERIDIONAL OVERTURNING CIRCULATION

4.1. Motivation

For most of this chapter, we will limit our attention to the aforementioned AMOC, which represents a circulation pattern most relevant for marine and terrestrial climates in many ways:

(i) The AMOC represents a mechanism of long-term 'memory' in the climate system.

(ii) The AMOC is the most important oceanic flow component for meridional redistribution of heat.

(iii) The AMOC is an important pathway for the oceans' uptake of anthropogenic greenhouse gases and for the ventilation of the deep ocean interior.

(iv) The vigour of the AMOC and the associated heat transport are thought to experience a reduction between 30% and 50% over the next century as a consequence of global warming.

Taken together, highly possible long-term changes in the AMOC are thought both to be indicative of climate change and to contribute to climate change. Because of this the AMOC represents a subject of active ongoing research involving observations and numerical modelling. In the following, we outline the AMOC's the relevance for climate, climate variations and climate change. This is preceded by a description of the underlying pattern of circulation and its potential driving mechanisms. The authors intend to convey that although our knowledge about the AMOC has advanced dramatically over the last few decades, many uncertainties remain yet to be solved.

4.2. Circulation, Driving Mechanisms

A striking feature of the temperature distribution in the oceans – Fig. 1 displays a section of temperature along the meridional extent of the Atlantic – is the strong vertical contrast in temperatures at low and mid-latitudes, with warm upper-ocean waters floating on top of cold deep and abyssal waters. The vertical layering of waters of different temperatures (densities) is referred to as stratification. It was already recognised as early as 1798 by Count Rumford that – in the absence of any deep-ocean heat sinks at low latitudes – those cold waters had to originate from high latitudes propagating equatorward at depth.[6] Today, it is well established that the observed temperature distribution is a consequence of the AMOC, that moves roughly 19 Sv of warm, saline waters northward throughout the Atlantic and the same amount of cold water back south at depth ([16,17]; Fig. 2). Carried northward within the Gulf Stream/North Atlantic Current system the near surface waters release heat to the atmosphere and thus become gradually denser. The waters eventually reach the Nordic Seas and the Labrador Sea. Here, deep-reaching wintertime convection (i.e. vertical mixing throughout the upper 2000 m of water column) can occur [4,18,19], when the vertical stratification has eroded after periods of excessive heat loss (Fig. 2). The bulk of the newly formed deep waters – that

---

6 Longworth and Bryden [15] give an exciting account of the history of the recovery of the Atlantic Meridional Overturning Circulation.

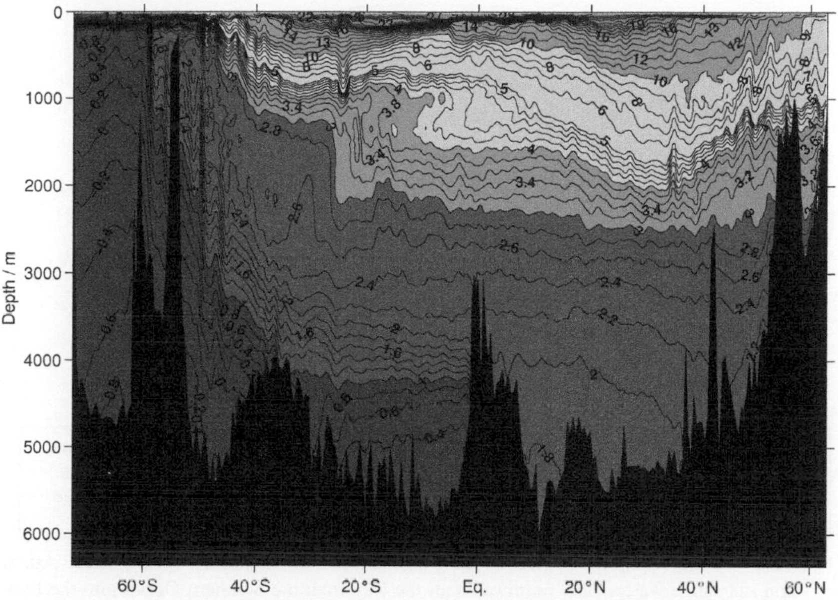

FIGURE 1 Section of potential temperature along the meridional extent of the Atlantic. For temperatures less than 5 °C and greater than 5 °C, the black contours have a spacing of 0.2 and 1 °C, respectively. The red, indian red, salmon, cyan, light blue and dark blue areas denote temperatures above 16 °C, from 10 to 16 °C, from 4 to 10 °C, from 3 to 4 °C, from 1 to 3 °C and below 1 °C, respectively. Lowered temperature measurements acquired during three research expeditions – aboard *RV Ronald H. Brown* in 2003 (section A16N, PI: Bullister [PMEL]) and in 2005 (section A16S; PIs: Wanninkhof [NOAA]/Doney [WHOI]) and aboard *RV James Clark Ross* in 1995 (section A23; PIs: Heywood/King [NOCS]) – were joined together to compile this figure. Data source: Clivar and Carbon hydrographic data office (http://whpo.ucsd.edu/atlantic.htm). Adapted from a figure of Lynne D. Talley (http://sam.ucsd.edu/vertical_sections/Atlantic.html#a16a23). (See Color Plate 22).

are subject to overflow and entrainment processes – constitute the North Atlantic Deep Water (NADW). The NADW is subsequently exported southward, partly confined to the deep western boundary current (DWBC) along the Americas below roughly 1000 m. The intensity of the strongly localised, buoyancy-loss induced formation of NADW at high latitudes (Fig. 2) 'pushing' surface waters downwards has long been thought to control the strength of the AMOC.

To close the circulation, the dense NADW needs to return to the upper ocean eventually. This is assumed to be accomplished mainly by two processes. The first process relates to winds and tides that represent the major sources of mechanical energy input into the ocean [20]. Ultimately, this energy input is balanced by dissipation into small scale motions, a process by which turbulent mixing occurs. Dissipation and mixing are ubiquitous in the open ocean; however, they seem most active in the vicinity of rough bathymetry such as exhibited by mid-oceanic ridges [21]. As a consequence deeper (denser) waters from below are mixed with overlying warmer (less

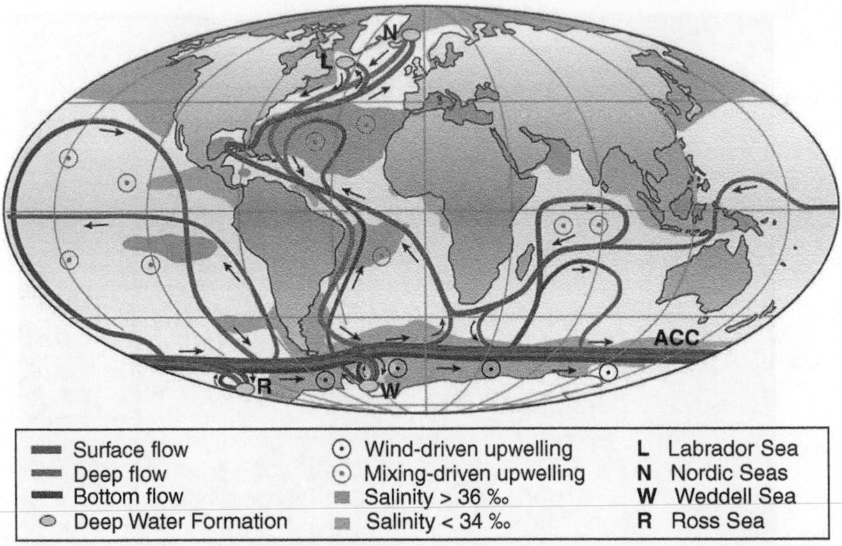

FIGURE 2 Strongly simplified sketch of the global overturning circulation system. In the Atlantic, warm and saline waters flow northwards all the way from the Southern Ocean into the Labrador and Nordic Seas. By contrast, there is no deep water formation in the North Pacific and its surface waters are fresher. Deep waters formed in the Southern Ocean are denser and thus spread in deeper levels than those from the North Atlantic. Note the strongly localised deep water formation areas in comparison with the wide-spread zones of mixing-driven upwelling. Wind-driven upwelling occurs along the Antarctic Circumpolar Current (ACC). This figure has been published by Kuhlbrodt et al. [17]. (See Color Plate 23).

dense) waters, thus making deep waters gradually lighter. This allows them to rise and to return to the upper ocean. The fact that as a direct consequence of vertical mixing even at the deep ocean below 1000 m exhibits a notable stable stratification (i.e. water becoming denser with depth, as shown in Fig. 1) has been used to argue that dissipation induced vertical mixing ‘pulling’ deep water upwards might ultimately have a stronger control on the vigour of the AMOC than the downwards ‘pushing’ at high latitudes. To move waters vertically across surfaces of constant density, vertical mixing is required and this cannot be generated by high latitude buoyancy forcing [22].

The second potentially powerful mechanism to ‘pull’ deep water back to the upper ocean to close the overturning circulation can be motivated by a careful inspection of Fig. 1. The NADW flows southward away from the regions of its formation and eventually partly reaches the Southern Ocean. While north of 40°S the deep surfaces of constant temperature show only a weak upwards slope towards the south, the situation changes dramatically south of 40°S. This is a direct result of 70% of the global wind energy input into the ocean taking place in this area. Due to the Ekman balance the strong westerly winds over the Southern Ocean push large amounts of near-surface waters northward, which are then replaced by waters being sucked upwards from the deep ocean. The manifestation of this process is the drastic increase in the upwards tilt of the deep temperature surfaces towards the south (Fig. 1). In this scenario, the transition from cold to warm waters (mixing across density surfaces) occurs near the sea surface in the Southern Ocean as suggested from model findings by Toggweiler and Samuels [23]. Whether the Southern Ocean's control on the vigour of the AMOC is stronger than that deep-ocean mixing mechanism is subject to current debate [17], as clear observational evidence is still lacking.

Besides the AMOC, a second major pattern of meridional overturning exists. This involves formation of deep waters by means of convection around Antarctica. The Antarctic Bottom Water (AABW) spreads northward and represents the coldest and therefore deepest water mass in the Atlantic, Pacific and Indian Oceans (roughly represented by the dark blue shaded part of the temperature field in Fig. 1).

In the Atlantic, the waters gradually mix into the lower parts of the overlying NADW, and eventually return southward. Even though the volume of NADW flowing southward and AABW moving northward are comparable in size [24] or possibly larger for the southern cell [25], the contribution of the AABW related meridional overturning cell to meridional heat transport is negligible, as the vertical temperature contrast between its upper and low branches is very small [26].

5. THE AMOC's ROLE IN HEAT TRANSPORT, OCEANIC UPTAKE OF CARBON AND VENTILATION OF THE DEEP OCEAN

The Earth's surface takes up heat by absorbing solar short-wave radiation. On a global average this is almost exactly balanced by the Earth's emission of long-wave radiation. Regional budgets of radiative energy fluxes, however, are unbalanced, as they show pronounced heat gain at low latitudes opposing to heat loss at high latitudes [27]. One-quarter of the 5 Pw[7] of maximum global heat transport – that the coupled ocean–atmosphere system is required to transport poleward in the Northern Hemisphere to approximately balance regional energy budgets – is carried by the AMOC in the Subtropical North Atlantic [27]. While most of the remaining three quarters of the heat transport is accomplished by the atmosphere, the AMOC is by far the most important oceanic component of meridional heat transport.

The steady increase in atmospheric CO_2 is widely regarded as one of the main drivers of the presently ongoing global warming. Exceeding the atmosphere in terms of carbon storage by more than a factor of 50 [28], the oceans exchange gases with the atmosphere. The CO_2 solubility in sea water

---

[7] 1 Pw = 1×10^{15} W; 5 Pw correspond to the output of 5 000 000 power stations.

increases with decreasing temperatures. The NADW formation at high latitudes, acting to increase carbon concentrations at depth is considered a major element in the ocean's carbon uptake ('solubility pump'). Changes in the strength and spatial structure of the AMOC might affect atmospheric CO_2 concentrations and thus global temperatures.

The flow of well oxygenated near-surface water to the deep ocean that goes along with the NADW formation at high latitudes and its subsequent export to the world ocean help to maintain the deep ocean basins as habitats for a diverse biota. This is probably the least known component of biodiversity on Earth. Substantial changes in the rate of deep-water ventilation by the AMOC are thus expected to have consequences for deep-ocean habitats.

5.1. Simultaneous Changes of the AMOC and Atlantic Climate in the Past

Especially in the Northern Hemisphere the amount of heat carried poleward by the oceans is very much tied to the strength of the AMOC. Analysis of ice cores from Greenland revealed more than 20 so-called 'Dansgaard-Oeschger events' during the last ice age (100 000–10 000 BC) over the course of which Greenland temperatures jumped by roughly 10 °C within a few decades subsequently followed by a gradual cooling on a millennium time scale [29,30]. Based on the analysis of ocean sediment cores these fluctuations are thought to be linked to abrupt changes in the deep-ocean circulation in the North Atlantic [31,32]. The observations are in qualitative agreement with numerical model simulations that associate the climate variations with temporal changes in the vigour of the AMOC [33,34]. The North Atlantic cold phases are generally thought to be linked with a very weak (or inactive) state of the AMOC that goes along with a near-cessation of NADW formation and northward heat transport in the North Atlantic. Warm phases are expected to coincide with a strong state of the AMOC. During the last ice age, the reduction in the NADW formation rates are likely to have arisen from events of massive input of freshwater from the Laurentide ice sheet (covering Canada) into the North Atlantic [35]. The subsequently fresher and thus less dense subpolar upper-ocean waters stabilised the vertical stratification of the water column, hence, heavily impeding deep-water formation.

To simulate the effect of freshwater input on deep water formation Vellinga and Wood [36] carried out a 'water hosing' experiment using a numerical model. They added a sufficient quantity of freshwater to the northern North Atlantic to cause the AMOC to switch off. This resulted in a strong cooling over the North Atlantic peaking at a temperature reduction of 8 °C around Greenland, standing out from patterns of moderate cooling over the entire Northern Hemisphere and warming over the Southern Hemisphere (Fig. 3). Thus, besides its strong importance for climate over the North Atlantic section the AMOC may also have a moderate impact on global climate patterns.

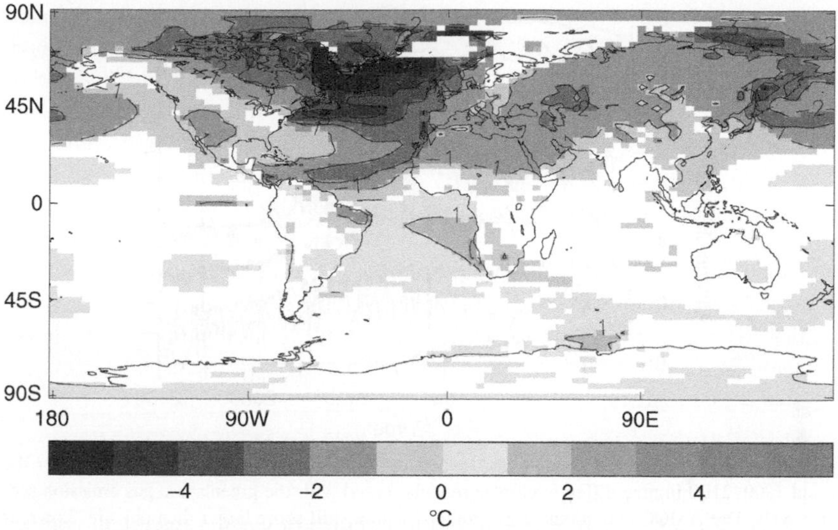

FIGURE 3 Change in surface air temperature during the years 1920–1930 after the collapse of the AMOC in a water hosing experiment using the HadCM3 climate model. Areas where the anomaly is not significant have been masked. This figure has been published by Vellinga et al. [36]. (See Color Plate 24).

Additionally, palaeoclimate records suggest that changes in the global circulation involving the AMOC during the early last deglacial period (19 000–14 500 a ago) went along with a significant net transfer of CO_2 from the ocean to the atmosphere, leading atmospheric CO_2 concentration to rise by about 50 ppmv [37].

5.2. Why should the AMOC Change as Part of Ongoing Climate Change?

In their fourth comprehensive climate assessment the Intergovernmental Panel of Climate Change (IPCC) considers it 'very likely' that the AMOC will have gradually slowed down by the end of the twenty-first century as a consequence of the Greenhouse climate [38]. Climate model projections – all of which are based on the greenhouse gas emission scenario A1B [38] – predict a reduction between 0 and 50% by the year 2100 [39], such that a complete (and possibly irreversible) AMOC shutdown is considered 'unlikely'. The future evolution of the AMOC in several selected climate model projections is shown in Fig. 4.

Future greenhouse gas emission scenarios carry a high level of uncertainty as they depend on parameters such as economic and population growth, technology development and basic political and social conditions, all of which are difficult to predict. Also, none of the present-day climate models have a sufficiently fine spatial resolution to resolve the processes that govern either the

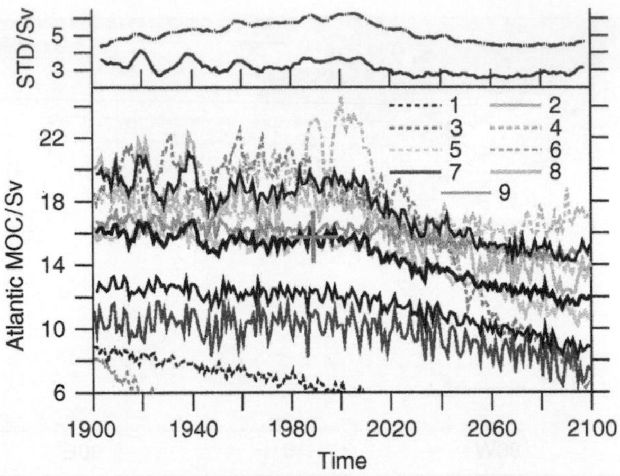

FIGURE 4 Evolution of the AMOC as defined by the maximum overturning at 24°N for the period 1900–2100 in nine different climate models forced with the greenhouse gas emission scenario A1B. The AMOC evolutions of integrations with a skill score larger than one are shown as solid lines, those from models with a smaller skill score as dashed lines [39]. The weighted ensemble mean is shown by the thick black curve together with the weighted standard deviations (thin black lines). This figure was published by Schmittner et al. [39]. (See Color Plate 25).

sinking or the rising, and have to rely on parameterisations instead. Both aspects may significantly add to the uncertainty in the prediction of the long-term AMOC evolution.

Reasons for a long-term greenhouse gas induced reduction of the strength of the AMOC include straightforward effects, such as warming of surface waters, melting of continental ice sheets acting to reduce high latitude salinity (a mechanism not included in many climate models), and intensification of the hydrological cycle [40]. All of these act to impede deep-water formation.

More complex feedbacks (that either stabilise or de-stabilise the AMOC) also involve wind field changes in the deep-water formation regions leading to buoyancy flux anomalies [41] and oceanic teleconnections driven by changes in the freshwater budget of the tropical Atlantic and South Atlantic [36,42–45].

Huang et al. [46] found a significant increase in wind stress and its energy input into the Southern Ocean between 1950 and 2000, which may have been caused by decreasing stratospheric ozone concentrations [47]. Using a climate model, Shindall and Schmidt [48] predicted the positive wind stress trend over the Southern Ocean, which will prevail until 2100 as a consequence of anthropogenic Greenhouse gas induced global warming. While oceanographers have not yet been able to establish a relationship between the multidecadal wind stress increase and changes in the ocean circulation – which may partly be a consequence of unavailability of a sufficient number of suitable observations over the last 50 years – one might speculate that in the future, beyond the end of this century, in a then different climate 'pulling', the AMOC in the Southern Ocean might gain in importance relative to 'pushing' it in the North Atlantic [49].

6. CAN WE DETECT CHANGES IN THE AMOC? IS THE AMOC CHANGING ALREADY?

Previously, direct estimates of the vigour of the AMOC have been obtained from transatlantic hydrographic (density profile) sections, assuming the geostrophic balance to hold on the ocean interior. Five such sections have been carried out along 24.5°N in the Atlantic over the last 50 years [50]. Taken together the five snapshots, each of which is assumed to be representative of the annual mean strength of the AMOC of the year in which they where taken (i.e. intra-seasonal variations assumed to be small), implies an AMOC slowdown of 30% (or 8 Sv) since 1957 [50]. Other measurements, focusing on single components of the AMOC gave rather inconclusive results regarding long-term AMOC changes. Using a combination of direct and indirect transport measurement techniques a gradual 1–2 Sv decrease in the amount of cold, dense inflow of deep waters from the Nordic Seas through the Faroe Bank Channel (feeding the NADW) has been found since 1970 [51], implying a long-term AMOC weakening. However, the continuation of the direct measurements showed an increase over the last few years back to the levels of the mid 1990s [52,53].

At the same time measurements in the Deep Labrador Current – which represents a major pathway for the export of NADW from the deep water formation regions – seems to have strengthened by 15% when comparing the 1996–1999 to 2000–2005 periods [54]. However, measurements in the DWBC further south off Grand Banks gave no significant change over roughly the same period [55]. It is uncertain how representative the strength of the DWBC off Grand Banks is, for the basin wide AMOC. Hydrographic measurements in the mid and high-latitude North Atlantic suggest that a substantial part of the southward export of NADW might be accomplished along a pathway in the ocean interior that feeds into the DWBC only in the subtropical North Atlantic [56]. Kanzow et al. [57] showed from observations and model simulations that fluctuations in the strength of the DWBC may not be a good indicator of AMOC changes in the tropical North Atlantic either, due to the presence of time-variable deep offshore recirculations.

A pilot system to measure the strength of the AMOC continuously at 26.5°N (i.e. the zonally integrated meridional transport profile between Florida and Morocco) has been operating since April 2004 [9,16,58]. Figure 5 shows a 1-year long time series of the AMOC between April 2004 and April 2005, exhibiting a time mean of 18.5 Sv and a rms variability of ±5.6 Sv [16]. The range of values the AMOC assumed within one year spans roughly 30 Sv (varying between 5 and 35 Sv). The observed intra-seasonal variability raises concerns

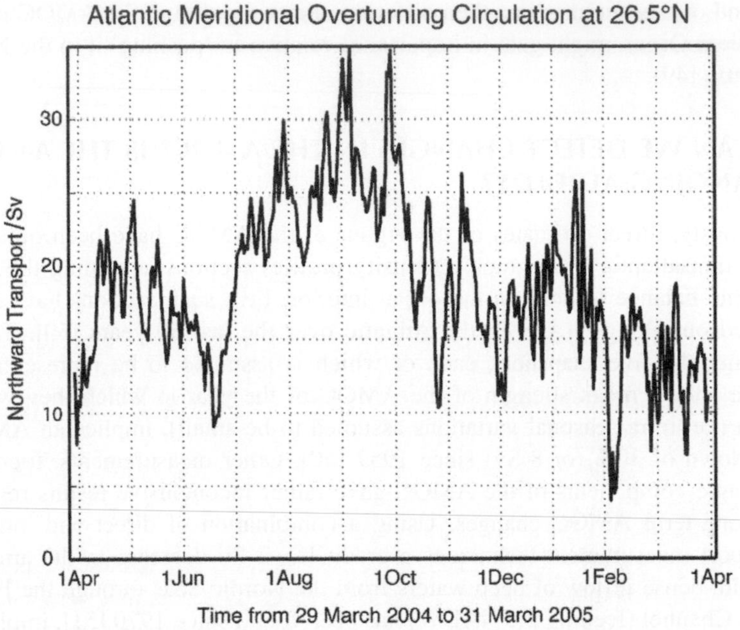

FIGURE 5 Time series of the strength of the Atlantic meridional overturning at 26.5°N, based on the continuous transport measurements within the RAPID/MOCHAexperiment [16], defined as the vertical integral of the transport per unit depth down to the deepest northward velocity (~1100 m) on each day. It represents the sum of the Florida Current, Ekman and upper mid-ocean transports [16]. This figure was published by Kanzow et al. [59].

whether the hypothesised 30% slowdown of the AMOC [50] may represent aliasing effects (as a consequence of not resolving the large intra-seasonal variations) rather than a sustained change of the ocean circulation [16,59].

While oceanographers have not yet been able to document a statistically significant trend in the strength of the AMOC, it is worth asking, how much time it would take to detect a possible long-term trend from continuous measurements at 26.5°N. Making assumptions about the short term noise level of the AMOC, Baehr et al. [60] concluded from the analysis of an AMOC future projection, that a 0.75 Sv per decade decline could be detected after three decades. A more abrupt (than currently expected) AMOC change would be detectable earlier. The detectability could most likely be shortened significantly if several continuously observing AMOC monitoring system were operated simultaneously at different latitudes.

7. CONCLUSION

Observations have revealed that patterns of present-day regional and large-scale ocean circulation may display strong changes on intra-seasonal to multi-decadal time scales. Physical oceanographers have developed a variety of tools to quantify circulation changes, which involve direct and indirect measurement techniques and numerical simulations. While most of the documented present-day circulation changes are believed to fall within the class of natural (ocean climate) variability even at decadal and longer time scales, it is a non-trivial task to disentangle climate variability from presently possibly ongoing climate shifts.

The AMOC has been in the focus of climate change research. The interpretation of palaeo-climate records in the light of findings from numerical climate models reveals that the AMOC has undergone large changes in the Earth's past and that these went along with climate shifts in the North Atlantic sector and beyond. In the present day climate, the AMOC represents the major oceanic mechanism of meridional heat transport. The AMOC moves volumes of cold waters (having sunk at high latitudes) southward throughout the Atlantic at depth and keeps them out of contact with the atmosphere for centuries, until the waters rise to the upper ocean eventually. Thereby the AMOC ventilates the deep ocean with oxygen rich waters. The sinking of waters in the Nordic Seas and the Labrador Sea (push) and their eventual rising (pull) are necessary ingredients for the existence of the AMOC, both of which are thought to change in a changing climate.

Model projections imply that the AMOC might slow down between 0 and 50% by the end of the twenty-first century. This is thought to be due to an increase in vertical density stratification at high latitudes (both due to warming and freshening of surface waters) as a result of global warming. However, none of the present-day climate models have a sufficiently fine spatial resolution to resolve the processes that govern either the sinking or the rising, and have to rely on parametrisations instead. Additionally, the climate model projections that produced the range of 0–50% in AMOC decline all rely on the same greenhouse gas forcing scenario, which will inevitably differ from the actual one. Thus, the true range of uncertainty of the future evolution of the AMOC is even larger. There is clearly a need to monitor the state of the AMOC continuously over coming decades.

To date there is no clear evidence that the AMOC has started to decrease in strength, partly because it has only very recently been a subject of continuous monitoring. Indeed a reliable time series of the strength of the AMOC spanning the last 50 year (or so) does not exist. The recent continuous measurements at 26.5°N suggest that the amplitude of intra-seasonal variations of the strength of AMOC is larger than previously thought. This makes it doubtful whether reliable estimates of long-term AMOC changes can be inferred from the few sporadic attempts to estimate the strength of the AMOC, that have been done in the past. Measurements that have focused on the observation of one particular aspect rather than the whole AMOC (such the strength of the DWBC or the deep Nordic Sea inflow into the Atlantic) do not show clear, uniform trends either. In addition it is difficult to assess how a change in one of the components translates into a change of the whole

AMOC. However, even with the recently started suitable continuous AMOC observations now well under way, the detection of a possible ongoing, global-warming-induced decline in the vigour of the AMOC may still be decades away, unless more observing systems are put into place. This will strongly depend on how fast the decline actually comes about, if at all.

The major difficulty for scientists to document ocean circulation changes on climate relevant time scales arises from the sparseness of historical in situ observations both in space and time. Over the last decades it has been (and partly still is) a technical, logistical and financial challenge to maintain ocean observatories at key locations continuously for more a few years and/or to repeat measurement campaigns at a frequency that is sufficient to detect trends with a high level of confidence. However, the awareness that understanding the processes that govern ocean circulation changes may be vital for present and future societies has triggered dedicated, internationally coordinated field programmes and along with them technical developments (such as autonomous in situ profilers or advances in remote sensing). As a consequence physical oceanography is currently undergoing a step change in capacity, capability and understanding, from which future generations will certainly profit.

REFERENCES

1. T.N. Lee, W.E. Johns, R. Zantopp, E.R. Fillenbaum, J. Phys. Oceanogr. 26 (1996) 962–963.
2. D. Zhang, M.J. McPhaden, Ocean Model. 15 (2006) 250–273.
3. S. Häkkinen, P. Rhines, Science 304 (2004) 555–559.
4. M. Bersch, I. Yashayaev, K.P. Koltermann, Ocean Dyn. 57 (2007) 223–235, doi:10.1007/s10236-007-0104-7.
5. J.W. Hurrell, Y. Kushnir, G. Ottersen, M. Visbeck, in: J.W. Hurrell, Y. Kushnir, G. Ottersen, M. Visbeck (Eds.), The North Atlantic Oscillation, Geophys. Monogr. Ser., vol. 134, American Geophysical Union, Washington, DC, 2003, pp. 1–36.
6. H. Hàtùn, A.B. Sandø, H. Drange, B. Hansen, H. Valdimarsson, Science 309 (2005) 1841–1844.
7. J.R. Knight, R.J. Allan, C.K. Folland, M. Vellinga, M.E. Mann, Geophys. Res. Lett. 32 (2005) L20708, doi:10.1029/2005GL024233.
8. J.W. Hurrell, M. Visbeck, A. Busalacchi, R.A. Clarke, T.L. Delworth, R.R. Dickson, W.E. Johns, K.P. Koltermann, Y. Kushnir, D. Marshall, C. Mauritzen, M.S. McCartney, A. Piola, C. Reason, G. Reverdin, F. Schott, R. Sutton, I. Wainer, D. Wright, J. Clim. 19 (2006) 5100–5121.
9. W.E. Johns, L.M. Beal, M.O. Baringer, J.R. Molina, S.A. Cunningham, T. Kanzow, D. Rayner, J. Phys. Oceanogr. 38 (2008) 605–623.
10. T.M. Joyce, D.S. Bitterman, K.E. Prada, Deep-Sea Res. 29 (1992) 903–913.
11. J. Fischer, M. Visbeck, J. Atm. Ocean Technol. 10 (1993) 764–773.
12. R. Davis, W. Zenk, in: G. Siedler, J. Church, J. Gould (Eds.), Ocean Circulation and Climate, Acadamic Press, New York, 2001.
13. V.W. Ekman,Arch. Math. Astron. Phys. 2 (1905) 1–53.

14. G. Wüst, Wissenschaftliche Ergebnisse der Deutschen Atlantischen Expedition. "Meteor," 1925–1927, vol. 6, 1935, pp. 109–288 (In English, W.J. Emer y (Ed.), The Stratosphere of the Atlantic Ocean, Amerind, New Delhi, 1978, 112 pp.).
15. H. Longworth, H.L. Bryden, in A. Schmittner, J.C.H. Chiang, S.R. Hemmings (Eds.), Geophys. Monogr. Ser., vol. 173, American Geophysical Union, Washington, DC, 2007.
16. S.A. Cunningham, T. Kanzow, D. Rayner, M.O. Barringer, W.E. Johns, J. Marotzke, H.R. Longworth, E.M. Grant, J.J.-M. Hirschi, L.M. Beal, C.S. Meinen, H.L. Bryden, Science 317 (2007) 935–938.
17. T. Kuhlbrodt, A. Griesel, M. Montoya, A. Levermann, M. Hofmann, S. Rahmstorf, Rev. Geophys. 45 (2007) RG2001, doi:10.1029/2004RG000166.
18. J.M. Lilly, P.B. Rhines, M. Visbeck, R. Davis, J.R.N. Lazier, F. Schott, D. Farmer, J. Phys. Oceanogr. 29 (1999) 2065–2098.
19. J. Marshall, F. Schott, Rev. Geophys. 37 (1999) 1–64.
20. C. Wunsch, R. Ferrari, Ann. Rev. Fluid Mech. 36 (2004) 281–314.
21. K.L. Polzin, J.M. Toole, J.R. Ledwell, R.W. Schmitt, Science 276 (1997) 93–96.
22. F. Paparella, W.R. Young, J. Fluid Mech. 466 (2002) 205–214.
23. J.R. Toggweiler, B. Samuels, J. Phys. Oceanogr. 28 (1998) 1832–1852.
24. A.H. Orsi, W.M. Smethie, J.L. Bullister, J. Geophys. Res. 107 (2002) 3122, doi:10.1029/2001JC000976.
25. G.C. Johnson, J. Geophys. Res. 113 (2008) C05027, doi:10.1029/2007JC004477.
26. G. Boccaletti, R. Ferrari, A. Adcroft, D. Ferreira, J. Marshall, Geophys. Res. Lett. 32 (2005) L10603, doi:10.1029/2005GL022474.
27. H.L. Bryden, S. Imawaki, in: G. Siedler, J. Church, J. Gould (Eds.), Ocean Circulation and Climate, Academic Press, San Diego, 2001, pp. 455–474.
28. J.L. Sarmiento, N. Gruber, Ocean Biogeochemical Dynamics, Princeton University Press, Princeton, 2006.
29. S.J. Johnson, H.B. Clause, W. Dansgaard, K. Fuhrer, N. Gundestrup, C.H. Hammer, P. Iversen, J. Jouzel, B. Stauffer, J.P. Steensen, Nature 311 (1992) 313.
30. W. Dansgaard et al., Nature 364 (1993) 218–220.
31. M. Sarnthein, K. Winn, S.J.A. Jung, J.-C. Duplessy, L. Labeyrie, H. Erlenkeuser, G. Ganssen, Paleoceanography 9 (1994) 209–267.
32. J.F. Adkins, E.A. Boyle, L. Keigwin, E. Cortijo, Nature 390 (1997) 154–155.
33. A. Ganopolski, S. Rahmsdorf, V. Petoukhov, M. Claussen, Nature 391 (1998) 351–356.
34. A. Ganopolski, S. Rahmsdorf, Nature 409 (2001) 153–158.
35. S. Hemming, Rev. Geophys. 42 (2004) RG1005, doi:10.1029/2003RG000128.
36. M. Vellinga, R.A. Wood, J.M. Gregory, J. Clim. 15 (2002) 764–780.
37. M. Sarnthein, P.M. Grootes, J.P. Kennett, M.-J. Nadeau, in: A. Schmittner, J. Chiang, S. Hemming (Eds.), Geophys. Monogr. Ser., vol. 173, American Geophysical Union, Washington, DC, 2007, pp. 175–196.
38. IPCC, Climate Change, The Scientific Basis: Contribution of Working Group I to the Fourth Assessment Report of the Intergovernmental Panel of Climate Change, Cambridge University Press, New York, 2007.
39. A. Schmittner, M. Latif, B. Schneider, Geophys. Res. Lett. 32 (2005) L23710, doi:10.1029/2005GL024368.
40. J.M. Gregory et al., Geophys. Res. Lett. 32 (2005) L12703, doi:10.1029/2005GL023209.
41. P. Gent, Geophys. Res. Lett. 28 (2001) 1023–1026.
42. M. Latif, E. Roeckner, U. Mikolajewicz, R. Voss, J. Clim. 13 (2000) 1809–1813.
43. R.B. Thorpe, et al., J. Clim. 14 (2001) 3102–3116.

44. A.X. Hu, G.A. Meehl, W.M. Washington, A. Dai, J. Clim. 17 (2004) 4267–4279.
45. U. Krebs, A. Timmermann, J. Clim. 20 (2007) 4940–4956.
46. R.X. Huang, W. Wang, L.L. Liu, Deep-Sea Res. II 53 (2006) 31–41.
47. D.W. Thompson, S. Solomon, Science 296 (2002) 895–899.
48. D.T. Shindell, G.A. Schmidt Geophys. Res. Lett. 31 (2004) L18209, doi:10.1029/2004GL020724.
49. M. Visbeck, Nature 447 (2007) 383.
50. H.L. Bryden, H.L. Longworth, S.A. Cunningham, Nature 438 (2005) 655–657.
51. B. Hansen, W.R. Turrell, S. Østerhus, Nature 411 (2001) 927–929.
52. B. Hansen, S. Østerhus, Prog. Oceanogr. 75 (2007) 871–865, doi:10.1016/j.pocean.2007.09.004.
53. S. Østerhus, T. Sherwin, D. Quadfasel, B. Hansen, in: R.R. Dickson, et al. (Eds.), Arctic-Subarctic Ocean Fluxes, Springer Science, Business Media B.V., Berlin, 2008, pp. 427–441.
54. M. Dengler, J. Fischer, F.A. Schott, R. Zantopp, Geophys. Res. Lett. 33 (2006) L21S06, doi:10.1029/2006GL026702.
55. F.A. Schott, J. Fischer, M. Dengler, R. Zantopp, Geophys. Res. Lett. 33 (2006) L21S07, doi:10.1029/2006GL026563.
56. S. Lozier, Science 277 (1997) 361–364, doi:10.1126/science.277.5324.361.
57. T. Kanzow, U. Send, M. McCartney, Deep-Sea Res. I 55 (2008) doi:10.1016/j.dsr.2008.07.011.
58. T. Kanzow, S.A. Cunningham, D. Rayner, J.J.-M. Hirschi, W.E. Johns, M.O. Baringer, H.L. Bryden, L.M. Beal, C.S. Meinen, J. Marotzke, Science 317 (2007) 938–941.
59. T. Kanzow, J.J.-M. Hirschi, C. Meinen, D. Rayner, S.A. Cunningham, J. Marotzke, W.E. Johns, H.L. Bryden, L.M. Beal, M.O. Baringer, J. Oper. Oceanogr. 1 (2008) 19–28.
60. J. Baehr, H. Haak, S. Alderson, S.A. Cunningham, J.H. Jungclaus, J. Marotzke, J. Clim. 20 (2007) 5827–5841.

Chapter 21

Ocean Acidification as an Indicator for Climate Change

Carol Turley and Helen S. Findlay
Plymouth Marine Laboratory, Prospect Place, The Hoe, Plymouth PL1 3DH, United Kingdom

1. **Introduction**
   1.1. Carbonate Chemistry
   1.2. Combined Impacts of Ocean Acidification and Climate Change
2. **Evidence from Observations**
   2.1. Evidence from Geological and Ice Core Records
   2.2. Evidence from Long-Term Oceanographic Time Series
   2.3. Evidence from Oceanographic Cruises
3. **Model Predictions of Future Change**
4. **Impacts**
   4.1. Past Observations
   4.2. Current Observations
   4.3. Experimental Observations
   4.4. Coral Ecosystems and Their Services
   4.5. Combined Impacts
5. **Biogeochemical Cycling and Feedback to Climate**
   5.1. Changes to the Ocean Carbon Cycle
   5.2. Changes to Ocean Nutrient Cycles
   5.3. Changes to Flux of Other Climate Reactive Gases from the Ocean
6. **Adaptation, Recovery and Mitigation**
   6.1. Adaptation
   6.2. Recovery
   6.3. Mitigation
7. **Conclusion**
References

1. INTRODUCTION

1.1. Carbonate Chemistry

Oceans have the capacity to absorb large amounts of Carbon dioxide (CO_2) because CO_2 dissolves and reacts in seawater to form bicarbonate (HCO_3^-) and protons (H^+). About a quarter to a third of the CO_2 emitted into the atmosphere from the burning of fossil fuels, cement manufacturing and land use changes has been absorbed by the oceans [1]. Over thousands of years, the changes in pH have been buffered by bases, such as carbonate ions (CO_3^{2-}). However, the rate at which CO_2 is currently being absorbed into the oceans is too rapid to be buffered sufficiently to prevent substantial changes in ocean pH and CO_3^{2-}. As a consequence, the relative seawater concentrations of CO_2, HCO_3^-, CO_3^{2-} and pH have been altered. Since pre-industrial times the oceans pH has decreased by a global average of 0.1 (compare Fig. 1a and b). The

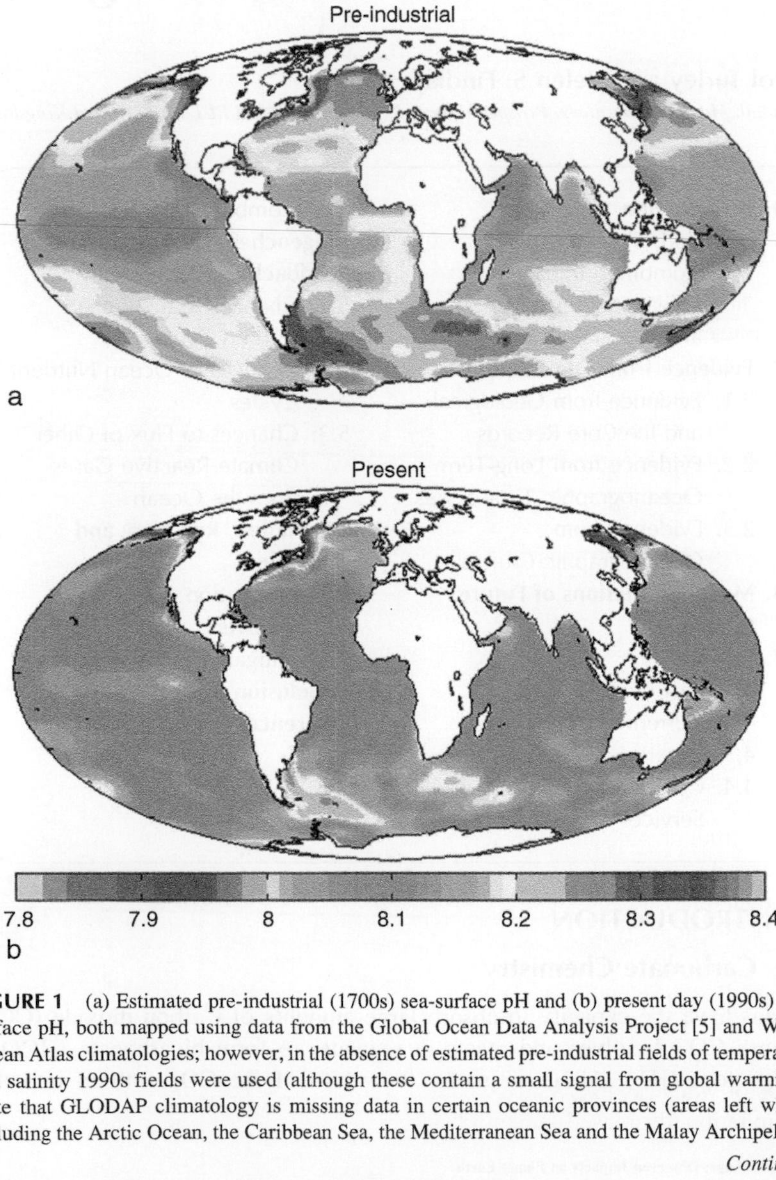

FIGURE 1 (a) Estimated pre-industrial (1700s) sea-surface pH and (b) present day (1990s) sea-surface pH, both mapped using data from the Global Ocean Data Analysis Project [5] and World Ocean Atlas climatologies; however, in the absence of estimated pre-industrial fields of temperature and salinity 1990s fields were used (although these contain a small signal from global warming). Note that GLODAP climatology is missing data in certain oceanic provinces (areas left white) including the Arctic Ocean, the Caribbean Sea, the Mediterranean Sea and the Malay Archipelago.

Continued

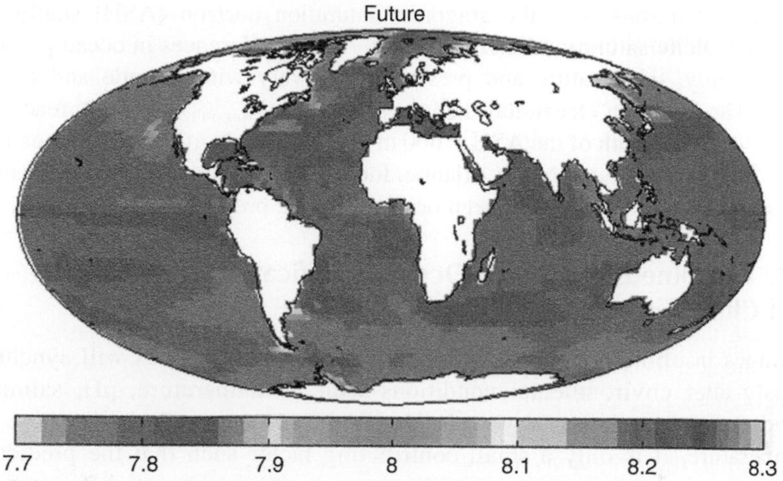

FIGURE 1 Cont'd (c) Predicted pH across the world's oceans for yr 2100 using the SOC model, which was part of the OCMIP-2 project [6] and used the IS92a CO_2 scenario. Note that the pH scale is different in (c). Courtesy of Andrew Yool (National Oceanography Centre, Southampton). (See Color Plate 26).

Intergovernmental Panel on Climate Change (IPCC) [2], using IS92 CO_2 emissions scenario, predicts that the pH of the surface ocean will decrease by as much as 0.4 by the year 2100 (Fig. 1c) and 0.77 by 2300 [3]. It will take tens of thousands of years for these changes in ocean chemistry to be buffered through neutralisation by calcium carbonate sediments and the level at which the ocean pH will eventually stabilise will be lower than it currently is [4].

The CO_3^{2-} concentration directly influences the saturation, and consequently the rate of dissolution, of calcium carbonate ($CaCO_3$) minerals in the ocean. The saturation state (Ω) is used to express the degree of $CaCO_3$ saturation in seawater:

$$\Omega = \left[Ca^{2+}\right]\left[CO_3^{2-}\right]/K_{sp}^*$$

where K_{sp}^* is the solubility product for $CaCO_3$ and $[Ca^{2+}]$ and $[CO_3^{2-}]$ are the *in situ* calcium and carbonate concentrations, respectively. When $\Omega > 1$, seawater is super-saturated with respect to mineral $CaCO_3$ and the larger this value the more suitable the environment will be for organisms that produce $CaCO_3$ (shells, liths and skeletons). When $\Omega < 1$, seawater is under-saturated and corrosive to $CaCO_3$. Currently, the vast majority of the surface ocean is super-saturated with respect to $CaCO_3$. The depth at which $\Omega = 1$ is known as the saturation horizon. The three main mineral forms of $CaCO_3$, in order of least soluble to most soluble, are calcite, aragonite and magnesium-calcite. Therefore, each mineral form has different saturation state profiles and saturation horizons with the aragonite saturation horizon (ASH) shallower than the calcite saturation horizon (CSH). Due to differences in ocean properties (salinity, temperature and pressure) both vary with latitude and ocean basin. The Southern Ocean has the lowest Ω, with $\Omega_{aragonite}$ currently reaching below 1.5. The depth of the ASH is 600 m or less in the North Pacific but can be over 2000 m deep in the North Atlantic. Increasing atmospheric CO_2 will cause Ω to decrease, as has already been occurring since pre-industrial times [6].

1.2. Combined Impacts of Ocean Acidification and Climate Change

Changes in climate resulting from anthropogenic influences will synchronously alter environmental conditions such as temperature, pH, salinity, wind strength and oxygen levels [7]. While seawater pH is sensitive to temperature, it is only a small contributing factor such that the predicted range for future temperatures will not make a significant difference to the pH decline [8]. However, organisms' responses may be different with increasing temperature depending on the level at which they adapted [9]. pH is also sensitive to changes in salinity, as a result of changes in total alkalinity and dissolved inorganic carbon, so organisms in coastal waters with riverine input, can experience larger variability in pH than in open oceans [10]. Both increasing temperature and decreasing salinity will also act to increase ocean stratification, which in turn will alter the nutrient supply that fuels primary production. A change in wind strength is also an important consideration for ocean acidification for two reasons. Firstly, wind strength determines the flux of CO_2 between the ocean and the atmosphere, so may reduce the ocean CO_2 sinks [11]. Secondly, wind strength drives ocean currents, mixes nutrients into the productive upper ocean and is particularly important for generating upwelling areas [12]. Upwelling areas, although rich in nutrients, are also rich in CO_2 and are therefore areas of natural low pH [13]. A reduction in oxygen (O_2), hypoxia, within the oceans occurs largely as a result of increased nutrient or organic matter input (e.g. caused by increased land-run off). An increase in nutrient load can substantially increase biological productivity and subsequent microbial decomposition of this excess productivity consumes large amounts of O_2 and releases CO_2 through respiration, causing hypoxia and low pH.

2. EVIDENCE FROM OBSERVATIONS

2.1. Evidence from Geological and Ice Core Records

Ice cores provide high resolution and accurate records of atmospheric CO_2 concentrations over the last 650 000 a and together with marine paleo-proxies (e.g. boron isotopes) serve to arrive at a reasonable estimate of ocean carbonate chemistry over millions of years [14].

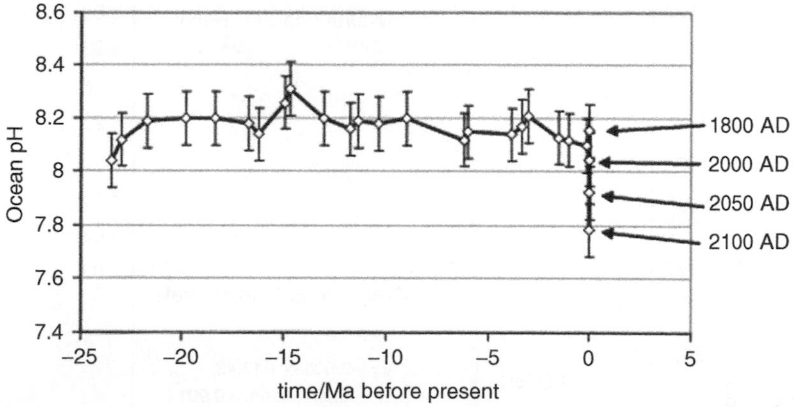

FIGURE 2 Past (white diamonds, data from Pearson and Palmer [14]) and contemporary variability of surface ocean pH (diamonds with dates). Future predictions are model-derived values based on IPCC mean scenarios. Adapted from Turley et al. [15].

Such measurements indicate no major undersaturation of the surface ocean for at least the last 65 Ma and that the current rate and magnitude of CO_2 induced chemical change occurring in the surface ocean are unprecedented for at least the past 25 Ma (Fig. 2).

Observations that CO_2 variations in the glacial and inter-glacial periods of the last 50 000 a correlated with the shell weights of fossil planktonic foraminifers [16] indicate that marine calcifiers are influenced by small fluctuations in atmospheric CO_2 values and those effects are likely to progressively intensify with increasing CO_2.

2.2. Evidence from Long-Term Oceanographic Time Series

The Pacific time-series station, off Hawaii (Hawaii Ocean Time-Series, HOTS), shows an increase in seawater CO_2 concurrent with the increase in atmospheric CO_2 recorded at Mauna Loa. The resultant decrease in surface ocean pH is $0.0019 \pm 0.00025\ a^{-1}$ (Fig. 3a) [17]. Aragonite and calcite saturation states also both show a decline over the last 20 a (Fig. 3b and c).

The other two major time series stations, the Bermuda Atlantic Time-Series (BATS) and the European Station for Time-Series in the Ocean at the Canary Islands (ESTOC), located either side of the North Atlantic, show a decrease in the seawater pH of around $0.0012 \pm 0.0006\ a^{-1}$ at BATS, and $0.0017 \pm 0.0004\ a^{-1}$ at ESTOC due to increased uptake of CO_2 [18]. These time-series data show that the Pacific and the subtropical gyre at both sites on the North Atlantic are becoming more acidic as predicted by ocean general-circulation models (OGCMs) (see below).

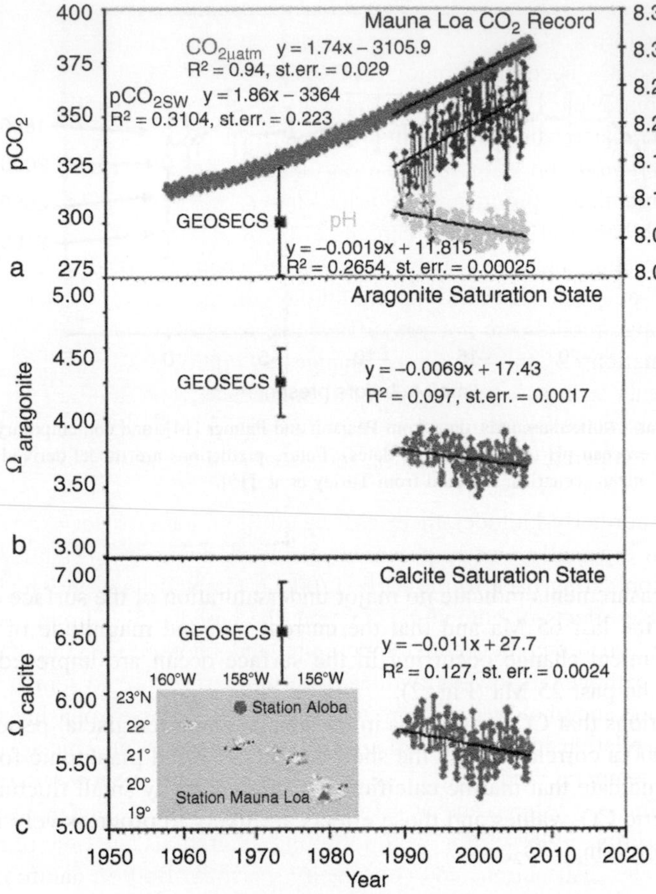

FIGURE 3 (a) The Mauna Loa records of atmospheric CO_2 over the last 50 a with the pCO_{2SW} and surface ocean pH recorder during the last two decades from the Hawaii Ocean Time-Series (HOTS) and the resultant changes to (b) Agaronite saturation state and (c) calcite saturation state over the same period. From Doney et al. [17]. (See Color Plate 27).

2.3. Evidence from Oceanographic Cruises

Sabine and colleagues [1] used inorganic carbon measurements from an international survey effort in the 1990s, consisting of 9618 hydrographic stations collected on 95 cruises in different oceans (pH data is mapped in Fig. 1b). They estimated a global oceanic anthropogenic CO_2 sink for the period from 1800 to 1994 of 118 ± 19 Pg of carbon, accounting for about 48% of the total fossil-fuel and cement-manufacturing emissions.

A hydrographic survey along the western coast of North America, from central Canada to northern Mexico, revealed upwelling of seawater undersaturated with respect to aragonite and with low pH (<7.75) onto large portions

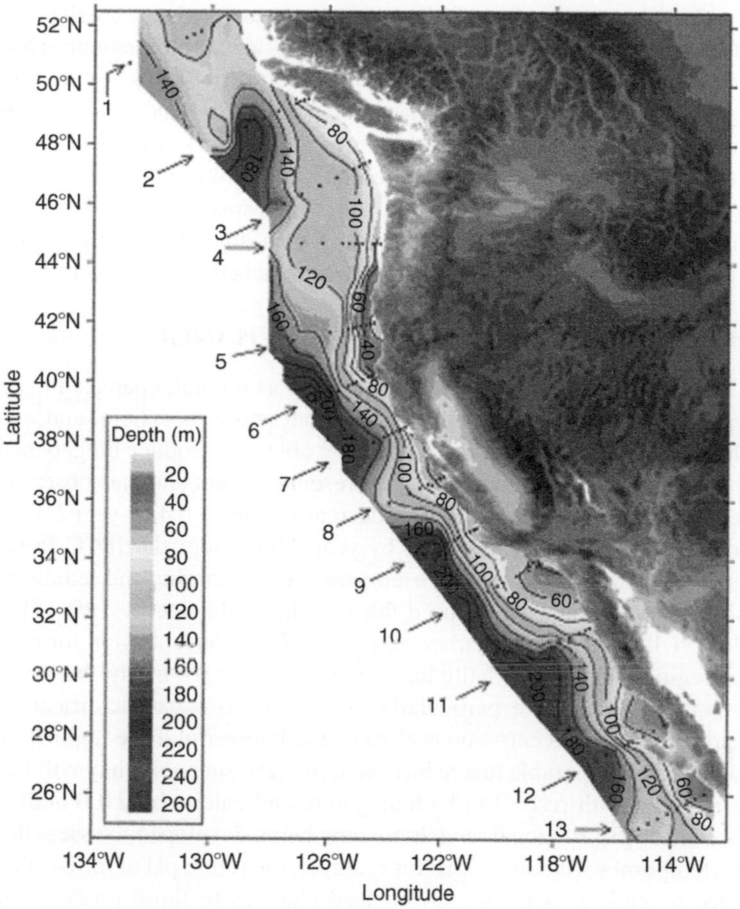

FIGURE 4 Distribution of the depths of the undersaturated water (aragonite saturation < 1.0; pH < 7.75) on the continental shelf of western North America from Queen Charlotte Sound, Canada to San Gregorio Baja California Sur, Mexico. On transect line 5 the corrosive water reaches all the way to the surface in the inshore waters near the coast. The black dots represent station locations. From Feely et al. [13]. (See Color Plate 28).

of the continental shelf [13] (Fig. 4). The areal extent of this natural phenomenon has been increased by the ocean uptake of anthropogenic CO_2. They estimated that during pre-industrial days the ASH would have been about 50 m deeper with no undersaturated waters reaching the surface. With the additional anthropogenic CO_2 signal the ASH has shoaled by around 1 $m \cdot a^{-1}$ bringing increasingly corrosive conditions with pH as low as 7.6 not just to the deeper benthic communities but also, increasingly, to the productive, shallower continental shelf ecosystems. How these ecosystems respond to this seasonal inflow of undersaturated waters from February to August, during the growing season, has not yet been reported but these coastal ecosystems may well represent the first shallow sea ecosystems that experience rapid and nonlinear undersaturation due to uptake of anthropogenic CO_2. In CO_2-rich vent waters in the Mediterranean, useful 'hot-spots' of waters with a range of pH that offer a useful natural laboratory to study the response of marine organisms to long term exposure to reduced pH [19]. A seawater pH of around 7.8 seemed a critical threshold to the growth and survival of many of the local calcifiers and should organisms on the west coast of North America exhibit similar vulnerability then an ocean acidification 'tipping point' may well have been or soon be reached in these waters.

3. MODEL PREDICTIONS OF FUTURE CHANGE

OGCMs have been used to reconstruct, as well as predict, changes in climate. Forced by the physical dynamics of the ocean and atmosphere, and coupled together with biological models, OGCMs are able to reproduce biogeochemical cycling within the oceans that closely represents present and past observations. These models predict a global average decrease in pH of 0.4 by year 2100 in the surface ocean (Fig. 1c) and of 0.77 by year 2300 under the IPCC IS92 CO_2 scenario [3]. Introducing changes in temperature, weathering and sedimentation into these simulations only reduced this maximum decline by 10% [3]. More detailed predictions of both carbonate ion and CO_2 concentration for different oceans regions and across latitudinal gradients strongly imply that the polar and sub-polar oceans are particularly vulnerable to ocean acidification [6]. The carbonate ion concentration is already much lower in these regions so they are particularly vulnerable to a reduction in pH [20] such that they will become undersaturated with respect to both aragonite and calcite by 2100 under IS92 CO_2 scenarios [6]. Regional models are now being developed to assess the spatial and temporal variability in pH; for example, the future pH of the North Sea is predicted to undergo similar CO_2-induced changes to those predicted in the open oceans although coastal and shelf sea pelagic and benthic activities and riverine input are important factors in contributing to a greater variability [10]. Continued uptake of CO_2 by the oceans is predicted to cause some areas in the ocean to be completely outside their natural ranges by the year 2050 [10].

4. IMPACTS

4.1. Past Observations

There have been several ocean acidification events in Earth's deep past, caused by massive input of carbon into the ocean, the best studied and most prominent of which occurred 55 Ma ago. The subsequent acidification led to the largest extinction of microscopic sea bed dwelling calcifiers [21]. In parts of the ocean, red clay instead of white carbonate was deposited for more than 100 000 a indicating the magnitude of the dissolution effect on global biogeochemical cycles and the duration of the recovery.

4.2. Current Observations

To date, there are limited observations of current changes in ocean biology as a result of ocean acidification. This is in part a result of a lack of chemical data with which observations can be correlated and a lack of research in this emerging area, but may also be an artefact of organisms' ability to cope with short-term variability in pH. Seawater pH in coastal and shelf sea water columns can fluctuate by up to 0.9 depending on time, season, position in water column and fresh water influence [10,22]. That is, there may be only short periods of low pH with the periods of high pH allowing the organisms to recover. Impacts may not become apparent until they are subjected to longer periods of lower pH or the whole pH range that they experience is reduced. In contrast, seasonal pH variation of open ocean surface waters is around 0.07 (Fig. 3a) which may make these regions more sensitive to current and future acidification. Indeed, the detected change in pH (−0.1) since the pre-industrial already exceeds the open ocean seasonal variation. Observed changes, for example, in species distribution which have been attributed to changes in climate, pollution, ecosystem detioration and so on, may have masked the role of ocean acidification. Further work at the international level needs to be carried out to explore current and future impacts of ocean acidification.

Observed differences in the cold water coral ecosystems between the North Atlantic and the North Pacific may be indicative of biological responses to changes in ocean chemistry. Cold-water corals in the North Pacific are found living close to the ASH, as it is much shallower than in the North Atlantic; however, they do not flourish or form large structures, such as are found in the North Atlantic [23]. Only 10% of all known stylasterid corals produce calcite instead of aragonite, yet in the North Pacific six out of seven stylasterid corals used calcite to form their spicules and skeletons [24]. Near the ASH, it may be less costly for these corals to produce calcite thereby reducing the affects of dissolution.

Coral on the Great Barrier Reef (GBR), Australia, have shown a 21% decrease in net calcification and 30% decrease in growth over the period 1988–2003 [25]. Sea surface temperature does not appear correlated to this decline, as might be expected if increasing temperature was causing bleaching events or decreasing health of the corals. The change in carbonate chemistry observed in our oceans (Fig. 1a and b) could be impacting the growth and net calcification of corals, but as yet there is no chemical data directly from the GBR to confirm this. However, reefs in the Red Sea have shown correlated responses in net calcification rate to natural fluctuations in Ω and temperature [26], providing observational evidence of a response of corals to changes to today's carbonate conditions.

Spatial variation in sea bed organisms has been observed across a large pH range at natural marine volcanic CO_2 vent sites. A number of key ecosystem changes are apparent, for example, calcareous algae were replaced by non-calcareous algae and sea-grasses with the latter increasing their primary production. There was a large reduction in biodiversity, particularly a loss of calcifying organisms at low pH levels. A number of taxa appear to be more susceptible to acidification impacts than others, for example, echinoderms (particularly sea urchins) did not appear below pH 7.6, whereas molluscs (limpets) and crustaceans (barnacles) were present until pH 6.5 [19].

Coccolithophores, microscopic plants that secrete $CaCO_3$ platelets called liths, occur over a variety of environmental conditions throughout the worlds oceans yet they are excluded from certain locations, for example, the Baltic Sea. Areas known to have an extremely large seasonal cycle of calcite saturation states, with wintertime values declining to ≤ 1, appear to be areas where coccolithophores are absent [27] implying that the saturation state may have a large influence on their distribution, although low salinity or differences in the magnitude of the spring bloom will also contribute [27,28].

4.3. Experimental Observations

Experimental approaches have, so far, been carried out in controlled laboratory experiments on single organisms or in larger volume sediment or seawater mesocosms enriched with CO_2 containing mixed populations. There are many important biological processes within the lifecycle of an organism or even more so in an ecosystem. Therefore an impact on a process, be it at the cellular level or ecosystem level, may have a negative impact on the ultimate successful functioning of the ecosystem. An experimental approach is a key tool in determining the weak links in these processes.

4.3.1. Primary Production

The Royal Society [29] concluded that unlike land plants, most marine phytoplankton are thought to have mechanisms to actively concentrate CO_2 so that changes in seawater pH and CO_2 have little (<10%) if any direct effect on their growth rate or their elemental composition. However, whilst taxon specific differences in CO_2 sensitivity have been observed in laboratory culture [30] it is currently unknown whether a reduction of the advantage of possessing a CO_2 concentrating mechanism will impact phytoplankton species diversity in the natural environment. This is a possibility and, should it occur, may impact the contribution of different functional groups, primary production, food web structure and marine biogeochemical cycles. The coccolithophore *Emiliania huxleyi* seem to be an exception to this generalisation, having low affinity for inorganic carbon such that it could be carbon limited in today's ocean, with increasing CO_2 resulting in increased productivity [31].

4.3.2. Calcification

Although there is variability amongst experiments, with some studies showing no change or even increased calcification [49,50], most calcifying species studied to date, representing the major marine calcifying groups (coccolithophores, pteropods, foraminifera, corals, calcareous macroalgae, mussels, oysters, echinoderms and crustacean), show reduced net calcification rates in response to elevated CO_2 [reviewed in Refs. 32-25]. For example, a mean decrease of 16% (double pre-industrial CO_2 concentration (2× CO_2)) and 20% (triple pre-industrial CO_2 concentration (3× CO_2)) for coccolithophores; 6% (2× CO_2) and 9% (3×CO_2) for foraminifera; 24% (2× CO_2) and 41% (3× CO_2) for Scleractinian corals; 25% (2× CO_2) for coralline red algae; 25% (2× CO_2) and 37% (3× CO_2) for mussels; and 10% (2× CO_2) and 15% (3× CO_2) for oysters.

This variability between major groups of organisms may result primarily from the different mechanisms used to carry out calcification. Coccolithophores, for example, carry out calcification in an intracellular compartment which may be buffered against external changes by their own homeostatic mechanisms. Foraminifera and corals carry out calcification in an enclosed yet extracellular space, relying on membrane transporters to regulate conditions. In more complex multicellular organisms, such as crustaceans and molluscs, metabolic energy balance as well as whole animal acid-base regulation (see below) may be more important in determining the responses of calcification to decreased seawater pH. To maintain a calcified structure, when exposed to a more acidic environment for a short time, an organism may have to divert energy from other metabolic processes in order to compensate for dissolution. Other metabolic processes may also be impacted by CO_2 so this compensation may not always be possible over longer time periods. Evidence strongly indicates that dissolution rates will, over the timescale associated with ocean acidification, become greater than the rate at which organisms can grow and calcify, resulting in an inevitable reduction in biogenic $CaCO_3$.

4.3.3. Acid–Base Regulation and Internal Physiology

Much is known about the short-term effects of very high concentrations of CO_2 (higher than we will see due to ocean acidification) on respiration and acid–base balance in marine invertebrates and fish [9]. These early experiments were important in the discovery that CO_2 in seawater readily diffuses across animal surfaces, lowering the pH of internal fluids and that many animals have developed compensation mechanisms to regulate their internal pH. We now know that for normal function of an organism, internal pH must be kept within relatively narrow ranges because processes such as enzyme function, protein phosphorylation, chemical reactions and the carrying capacity of haemoglobin for O_2 are all influenced by pH and that these can be regulated for short periods of exposure to high CO_2. Evidence so far indicates that fish are tolerant to these short-term high CO_2 exposures but organisms such as squid, may be more vulnerable (reviewed in [9,33]). However, we do not yet know the impact of long term exposure to the relatively lower levels of CO_2 they will experience in the future from ocean acidification.

4.3.4. Fertilisation, Embryo Development, Larval Development and Settlement

Physiological impacts induced by lowered pH have the potential to affect an animal at any stage in its life cycle; however, adults tend to have more protection as well as better mechanisms to deal with a fluctuating environment with early life stages tending to be more vulnerable. Many benthic marine invertebrates produce free-swimming larvae, which spend time developing through several larval stages in the plankton before settling into the adult form (Fig. 5). Large numbers of larvae are often produced because of high rates of mortality, for example, coastal estuarine bivalves experience more than 98% mortality during settlement [36]. Oyster [37], echinoderm [38,39] and fish larvae [40] as well as barnacle, tube worm and copepod eggs [41, personal observation] have all been found to either be increasingly malformed or have slower rates of development at high CO_2. Barnacle settlement has also been affected [42].

Assuming that some larvae are still viable and go on to settle on the shore, delayed development could leave juveniles susceptible to additional stresses such as wave impact and temperature and salinity variations. In addition, if they settle later, they may miss their survival window.

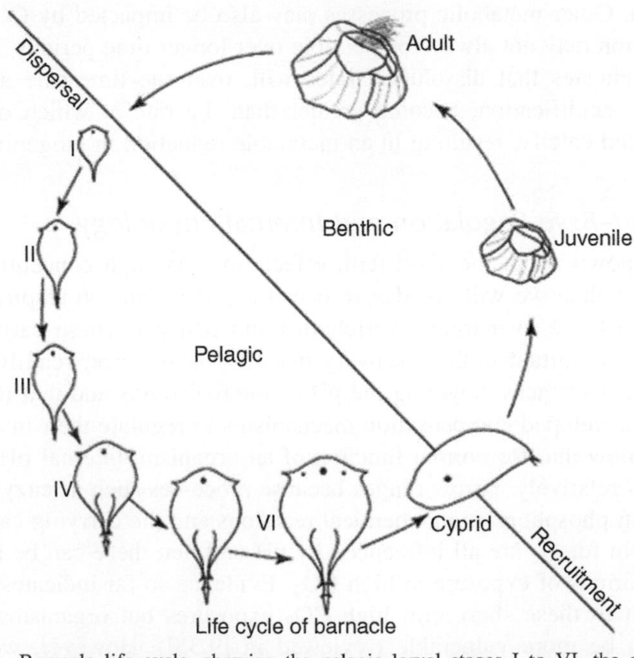

FIGURE 5 Barnacle life cycle, showing the pelagic larval stages I to VI, the cyprid larvae settling to become a benthic juvenile and finally an adult. From Desai and Anil [43].

4.3.5. Communication

Chemical cues are used for marine communication and can have strong influences on habitat selection and predator–prey interactions as well as courtship and mating, species recognition, and symbiotic relationships [44]. Some of these cues are known to be susceptible to changes in pH during formation and detection or within the seawater itself. Settlement of oyster larvae can be induced or inhibited by the presence of weak bases or acids, respectively, possibly as a mechanism for suitable habitat selection (e.g. [45–47]). Weakly acidic environments also impaired the ability of juvenile salmon to detect and respond to alarm cues [48]. The normal response to predators by littorinid snails on rocky shores is to thicken their shells. However, under CO_2-induced acidification the snails switched from thickening shells in the presence of predators to increased avoidance behaviour [49].

4.3.6. Interactions

The responses that occur within one individual can lead to changes in how it interacts with others and its environment. For example, burrowing brittlestars were found to have increased muscle wastage in their arms as compensation for increasing their calcified material under low pH conditions [50]. These brittlestars are important prey for commercial fish and aid nutrient and oxygen cycling between the sediment and the overlying water [51]. Reduced muscle may lead to reduced ability to feed themselves, lower quality of food for predators and reduce nutrient flow.

The importance of microbial and viral activity in the oceans is becoming ever more apparent; recent experiments show that bleaching of corals can be induced by increased viral activity [52], however, some evidence suggests that viral activity may decrease with increasing CO_2 [53]. The ability of a host to have an immune response to viral attack is critical for its health. Preliminary evidence suggests that at lowered pH mussels are unable to induce normal immune responses [54].

Organisms that occupy the same ecosystem space or function, may be outcompeted if they are less suited to surviving in a high CO_2 ocean. This may lead to an overall loss of biodiversity or even regime shift but may not lead to a complete breakdown in all functions of the ecosystem [19]. Changes to populations and their interactions within communities could well influence the relative composition, productivity, timing, location and predominance of the major functional groups and thereby impact the rest of the food web.

4.4. Coral Ecosystems and Their Services

Corals are the most studied organisms in relation to impacts of ocean acidification. Should these impacts occur in the natural environment they will have a large impact on the ecosystems they support. Corals are, therefore, a useful example of how ecosystems and their services may be impacted in the future.

4.4.1. Tropical Coral Reef Ecosystems

Tropical corals have adapted, over millions of years, to live in warm, sunlit waters highly saturated in aragonite. They are among the most diverse marine ecosystems, supporting about a quarter of all marine biodiversity. They are very important in local shore protection, important to tourism, and supply a critical level of subsistence protein as well as providing an income source in the developing world through fishing and tourism [7]. Unabated CO_2 emissions will result in suboptimal aragonite saturation states for coral growth by 2070 such that many reefs could be threatened resulting in reduced coral cover [6,32,55–57]. At this time erosion will outpace calcification so that reef structures will not be able to withstand the waves nor rebuild sufficiently after a storm.

Indeed, coral reefs in the waters off Panama and Galapagos, which live in a naturally more acidic and high CO_2 environment, suffer some of the highest erosion rates measured. They contain extremely low percentages of interskeletal pore cement to hold them in place compared to the coral reefs off the Bahamas that live in waters with less CO_2 and higher pH [58]. These reefs may be a vision into the future of reefs worldwide, since the Panama and Galapagos environments replicate the expected increased in acidity and CO_2.

4.4.2. Cold-Water Corals

Scleractinian cold-water corals, often referred to as deep-water or deep-sea corals, are long-lived (hundreds of years old), are found around 200–1000+ m depth throughout the worlds' oceans and can form large (100 km^2) reef frameworks that persist for millennia. They are biodiversity hotspots and play an important role as a refuge, feeding ground and nursery for deep-sea organisms, including commercial fish [23,59,60]. However, they may be the most vulnerable marine ecosystems to ocean acidification [23,61]. Future projections of global aragonite saturation state indicate that 70% of cold-water corals are likely to experience undersaturation this century through the shoaling of the ASH and in some instances this could be as early as 2020 [6,23,61]. It would seem unlikely that scleractinian cold-water corals would be able to calcify under these conditions; it would be more likely that aragonitic structures would experience dissolution in these corrosive waters. As yet there have been no experiments on their reaction at high CO_2 but if they respond in the same way as their warm-water cousins their calcification rates may decrease well before aragonite under-saturation occurs.

4.5. Combined Impacts

Temperature already provides limits to the survival of organisms; it alters many physiological processes by acting on the rates at which these processes occur (e.g. speeding up metabolism, enzyme activity, etc.). However, organisms are acclimatised to a certain temperature range. Acidification may act to narrow these ranges [9]. Increasing temperature will also drive many species polewards, either as a result of biogeographic range expansion (by temperate and tropical species) or as a result of contraction (by boreal and polar species). However, ocean acidification may act in the opposite direction, as the polar waters will be most affected by increasing CO_2 [6]. This could lead to a complete disappearance of boreal and polar species and may restrict the ability of temperate and tropical species to migrate.

Available oxygen is also a significant factor in controlling the distribution of organisms in marine environments. Eutrophication events and warming of waters decreases the oxygen content causing hypoxia. As mentioned previously, hypoxia is nearly always accompanied by an elevation of CO_2 (and thus a decrease in pH) and will compound the impacts [62].

Corals are again a good example of the effects of multiple stresses. They are affected by both ocean acidification and by warming of ocean surface waters leading to declining calcification and increase in bleaching [7,63]. Other climate change factors (sea-level rise, storm impact, aerosols, ultra-violet irradiation) and non-climate factors (over-fishing, invasion of non-native species, pollution, disease, nutrient and sediment load) add multiple impacts on coral reefs, increasing their vulnerability and reducing their resilience [7,32,63–65]. A recent report shows that about half of the coral reef ecosystem resources within the United States and Pacific Freely Associated States jurisdiction are considered by scientists to be in 'poor' or 'fair' condition and have declined over time due to several natural and anthropogenic threats [66]. Another consensus of opinion is that one-third of reef-building corals face elevated risk of extinction from climate change and local impacts and that the loss of reef ecosystems would lead to large-scale loss of global biodiversity [67].

5. BIOGEOCHEMICAL CYCLING AND FEEDBACK TO CLIMATE

5.1. Changes to the Ocean Carbon Cycle

Over several thousands of years, around 90% of the anthropogenic CO_2 emissions will end up in the ocean [4]. Because of the slow mixing time of the ocean the current oceanic uptake fraction is only about one-third of this value [1], without which atmospheric CO_2 would be about 55 ppm higher today than what is currently observed (385 ppm).

The Southern Ocean is estimated to account for around 25% of the anthropogenic CO_2 taken up by all the oceans while the North Atlantic is estimated to account for 40% [1]. Unlike the Southern Ocean which has a strong biological pump, the North Atlantic CO_2 sink is thought to be mainly due to the physical pump, with the 'biological pump' contributing only around 10% [68]. As the surface ocean CO_2 concentrations continue to increase the ocean's ability to absorb more CO_2 from the atmosphere will slow down.

Whilst there were indications that this might be occurring in the analysis of 1990s oceanographic cruises by Sabine et al. [1], more recent analysis of CO_2 in the NE Atlantic [69] and Southern Ocean [11] show a decrease in CO_2 uptake over the last 1–2 decades. Whether this decrease in the efficiency of the ocean sink for anthropogenic CO_2 is decadal variation awaits further long time series study. If the ocean CO_2 sink is becoming less efficient then more CO_2 will remain in the atmosphere exacerbating global warming.

The 'biological pump' removes carbon from surface waters to the deep ocean via the organic or 'soft' tissue pump (which decreases CO_2 of surface water, increasing its ability to absorb atmospheric CO_2) while the inorganic or 'hard' $CaCO_3$ pump increases CO_2 of the surface water and decreases its ability to absorb atmospheric CO_2. Decreasing calcification and $CaCO_3$ export rates could therefore play a direct role in ameliorating future global change. However, decreasing primary production and export rates (the soft tissue pump) would have the opposite effect, resulting in less atmospheric CO_2 draw down by surface waters. To add to the complexity of these key mechanisms in the carbon cycle, there may be strong association between 'soft' and 'hard' pumps with a 'ballasting' of organic matter by carbonate particles, making the organic matter sink faster than it would on its own. A decrease in $CaCO_3$ production [70] would then lead to a reduction in the efficiency with which organic matter is transported to depth, weakening the biological pump and resulting in higher surface ocean CO_2. This would reduce fossil fuel CO_2 uptake by the ocean and exacerbate future climate change [71]. Although we have a poor understanding of the importance of these two mechanisms experiments looking at the calcification and primary production of coccolithophores in 27 m^3 seawater enclosures (mesocosms) found a shift in the ratio of organic carbon to calcium carbonate production and vertical flux with rising atmospheric CO_2 [72].

5.2. Changes to Ocean Nutrient Cycles

Another experiment maintaining natural plankton communities in mesocosms at 1× pre-industrial CO_2, 2× pre-industrial CO_2 and 3× pre-industrial CO_2 showed that primary productivity increased by as much as 39% under high CO_2 while nutrient uptake remained the same. This excess carbon consumption was associated with a more efficient biological pump and increasing C:N ratios [73]. If these findings were transferrable to the natural environment this could lead to an expansion of deep ocean oxygen minimum zones. Increasing C:N ratios would also lower the nutritional value of organic matter produced by primary producers thereby having further implications for marine ecosystem dynamics.

Nutrients such as nitrogen, phosphorus and iron often limit phytoplankton growth in major parts of the worlds' oceans. The lower pH expected over the next hundred years can theoretically impact the speciation of many elements

[15,29,74]. These include biologically important nutrients (nitrogen, phosphate and silica) and micronutrients (iron, cobalt, manganese, etc.). For instance, a decrease in pH of 0.3 could reduce the fraction of NH_3 by around 50% [75]. In addition, the key process of nitrification is sensitive to pH with rates reduced by ~50% at pH 7 [76]. This may result in a reduction of ammonia oxidation rates and the accumulation of ammonia instead of nitrate. Using this data to parameterise a shelf sea ecosystem model about a 20% decrease in pelagic nitrification by 2100 was predicted [10]. *Trichodesmium* cyanobacteria play a key role in sustaining primary production in the large low nutrient areas of the worlds' oceans through nitrogen fixation and show a >35% increase in rates of nitrogen fixation under elevated CO_2 of 750 ppm [77]. In addition, the proportion of soluble iron may increase which might be beneficial to the 10% of the oceans where iron is thought to limit primary production.

Depending on their nutrient requirements and uptake abilities, primary producers may respond differently to the effects of ocean acidification and nutrient speciation. Each response has the potential to impact the biodiversity and nutritional value of phytoplankton and the food webs and biogeochemical cycles that depend on them. Clearly, unravelling the combined impacts of declining pH on critical seawater constituents, such as nutrients and key biogeochemical processes such as nitrification, denitrification, nitrogen fixation and nutrient uptake will be a challenge.

5.3. Changes to Flux of Other Climate Reactive Gases from the Ocean

As well as their important role in calcification, coccolithophores are also major producers of dimethyl sulphide (DMS) which may have a role in climate regulation via the production of cloud condensation nuclei [78]. A reduction in the occurrence of coccolithophore blooms that occur in large areas of the global oceans, often as large as 10^5 km^2, could lead to a reduced flux of DMS from the oceans to the atmosphere and hence to further increases in global temperatures via cloud changes [78,79]. As the oceans, and organisms within them, are a major source of other atmosphere changing gases [80,81] changes to the biology could also alter their production and cycling.

6. ADAPTATION, RECOVERY AND MITIGATION

It is difficult to predict if marine organisms and ecosystem will adapt to or recover from the rapid changes to ocean carbonate chemistry. An optimistic view may be that for organisms with short generation times micro-evolutionary adaptation could be rapid and that species adversely affected by high CO_2 could be replaced by more CO_2-tolerant strains or species, with minimal impacts up the food chain. The less optimistic view is that CO_2-sensitive groups, such as the marine calcifiers, will be unable to compete ecologically, resulting in widespread extinctions with profound ramifications up the food chain.

6.1. Adaptation

There are periods within a coccolithophore life cycle that are non-calcifying. In addition, there are some species that appear to have lost the ability to form $CaCO_3$ liths [82]. This suggests that the biochemical pathways involved in calcification in coccolithophores can be turned on and off. Should coccolithophores struggle to form their coccoliths in future high CO_2 scenarios, as is suggested by experimental data, they may have the genetic diversity and capability to adapt. Indeed, this may have happened several times throughout the course of evolution [83] although they would have had more time to do this then than is available during the current acidification event.

Although tropical Scleractinian corals have adapted, over millions of years, to live in warm, sunlit waters highly saturated in aragonite they have survived, and even retained their algal symbionts and completed gametogenesis, for a year in experiments at pH 7.4 although in a 'naked', decalcified form [84]. When transferred back to ambient pH conditions of 8.2, the soft-bodied corals calcified and reformed colonies. However, it should be noted that if this occurred in the wild the naked corals would be prone to greater grazing and they could not build reef structures which create important biodiversity hotspots.

A fossil coral from ~70 Ma ago had skeletal features identical to those observed in present-day Scleractinians but was made entirely of calcite rather than the aragonite of today's Scleractinian coral skeletons [85]. This implies that in geological times, some corals may have been able to switch between different carbonate forms to make their skeletons. However, the estimated rate of change during even the largest of these previous acidification events was an order of magnitude lower (over several thousand years) than our predicted current change (over a few hundreds of years) [86] so current corals may not have sufficient time to adapt.

Tropical coral migration to higher latitudes with more optimal sea surface temperature is unlikely, due both to latitudinally decreasing aragonite concentrations and projected atmospheric CO_2 increases [6,57,87]. Coral migration is also limited by lack of available substrate.

It would therefore seem unlikely that coral reefs would be able to adapt to a high CO_2 ocean sufficiently quickly in this current rapid anthropogenic perturbation, neither through switching to another carbonate form nor through migration.

The changes in current ecosystem composition caused by a natural CO_2 vent systems emitted by a volcano have shown a lack of many calcifying organisms in the lower pH areas ($pH < 7.8$) and a shift to predominance of sea grass beds or invasive alien species [19]. This study demonstrates the inability of many calcifiers to adapt to longer term decline in pH and gives an unattractive *in situ* insight into future ecosystems in a high CO_2 ocean.

6.2. Recovery

Ocean carbon models and the sediment record both indicate that chemical recovery from projected CO_2 emissions will require thousands of years (chemical equilibration with carbonate minerals) to hundreds of thousands of years (equilibration with the carbonate-silicate cycle) [4]. This means that the chemical effects of CO_2 released from anthropogenic sources are not confined to a century time scale.

Diversity of the sea bed dwelling organisms after the acidification event 55 Ma ago took several hundreds of thousands of years to recover. In contrast, there is evidence that planktonic calcifiers tracked their habitat during this event (e.g. tropical species migrated towards the poles), thereby avoiding extinction [88]. The geological record also shows that Scleractinian corals have survived several mass extinction events, likely due to perturbations in the carbon cycle, but they took several millions of years to recover [89–91]. These lessons from the past indicate that should increasing ocean acidification lead to significant loss of biodiversity and even extinction, biological systems may not 'recover' to pre-industrial ecosystems, but rather may 'transition' to a new state.

6.3. Mitigation

As concerns over climate change grow there are increasing numbers of geoengineering solutions proposed. However, they often do not take into account or resolve the issue of ocean acidification (e.g. addition of sulphur dioxide into the stratosphere to deflect some of the sun's energy or ocean pumps of deep water rich in nutrients to increase productivity and drawdown CO_2) nor do they look at potential deleterious impacts on the marine environment (adding quicklime to the oceans to soak up CO_2, iron or urea fertilisation to increase ocean productivity and drawdown CO_2).

Currently, expert opinion is that the only method of reducing the impacts of ocean acidification on a global scale is through urgent and substantial reductions in anthropogenic CO_2 emissions [7,15,29]. A threshold of no more than a 0.2 pH decrease has been recommended to avoid aragonite undersaturation in surface waters [93]. In terms of atmospheric CO_2 concentration this would be just above the 450 ppm stabilisation scenario (Fig. 6). However some polar waters would experience aragonite undersaturation even at this stabilisation level.

7. CONCLUSION

The oceans have been buffering climate change by absorbing about a quarter to a third of the CO_2 emitted into the atmosphere from anthropogenic sources. This has resulted in the measurable alteration of surface ocean concentrations of CO_2, HCO_3^-, CO_3^{2-} and pH as well as the reduction of the

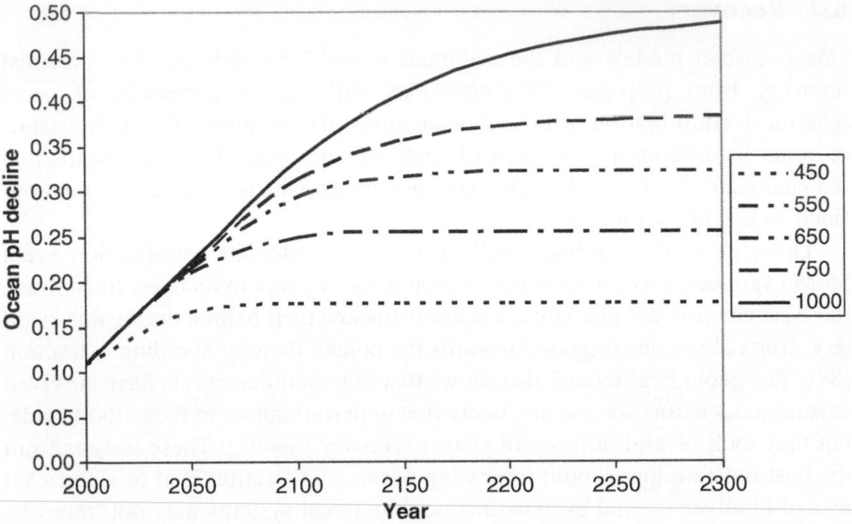

FIGURE 6 Trajectories for surface ocean pH decrease calculated for different atmospheric CO_2 concentration profiles leading to stabilisation from 450 to 1000 ppm. From Turley [92]. (See Color Plate 29).

saturation state and shoaling of the saturation horizons of $CaCO_3$ minerals. Since pre-industrial times ocean pH has decreased by a global average of 0.1 and it has been estimated that unmitigated CO_2 emissions will cause ocean pH to decrease by as much as 0.4 by the year 2100 and 0.77 by 2300. These will be the most rapid and greatest changes in ocean carbonate chemistry experienced by marine organisms over the past tens of millions of years. Laboratory experiments, field observations of natural CO_2-rich seawater 'hot spots' and studies of previous ocean acidification events in Earth's history, indicate that these changes are a threat to the survival of many marine organisms but particularly organisms that use $CaCO_3$ to produce shells, tests and skeletons (e.g. coccolithophores, pteropods, foraminifera, corals, calcareous macroalgae, mussels, oysters, echinoderms and crustacean). The ASH is already shoaling, bringing increasingly corrosive waters to the productive, shallower shelf seas along the western coast of North America and models predict that polar and some sub polar waters will be undersaturated this century while saturation states in tropical surface oceans will be substantially reduced. Recent experiments reveal that other important biological processes (productivity, internal physiology, fertilisation, embryo development, larval settlement and communication) are also vulnerable to future changes in ocean chemistry. There could also be changes to ocean carbon and nutrient cycles but, because of their complexity, it is hard to predict what the implications of the changes to biology will be on marine food webs, ecosystems and the services they provide. However, examination of previous episodes in Earth's history indicates that unmitigated CO_2 emissions are likely to result in widespread extinctions. It will take tens of thousands of years for the changes in ocean chemistry to be buffered through neutralisation by calcium carbonate sediments and the level at which ocean pH will eventually stabilise will be lower than it currently is. The only way of reducing the impacts of ocean acidification on a global scale is through urgent and substantial reductions in anthropogenic CO_2 emissions. Ocean acidification is a key argument for united global societal action in future climate change negotiations.

REFERENCES

1. C.L. Sabine, R.A. Feely, N. Gruber, R.M. Key, K. Lee, J.L. Bullister, R. Wanninkhof, C.S. Wong, D.W.R. Wallace, B. Tilbrook, F.J. Millero, T.H. Peng, A. Kozyr, T. Ono, A.F. Rios, Science 305 (2004) 367–371.
2. IPCC, in: J.T. Houghton, Y. Ding, D.J. Griggs, M. Noguer, P.J. van der Linden, X. Dai, K. Maskell, C.A. Johnson (Eds.), Climate Change 2001: The Scientific Basis. Contribution of Working Group 1 to the Third Assessment Report of the Intergovernmental Panel on Climate Change, Cambridge University Press, Cambridge, UK and New York, NY, USA, 2001.
3. K. Caldeira, M.E. Wickett, Nature 425 (2003) 365.
4. D.E. Archer, H. Kheshgi, E. Maier-Reimer, Glob. Biogeochem. Cycles 12 (1998) 259–276.
5. N. Gruber, J.L. Sarmiento, T.F. Stocker, Glob. Biogeochem. Cycles 10 (1996) 809–837.
6. J.C. Orr, V.J. Fabry, O. Aumont, L. Bopp, S.C. Doney, R.A. Feely, A. Gnanadesikan, N. Gruber, A. Ishida, F. Joos, R.M. Key, K. Lindsay, E. Maier-Reimer, R. Matear, P. Monfray, A. Mouchet, R.G. Najjar, G.-K. Plattner, K.B. Rodgers, C.L. Sabine, J.L. Sarmiento, R. Schlitzer, R.D. Slater, I.J. Totterdell, M.-F. Weirig, Y. Yamanaka, A. Yool, Nature 437 (2005) 681–686.
7. A. Fischlin, G.F. Midgley, J. Price, R. Leemans, B. Gopal, C. Turley, M. Rounsevell, P. Dube, J. Tarazona, A. Velichko, Climate Change 2007: Climate Change Impacts, Adaptation and vulnerability, Chapter 4: Ecosystems, their properties, goods and services, Forth Assessment Report of the Intergovernmental Panel on Climate Change, Cambridge University Press, Cambridge, 2007, pp. 211–272.
8. L. Cao, K. Caldeira, A.K. Jain, Geophys. Res. Lett. 34 (2007) L05607, doi:10.1029/2006GL028605.
9. H.O. Pörtner, M. Langenbuch, B. Michaelidis, J. Geophys. Res. Oceans 110 (2005) C09S10, doi:10.1029/2004JC002561.
10. J.C. Blackford, F.J. Gilbert, J. Mar. Syst. 64 (2007) 229–241.
11. C. Le Quéré, C. Rödenbeck, E.T. Buitenhuis, T.J. Conway, R. Langenfelds, A. Gomez, C. Labuschagne, M. Ramonet, T. Nakazawa, N. Metzl, N. Gillett, M. Heimann, Science 316 (2007) 1735–1738.
12. R. Torres, D.R. Turner, N. Silva, J. Rutllant, Deep-Sea Res. II 46 (1999) 1161–1179.
13. R.A. Feely, C.L. Sabine, J.M. Hernandez-Ayon, D. Ianson, B. Hales, Science 22 May (2008) 1–4, doi:10.1126/science.1155676.
14. P.N. Pearson, M.R. Palmer, Nature 406 (2000) 695–699.
15. C. Turley, J. Blackford, S. Widdicombe, D. Lowe, P.D. Nightingale, A.P. Rees, in: H.J. Schellnhuber, W. Cramer, N. Nakicenovic, T. Wigley, G. Yohe (Eds.), Avoiding Dangerous Climate Change, Cambridge University Press, Cambridge, 2006, pp. 65–70.
16. S. Barker, H. Elderfield, Science 297 (2002) 833–836.
17. S.C. Doney, V.J. Fabry, R.A. Feely, J.A. Kleypas, Ann. Rev. Mar. Sci. 1 (2009), 1: 169–192.

18. J.M. Santana-Casiano, M. González-Dávila, M.-J. Rueda, O. Llinás, E.-F. González-Dávila, Global Biogeochem. Cycles 21 (2007) GB1015, doi:10.1029/2006GB002788.
19. J.M. Hall-Spencer, R. Rodolfo-Mpi, S. Martin, R. Ransome, M. Fine, S.M. Turner, S.J. Rowley, D. Tedesco, M.-C. Buia, Nature (2008) 454, 96–99.
20. R.G.J. Bellerby, A. Olsen, T. Furevik, L.G. Anderson, in: H. Drange (Ed.), The Nordic Seas: An Integrated Perspective, Geophysical Monograph Series, vol. 158, American Geophysical Union, Washington, DC, 2005, pp. 189–197.
21. E. Thomas, in: S. Monechi, R. Coccioni, M.R. Rampino (Eds.), Large Ecosystem Perturbations: Causes and Consequences, vol. 424, Geological Society of America, Special Paper, Boulder, CO, 2007, pp. 1–23.
22. K.R. Hinga, Mar. Ecol. Prog. Ser. 238 (2002) 281–300.
23. J.M. Guinotte, J. Orr, S. Cairns, A. Freiwald, L. Morgan, R. George, Front. Ecol. Environ. 4 (2006) 141–146.
24. S.D. Cairns, I.G. Macintyre, Palaios 7 (1992) 96–107.
25. T.F. Cooper, G. De'Ath, K.E. Fabricius, J.M. Lough, Glob. Change Biol. 14 (2008) 539–538.
26. J. Silverman, B. Lazar, J. Erez, J. Geophys. Res. 112 (2007) C05004, doi:10.1029/2006JC003770.
27. T. Tyrrell, B. Schneider, A. Charalampopouloul, U. Riebesell, Biogeosciences 5 (2008) 485–494.
28. H.S. Findlay, T. Tyrrell, R.G.J. Bellerby, A. Merico, I. Skjelvan, Biogeosci. 5 (2008) 1395–1410.
29. The Royal Society. Ocean Acidification Due to Increasing Atmospheric Carbon Dioxide, Policy document 12/05, The Royal Society, London, 2005.
30. B. Rost, U. Riebesell, S. Burkhardt, D. Sültemeyer, Limnol. Oceanogr. 48 (2003) 55–67.
31. B. Rost, U. Riebesell, in: H.R. Thierstein, J.R. Young (Eds.), Coccolithophores: From Molecular Processes to Global Impact, Springer, Berlin, 2004, pp. 99–125.
32. J.A. Kleypas, R.A. Feely, V.J. Fabry, C. Langdon, C.L. Sabine, L.L. Robbins, Impacts of Ocean Acidification on Coral Reefs and Other Marine Calcifiers: A Guide for Future Research. Report of a workshop held 18–20 April 2005, St. Petersburg, FL, sponsored by NSF, NOAA, and the US Geological Survey, 2006.
33. P.M. Haugan, C. Turley, H.O. Pörtner, DN-utredning 1 (2006) 1–36.
34. J.M. Guinotte, V.J. Fabry, Ann. N. Y. Acad. Sci. 1134 (2008) 320–342.
35. F. Gazeau, C. Quibler, J.M. Jansen, J.-P. Gattuso, J.J. Middelburg, C.H.R. Heip, Geophys. Res. Lett. 34 (2007) L07603, doi:10.1029/2006GL028554.
36. M.A. Green, M.E. Jones, C.L. Boudreau, P.L. Moore, B.A. Westman, Limnol. Oceanogr. 49 (2004) 727–734.
37. H. Kurihara, S. Kato, A. Ishimatsu, Aquat. Biol. 1 (2007) 91–98.
38. H. Kurihara, Y. Shirayama, Mar. Ecol. Prog. Ser. 274 (2004) 161–169.
39. S. Dupont, J. Havenhand, W. Thorndyke, L. Peck, M. Thorndyke, Mar. Ecol. Prog. Ser. 373 (2008) 285–294.
40. T. Kikkawa, J. Kita, A. Ishimatsu, Mar. Pollut. Bull. 48 (2004) 108–110.
41. H. Kurihara, S. Shimode, Y. Shirayama, Mar. Pollut. Bull. 49 (2004) 721–727.
42. H.S. Findlay, M.A. Kendall, J.I. Spicer, C. Turley, S. Widdicombe, Aquat. Biol. 3 (2008) 51–62.
43. D.V. Desai, A.C. Anil, J. Mar. Biol. Assoc. U.K. 85 (2005) 909–920.
44. A.W. Decho, K.A. Browne, R.K. Zimmer-Faust, Limnol. Oceanogr. 43 (1998) 1410–1417.
45. S.L. Coon, M. Walch, W.K. Fitt, R.M. Weiner, D.B. Bonar, Biol. Bull. 179 (1990) 297–303.
46. D.B. Bonar, S.L. Coon, M. Walch, R.M. Weiner, W. Fitt, Bull. Mar. Sci. 46 (1990) 484–498.
47. M.J. Anderson, Biol. Bull. 190 (1996) 350–358.

48. A.O.H.C. Leduc, E. Roh, M.C. Harvey, G.E. Brown, Can. J. Fish. Aquat. Sci. 63 (2006) 2356–2363.
49. R. Bibby, P. Cleal-Harding, S. Rundle, S. Widdicombe, J.I. Spicer, Biol. Lett. 3 (2007) 699–701.
50. H.L. Wood, J.I. Spicer, S. Widdicombe, Proc. R. Soc. B 275 (2008) 1767–1773.
51. K. Vopel, D. Thistle, R. Rosenberg, Limnol. Oceanogr. 48 (2003) 2034–2045.
52. R. Danovaro, L. Bongiorni, C. Corinaldesi, D. Giovannelli, E. Damiani, P. Astolfi, L. Greci, A. Pusceddu, Environ. Health Perspect. 116 (2008) 441–447.
53. J.B. Larsen, A. Larsen, R. Thyrhaug, G. Bratbak, R.-A. Sandaa, Biogeosciences 5 (2008) 523–533.
54. R. Bibby, S. Widdicombe, H. Parry, J.I. Spicer, R. Pipe, Aquat. Biol. 2 (2008) 67–74.
55. C. Langdon, T. Takahashi, C. Sweeney, D. Chipman, J. Goddard, F. Marubini, H. Aceves, Glob. Biogeochem. Cycles 14 (2000) 639–654.
56. J.M. Guinotte, R.W. Buddemeier, J.A. Kleypas, Coral Reefs 22 (2003) 551–558.
57. R.A. Feely, C.L. Sabine, K. Lee, W. Berelson, J. Kleypas, V.J. Fabry, F.J. Millero, Science 305 (2004) 362–366.
58. P.D. Manzello, J.A. Kleypas, D.A. Budd, C.M. Eakin, P.W. Glynn, C. Langdon, Proc. Natl. Acad. Sci. USA (2008), 105, 10450–10455.
59. A. Freiwald, J.M. Roberts, Cold-Water Corals and Ecosystems, Springer, Heidelberg, 2005.
60. J.M. Roberts, A.J. Wheeler, A. Freiwald, Science 312 (2006) 543–547.
61. C.M. Turley, J.M. Roberts, J.M. Guinotte, Coral Reefs 26 (2007) 445–448.
62. J.S. Gray, R.S.-S. Wu, Y.Y. Or, Mar. Ecol. Prog. Ser. 238 (2002) 249–279.
63. O. Hoegh-Guldberg, P.J. Mumby, A.J. Hooten, R.S. Steneck, P. Greenfield, E. Gomez, C.D. Harvell, P.F. Sale, A.J. Edwards, K. Caldeira, N. Knowlton, C.M. Eakin, R. Iglesias-Prieto, N. Muthiga, R.H. Bradbury, A. Dubi, M.E. Hatziolos, Science 318 (2007) 1737–1742.
64. C. Langdon, M.J. Atkinson, J. Geophys. Res. 110 (2005) C09S07, doi:10.1029/2004JC002576.
65. C. Langdon, W.S. Broecker, D.E. Hammond, E. Glenn, K. Fitzsimmons, S.G. Nelson, T.-H. Pend, Glob. Biogeochem. Cycles 17 (2003) 1011, doi:10.1029/2002GB00.
66. NOAA, The State of Coral Reef Ecosystems of the United States and Pacific Freely Associated States. NOAA Technical Memorandum NOSNCCOS73, 2008.
67. K.E. Carpenter, M. Abrar, G. Aeby, R.B. Aronson, S. Banks, A. Bruckner, A. Chiriboga, J. Cortés, J.C. Delbeek, L. DeVantier, G.J. Edgar, A.J. Edwards, D. Fenner, H.M. Guzmán, B.W. Hoeksema, G. Hodgson, O. Johan, W.Y. Licuanan, S.R. Livingstone, E.R. Lovell, J.A. Moore, D.O. Obura, D. Ochavillo, B.A. Polidoro, W.F. Precht, M.C. Quibilan, C. Reboton, Z.T. Richards, A.D. Rogers, J. Sanciangco, A. Sheppard, C. Sheppard, J. Smith, S. Stuart, E. Turak, J.E.N. Veron, C. Wallace, E. Weil, E. Wood, Science 321 (2008) 560–563.
68. B. Pasquer, G. Laruelle, S. Becquevort, V. Schoemann, H. Goosse, C. Lancelot, J. Sea Res. 53 (2005) 93–108.
69. U. Schuster, A.J. Watson, J. Geophys. Res. 112 (2007) C11006, doi:10.1029/2006JC003941.
70. U. Riebesell, I. Zondervan, B. Rost, P.D. Tortell, R. Zeebe, F.M. Morel, Nature 407 (2000) 364–367.
71. A. Ridgwell, R.E. Zeebe, Earth Planet. Sci. Lett. 234 (2005) 299–315.
72. B. Delille, J. Harlay, I. Zondervan, S. Jacquet, L. Chou, R. Wollast, R.G.J. Bellerby, M. Frankignoulle, A.V. Borges, U. Riebesell, J.-P. Gattuso, Glob. Biogeochem. Cycles 19 (2005) GB2023, doi:10.1029/2004GB002318.
73. U. Riebesell, K.G. Schulz, R.G.J. Bellerby, M. Botros, P. Fritsche, M. Meyerhöfer, C. Neill, G. Nondal, A. Oschlies, J. Wohlers, E. Zöllner, Nature 450 (2007) 545–548.
74. R.E. Zeebe, D. Wolf-Gladrow, CO_2 in Seawater: Equilibrium, Kinetics, Isotopes, Elsevier Oceanography Series, Elsevier, Amsterdam, 2001.

75. J.A. Raven, in: T. Platt, W.K.W. Li (Eds.), Photosynthetic Picoplankton, vol. 214, Canad. Bull. Fish. Aquat. Sci., Ottawa, 1986, pp. 1–583.
76. M.H. Huesemann, A.D. Skillman, E.A. Crecelius, Mar. Pollut. Bull. 44 (2002) 142–148.
77. D.A. Hutchins, F.-X. Fu, Y. Zhang, M.E. Warner, Y. Feng, K. Portune, P.W. Berhardt, Limnol. Oceanogr. 52 (2007) 1293–1304.
78. R.J. Charlson, J.E. Lovelock, M.O. Andreae, S.G. Warren, Nature 326 (1987) 655–661.
79. G. Malin, S. Turner, P. Liss, P. Holligan, D. Harbour, Deep-Sea Res. Part I – Oceanogr. Res. Pap. 40 (1993) 1487–1508.
80. P.D. Nightingale, P.S. Liss, in: H. Elderfield (Ed.), The Oceans and Marine Geochemistry, A Treatise on Geochemistry, vol. 6, Elsevier, New York, 2003.
81. C. Hughes, G. Malin, C.M. Turley, B.J. Keely, P.D. Nightingale, P.S. Liss, Limnol. Oceaongr. 53 (2008) 867–872.
82. J. Henderiks, R.E.M. Rickaby, Biogeosciences 4 (2007) 323–329.
83. C. de Vargas, I. Probert, Micropaleontology 50 (2004) 45–54.
84. M. Fine, D. Tchernov, Science 315 (2007) 1811.
85. J. Stolarski, A. Meibom, R. Przenioslo, M. Mazur, Science 318 (2007) 92–94.
86. K.A. Panchuk, A. Ridgwell, L.R. Kump, Geology 36 (2008) 315–318.
87. J.A. Kleypas, R.W. Buddemeier, J.-P. Gattuso, Int. J. Earth Sci. 90 (2001) 426–437.
88. S.J. Gibbs, P.R. Bown, J.A. Sessa, T.J. Bralower, P.A. Wilson, Science 314 (2006) 1770–1773, doi:10.1126/science.1133902.
89. G.D. Stanley, D.G. Fautin, Science 291 (2001) 1913–1914.
90. G.D. Stanley, Earth Sci. Rev. 60 (2003) 195–225.
91. J.E.N. Veron, A Reef in Time: The Great Barrier Reef from Beginning to End, Belknap Press, USA, 2008, pp. 1–282.
92. C. Turley, Mineral. Mag. 72 (2008) 363–366.
93. German Advisory Council on Global Change, The Future Oceans – Warming up, Rising High, Turning Sour Special Report, WBGU, Berlin, http://www.wbgu.de, ISBN 3-936191-14-X, 1-110, 2006.

Chapter 22

Ice Sheets: Indicators and Instruments of Climate Change

David G. Vaughan
British Antarctic Survey, Natural Environment Research Council, Madingley Road, Cambridge CB3 0ET, United Kingdom

1. Introduction
2. Sea-Level and Ice
3. How Ice Sheets Work
4. Summary
References

1. INTRODUCTION

Ice sheets of Greenland and Antarctica are uniquely arresting and captivating features of the Earth's natural environment. The hyperbole attached to their description, their sheer size and remoteness from the normal lives of most of us, guarantees their iconic status, but the emerging understanding that ice sheets contain a threat, which cannot be fully evaluated mean, that they have become a central issue in the climate change debate. However, while global climate has undoubtedly warmed during the recent past, and human activity has been a major factor in this change, the role of ice sheets as indicators of climate change and as influential components in the planets climate engine, is a complex one (Fig. 1).

This chapter will discuss the position of the great ice sheets within the climate change debate, contrasting the differing risks posed to sea-level rise by the Greenland and Antarctic ice sheets as likely contributors to future sea-level rise, and how they may differently influence the wider debate on limiting greenhouse-gas emissions.

2. SEA-LEVEL AND ICE

Although there are regional differences due, in part, to local subsidence and emergence rates of coastal land, global sea level is rising. We see this rise both in the century-long record from tide gauges around the world, and in

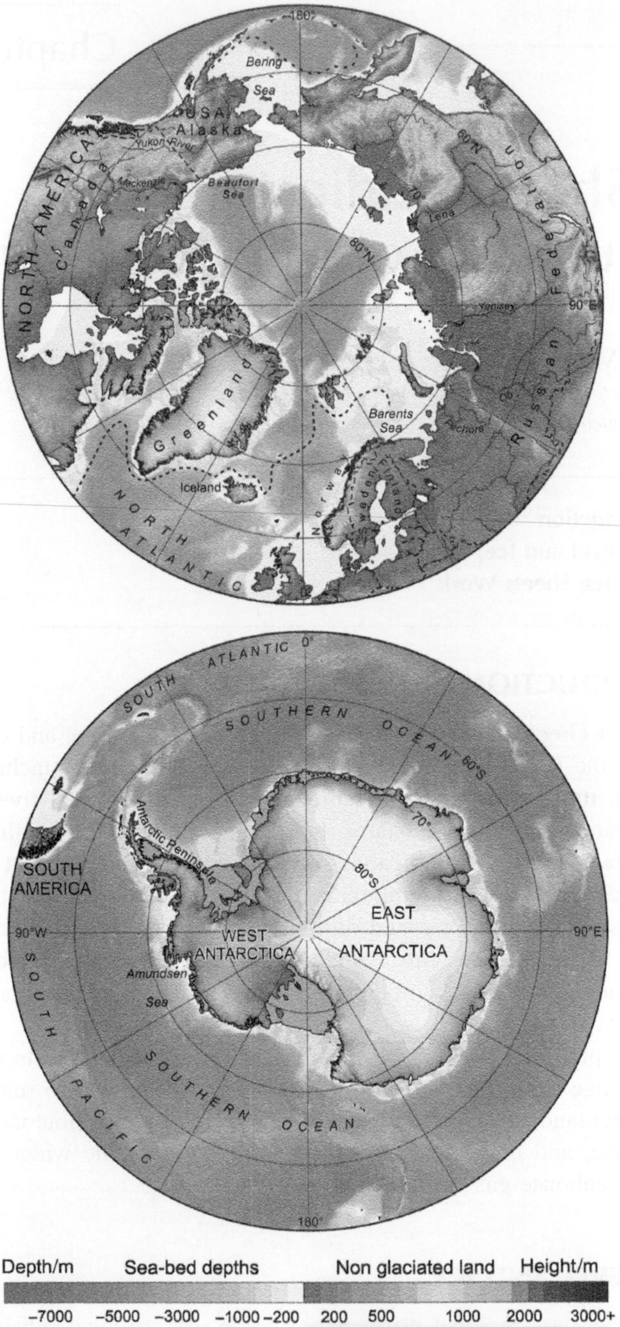

FIGURE 1 Maps of the north and south polar regions showing ice sheets and place names used in the text. Modified from Ref. [19]. (See Color Plate 30).

the shorter record from satellite monitoring of the ocean surface elevation. In recent decades, the rate of global sea-level rise has been more than 3 mm/year. This rise is made up from various contributions: thermal expansion of ocean waters; changes in the mass of water contained in mountain glaciers, reservoirs and ground-water acquifers and changes in the ice-sheets of Antarctica and Greenland.

This rate of sea-level rise may not sound serious but unlike some other climate variables, sea-level tends to change smoothly, and the current rate of rise is likely to continue and most probably grow in the future. The cumulative sea-level rise over coming decades will have surprising and profound and impacts on coastal ecosystems, human populations and the stability of some economies. Climate change is very likely to accelerate most of the individual contributions to sea-level rise, and thus accelerate the rise in global sea-level. It is still not entirely clear that accelerating rates of sea level rise in the late twentieth century indicate that this acceleration has already begun [1], but there is very little doubt that sea-level rise will accelerate substantially during the twenty-first century.

The Fourth Assessment Review of the Intergovernmental Panel on Climate Change[1] (IPCC 4AR) [2] contains the most authoritative assessment and projection of sea-level rise so far undertaken. The review includes discussion of all the major contributors to sea-level rise, including, the contribution of the Antarctic and Greenland ice sheets. In summary, it predicts that by 2090–2099, sea-level will have risen 20–60 cm compared to 1980–1999. However, there are strong statements within the IPCC report that retain the possibility that the IPCC-4AR projections of sea-level rise maybe incomplete and potentially too conservative, and that the potential contribution from the ice sheets holds substantial uncertainty. These statements are most succinctly summarised in the Summary for Policymakers [3],

> *Models used to date [within the IPCC-4AR review] do not include uncertainties in climate-carbon cycle feedback nor do they include the full effects of changes in ice sheet flow, because a basis in published literature is lacking. The projections include a contribution due to increased ice flow from Greenland and Antarctica at the rates observed for 1993 to 2003, but these flow rates could increase or decrease in the future. For example, if this contribution were to grow linearly with global average temperature change, the upper ranges of sea level rise for SRES scenarios shown in Table SPM.3 would increase by 0.1 to 0.2 m. Larger values cannot be excluded, but understanding of these effects is too limited to assess their likelihood or provide a best estimate or an upper bound for sea level rise.*

The IPCC assessment has already been widely criticised by studies that range from the scientific [4] to those that are almost ideological in approach [5]. To fully understand the difficulties that the IPCC have faced, and the potential for

---

[1] The IPCC is a group of largely government-nominated specialists who are tasked with producing the most complete assessment of the science, impacts and potential responses to anthropogenic climate change.

resolution of this important question, we must first discuss the workings of the ice sheets, and the recent observations that have led the IPCC to be so cautious.

3. HOW ICE SHEETS WORK

To understand the influence that ice sheets have on the Earth system, it is important to have some insight into how an ice sheet, indeed any glacier, operates as a natural self-regulating entity.

Even during the warmest months, the inland parts of the Antarctic and Greenland ice sheets are too cold for significant melting of the snow surface to occur, and so, year-by-year, snow accumulates. If unchecked, this accumulation of snow would cause endless thickening, and in the absurd limit, this would drain the oceans of water. But snow eventually turns to ice, and this ice can, under great pressure, deform. Thus, within the ice sheets, ice is constantly moving, from the interior, into the glaciers and ice streams that take it back towards the coast. Eventually, ice re-enters the ocean either through iceberg calving, or though melting from the glacier surface or directly into the ocean.

The key mechanism that allows ice sheets to achieve apparently steady configurations is that glacial flow is driven by gravity and the driving force is proportional to the slope on the surface of the ice sheet. If snow accumulates more quickly than it is being removed by the ice flow, the ice sheet will get thicker, the surface slope from the interior to the coast will increase, and the rate of ice flow will increase – this will reverse the thickening. Thus the ice sheet has an inbuilt negative feedback, a system of self-regulation, which can produce equilibrium between accumulation and loss. In fact, the nonlinear mechanical properties of ice and its potential to slide over its bed, mean that even modest changes in the slope can have an enormous impact on ice-flow speed, and the ice sheet can be surprisingly quick to regain equilibrium after any disturbance.[2]

However, since the climatic variables that influence ice sheets are never entirely stable, no ice sheet is entirely in balance. Indeed, the geological records contain abundant evidence that ice-sheet configurations change dramatically with changing climate: twenty thousand years ago, around the Last Glacial Maximum (LGM), unbroken ice sheets covered vast land masses across the Earth. These ice sheets did not only cover enormous areas, they were also very thick, and they contained sufficient ice that global sea levels were some 120 m lower than they are today.

[2] I am sometimes asked if the ice sheets will melt. And in answer, I am sometimes tempted to reply in positive terms, 'yes, all the ice in the ice sheets will eventually melt'. This would, of course, be unfair, without also noting that, so long as the ice that melts is balanced by new snow replacing it, the ice sheets will remain the same size. Although glib, this point explains why glaciologists refer to ice-sheet growth or retreat, and rarely use 'melt', so much favoured by journalists.

It took almost a hundred thousand years for the ice sheets to grow to this size, by the year-on-year addition of snowfall, but as climate warmed and melting took hold, it took little more than 15 000 years for most of the LGM ice sheets to melt, and for sea-level to rise. This asymmetry in rates of ice sheet loss and growth, is a feature of the geological record, and means that while global sea-level fall is only ever a rather sedate process, whereas sea-level rise can be comparitively rapid. Records from coral reefs show that the peak periods of ice-sheet retreat since the LGM, caused sea-level rise at rates of around 4 m/century.

Such rates of sea-level rise sound dramatic, even frightening, but just because such rates of change occurred in the past, should we expect similar rates in the future? Although several authors have used this analogue to infer the possibility of substantial rates of future sea-level rise [5], I believe that this is not a useful analogue for two reasons. Firstly, these estimates could be argued to be substantially too high because the highest rates of sea-level rise since the LGM occurred when there was vastly more ice in the world available to melt, and the total length of melting margin of those ice sheets was much longer. Secondly, these estimates could be argued to be too low, because those rates of sea-level rise occurred at times when climate change was many times slower than is predicted for the coming one or two centuries.[3] The fact that these are opposing, is no assurance that they will in any substantial way, 'cancel out'. A much better analogue for future conditions might be the last inter-glacial, for which there is emerging, but as yet equivocal, evidence for substantial rates of sea-level change. It is thus arguable that behaviour cannot be taken as a sufficient basis for predicting the future – we need to consider the ice sheets in much greater detail and predict their future on the basis of understanding their configuration and possible responses from a more mechanistic approach.

Today, there are two great ice sheets left on the planet, one on Greenland and the other on Antarctica. The Antarctic ice sheet is by far the larger covering an area similar to that of the contiguous states of the USA. By comparison, Greenland covers an area a little greater than that of Mexico. Trying to measure the degree of imbalance that exists in these ice sheets, has been a preoccupation of glaciologists for many decades, even before it was postulated that human activities were having an influence on climate, or that they could potentially also influence the ice sheets. These efforts were dogged by uncertainty for decades, until the late 1990s, when a new generation of satellites that could measure the volume, and flow or the ice sheets was launched.

---

[3] The IPCC predicts global warming temperatures for the decade 2090–2099, that are 1.8–6.4 °C higher than 1980–1999; where the range expresses uncertainty due to the future emissions of greenhouse gases and uncertainty in model projections. This should be compared to the global temperature difference between glacial and interglacial conditions which is 4–7 °C, and this change took thousands of years to complete.

Since then assessment of the rates and patterns of change have become common and now show broad agreement [6,7].

Not only the area, but also the volume of ice contained in the ice sheets is substantially different. Antarctica contains sufficient ice to raise global sea level by 60 m, while loss of Greenland could raise it by around 7 m. However, these figures are to some extent misleading, since the risks associated with the Greenland and Antarctic ice sheets are closer in magnitude, but might be said to have a subtly different 'flavour'.

The risk associated with the Greenland ice sheet can be characterised as the possibility that climate change will push Greenland into a state of imbalance, where it loses most of its ice (raising global sea-level by a little more than 5 m) over a period of several centuries to millennia. Because most of the Greenland ice sheet rests on rock that is close to, or above, current sea-level, for ice to be lost it must be transported to the sea by the glaciers. The rate at which this could occur is likely to be limited by the rate at which those glaciers could conceivably flow, and this is unlikely to be more than a few times the present rates. However, perhaps the absolute rate of loss is not the most significant point; lowering of the interior of the ice sheet would mean increasing areas of summer melt, once this process of retreat is fully underway, it would push the ice sheet as a whole further into imbalance, and the process of retreat could become self-sustaining. Continuing retreat would become inevitable and irreversible [8]. The critical threshold in atmospheric warming, that is required to begin this process could, if the higher projections of climate warming are to be believed, be exceeded by the end of the twenty-first century.

This argument frames the role of the Greenland ice sheet in the climate change debate. The hypothesis that human activities could push Greenland into a state where it continues to retreat for many centuries implies committing many, as yet unborn, generations to cope with the impacts of a rising sea-level that earlier generations have caused. Rightly, or wrongly, this has been used by many campaigners as a potent argument for early and vigorous reduction in greenhouse gas emissions to prevent this 'tipping point' being exceeded [9].

In contrast, large parts of the Antarctic ice sheet are considered to be relatively immune to rising atmospheric temperatures. Current climate is sufficiently cold that even under the most extreme projections year-round the ice sheet will remain frozen and generally unaffected. Indeed, for large parts of the East Antarctic ice sheet, the most likely impact of rising temperatures is an increase in snowfall rates. Perhaps, counter-intuitively, this may be a mechanism that slows the rate of future sea-level rise, although this should not be taken as a possible fix for the sea-level rise problem. Even if this hypothesis proves to be correct, its magnitude would probably reduce rates of sea-level rise by only 1 $mm \cdot a^{-1}$ [10]. Furthermore, a recent analysis of ice-core data from East Antarctica failed to show any evidence that increases in snowfall rates has yet begun [11]. The conclusion that the Antarctic ice sheet may be largely immune to atmospheric climate change,[4] certainly does not mean that it is insignificant in the sea-level rise debate. There is a part of the Antarctic ice sheet that is changing more rapidly than any other on the planet. The Amundsen Sea embayment of West Antarctica, is an area of ice sheet about the size of Texas, across which the ice sheet has been thinning for more than a decade at rates of a few centimetres per year on the interior to rates of several metres per year near the coast.

The Amundsen Sea embayment is a part of the West Antarctic ice sheet that has long been the focus of concern. In a series of traverses across Antarctica begun during the last International Geophysical Year (1957–1958), glaciologists discovered that much of the West Antarctic Ice Sheet rests on a bed that is below sea level. Indeed, the bed topography beneath the ice is like a bowl, sloping down from the edge of ice sheet into a deep tectonic rift beneath its interior. If the ice sheet were removed, most of the area would be inundated by the sea. For this reason, those early glaciologists described WAIS as a "marine ice sheet", and they were quick to see a particular significance in this configuration. WAIS is the only significant marine ice sheet left on Earth, and it was suggested that this is because marine ice sheets are intrinsically unstable. With a bed below sea level, the ice sheet is anchored to its bed only because it is too thick to float. If areas around the margin of a marine ice sheet were to loose contact with the bed, there would be a reduction in the force that restrains ice-flow. Ice-flow could then accelerate and leave an imbalance between outflow and replenishment by snowfall. This imbalance would in turn cause thinning of the ice sheet at the point at which it begins to float, allowing the so-called, grounding line, to retreat inland. It has been frequently argued [13,14] that this could set up a positive-feedback cycle that would be sufficiently strong to overcome the negative feedbacks that act to keep ice sheets in a state of balance. Once this positive-feedback cycle was set in motion, ice-sheet 'collapse' or 'disintegration' would inevitably follow and with severe global consequences since WAIS contains sufficient ice to raise global sea level by about 5 m.

This idea, remained relatively dormant for many years, but has been recently reinvigorated by observations, that the ice sheet in the Amundsen Sea embayment, which has often been cited that the most vulnerable part of WAIS [15], is actually the area that is thinning most rapidly [16].

Furthermore, the thinning of the ice sheet in the Amundsen Sea embayment has been accompanied by indications that the some parts of the ice sheet are lifting away from their beds and beginning to float, which is another feature expected if a retreat of the type expected to result from the marine ice

---

[4] This is not universally true, since a small part of Antarctica, the Antarctic Peninsula, has experienced very rapid recent rises in atmospheric temperature, and this does appear to be causing glacier and ice-sheet retreat in this area [12]. However, this area is only a small player in its contribution to sea-level rise.

sheet instability was beginning. Finally, it is now clear that the rate of ice loss from the Amundsen Sea embayment is accelerating, at a surprising rate. The loss from the ASE currently accounts for only 0.5 mm/year of global sea-level rise [17], but it does appear to be growing in an exponential way (around 1.3% per year).

The cause of these impressive, and concerning changes, is not yet entirely clear, but there is strong evidence that it arises from some change in the ocean surrounding the ice sheet – the Amundsen Sea itself. Essentially, it appears that the ocean is delivering more heat to the ice sheet than it once did, and that is melting and thinning the coastal margin of the ice sheet. Although it cannot be proved equivocally many researchers believe that this change in the ocean is has been driven by changes in the atmospheric circulation associated with greenhouse warming.[5]

To return to the comparison with Greenland, and the significance of these Antarctic changes to the policy debate; the fact that WAIS is a marine ice sheet means that, unlike the Greenland sheet, its response is not constrained by the speed at which its glaciers can deliver ice to the ocean. In effect, the sea could eat into the margin of the ice sheet and remove at whatever rate the ocean could remove the iceberg debris. Thus, the worst rates of sea level rise that could occur from WAIS are likely to be more rapid than that from Greenland. However, there is further complication, which is that current changes in the ASE, as mentioned earlier, are not likely to be due to changing atmospheric temperature, and making a clear attribution between ice-sheet change and anthropogenic influences on climate is significantly harder in Antarctica than it is for Greenland. Indeed, if there is proved to be an inherent instability in marine ice sheets, it may be argued that its retreat could be driven by a relatively small anthropogenic or natural variation. So, rapid sea-level rise from Antarctica could have a human, or a natural origin. Either way, the risks to unprotected coastal populations, and the cost of improving coastal defences to maintain protection, are the same, but the correct framing of the issues within the debate concerning reduction of greenhouse gas emissions is quite different.

4. SUMMARY

While we may debate the finer points of how climate change may interact with ice sheets, the general understanding that if our planet warms, ice sheets and glaciers will retreat and sea levels will rise, is not widely contested. It is clear from the geological records that this close connection between

[5] Essentially, their hypothesis is that increasing circulatory winds in the Southern Ocean draw more warm water onto the Antarctic continental shelf, and provide more heat for ice-sheet melting. However, a shortage of marine monitoring sites means that this is difficult to pin down with any real confidence.

temperature and sea-level has been maintained through many glacial cycles. And it is also clear that ice sheets contain the potential to raise sea-level at rates that are many times higher than those we have observed in recent decades. However, the overriding questions regarding the rate that ice sheets will contribute to sea level in coming centuries will remain unresolved until we attain a substantially improved understanding of ice sheet behaviour. The geological record of past changes is a guide in this regard, but today's ice sheets cannot be expected to respond to future anthropogenic change exactly as it did to past natural variations. The contrasting configurations of the Greenland and Antarctic ice sheets mean that they occupy subtly different positions within the climate change debate, both of which have been described as representing tipping points.

The way that the risk that ice sheets pose is perceived is an area that is also worth brief consideration. Studies of risk perception show that they many identifiable factors amplify the social importance and fear surrounding particular risks [18]. These include: the longevity of the risk; its apparent invisibility; and its potential to cause catastrophic events (rather than impacting a few individuals at any one time). As described in this chapter, each of these factors applies strongly to the threat of sea-level rise from ice sheet. Ice sheets may produce effects that, as with nuclear waste, many future generations will have to live with; they are so remote that they can be viewed as essentially invisible, and recent coastal flooding events, for example in New Orleans, have shown the huge scale and severity of the impacts that may become all the more frequent in future. All these factors may serve amplify the perception of the risk and to exert a powerful influence on the public and policy-makers alike, but this does not mean that the risk does not exist, or can be ignored, and it is clear that a substantial number of scientists see sea-level rise as a worrying adjunct to the wider climate change debate, and an area where the scientific understanding that underpins predictions is particularly lacking.

We know immeasurably more about current changes in the ice sheets of Antarctica and Greenland, than we did even a decade ago, but it is arguable that this increase understanding has actually increased our uncertainty. Whereas the ice sheets could once be ignored as sleeping giants, there is now evidence they are becoming restless.

REFERENCES

1. R.S. Nerem, E. Leuliette, A. Cazenave, Comptes Rendus Geosci. 338 (14–15) (2006) 1077–1083.
2. P. Lemke, J. Ren, R.B. Alley, I. Allison, J. Carrasco, G. Flato, Y. Fujii, G. Kaser, P. Mote, R.H. Thomas and T. Zhang. 2007: "Observations: Changes in Snow, Ice and Frozen Ground." pp. 339–383 in *Climate Change 2007: The Physical Science Basis. Contribution of Working Group I to the Fourth Assessment Report of the Intergovernmental Panel on Climate Change*, edited by

S. Solomon, D. Qin, M. Manning, Z. Chen, M. Marquis, K.B. Avery, M. Tignor and H.L. Miller. Cambridge, United Kingdom and New York, NY, USA: Cambridge University Press.

3. S. Solomon, D. Qin, M. Manning, Z. Chen, M. Marquis, K.B. Avery, M. Tignor and H.L. Miller. 2007. *Summary for Policymakers.* Cambridge, United Kingdom and New York, USA: Cambridge University Press.
4. S. Rahmstorf, A. Cazenave, J.A. Church, J.E. Hansen, R.F. Keeling, D.E. Parker and R.C.J. Somerville. 2007. "Recent climate observations compared to projections." *Science* 316 (5825): 709–709.
5. J.E Hansen, Environ. Res. Lett. 2 (2) (2007).
6. D.J. Wingham, A.J. Ridout, R. Scharroo, R.J. Arthern, C.K. Schum, Science 282 (1998) 456–458.
7. H.J. Zwally, et al., J. Glaciol. 51 (175) (2005) 509–527.
8. J.K. Ridley, P. Huybrechts, J.M. Gregory, J.A. Lowe, J. Clim. 18 (17) (2005) 3409–3427.
9. T.M. Lenton, et al., Proc. Natl. Acad. Sci. 105 (6) (2008) 1786–1793.
10. D.G. Vaughan, J.W. Holt, D.D. Blankenship, Eos. Transact. Am. Geophys. Union 88 (46) (2007) 485–486.
11. A.J. Monaghan, et al., Science 313 (5788) (2006) 827–831.
12. A. Cook, A.J. Fox, D.G. Vaughan, J.G. Ferrigno, Science 22 (2005) 541–544.
13. J. Weertman, J. Glaciol. 13 (67) (1974) 3–11.
14. C. Schoof, J. Fluid Mech. 573 (2007) 27–55.
15. T.J. Hughes, J. Glaciol. 27 (97) (1981) 518–525.
16. D.G. Vaughan, Clim. Change 91 (1–2) (2008) 65–79.
17. E. Rignot, et al., Nat. Geosci. 1 (2) (2008) 106–110.
18. R.E. Kasperson, J.X. Kasperson, Ann. Am. Acad. Pol. Soc. Sci. 545 (1996) 95–105.
19. O.A. Anisimov, D.G. Vaughan, T.V. Callaghan, C. Furgal, H. Marchant, T.D. Prowse, H. Viljalmasson and J.E. Walsh. 2007. "Chapter 15 – Polar Regions (Arctic and Antarctic)." pp. 653-685 in *Climate Change 2007: Impacts, Adaptation and Vulnerability. Contribution of Working Group II to the Fourth Assessment Report of the Intergovernmental Panel on Climate Change*, edited by M.L. Parry, O.F. Canziani, J.P. Palutikof, P.J.v. d. Linden and C.E. Hanson. Cambridge, UK: Cambridge University Press.

Chapter 23

Lichens as an Indicator of Climate and Global Change

Andre Aptroot
ABL Herbarium, Gerrit van der Veenstraat 107, NL-3762 XK Soest, The Netherlands

1. Introduction
2. Predicted Effects
3. Observed Effects
4. Uncertain Effects
5. Habitats with Vulnerable Lichens
5.1. Low Level Islands with Endemic Lichens
5.2. The Extended Regions with Similar Climate but Local Endemism
5.3. The (Ant-)Arctic and Tundra Regions
5.4. High Ground in the Tropics
6. Conclusion
Acknowledgement
References

1. INTRODUCTION

Lichens have been observed to respond rapidly to climate change. So far, the changes are as expected with a rather rapid increase of (sub)tropical species in temperate areas, and a gradual decrease of some boreo-alpine elements [1]. So far, comparatively few publications have addressed the issue of lichens in connection with global warming [2]. No lichens have, so far, been reported to be seriously threatened by climate change. Marked shifts in occurrence and distribution have been predicted based on known habitat preferences and projected climate change [3].

Lichens, like most cryptogams, tend to be widespread, much more so than phanerogams or land animals. Also, many of the species seem to be capable of rather rapid dispersal, as shown by the recent arrival of some (sub)tropical species in a temperate area [1].

In this chapter, predicted, observed and uncertain effects related to lichen and climate change are discussed together with the habitats of vulnerable lichens, with special attention to mountain tops in the tropics – the most likely place for possible extinction of lichens as a result of global warming.

Climate Change: Observed Impacts on Planet Earth

2. PREDICTED EFFECTS

As a result of the attention paid to the effects of global warming on various groups of organisms and various ecosystems, some lichenologists have addressed the question of what effects global warming might have or have had on lichens.

Nash and Olafsen [4] predict that global warming in arctic areas may have a positive effect on lichens with cyanobacteria as photobiont, because the conditions for nitrogen fixation will improve. They reasoned that under field conditions of optimal water hydration, lichen photosynthesis is primarily light-limited and nitrogen fixation is temperature-limited in both *Peltigera canina* and *Stereocaulon tomentosum* at Anaktuvuk Pass, Alaska. Thus, they continued, 'where duration of optimal hydration conditions remains unchanged from the present-day climate, the anticipated temperature increases in the Arctic may enhance nitrogen fixation in these lichens more than carbon gain. Because nitrogen frequently limits productivity in Arctic ecosystems, the results are potentially important to the many Arctic and subarctic ecosystems in which such lichens are abundant'. The expected effect will be a spread of these species at the cost of other lichens and/or plants. So far, this has not been unequivocally observed; rather the contrary: lichens have recently decreased in arctic regions, probably due to the increase in phanerogams [5].

Insarov and Schroeter [6] and Insarov and Insarova [7] predict that lichens might, like other groups of organisms, show a response to global warming. As lichens are generally swift colonisers that disperse well, not only negative changes (extinctions) might be observed but also new invasions of more warmth-loving species in areas where they have not occurred before. In order to detect such changes, they installed some base-line monitoring transects across steep climatic gradients, but so far, no results have been reported.

Ellis and co-workers [3,8] predict the response, in terms of changed distribution on the British Isles, of groups of lichens with different current distribution patterns and known ecological preferences, based on the current distribution and on several different climate scenarios. Although numerous historic data are also available, no unequivocal correlation between global warming and past changes in the lichen flora of the British Isles has been shown.

Zotz and Baader [9] describe the different projected scenarios as regards lichens and bryophytes in the different biomes in the world.

Finally, as a result of widespread melting of glaciers, new habitats for (especially) stone-inhabiting lichen are being formed. However, only the pioneer species can be expected to benefit from this.

3. OBSERVED EFFECTS

So far, few studies have demonstrated a correlation between global change and change in lichen habitat. The study by van Herk et al. [1] was the first and only one reported in the meta-analysis by Parmesan and Yohe [10] in their study of 'globally coherent fingerprint of climate change impacts across natural systems'. The lichen study was based on a long-term (22 a) monitoring involving all the 329 epiphytic and terrestrial lichen species occurring in the Netherlands and were considered in relation to their world distribution. The investigation focussed on the exposed wayside trees in the province Utrecht in the Netherlands. The research was initially started to document changes resulting from changes in sulphur dioxide air pollution levels. When the levels dropped, the effects on the lichens were clearly visible. However, the pattern was disturbed by a new emergent air pollution problem – ammonia from increasingly intensive cattle farming. As different lichens show different responses to this pollutant, the lichen monitoring was continued for a different purpose, viz. a detailed mapping of the areas with problematic ammonia pollution. Changes between 1995 and 2001, however, could not be explained in terms of air pollution variables alone. Analysis, however, showed a positive correlation with temperature, oceanity and nutrient demand, indicating a recent and significant shift towards species preferring warmer circumstances, independent from, and concurrent with changes due to nutrient availability. In short, warmth-loving, oceanic lichens are expanding and boreal lichens are diminishing.

The lichens that are expanding most dramatically are those with the green algae *Trentepohlia* as their photobiont. As these lichen species (i.e., the mycobiont) belong to different unrelated taxonomic groups and the effect has been observed in different ecosystems (exposed trees, forests), Aptroot and van Herk [11] argue that it seems likely that the effect of the global warming is, in fact, directly related to the alga, and all lichens with this alga can profit from the expansion of their photobiont. The process as described here is continuing and probably even accelerating. A recent study by van Herk [12] shows that most of the recent changes can be now attributed to global warming (see Figs. 1 and 2).

4. UNCERTAIN EFFECTS

Some observed and reported changes in the lichen flora cannot be unequivocally attributed to global warming. There are several reasons for this but the most common one is that comparable historic or background data are wanting. Also climate change is not the only change taking place and some of the changes occurring locally may interact or even counteract. Examples are isolated finds of warmth-loving species in more boreal countries, like *Flavoparmelia caperata* in Denmark, reported by Søchting [13] and attributed by him to global warming.

Another type of uncertainty is the intermittent and sometimes devastating effects of El Nino on coastal lichens along the Pacific coast of South America and on the Galapagos Islands. These have been documented, for example, by Follmann [14] and attributed directly to El Nino. The question remains

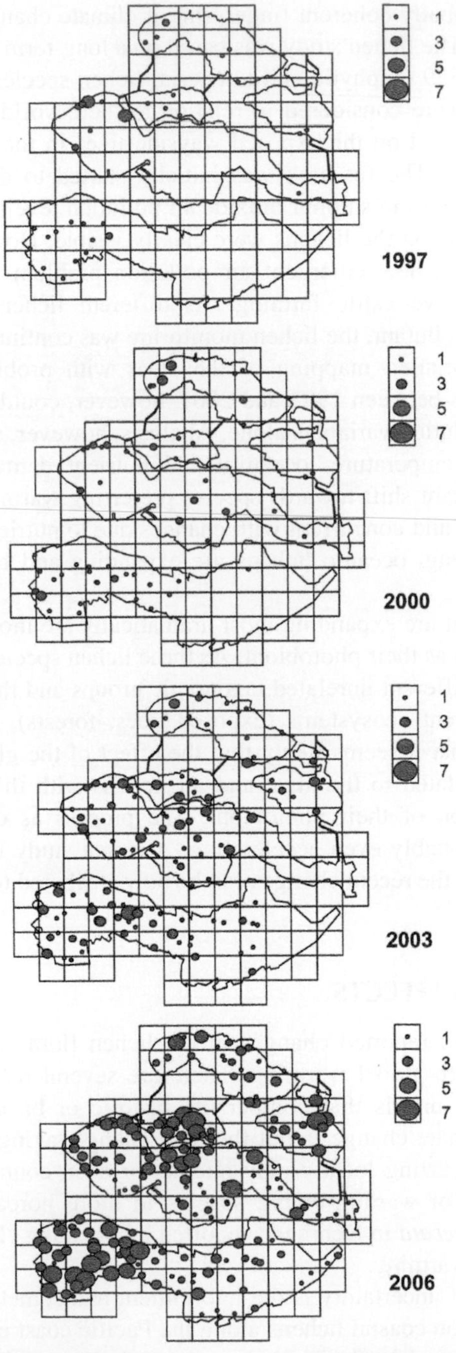

FIGURE 1 The distribution of lichen species with *Trentepohlia* phycobiont in the Netherlands province of Zeeland, in 1997, 2000, 2003 and 2006. The dot size refers to the number of species per site [12].

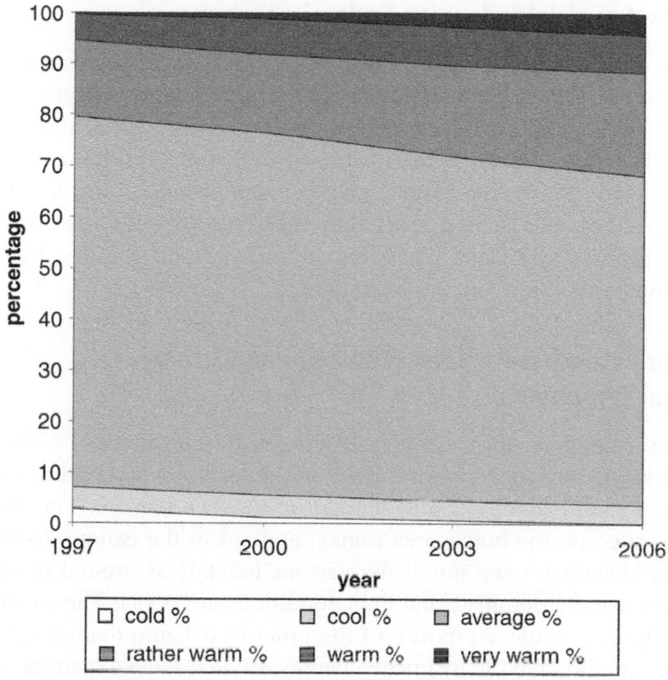

FIGURE 2 Changes of the epiphytic lichen composition in the Netherlands province of Zeeland in relation to temperature preference. The percentages are derived from species' frequencies per year. The total number of species for which a temperature preference is known is given as 100% [12].

whether the intensity of the El Nino effects is changing due to global warming, or not. In any event, lichens appear to have suffered more during the past few decades than ever before.

In some cases the patterns and processes are confused. An example is the reported work by Cezanne and co-workers [15] claiming that changes in lichen were indicators of climate change. However, all the observations were made within a year and the various stations were visited only once; the paper describes the correlation of the lichen vegetation with climatic parameters, but only a spatial pattern is shown. In summary the conclusion made, does not bear up to scientific scrutiny.

5. HABITATS WITH VULNERABLE LICHENS

There are four main habitats where lichens are potentially most vulnerable to climate change in the form of global warming, changes in precipitation and changes in the incidence of fog.

5.1. Low Level Islands with Endemic Lichens

Examples of such islands include Porto Santo [16–18] and Bermuda [19]. The fact that some of the lichens on these islands are endemic suggests that they are either not capable of dispersal and/or their ecological requirements are not met elsewhere. In the event of a marked temperature rise or a change in the incidence of fog, the climate may become unsuitable for these species, and the chances of reaching a suitable substitute location are remote. The risks are highest at islands without mountains, as no suitable habitat will become available higher up the mountains.

5.2. The Extended Regions with Similar Climate but Local Endemism

The main examples are the extensive tropical rain forests. Although the climate and the physiognomy of the vegetation can be very similar over large areas, there can be a considerable amount of local endemism. This endemism is concentrated on the higher tree trunks, and not in the canopy (where wind moves the diaspores) and not at the various habitats at ground level (where bryophytes usually dominate and light conditions are poor). The endemic species usually have large acospores, of the order of 0.1 mm diameter. The risks to these lichen involved in a climate change, are that large expanses of habitat will change simultaneously, and the species with large diaspores have little chance of reaching a relatively remote suitable habitat. Incidentally, this risk may be small compared to the more direct and imminent risk of habitat destruction by logging. Furthermore, it has been pointed out by Zotz and Baader [9] that if tropical coastal regions become warmer, no species may continue to exist that are capable of occupying the habitats that become available as a result of other species shifting to higher elevations.

5.3. The (Ant-)Arctic and Tundra Regions

These areas are very rich in lichens, which often dominate the vegetation, both in biomass and in species diversity. Some lichens have been shown to decline, possibly indirectly as a result of global warming, due to increases in vascular plant biomass [5]. This is a potential threat to the rich (ant-)arctic lichen flora, but cannot be considered as an immediate one, as most (ant-) arctic lichens are relatively abundant and widespread, and a major impact will only occur in the unlikely event of the whole (ant-)arctic biome collapsing.

5.4. High Ground in the Tropics

High mountains in tropical areas sustain a rather depauperate lichen flora consisting predominantly of species widespread in boreo-alpine areas elsewhere in the world, but also including local endemics. These species have nowhere to go, other than literally in air, in the case of global warming. The mountains in New Guinea are examples of this group. They are among the most isolated biomes, as they are not connected to temperate regions, as, for example, the Andes are. Mount Wilhelm, reaching about 4500 m, is the highest mountain in Oceania, and from a lichenological point of view is the best investigated high mountain in New Guinea. It is also the richest in lichen species, as several other mountain tops are grass-covered. This is an isolated mountain, of which only less than 100 km^2 lies above the tree line and is at least partly suitable for boreo-alpine terricolous and saxicolous lichen growth. Among these are many cosmopolitan species [20,21]. The species, virtually on the equator, must be considered as 'boreo-alpine' or 'temperate' in a climatic (not geographic) sense. They cannot be considered as 'circumpolar' or 'bipolar' as is often stated [22]. For these New Guinean lichens their next closest localities are in Taiwan, over 4000 km away, and in the Himalayas, more than 5000 km away. How the species actually arrived remains unknown, although the presence of relatively many species that are associated with bird perching suggests that birds may have played an important role as vector of lichen diaspores, next to or even instead of wind and air currents.

The alpine lichen zone on Mount Wilhelm is restricted to a narrow altitudinal belt, above the tree limit at 3900 m to about 4300 m. This belt is known in botanical and tourism descriptions of the vegetation and the climb, as the 'dead lichen zone', because the abundant *Thamnolia* is mistaken for dead lichens. The area consists of a granite bedrock with large boulders, vertical cliffs and horizontal stretches with some soil compaction supporting heath-like dwarf shrub vegetation. This is a small zone where the recently described endemic *Sticta alpinotropica* [23] occurs on rocks, and the equally endemic *Thamnolia juncea* [24] is found in the (sub-)alpine grassland. The known world populations of both species amount to only a few square metre. Below the tree limit, the availability of various susbstrates for lichens is much wider, and the lichen diversity in the cloud forest belt is very high. This is the zone where numerous endemic species occur, for example, of the genera *Anzia* [25] and *Menegazzia* [26].

6. CONCLUSION

Lichens are unequivocally responding to global change. The effects are, so far, apparent only in the last two decades (since ca. 1990) and in the temperate region only. Lichens have indirectly suffered from global change effects in arctic regions. The most severe effects of climate change, leading to probable extinctions, is expected (but has not been observed as yet) on high mountains in tropical regions.

ACKNOWLEDGEMENT

Kok van Herk is warmly thanked for comments on a draft of this paper.

REFERENCES

1. C.M. van Herk, A. Aptroot, H.F. van Dobben, Lichenologist 34 (2002) 141–154.
2. A. Aptroot, Bull. Br. Lichen Soc. 96 (2005) 14–16.
3. C.J. Ellis, B.C. Coppins T.P. Dawson, Biol. Conserv. 135 (2007) 396–404.
4.. T.H. Nash, A.G. Olafsen, Lichenologist 27 (1995) 559–565.
5. J.H.C. Cornelissen, T.V. Callaghan, J.M. Alatalo, A. Michelsen, E. Graglia, A.E. Hartley, D.S. Hik, S.E. Hobbie, M.C. Press, C.H. Robinson, G.H.R. Henry, G.R. Shaver, G.K. Phoenix, D.G. Jones, S. Jonasson, F.S. Chapin III, U. Molau, C. Neill, J.A. Lee, J.M. Melillo, B. Sveinbjörnsson R. Aerts, J. Ecol. 89 (2001) 984–994.
6. G. Insarov, H. Schroeter, in: P.L. Nimis, C. Scheidegger, P.A. Wolseley (Eds.), Monitoring with Lichens – Monitoring Lichens, Kluwer, Dordrecht, 2002, pp. 183–202.
7. G. Insarov, I. Insarova, Bibl. Lichenol. 82 (2002) 209–220.
8. C.J. Ellis, B.C. Coppins, T.P. Dawson, M.R.D. Seaward, Biol. Conserv. 140 (2007) 217–235.
9. G. Zotz, M.Y. Baader, Prog. Bot. (2009) 70, 147–170.
10. C. Parmesan, G. Yohe, Nature 421 (2003) 37–42.
11. A. Aptroot, C.M. van Herk, Environ. Pollut. 146 (2007) 293–298.
12. C.M. van Herk, Bibl. Lichenol. 99 (2008) 207–225.
13. U. Søchting, Graph. Scr. 15 (2004) 53–56.
14. G. Follmann, Cryptogamic Bot. 5 (1995) 224–231.
15. R. Cezanne, M. Eichler, U. Kirschbaum, U. Windisch, Sauteria 15 (2008) 159–174.
16. H. Krog, H. Østhagen, Norweg. J. Bot. 27 (1980) 185–188.
17. H. Krog, Lichenologist 22 (1990) 241–247.
18. R. Haugan, Mycotaxon 44 (1992) 45–50.
19. L.W. Riddle, Bull. Torrey Bot. Club 43 (1916) 145–160.
20. A. Aptroot, P. Diederich, E. Sérusiaux, H.J.M. Sipman, Bibl. Lichenol. 64 (1997) 1–220.
21. H. Streimann, Bibl. Lichenol. 22 (1986) 1–145.
22. D.J. Galloway, A. Aptroot, Cryptogamic Bot. 5 (1995) 184–191.
23. A. Aptroot, Lichenologist 40 (2008).
24. R. Santesson, Symb. Bot. Ups. 34 (1) (2004) 393–397.
25. I. Yoshimura, H.J.M. Sipman, A. Aptroot, Bibl. Lichenol. 58 (1995) 439–469.
26. P.W. James, A. Aptroot, P. Diederich, H.J.M. Sipman, E. Sérusiaux, Bibl. Lichenol. 78 (2001) 91–108.

Chapter 24

Coastline Degradation as an Indicator of Global Change

Robert J. Nicholls
School of Civil Engineering and the Environment and the Tyndall Centre for Climate Change Research, University of Southampton, Southampton SO17 IBJ, UK

Colin Woodroffe
School of Earth and Environmental Sciences, University of Wollongong, NSW 2522, Australia

Virginia Burkett
U.S. Geological Survey, 540 North Courthouse street, Many, LA 71449, USA

1. Introduction
2. Sea-Level Rise and Coastal Systems
3. Climate Change and Global/ Relative Sea-Level Rise
4. Increasing Human Utilisation of the Coastal Zone
5. Climate Change, Sea-Level Rise and Resulting Impacts
6. Recent Impacts of Sea-Level Rise and Climate Change
7. Global Warming and Coasts at Latitudinal Extremes
8. The Challenge to Understand Contemporary Impacts
9. Concluding Remarks
Acknowledgements
References

1. INTRODUCTION

Coastal degradation has been widely reported around the world's coasts over the past century, and especially in recent decades as discussed later in this chapter [1,2]. This degradation can be attributed to the intensification of a wide range of drivers of coastal change that are linked directly and indirectly to an expanding global population and economy. The twentieth century was also characterised by recognition of human-induced climate change and sea-level rise, which constitutes an additional set of coastal drivers [3]. This chapter explores the relative contribution of climate change to observed coastal changes, focusing particularly on the extent to which climate change can be attributed as a significant driver of the change.

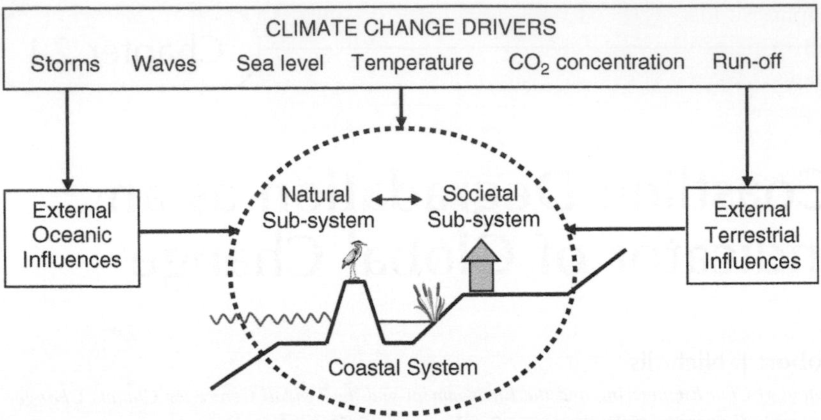

FIGURE 1 The coastal system showing how it is impacted by climate change. The natural environment and coastal inhabitants interact directly, and are affected by external terrestrial and marine issues. Climate change, including sea-level rise, can directly or indirectly effect the coastal system (as can non-climate drivers of change). (Adapted from Ref. [3]).

An analytical framework is adopted, based on a systems view of coasts as defined in Fig. 1. Comprising the narrow interface between land and sea, coastal systems are influenced by both marine and land surface processes. Coastal systems include intertidal zones and adjacent coastal lowlands and bays, lagoons, estuaries and nearshore waters. The connectivity of coasts with both marine and terrestrial systems is responsible, in part, for the high variability and complexity among coastal system types. In contrast to terrestrial systems that have physical gradients that can stretch over tens or thousands of kilometres, coastal biotic and abiotic gradients are often relatively short, particularly along steep rocky shores. Many coastal areas support large and growing populations and high economic activity [4,5], which are changing coastal environments. River catchments feeding to the coast are increasingly modified, such that coastal systems are also influenced by these external changes [6]. Hence, few of the world's coastlines are now beyond the influence of human pressures [7], with many being dominated by human activities [8] and most coastal systems include elements of human development that interact with environmental changes associated with a warming climate.

Global warming through the twentieth century has caused a series of changes with important implications for coastal areas (Fig. 1). These include rising temperatures (both air and sea surface temperatures), rising sea level, increasing CO_2 concentrations with an associated reduction in seawater pH, and more intense precipitation on average (with substantial regional variation). It has also been argued that tropical storms have become more intense [9]. The tragic impacts of Hurricane Katrina on the Gulf of Mexico coast of the United States in 2005 and of Cyclone Nargis on Myanmar in 2008 emphasise the enormous devastation that these events cause, but it cannot be shown that these individual events were more intense as a result of climate change, and no firm conclusions on intensification of storms can be drawn at present.

Sea-level rise is one of the most widely cited outcomes of global warming. Rising global sea level due to thermal expansion and the melting of land-based ice is already being observed with a global-mean rise of 17 ± 5 cm during the twentieth century [9] and a slow accelerating trend [10]. Higher sea level will directly impact coastal areas, including some of the most densely-populated and economically active land areas on Earth.

In this chapter, we outline historical climate and sea-level change and discuss how this impacts coasts, but we also recognise that coastal systems are subject to many other drivers, most especially the impacts of human development. We further discuss the need to discriminate whether coastal degradation can be attributed to the effects of climate or to what degree they are related to non-climate drivers.

2. SEA-LEVEL RISE AND COASTAL SYSTEMS

Since the peak of the last glacial maximum about 20 000 a (years) ago, sea level has risen ~125 m [11]. Geologic evidence indicates inundation of coastal lowlands and retreat of shorelines during periods of rapid sea-level rise, such as major meltwater pulses (Fig. 2). This pattern of sea-level rise was experienced around the world, driven by the melt of the large ice sheets

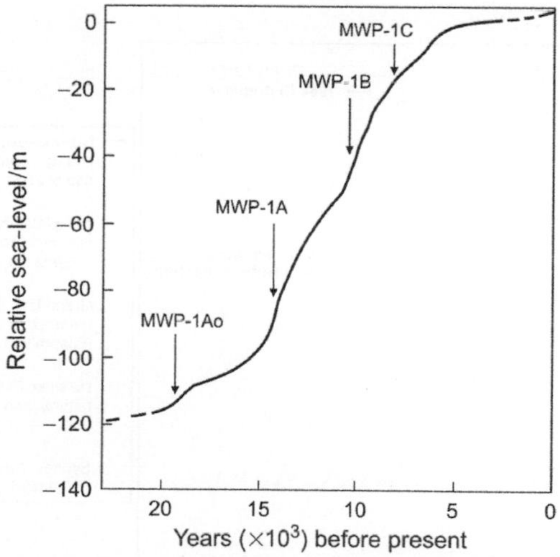

FIGURE 2 Sea-level history since the peak of the last glacial maximum with arrows indicating the timing of meltwater pulses. Abbreviations: MWP = meltwater pulse. MWP-1A0, c. 19000 a ago, MWP-1A, 14600–13500 a ago, MWP-1B, 11500–11000 a ago, MWP-1C, ~8200–7600 a ago (Source: Ref. [12]).

which appears to have ceased 7500–6000 a ago. The level of the sea has risen less than 3 m over the past 6000 a and regional variations of sea level on time scales of a few 100 a or longer are likely to have been less than 0.3–0.5 m [13]. As sea level stabilised extensive coastal plains were formed, and the first evidence of early civilisations appeared on the plains [14,15].

Coastline location and stability is intimately linked with changes in mean sea level. However, even under conditions of relatively stable mean sea level, coasts are extremely dynamic systems, involving co-adjustment of form and process at different time and space scales, termed morphodynamics [16,17]. Hence, erosion and deposition of coasts are naturally occurring due to short-term wave and tide conditions, as well as seasonal and longer-term climatic variability. The El Niño phenomenon, for example, has been shown to influence wave processes that shape beaches in the southwest Pacific [18] and cliffs in the eastern Pacific [19].

3. CLIMATE CHANGE AND GLOBAL/RELATIVE SEA-LEVEL RISE

The impacts and responses of coasts to sea-level rise are a product of relative (or local) sea-level rise rather than global changes alone. Relative sea-level rise takes into account global-mean sea-level rise, regional trends in the absolute elevation of the ocean surface, and geological uplift or subsidence and related processes which change the position of the land/sea boundary. Relative sea-level rise is only partly a response to climate change and can vary significantly among coastal systems (Fig. 3). Abrupt changes may occur, for example, where an earthquake causes rapid vertical displacement of a part

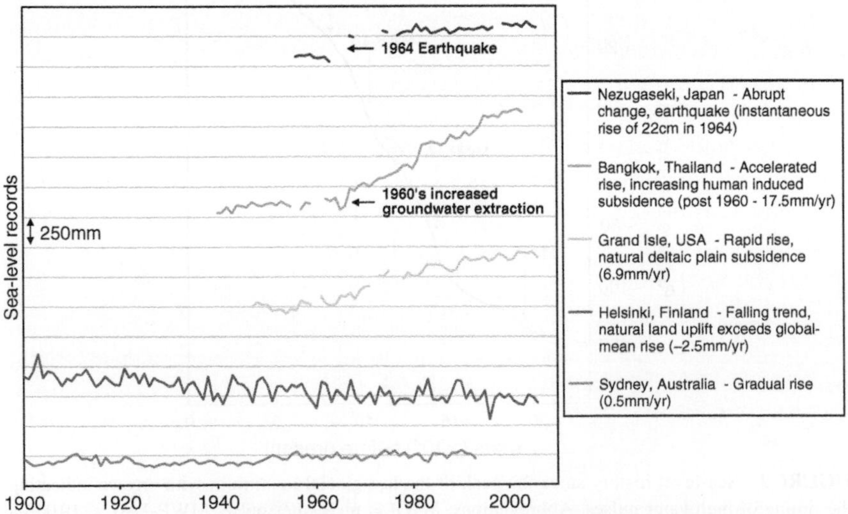

FIGURE 3 Selected relative sea-level records for the twentieth century, illustrating different types of trend. The records are offset for display purposes. Source: http://www.pol.ac.uk/psmsl/.

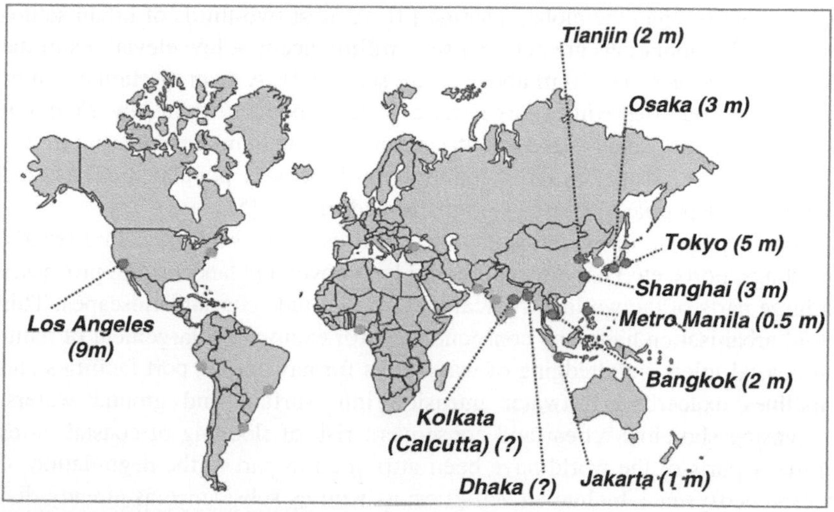

FIGURE 4 Subsiding coastal megacities with the maximum observed subsidence (in m) (adapted from Ref. [23]). Subsidence in Los Angeles was very localised (about 1 km^2) and due to oil extraction. Dhaka and Kolkata are thought to be subsiding, but data is limited.

of the Earth's surface (see Nezugaseki; Fig. 3). Sea level is presently falling due to ongoing glacial isostatic adjustment (rebound) in some high-latitude locations that were formerly sites of large (kilometre-thick) glaciers, such as Hudson Bay and the northern Baltic (see Helsinki, Fig. 3). In contrast, sea level is rising more rapidly than global-mean trends on subsiding coasts, including deltas such as the Mississippi delta (see Grand Isle, Fig. 3), the Nile delta, and the large deltas of south and east Asia [20,21]. Most dramatically, human-induced subsidence of susceptible areas due to drainage of organic soils and withdrawal of groundwater can produce dramatic rises in relative sea level, especially in susceptible coastal areas and cities built on recently-deposited deltaic landforms [22]. Four noteworthy examples over the twentieth century are parts of Tokyo and Osaka which subsided up to 5 and 3 m, respectively, most of Shanghai which subsided up to 3 m, and nearly all of Bangkok which subsided up to 2 m (see Bangkok; Figs. 3 and 4). As a management response to human-induced subsidence, stopping shallow sub-surface fluid withdrawals can reduce subsidence.

4. INCREASING HUMAN UTILISATION OF THE COASTAL ZONE

Human use of the coast increased dramatically during the twentieth century. It has been estimated that 37% of the world's population lives within 100 km, and 49% lives within 200 km, of the coast [24]; the greatest number of people live at low elevations and population densities in coastal regions are about

3 times higher than the global average [4]. Almost two-thirds of urban settlements with populations greater than five million occur at low elevations in the coastal zone (less than 10 m above mean sea level). A disproportionate number of the countries with a large share of their population in low elevation coastal zones are small island countries. Most of the human population in this zone, however, resides in large countries with densely populated deltas, and migration of people to coastal regions is widespread [5].

The expansion of human settlements and associated infrastructure (roads, buildings, ports, etc.) has directly altered land cover and land surface processes in large parts of the world's tropical and mid-latitude coastal landscapes. This rapid urbanisation has many consequences; for example, enlargement of natural coastal inlets and dredging of waterways for navigation, port facilities and pipelines exacerbate saltwater intrusion into surface and ground waters. Increasing shoreline retreat and consequent risk of flooding of coastal cities in many parts of the world have been attributed in part to the degradation of coastal ecosystems by human activities, as well as subsidence as already discussed [3]. As a result of this, cities often progressively move to artificial defensive and drainage systems as they develop/expand and their influence on their environs increases.

The natural ecosystems within watersheds have been fragmented and the downstream flow of water, sediment and nutrients to the coast disrupted [25]. Land-use change and hydrological modifications have had downstream impacts, in addition to localised influences, including human development on the coast. Hillslope erosion has increased the sediment load reaching the coast; for example, suspended loads in the Huanghe (Yellow) River have increased 2–10 times over the past 2000 a [26]. In contrast, damming and channelisation has greatly reduced the supply of sediments to the coast on other rivers through retention of sediment in dams [6], and this effect has dominated through the twentieth century [27,28].

The structure and ecological functions of natural systems are altered as a result of population growth, and ecological services provided by coastal systems are often disrupted directly or indirectly by human activities. For example, tropical and subtropical mangrove forests provide goods and services because they accumulate and transform nutrients, support rich ecological communities of fish and crustaceans, attenuate waves and storm surge impacts, and their root systems trap and bind sediments [29,30]. Large-scale conversions of coastal mangrove forests to shrimp aquaculture have occurred during the past three decades along the coastlines of Asia and Central America [31], and the decline or loss of mangrove forests reduces all of these ecosystem services [32]. Similar reductions of temperate salt marshes and wetlands in deltas are often linked to direct land use change [33,34]. Hence, on those developed coasts that have experienced disproportionately rapid expansion of settlements, urban centres, and tourist resorts, the direct impacts of human activities on the coastal zone are profound, with more widespread indirect effects of human activities.

5. CLIMATE CHANGE, SEA-LEVEL RISE AND RESULTING IMPACTS

Relative sea-level rise has a wide range of effects on the natural system; the main effects are summarised in Table 1. Flooding/submergence, ecosystem change and erosion have received significantly more attention than salinisation and rising water tables. Rising sea level alters all coastal processes. The immediate effect is submergence and increased flooding of coastal land, as well as saltwater intrusion into surface waters. Longer term effects also occur as the coast adjusts to the new environmental conditions, including

TABLE 1 The main natural system effects of relative sea-level rise, including climate and non-climate interacting factors

Natural system effect			Interacting factors: Climate	Interacting factors: Non-climate
1. Inundation (including flood and storm damage)	a. Surge (from the sea)		Wave/storm climate, erosion, sediment supply	Sediment supply, flood management, erosion, land reclamation
	b. Backwater effect (from rivers)		Run-off	Catchment management and land use
2. Morphological Change	a. Wetland loss (and change)		CO_2 fertilisation of biomass production, sediment supply, migration space	Sediment supply, migration space, land reclamation (i.e., direct destruction)
	b. Erosion (of beaches and soft cliffs)		Sediment supply, wave/storm climate	Sediment supply
3. Hydrological change	a. Saltwater intrusion	i. Surface waters	Run-off	Catchment management (over-extraction), land use
		ii. Ground-water	Rainfall	Land use, aquifer use (over-pumping)
	b. Rising water tables/ impeded drainage		Rainfall, run-off	Land use, aquifer use, catchment management

Some interacting factors (e.g., sediment supply) appear twice as they can be influenced both by climate and non-climate factors.

wetland loss and change in response to higher water tables and increasing salinity, erosion of beaches and soft cliffs and saltwater intrusion into groundwater. These lagged changes interact with the immediate effects of sea-level rise and generally exacerbate them. For instance, coastal erosion will tend to degrade or remove natural protective features (e.g. saltmarshes, mangroves and sand dunes) that in turn increase extreme water levels and hence the risk of coastal flooding.

A rise in mean sea level also has a net effect of intensifying flooding during extreme storm events [35]. Changes in storm characteristics could have also influenced extreme water levels. Increases in tropical cyclone intensity in the North Atlantic over the past three decades are consistent with the observed changes in sea surface temperatures [9] and wave data in the North Atlantic support this observation [36]. However, it is difficult to prove if this is a systematic change or a component of cyclic variations in the frequency and intensity of tropical storms. Changes in storm tracks might also result from global climate change; in this context, Cyclone Catarina was the first documented hurricane in the South Atlantic, striking the coast of Brazil in March 2004 as a Category 2 storm on the Saffir–Simpson Hurricane Scale [37,38]. The cyclone killed at least three people and caused an estimated US $\$350 \times 10^6$ in damage in Brazil, and it is unclear whether this indicates an extremely unusual event, or the beginning of a new trend under global warming.

Changes in the natural system due to sea-level rise have many important direct socio-economic impacts on a range of sectors with the effect being overwhelmingly negative. For instance, flooding can damage key coastal infrastructure, the built environment, and agricultural areas, and in the worst case lead to significant mortality as occurred in 2008 when Cyclone Nargis devastated southern Myanmar. Erosion can lead to losses of the built environment and related infrastructure and have adverse consequences for sectors such as tourism and recreation. In addition to these direct impacts, there are indirect impacts such as negative effects on human health. For example, mental health problems increase after a flood [39], or the release of toxins from eroded landfills and waste sites which are commonly located in low-lying coastal areas, especially around major cities (e.g. Ref. [40]). Thus, sea-level rise has the potential to trigger a cascade of direct and indirect human impacts.

6. RECENT IMPACTS OF SEA-LEVEL RISE AND CLIMATE CHANGE

Sea level was relatively stable in the sixteenth to eighteenth centuries; it started to rise in the nineteenth century and rose about 20 cm by the end of the twentieth century, with a global rise of 17 ± 5 cm rise in that century [9,41]. Although this change may seem small, it has had many significant effects, most particularly in terms of the return periods of extreme water levels [35,42]. Worldwide there are many coasts that have been observed to be eroding [43]. However, attributing particular impacts such as erosion to sea-level rise is difficult as erosion can be promoted by processes other than sea-level rise (Table 1). As already discussed, many of these non-climate drivers of change operated over the twentieth century. While sea-level rise is often inferred as an underlying cause of widespread retreat of sandy shorelines [44], negative sediment budgets also lead to erosion [17]. Human reduction in sediment supply to the coast has contributed to observed changes through activities such as construction of levees, dikes and dams on rivers that drain to the coast [6,45]. Equally, changes in flooding and flood risk are difficult to attribute to global sea-level rise. For instance, flood defences have often been upgraded substantially through the twentieth century, especially in those (wealthy) places where there are sea-level measurements. Most of this defence upgrade reflects expanding populations on the coastal plains and changing attitudes to risk. In many places, relative sea-level rise has rarely even been considered in the design of past coastal infrastructure.

The accelerated rate of sea-level rise observed since the late 1800s has been accompanied by coastal erosion and rapid wetland losses in many low-lying coastal regions. On the US east coast, relative sea levels have risen at rates of between 2 and 4 $mm \cdot a^{-1}$ over the twentieth century due to varying patterns of subsidence caused by glacial isostatic adjustment. Both rates of sea-level rise and coastal retreat have been measured, providing the opportunity to explore shoreline response to sea-level rise. Away from inlets and engineered shores, the shoreline retreat rate is 50–100 times the rate of sea-level rise, as might be anticipated using the concept of the Bruun Rule [46]. Near inlets, the indirect effects of sea-level rise which cause the associated estuary/lagoon to trap beach sediment can have much larger erosional effects on the neighbouring open coasts than predicted by the Bruun Rule [47]. So, whereas a simple heuristic like the Bruun rule describes the relationship for some shores, more general relationships are required to fully understand coastal change, taking account of sea-level change, sediment supply and coastal morphology [17].

In coastal Maryland and Louisiana, for example, wetland losses and shoreline retreat have led to a rapid restructuring of coastal ecosystems [33,48,49]. In Florida, a decline in coastal cabbage palm forests since the 1970s has been attributed to salt water intrusion associated with sea-level rise [50,51]. Due to extensive human development along these coastlines, it is not possible to quantitatively isolate climate change effects versus changes due to other human development activities.

Human responses to sea-level rise are even more difficult to document. A rare example is human abandonment of low-lying islands in Chesapeake Bay, USA, during the late nineteenth/early twentieth century which seems to have been triggered by the acceleration of sea-level rise and resulting land loss [52].

There have certainly been impacts from relative sea-level rise resulting from large rates of subsidence, such as the Mississippi delta where relative

FIGURE 5 A line of telegraph poles south of Bangkok, Thailand: built on subsiding land, they are now up to 1 km out to sea.

sea-level rise approaches 1 $cm{\cdot}a^{-1}$! (see Grand Isle, Fig. 3). Between 1978 and 2000, 1565 km^2 of intertidal coastal marshes and adjacent lands were converted to open water, due to sediment starvation and increases in the salinity and water levels of coastal marshes as a result of human development activities coupled with high rates of relative sea-level rise [53]. The flooding in New Orleans during Hurricane Katrina was significantly exacerbated by subsidence compared to earlier flood events such as Hurricane Betsy in 1965 [54]. Coastal retreat has occurred due to subsidence, such as south of Bangkok where shoreline retreat has been more than 1 km (Fig. 5). However, all the major cities that were impacted by relative sea-level rise have been defended, even when the change in relative sea-level rise was several metres.

Hence, while global sea-level rise has been a pervasive process, it is difficult to unambiguously link it to impacts, except in some special cases; most coastal change in the twentieth century was a response to multiple drivers of change. However, changes in two contrasting environments, polar coasts and tropical reefs, do appear to be directly exacerbated by warmer temperatures.

7. GLOBAL WARMING AND COASTS AT LATITUDINAL EXTREMES

Global warming poses a particular threat to coasts at the latitudinal extremes, polar coasts and coral reefs. Polar coasts are experiencing permafrost melt and a decrease in the extent of sea ice as result of warming which is leading to a significant acceleration in erosion rates. Rapid shoreline erosion has been occurring on parts of the Arctic coast over recent decades, attributed in part to reduced sea ice cover allowing more wave activity to reach the shoreline [55]. Reduction in thickness of near-coastal ice, more rapid ice movement and retreat of the glacier fronts in Greenland appears related to warmer temperatures [56,57]. Similar trends to Greenland have been reported from the Antarctic Peninsula [58,59].

Parts of the Alaska coastline on the Beaufort Sea have retreated as much as 0.9 km in the past 50 a (Fig. 6). This coastal region is exposed to a combination of factors relating to climate change – sea-level rise, the thawing of permafrost and the reduction in sea ice that protects that coastline from

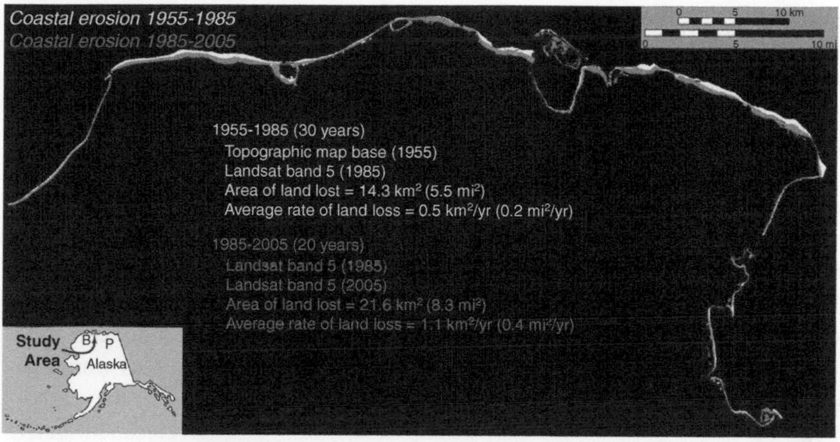

FIGURE 6 An example of land loss due to coastal erosion in northern Alaska over 50 years (1955–1985 and 1985–2005) based on ground survey and satellite (Landsat) measurements (Source: J.C. Mars and D.W. Houseknech, Unpublished work, U.S. Geological Survey, Reston, VA, USA, 2007) (see Ref. [60]).

erosion during part of the year – all of which are contributing to rapid shoreline retreat. Erosion at the coastline has led to the breaching of thermokarst lakes, causing initial draining followed by an increase in marine flooding that alters plant and animal community structure [60]. Similar retreat is occurring at sites in Arctic Canada [61], and evidence documented from traditional ecological knowledge also points to widespread change of coastlines across the North American Arctic from the Northwest Territories, Yukon and Alaska in the west to Nunavut in the east [62]. However, the impacts associated with human settlement along polar coasts are relatively very low due to the low population in these regions.

Within the tropics, widespread coral bleaching was detected on an unprecedented scale around the globe in response to El Niño-related warming in 1998 [63,64]. Further bleaching occurred across much of the Great Barrier Reef off northeastern Australia in 2002 [65] and in the Caribbean in 2005 [66]. Bleaching occurs when warmer than usual sea surface temperatures lead to expulsion of the symbiotic zooxanthellae from within the coral tissue; the coral surface becomes pale, in many cases leading to mortality. It seems that temperatures ~1 °C above the monthly maximum experienced by the coral result in bleaching, and that persistently high temperatures, or temperatures more than 2 °C above this threshold, can cause the coral to die. Threshold temperatures above which corals bleach have evidently been occurring more frequently [67–69], and the prospect of further global warming implies that reefs may bleach with a frequency that exceeds their ability to recover between events.

Coral reefs are also susceptible to many other stresses, and there are many reefs that are severely degraded as a consequence of human activities, particularly overfishing and pollution [70]. As with other considerations of coastal degradation it is difficult to disentangle the effects of human-induced pressures from those that result directly from climate change. The synergistic effects of various pressures combine to affect reefs, but the occurrence of bleaching on remote reefs well away from direct human development, and its incontrovertible association with increased sea surface temperature provides a salutary warning of the likely consequences should global warming continue unabated. Human impacts, such as overfishing, appear to be exacerbating the stresses on reef systems and, at least on a local scale, exceeding the thresholds beyond which coral is replaced by other organisms [71]. Nevertheless, as with polar coasts, it is difficult to avoid the conclusion that these remote coastlines are changing for the worst as a consequence of rising sea surface temperatures.

8. THE CHALLENGE TO UNDERSTAND CONTEMPORARY IMPACTS

While significant coastal degradation has occurred over the twentieth century it is difficult to unambiguously attribute the relative role of climate change. Most degradation has occurred on coasts that are influenced by one or more non-climate related drivers such as ongoing tectonic or isostatic adjustments, or, increasingly often, as a result of human activities. Further, the magnitude of climate change to date remains relatively small. In the next few decades, global warming will continue and is expected to accelerate, resulting in climate-induced impacts becoming more apparent.

In some coastal regions it is possible to discriminate between those effects that can already be attributed to climate change. Rising air and sea surface temperatures have resulted in detectable impacts on polar and tropical coasts. There is an emerging consensus that the increased frequency of bleaching on coral reefs is related to higher sea surface temperatures. Melting of sea ice and permafrost in high latitudes results from increased temperatures, and this is related to rapid erosion of polar coasts. However, these coasts were already experiencing extensive erosion, and there is no clear procedure for differentiating how much erosion would have been occurring because of ongoing factors, such as isostatic adjustments of the land, and how much additional retreat has occurred because of climate change.

A significant component of global-mean sea-level rise also results from global warming, primarily because of thermal expansion, but with a component from ice melt. Discriminating the impacts of the global-mean sea-level component at regional and local scales where other contributions to relative sea-level change are of variable importance remains problematic. This presents a challenge to further test and refine our understanding about the impacts of climate on coasts, so that better predictions can be made and management plans put in place to respond to the anticipated impacts.

To meet this challenge, it will be necessary to continue and expand monitoring of coastlines, including both the climate and non-climate drivers, and the responses of coastal systems. Climate change is a global phenomenon, and therefore this monitoring and analysis needs to consider changes over broad scales. There will be an increasing role for more sophisticated remote sensing which will be an important tool [34,60]. Comparative studies offer the opportunity to assess sensitivity, comparing those coasts with intense human pressures with more pristine counterparts in less densely populated regions. However, as indicated above, the indirect effects of human modification of the Earth are leaving a pervasive signal in even these remote places; global sea-level change effects those coasts that are uninhabited as well as those that are intensively developed. Studies of analogues of climate change and sea-level rise are also relevant, such as relative sea-level rise on subsiding coasts which can provide insights into outcomes expected more widely in response to global warming induced sea-level rise.

9. CONCLUDING REMARKS

Finding a climate change signal on coasts is more problematic than often assumed. Coasts undergo natural dynamics at many scales, with erosion and recovery in response to climate variability such as El Niño, or extreme events such as storms and infrequent tsunamis. Additionally, humans have had enormous impacts on most coasts, overshadowing most changes that we can presently attribute directly to climate change.

Using the geographic examples cited in this paper, various impacts can be inferred on coasts as a consequence of changes in climate. However, each area of coast is experiencing its own pattern of relative sea-level change and climate change, making discrimination of the component of degradation that results from climate change problematic. The best examples of a climate influence are related to temperature rise at low and high latitudes, as seen by the impacts on coral reefs and polar coasts, respectively. Observations through the twentieth century demonstrate the importance of understanding the impacts of sea-level rise and climate change in the context of multiple drivers of change; this will remain a challenge under a more rapidly changing climate.

Nevertheless, there are emerging signs that climate change provides a global threat – sea ice is retreating – permafrost in coastal areas is widely melting – reefs are bleaching more often – and the sea is rising, amplifying widespread trends of subsidence and threatening low-lying areas. From this analysis some important lessons about the response to these challenges become evident. To devise successful response strategies for coastal degradation it will be important to understand coastal changes in the context of integrated assessment and multiple drivers of change, with climate only being part of the problem [72]. To enhance the sustainability of coastal systems, management strategies will also need to address this challenge, focusing on the drivers that are dominant at each section of coast.

ACKNOWLEDGEMENTS

We thank Tom Doyle, USGS for his constructive comments on an earlier draft. Susan Hanson drew Fig. 3.

REFERENCES

1. C.J. Crossland, H.H. Kremer, H.J. Lindeboom, J.I. Marshall Crossland, M.D.A. Le Tissier, Coastal fluxes in the anthropocene.The Land-Ocean Interactions in the Coastal Zone Project of the International Geosphere–Biosphere Programme Series: Global Change – The IGBP Series, 2005.
2. I. Valiela, Global Coastal Change. Blackwell, Malden, MA, USA, 2006.
3. R.J. Nicholls, P.P. Wong, V.R. Burkett, J.O. Codignotto, J.E. Hay, R.F. McLean, S. Ragoonaden, C.D. Woodroffe, in: M.L. Parry, O.F. Canziani, J.P. Palutikof, P.J. van der Linden, C.E. Hanson (Eds.), Coastal Systems and Low-Lying Areas. Climate Change 2007: Impacts, Adaptation and Vulnerability. Contribution of Working Group II to the Fourth Assessment Report of the Intergovernmental Panel on Climate Change, Cambridge University Press, Cambridge, UK, 2007, pp. 315–356.
4. C. Small, R.J. Nicholls, J. Coast. Res. 19 (2003) 584–599.
5. G. McGranahan, D. Balk, B. Anderson, Environ. Urban. 19 (2007) 17–37.
6. J.P.M. Syvitski, C.J. Vörösmarty, A.J. Kettner, P. Green, Science 308 (2005) 376–380.
7. R.W. Buddemeier, S.V. Smith, D.P. Swaaney, C.J. Crossland The role of the coastal ocean in the disturbed and undisturbed nutrient and carbon cycles. LOICZ Reports and Studies Series No. 24, 2002.
8. K.F. Nordstrom, Beaches and Dunes of Developed Coasts, Cambridge University Press, Cambridge, UK, 2000.
9. G.A. Meehl, T.F. Stocker, W. Collins, P. Friedlingstein, A. Gaye, J. Gregory, A. Kitoh, R. Knutti, J. Murphy, A. Noda, S. Raper, I. Watterson, A. Weaver, Z.-C. Zhao, in: S. Solomon, D. Qin, M. Manning (Eds.), Global climate projections Climate Change 2007: The Physical Science Basis. Contribution of Working Group I to the Fourth Assessment Report of the Intergovernmental Panel on Climate Change, Cambridge University Press, 2007, pp. 747–845.
10. P.L. Woodworth, N.J. White, S. Jevrejeva, S.J. Holgate, J.A. Church, W.R. Gehrels, Int. J. Clim. (in press).
11. K. Lambeck, J. Chappell, Science 292 (2001) 679–686.
12. V. Gornitz, Sea level rise, after the ice melted and today. U.S. National Aeronautics and Space Administration, Science Brief, accessed September 18, 2007, at http://www.giss.nasa.gov/research/briefs/gornitz_09/.
13. P.A. Pirazzoli, Sea-Level Changes: The Last 20000 Years, Wiley, Chichester, 1996.
14. D.J. Stanley, A.G. Warne, Nature 363 (1993) 435–438.
15. J.W. Day, J.D. Gunn, J. Folan, A. Yáñez-arancibia, B.P. Horton, EOS Trans. 88 (2007) 169–170.
16. C.D. Woodroffe, Coasts, Form, Process and Evolution, Cambridge University Press, Cambridge, 2003.
17. M.J.F. Stive, P.J. Cowell, R.J. Nicholls, in: Slaymaker, O., Spencer, T., Embleton-Hamann, C. (Eds.), Geomorphology and Global Environmental Change. International Association of Geomorphologists, Cambridge University Press, in press.
18. R. Ranasinghe, R. McLoughlin, A.D. Short, G. Symonds, Mar. Geol. 204 (2004) 273–287.
19. C.D. Storlazzi, G.B. Griggs, Geol. Soc. Am. Bull. 112 (2000) 236–249.
20. J.P. Ericson, C.J. Vorosmarty, S.L. Dingman, L.G. Ward, M. Meybeck, Glob. Planet. Change 50 (2005) 63–82.

21. C.D. Woodroffe, R.J. Nicholls, Y. Saito, Z. Chen, S.L. Goodbred, in: Harvey, N. (Ed.) Global Change and Integrated Coastal Management: The Asia-Pacific Region, Springer, New York, pp. 277–314, 2006.
22. R.J. Nicholls, S.E. Hanson, J.A. Lowe, R.A. Warrick, X. Lu, A.J. Long, T.R. Carter Draft Guidelines on Constructing Sea-Level Scenarios for Impact and Adaptation Assessment. Guidance prepared for the Intergovernmental Panel on Climate Change (IPCC) Technical Group on Climate Impact Assessment (TGCIA), 2008.
23. R.J. Nicholls, Geojournal 37 (1995) 369–379.
24. J.E. Cohen, C. Small, A. Mellinger, J. Gallup J. Sachs. Sci. 278 (1997) 1211.
25. C. Nilsson, C.A. Reidy, M. Dynesius, C. Revenga, Science 308 (2005) 405–408.
26. X. Jiongxin, Environ. Manage. 31 (2003) 328–341.
27. P.H. Gleick, H. Cooley, D. Katz, E. Lee, G. H. Wolff, M. Palaniappan, J. Morrison, A. Samulon The World's Water 2006–2007: The Biennial Report on Freshwater Resources. Island Press, Washington, 2006, pp. 368.
28. J.P.M. Syvitski, Y. Saito, Glob. Planet. Change 57 (2007) 261–282.
29. D.R. Cahoon, P. Hensel, Hurricane Mitch: a regional perspective on mangrove damage, recovery and sustainability, USGS Open File Report 03-183, 2002.
30. B.B. Lin, J. Dushoff, Manage. Environ. Quality 15 (2004) 131–142.
31. M. Spalding, D. Blasco, C. Field, World Mangrove Atlas, The International Society for Mangrove Ecosystems, The University of the Ryukyus, Okinawa, Japan, 1997.
32. D.M. Alongi, Est. Coast. Shelf Sci. 76 (2008) 1–13.
33. V.R. Burkett, D.A. Wilcox, R. Stottlemeyer, W. Barrow, D. Fagre, J. Baron, J. Price, J.L. Neilsen, C.D. Allen, D.L. Peterson, G. Ruggerone, T. Doyle, Ecol. Complexity 2 (2005) 357–394.
34. J.M. Coleman, O.K. Huh, D. Braud, J. Coast. Res. 24 (2008) 1–14.
35. P.L. Woodworth, D.L. Blackman, J. Clim. 17 (2004) 1190–1197.
36. P.D. Komar, J.C. Allan, J. Coast. Res. 24 (2008) 479–488.
37. A. B. Pezza, I. Simmonds, Geophys. Res. Lett. 32 (2005) L15712.
38. R. McTaggart-Cowan, L.F. Bosart, C.A. Davis, E.H. Atallah, J.R. Gyakum, K. A. Emanuel, Mon. Weather Rev. 134 (2006) 3029–3053.
39. R. Few, M. Ahern, F. Matthies, S. Kovats Floods, health and climate change: a strategic review. Working Paper 63, Tyndall Centre for Climate Change Research, University of East Anglia, 2004, pp. 138.
40. T.J. Flynn, S.G. Walesh, J.G. Titus, M.C. Barth, in: M.C. Barth, J.G. Titus (Eds.), Greenhouse Effect and Sea Level Rise: A Challenge for this Generation, Van Nostrand Reinhold, New York, USA, 1984.
41. J.M. Gregory, J.A. Lowe, S.B.T Tett, J. Clim. 19 (2006) 4576–4591.
42. K. Zhang, B.C. Douglas, S.P. Leatherman, J. Clim. 13 (2000) 1748–1761.
43. E.C.F. Bird, Coastline Changes, Wiley Interscience, Chichester, 1985.
44. S.P. Leatherman, B.C. Douglas, M.S. Kearney, S.P. Leatherman (Eds.), Sea Level Rise, History and Consequences, Academic Press, London, 2001, pp. 181–223.
45. E.C.F. Bird, Submerging Coasts: The Effects of a Rising Sea Level on Coastal Environments, Wiley, Chichester, 1993.
46. K. Zhang, B.C. Douglas, S.P. Leatherman, Clim. Change 64 (2004) 41–58.
47. M.J.F. Stive, Clim. Change 64 (2004) 27–39.
48. K.W. Krauss, J.L. Chambers, J.A. Allen, D.M. Soileau Jr, A.S. DeBosier, J. Coast. Res. 16 (2000) 153–163.
49. USGS Synthesis of U.S. Geological Survey Science for the Chesapeake Bay Ecosystem and Implications for Environmental Management. US Geological Survey, Reston, Virginia, Circular 1316, 2007, pp. 71.

50. K. Williams, K.C. Ewel, R.P. Stumpf, F.E. Putz, T.W. Workman, Ecology 80 (1999) 2045–2063.
51. K. Williams, M. MacDonald, L. da Silveira Lobo Sternberg, J. Coast. Res. 19 (2003) 1116–1121.
52. S.J.A. Gibbons, R.J. Nicholls, Glob. Environ. Change 16 (2006) 40–47.
53. J. Barras, S. Beville, D. Britsch, S. Hartley, S. Hawes, J. Johnston, P. Kemp, Q. Kinler, A. Martucci, J. Porthouse, D. Reed, K. Roy, S. Sapkota, J. Suhayda, Historical and projected coastal Louisiana land changes: 1978–2050, Open File Report 03-334, U.S. Geological Survey, 2003, pp. 39.
54. P. Grossi, R. Muir-Wood, Flood Risk in New Orleans: Implications for Future Management and Insurability, Risk Management Solutions (RMS), London, UK, 2006.
55. O.M. Johannessen, L. Bengtsson, M.W. Miles, S.I. Kuzmina, V.A. Semenov, G.V. Aleekseev, A.P. Nagurnyi, V.F. Zakharov, L. Bobylev, L.H. Pettersson, K. Hasselmann, H.P. Cattle, Arctic Climate Change – Observed and Modeled Temperature and Sea Ice Variability, Technical Report 218, Nansen Environmental and Remote Sensing Centre, University of Bergen, Norway, 2002.
56. W. Krabill, E. Hanna, P. Huybrechts, W. Abdalati, J. Cappelen, B. Csatho, E. Frederick, S. Manizade, C. Martin, J. Sonntag, R. Swift, R. Thomas, J. Yungel, Geophys. Res. Lett. 31 (2004) L24402.
57. E. Rignot, D. Braaten, P. Gogineni, W. Krabill, J.R. McConnell, Geophys. Res. Lett. 31 (2004) L10401.
58. E. Rignot, G. Casassa, P. Gogineni, W. Krabill, A. Rivera, R. Thomas, Geophys. Res. Lett. 31 (2004) L18401.
59. A.J. Cook, A.J. Fox, D.G. Vaughan, J.G. Ferrigno, Science 308 (2005) 541–544.
60. J.C. Mars, D.W. Houseknecht, Geology 35 (2007) 583–586.
61. G.K. Manson, S.M. Solomon, D.L. Forbes, D.E. Atkinson, M. Craymer, Geo-Marine Lett. 25 (2005) 138–145.
62. S. Fox, When the weather is uggianaqtuq: inuit observations of environmental change, Cooperative Institute for Research in Environmental Sciences. CD-ROM, University of Colorado, 2003.
63. T. Spencer, K.A. Teleki, C. Bradshaw, M.D. Spalding, Mar. Pollut. Bull. 40 (2000) 569–586.
64. J.M. Lough, Geophys. Res. Lett. 27 (2000) 3901–3904.
65. R. Berkelmans, G. De'ath, S. Kininmouth, W.J. Skirving, Coral Reefs 23 (2004) 74–83.
66. J.P. McWilliams, I.M. Cote, J.A. Gill, W.J. Sutherland, A.R. Watkinson, Ecology 86 (2005) 2055.
67. T.P. Hughes, A.H. Baird, D.R. Bellwood, M. Card, S.R. Connolly, C. Folke, R. Grosberg, O. Hoegh-Guldberg, J.B.C. Jackson, J. Kleypas, J.M. Lough, P. Marshall, M. Nystrom, S.R. Palumbi, J.M. Pandolfi, B. Rosen and J. Roughgarden, Science 301 (2003) 929–933.
68. O. Hoegh-Guldberg, Symbiosis 37 (2004) 1–31.
69. O. Hoegh-Guldberg, J. Geophys. Res. 110 (2005) C09S06.
70. J.M. Pandolfi, R.H. Bradbury, E. Sala, T.P. Hughes, K.A. Bjorndal, R.G. Cooke, D. McArdle, L. McClenachan, M.J.H. Newman, G. Paredes, R.R. Warner and J.B.C. Jackson, Science 301 (2003) 955–958.
71. R.W. Buddemeier, J.A. Kleypas, R.B. Aronson, Coral Reefs and Global Climate Change: Potential Contributions of Climate Change to Stresses on Coral Reef Ecosystems, Pew Center on Global Climate Change, Arlington, VA, USA, 2004.
72. R. J. Nicholls, P.P. Wong, V.R. Burkett, C.D. Woodroffe, J.E. Hay, Sustainability Sci. 3 (2008) 89–102.

Chapter 25

Plant Pathogens as Indicators of Climate Change

K.A. Garrett, M. Nita, E.D. De Wolf, L. Gomez and A.H. Sparks
Department of Plant Pathology, Kansas State University, Manhattan, Kansas 66506

1. Introduction
2. Climatic Variables and Plant Disease
3. Evidence that Simulated Climate Change Affects Plant Disease in Experiments
4. Evidence that Plant Disease Patterns have Changed due to Climate Change
Acknowledgements
References

1. INTRODUCTION

Plant disease risk is strongly influenced by environmental conditions [1]. While some animal hosts may provide their pathogens with a consistent range of body temperatures, plant pathogens are generally much more exposed to the elements. Plant disease will tend to respond to climate change, though a number of interactions take place among host, pathogen, potential vectors. In some cases, the actions of land managers may also complicate interpretation of climate change effects. In this chapter, we present a brief introduction to plant disease and a synthesis of research in plant pathology related to climate change. We discuss the types of evidence for climate change impacts ('climate change fingerprints') that might be observed in plant disease systems and evaluate what evidence of climate change fingerprints currently exists.

The battle against plant disease is not a new one, and plant disease management is essential for our continued ability to feed a growing human population. The Great Irish Hunger is one striking example of the impact of plant disease: in 1845 more than a quarter million Irish people starved as the result of an epidemic of potato late blight [2]. Plant diseases continue to cause serious problems in global food production. Currently more than 800 million people do not have adequate food and at least 10% of global food production is lost to plant disease [3]. Not only does plant disease affect human food production, it also impacts natural systems [4]. Introduced diseases such as chestnut blight in the Eastern US, and more recently the increasing occurrence of sudden oak death, have resulted in the rapid decline of dominant tree species and triggered major impacts on forest systems [5].

Plant pathogen groups include fungi, prokaryotes (bacteria and mycoplasmas), oomycetes, viruses and viroids, nematodes, parasitic plants and protozoa. The very different life histories of this diverse group of organisms and their different interactions with host plants produce a wide range of responses to environmental and climatic drivers. For example, viruses may be present in hosts while symptom expression is dependent on temperature [6]; thus, even the difficulty of detection of these pathogens varies with climate. Fungal pathogens are often strongly dependent on humidity or dew for plant infection [7], so changes in these environmental factors are likely to shift disease risk. Genetic variation in pathogen populations often makes plant disease management more complicated when pathogens overcome host disease resistance [3]. Pathogen species may quickly develop resistance to pesticides or adapt to overcome plant disease resistance, and may also adapt to environmental changes, where the rate of adaptation depends on the type of pathogen [8]. Pathogen populations may explode when weather conditions are favourable for disease development [9,10]. The potentially rapid onset of disease makes it difficult to anticipate the best timing of management measures, especially in areas with high levels of interannual variability in climatic conditions.

2. CLIMATIC VARIABLES AND PLANT DISEASE

Understanding the factors that trigger the development of plant disease epidemics is essential if we are to create and implement effective strategies for disease management [11]. This has motivated a large body of research addressing the effects of climate on plant disease [11,12]. Plant disease occurrence is generally driven by three factors: a susceptible host, the presence of a competent pathogen (and vector if needed) and conducive environment [9,10]. All three of these factors must be in place, at least to some degree, for disease to occur (Fig. 1). A host resistant to local pathogen genotypes or unfavourable weather for pathogen infection will lessen disease intensity. The synchronous interaction between host, pathogen and environment governs disease development. These interactions can be conceptualised as a continuous sequence of cycles of biological events including dormancy, reproduction, dispersal and pathogenesis [1]. In plant pathology this sequence of events is commonly referred to as a disease cycle. Although plant pathologists have long realised the importance of the disease cycle and its component events and the apparent relationships with environment, the quantification of these interactions did not begin in earnest until the 1950s [11]. The past five decades of research have

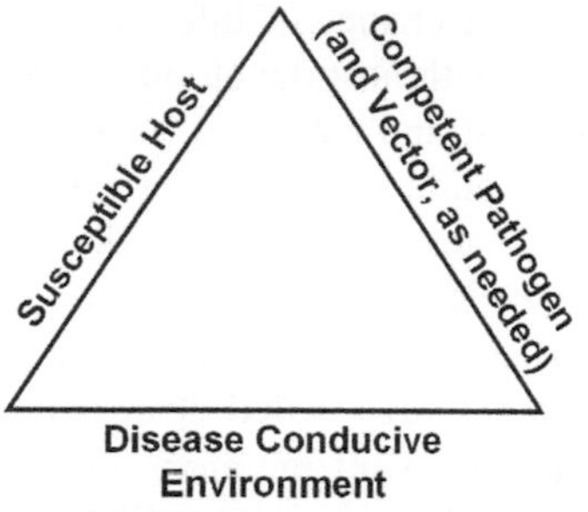

FIGURE 1 Plant disease results from the interaction of host, pathogen and environment. Climatic features such as temperature, humidity and leaf surface wetness are important drivers of disease, and inappropriate levels of these features for a particular disease may be the limiting factor in disease risk.

established a vast body of literature documenting the impact of temperature, rainfall amounts and frequency and humidity, on the various components of the disease cycle [11].

The quantification of the relationship between the disease cycle of a given plant disease and weather is also the foundation of many prediction models that can be used to advise growers days or weeks before the onset of an increase in disease incidence or severity [1]. Such prediction tools can allow a grower to respond in a timely and efficient manner by adjusting crop management practices. Given enough time to respond, a disease prediction might allow a grower to alter the cultivar they select for planting, the date on which the crop is sown, or the scheduling of cultural practices such as fertilisation or irrigation. A prediction of a low disease risk may also result in reduced pesticide use with positive economic and environmental outcomes. Larger scale predictions of disease risk, such as the typical risk for regions or countries based on climatic conditions, can be used to form policy and priorities for research (e.g. [13]).

Interestingly, the quantification of these relationships and application of this information as part of disease prediction models has also facilitated the simulation of potential impacts of climate change. For example, Bergot et al. [14] have used models of the impact of weather variables on the risk of infection by *Phytophthora cinnamomi* to predict the future distribution of disease caused by this pathogen in Europe under climate change scenarios. As more detailed climate change predictions are more readily available, many plant disease forecasting systems may be applied in this context.

Some relationships between climate and disease risk are obvious, such as some pathogens' inability to infect without sufficient surface moisture (i.e. dew or rain droplets) [7] or other pathogens' or vectors' inability to overwinter when temperatures go below a critical level. Other effects of climate may be more subtle. For example, a given pathogen may only be able to infect its host(s) when the plants are in certain developmental stages. This also means that in order to maximise their chance of infection, the life cycle of pathogen populations must be in sync with host development. Since climate change can influence the rate of both host and pathogen development, it could affect the development and impact of plant diseases. Here, we discuss a few examples where host phenology is the key to disease development.

Some pathogens depend on flower tissues as a point of entry to the host. For example, *Botrytis cinerea*, which causes gray mold of strawberry and other fruits (producing a gray fuzz-balled strawberry, which you may have seen at a grocery store or in your refrigerator), infects strawberry at the time of flowering [15]. It stays in flower parts until the sugar level of the berry increases, and then causes gray mold disease. Another example is Fusarium head blight of wheat and barley, which causes large yield losses, reductions in grain quality and contamination with mycotoxins (toxic substances created by the fungi) [16,17]. Several fungal species including *Fusarium graminearum* (teleomorph: *Gibberella zeae*) cause this disease, and anthesis (flowering) period seems to be the critical time for infection [17,18]. An important bacterial disease of apple and pears, called fire blight, also utilises flowers as a major point of entry [19]. The causal agent (*Erwinia amylovora*) can be disseminated by pollinating insects such as bees and moves into flowers to cause rapid wilting of branch tips.

Certain hosts become more resistant after a particular developmental stage, some exhibiting a trait referred to as adult plant resistance. There are many examples of genes that follow this pattern in wheat, including leaf rust (caused by the fungus *Puccinia triticina*) resistance genes *Lr13* and *Lr34* [20] and stripe rust (caused by *Puccinia striiformis* f. sp. *tritici*) resistance gene *Yr39* [21]. These genes are activated by a combination of wheat developmental stage and temperature changes. In grape, there are many cases of ontogenic (or age-related) resistance against pathogens. Once grape fruit tissue matures, certain fungal pathogens such as *Erysiphe necator* (formerly *Uncinula necator*, causing powdery mildew) [22], or *Guignardia bidwellii* (causing black rot) [23], or the oomycete pathogen *Plasmopara viticola* (causing downy mildew) [2] are less successful at infecting plants.

With changes in climate, host development patterns may be altered. For the examples above, the timing and duration of flowering in wheat are a function of the average daily temperature. Heavy rain and/or strong wind events can shorten flowering duration in strawberry and apple through flower damage. Some pathogen species may be able to maintain their synchrony with target host tissue, and others may become out of sync. Thus, there are some efforts to modify disease prediction systems to accommodate potential impacts from climate change. For example, in efforts to predict the risk of apple scab (caused by the fungus *Venturia inaequalis*), the concept of ontogenic resistance was utilised along with inoculum production [24] because tissues become less susceptible as the rate of tissue expansion decreases.

There is no doubt that weather influences plant disease; that relationship is fundamental to the modelling of plant disease epidemiology. Thus, it is fairly straightforward to predict that where climate change leads to weather events that are more favourable for disease, there will be increased disease pressure. But the relationship between climate change and associated weather events, and resulting changes in disease development will generally not be a simple one-to-one relationship (Fig. 2). The impacts will tend to be most dramatic when climatic conditions shift above a threshold for pathogen reproduction, are amplified through interactions, or result in positive feedback loops that decrease the utility of disease management strategies [25]. For example, the Karnal bunt pathogen, *Tilletia indica*, which reduces wheat quality, will tend to have lower reproductive rates per capita when populations are low because individuals of different mating types must encounter each other for reproductive success [26]. If climatic conditions change to favour pathogen reproduction, the pathogen will be released from this constraint and show a larger response to the change than would otherwise have been anticipated. The trend toward greater global movement of humans and materials also produces new types of interactions as pathogens are introduced to new areas and may hybridise to produce new pathogens [27,28].

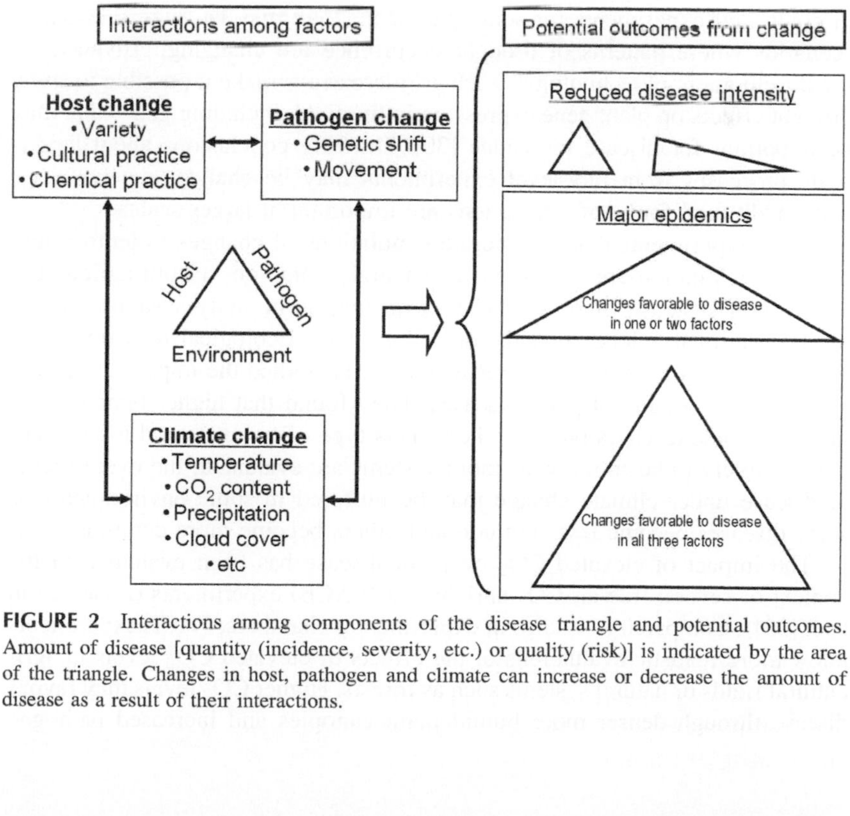

FIGURE 2 Interactions among components of the disease triangle and potential outcomes. Amount of disease [quantity (incidence, severity, etc.) or quality (risk)] is indicated by the area of the triangle. Changes in host, pathogen and climate can increase or decrease the amount of disease as a result of their interactions.

3. EVIDENCE THAT SIMULATED CLIMATE CHANGE AFFECTS PLANT DISEASE IN EXPERIMENTS

Next we consider two types of evidence for effects of changes in climate on plant disease. The first is evidence that simulated climate change affects plant disease in experimental settings. The effect of simulated climate change has been studied in experiments with altered heat treatments, altered precipitation treatments and carbon enrichment treatments. Where there are apparent effects from these treatments, this implies that, to the extent that the simulations do effectively represent future climate scenarios, plant disease will respond. The second type of evidence is for changes in patterns of plant disease in agricultural or wildland systems that can be attributed to climate change with some level of confidence, discussed in Section 4. In this case, the changes in plant disease might be taken as fingerprints of climate change. We also discuss what types of plant disease scenarios might qualify as fingerprints of climate change in this sense.

The range of possibilities for climate change simulations can be characterised in terms of the scale of the effect being considered [29]. For many well-studied pathogens and vectors, the temperature ranges that support single infection events or survival are fairly well characterised. The effects of plant water stress and relief from water stress on disease risk have also been studied in controlled experiments for some pathogens, and may be quite relevant to scenarios where patterns of drought occurrence are changing. Advances in the development of technologies such as microarrays make it possible to study drought effects on plant gene expression in the field, including genes that may be important for disease resistance [30]. Drawing conclusions about larger-scale processes from plot-level experiments may be challenging, however, since additional forms of interactions are important at larger scales.

Field experiments that incorporate simulations of changes in temperature and/or precipitation are becoming increasingly common in both agricultural and natural systems, often associated with long-term study systems such as the US National Science Foundation's Long-term Ecological Research sites. For example, in Montane prairie Roy et al. [31] studied the impact of heating treatments on a suite of plant diseases. They found that higher temperatures favoured some diseases but not others. This type of 'winners and losers' scenario is likely to be common as more systems are evaluated; the overall level of disease under climate change may be buffered in some environments as some diseases become less common and others become more common.

The impact of elevated CO_2 on plant disease has been evaluated in the context of several free-air CO_2 enrichment (FACE) experiments (reviewed in Ref. [32]). Compared to studies in experimental chambers, FACE experiments allow more realistic evaluations of the effects of elevated CO_2 levels in agricultural fields or natural systems such as forests. Higher CO_2 levels may favour disease through denser more humid plant canopies and increased pathogen reproduction but may reduce disease risk by enhancing host disease resistance [33], so the outcome for any given host-pathogen interaction is not readily predictable. Elevated ozone levels can also affect plant disease risk (reviewed in Ref. [32]).

In addition to the more direct influences of the abiotic environment on plant disease, climate change may also affect plant disease through its impact on other microbes that interact with pathogens. While certain microbes affect plant pathogens strongly enough to be used as biocontrol agents, a number of microbial interactions probably also have more subtle effects. As the effect of climate change on microbial communities is better understood [34], this additional form of environmental interaction can be included in models of climate and disease risk.

4. EVIDENCE THAT PLANT DISEASE PATTERNS HAVE CHANGED DUE TO CLIMATE CHANGE

If patterns of plant disease in an area have shifted at the same time that changes in climate are observed, when can this correlation be taken as evidence of climate change impacts on disease? Such an analysis is complicated by the number of factors that interact to result in plant disease. For example, if a disease becomes important in an area in which it was not important in the past, there are several possible explanations. The pathogen populations may have changed so that they can more readily infect and damage hosts. The pathogen species or particular vectors of the pathogen may be newly introduced to the area. In agricultural systems, host populations may have changed as managers have selected new cultivars based on criteria other than resistance to the disease in question. Management of the abiotic environment may have changed, such as changes in how commonly fields are tilled (tillage often reduces disease pressure), or changes in planting dates (which may result in more or less host exposure to pathogens). To rule out such competing explanations for changes in plant disease pattern, the argument for climate change as an important driver is strongest when (a) the pathogen is known to have been present throughout the area during the period in question, (b) the genetic composition of the pathogen and host populations has apparently not shifted to change resistance dynamics, (c) management of the system has not changed in a way that could explain the changes in disease pattern, (d) the climatic requirements of the pathogen and/or vector are well-understood and better match the climate during the period of greater disease pressure and (e) the change in disease pattern has been observed long enough to establish a convincing trend beyond possible background variation.

Even though the impact of changes in temperature, humidity and precipitation patterns has been quantified, the simulations of the potential impact of climate change remain just that, simulations. By their very nature these simulations depend on the best available projections of meteorological models.

Real evidence for the impact of climate change on plant disease could come from verification of the accuracy of these projections. This would require long-term records of disease intensity for the regions where impacts are projected and for control regions. Long-term monitoring of pathogens and other plant-associated microbes is necessary in general to understand their ecology, and to develop predictions of their impact on plant pathology [35]. The lack of availability of long-term data about disease dynamics in natural systems, and even in agricultural systems, limits opportunities for analysis of climate change effects on plant disease [36,37].

Interannual variation in climatic conditions can have important effects on disease risk. For wheat stripe rust (caused by *P. striiformis* Westend. f. sp. *tritici* Eriks.) in the US Pacific Northwest, disease severity was lower in El Niño years than in non-El Niño years [38]. If climate change alters the frequency and/or the intensity of El Niño events [39] or other extreme weather events, it will also alter patterns of disease risk; knowledge of the associations between disease and climate cycles is needed to inform predictions about plant disease epidemics under climate change [38].

Some general historical analyses of the relationship between disease and environmental factors have been developed. For example, the first annual appearance of wheat stem rust (caused by *Puccinia graminis* Pers.:Pers. f. sp. *tritici* Eriks. and E. Henn.) was compared for cool (1968–1977) and warm (1993–2002) periods in the US Great Plains, but a significant difference in arrival date was not observed [40]. In the UK, the abundance of two different wheat pathogens shifted in close correlation with patterns of SO_2 pollution during the 1900s [41,42]. For potato light blight, Zwankhuizen and Zadoks [43] have analysed epidemics in the Netherlands from 1950 to 1996 using agronomic and meteorological variables as predictors of disease severity. They found that some factors were associated with enhanced disease, such as greater numbers of days with precipitation, greater numbers of days with temperatures between 10 and 27 °C, and a relative humidity >90% during the growing season. Temperatures above 27 °C and higher levels of global radiation in the Netherlands appeared to reduce disease risk [43]. Baker et al. [44] evaluated late blight risk in central North America and found that the trends in climatic conditions should result in increased risk. Hannukkala et al. [45] evaluated late blight incidence and first appearance in Finland 1933–2002, concluding that there was higher risk in more recent years. The comparison of years is complicated in this case by changes in the pathogen population and management practices. Increases in fungicide use were consistent with increased disease risk; records of pesticide use or other management change are one potential form of evidence for climate change impacts.

Pathogens and insect pests of lodgepole pine (*Pinus contorta*) have been well-studied and offer an interesting example of a potential climate change fingerprint. Lodgepole pine is the most widely distributed pine species in natural (unmanaged) forests in western North America [46], including forests in British Columbia where there are more than 14 million ha of lodgepole pine [47]. Due to a lack of natural or human mediated disturbances, lodgepole pine has been increasing in abundance in British Columbia since the 1900s [47,48]. Recently, there have been increased cases of decline of lodgepole pines in these forests and researchers are evaluating the potential effects of climate change on these events.

Mountain pine beetle (*Dendroctonus ponderosae*) is a bark beetle native to western North American forests [49]. This beetle can infest many pine species, and lodgepole pine is a preferred host [46,48]. The distribution range has not been limited by availability of the host but by the temperature range required for beetle survival through the winter [46,50]. The beetle causes physiological damage to the host trees by creating tunnels (insect galleries) underneath the bark, and in addition, microorganisms, such as the blue-stain fungi complex, can take advantage of these wounds to cause secondary infestation that may further reduce plant health [46,49]. Dead pines are not marketable and also can facilitate the spread of wild fire [51]. Beetle populations can be very low for many decades, but when there is an outbreak, a large area of susceptible hosts may be killed. The beetle has been known to be native to British Columbia [48], but, probably due to low winter temperatures, outbreak events were not common. However, there have been a series of outbreaks in recent years, and 8 million hectares in British Columbia were affected in 2004 [48,51]. Carroll et al. [50] evaluated the shift in infestation range and concluded that the trend toward warmer temperatures more suitable for the beetle is part of the reason for this series of outbreaks. Further, in a study by Mock et al [48], genetic markers did not reveal any significant differences among beetle genotypes from inside and outside of British Columbia, indicating the beetle population had not changed. Thus, other factors including climate change are likely to be the reason why there have been more outbreaks in northern areas.

Dothistroma needle blight is a fungal disease (causal agent *Dothistroma septosporum*) of a variety of pine species worldwide [52], including lodgepole pines. The disease is associated with mild temperature ranges (18 °C is the optimum temperature for sporulation [53]) and rain events [52,54], and causes extensive defoliation, mortality and a reduced growth rate in pine [52,55]. As with the mountain pine beetle, Dothistroma needle blight has been found in British Columbia in the past, but damage due to this disease was relatively minor. However, the number of cases and intensity of epidemics in this region has increased since the late 1990s [55]. A study by Woods et al. [55] evaluated the relationship between these disease outbreaks and (i) regional climate change and (ii) long-term climate records (utilising the Pacific Decadal Oscillation, PDO, as an indicator variable). Although they did not find a substantial increase in regional temperature nor a significant correlation between PDO and directional increase of precipitation or temperature, increased mean summer precipitation in the study area was observed. The authors also found that in some locations, up to 40% of forest stands became dominated by lodgepole pine due to plantation development, and they hypothesised that a combination of increased rain events and the abundance of the favoured host were the probable cause of increased disease occurrence.

For both mountain pine beetle and Dothistroma needle blight, it is reasonable to assume that climate has influenced pathogen and pest behaviour; however, at the same time, there has been a substantial increase in the abundance of the host (lodgepole pine) in British Columbia [47,48]. Widely available and genetically similar hosts generally increase plant disease risk [56], and these factors may also explain at least part of the change in risk observed for lodgepole pine.

Another important disease that has exhibited recent changes in its pattern of occurrence is wheat stripe rust (or yellow rust, caused by the fungus *P. striiformis* f. sp. *tritici*). This disease decreased and then increased in importance in the US during the past century. Stripe rust was economically important in the 1930s–1960s, but the development of resistant wheat varieties successfully reduced the number of epidemic events. However, several epidemic events have been observed since 2000 [57,58]. The disease can cause 100% yield loss at a local scale [58], and epidemics in 2003 in the US resulted in losses estimated to total $300 million. Are these changes related to climate change?

Historically, *P. striiformis* f. sp. *tritici* was known to be active at relatively lower temperature ranges. Under favourable conditions (i.e. with dew or free water on plant surfaces), its spores can germinate at 0 °C [59], and the temperature range for infection was measured as between 2 and 15 °C with an optimum temperature of 7–8 °C [60,61]. And it could produce spores between 0 and 24.5 °C [59]. This pathogen species was not well adapted for higher temperature conditions and disease development declined at temperatures above 20 °C [60–62], while spores produced at 30 °C were shown to be non-viable [59].

However, more recent populations of *P. striiformis* f. sp. *tritici* were adapted to warmer temperature ranges [63]. Isolates from the 1970s to 2003 were compared, and newer (post-2000) isolates had a significantly ($P < 0.05$) higher germination rate and shorter latent period (period between infection and production of spores) than older isolates when they were incubated at 18 °C, whereas isolate effects were not different when incubation took place at 12 °C. In a follow-up study, Markell and Milus [64] examined isolates from the 1960s to 2004 with genetic markers and morphological comparisons, and found that isolates collected pre- and post-2000 could be classified into two different groups. Although within a population group less than nine polymorphic markers were identified, when pre-and post-2000 populations were compared there were 110 polymorphic markers [64]. The large difference between pre- and post-2000 groups led the authors to conclude that post-2000 isolates were introduced from outside of the US, rather than resulting from mutations in pre-2000 isolates.

Results from annual race surveys conducted by the United States Agricultural Research Service of Pullman, WA, indicated that pre-2000 isolates were not commonly collected in surveys after 2000 [64]. Thus, it seems that post-2000 isolates took the place of pre-2000 isolates. The question remains whether the success of post-2000 isolates is due to the change in climatic conditions (i.e. increase in overall temperature) or something else. Since post-2000 isolates were better adapted to a warmer temperature range, climate change might have played a role in selection for the new isolates, but there is another important factor for post-2000 isolates. All post-2000 isolates examined were able to cause disease on wheat plants with resistance genes *Yr8* and *Yr9*, while these resistance genes were effective at preventing disease for pre-2000 isolates [57,64]. There are other wheat varieties that are resistant to post-2000 isolates, but these varieties were less commonly grown since they were not effective against older isolates. Thus, the ability of new isolates to overcome these resistance genes was most likely the major factor behind the drastic change in populations of *P. striiformis* f. sp. *tritici* and recent epidemic events.

In summary, there is no doubt that plant disease responds to weather and that changes in weather events due to climate change are likely to shift the frequency and intensity of disease epidemics. Simulated climate change experiments reveal changes in plant disease intensity and the profile of plant diseases. When evidence for climate change is sought in observed changes in plant disease patterns, conclusions are less clear. Since the search for fingerprints of climate change is correlative by nature, there may always be alternative predictors for the changes, but this seems particularly true for plant disease. It is a typical biological irony that, while plant disease risk may be particularly sensitive to climatic variables and climatic shifts, plant disease may also be particularly difficult to use as an indicator of climate change because of the many interactions that take place to result in disease. However, as more data sets are collected and synthesised [37], and climate patterns exhibit greater changes over a longer period, the impacts of climate change on plant disease are likely to become clearer.

ACKNOWLEDGEMENTS

We appreciate support by the U.S. National Science Foundation (NSF) through Grant DEB-0516046 and NSF Grant EF-0525712 as part of the joint NSF-National Institutes of Health (NIH) Ecology of Infectious Disease program, by the U.S. Agency for International Development (USAID) to the Office of International Research, Education, and Development (OIRED) at Virginia Tech for the Sustainable Agriculture and Natural Resource Management (SANREM) Collaborative Research Support Program (CRSP) under Award No. EPP-A-00-04-00013-00 and for the Integrated Pest Management (IPM) CRSP under Award No. EPP-A-00-04-00016-00. This is contribution 09-116-B of the Kansas Agricultural Experiment Station.

REFERENCES

1. E.D. De Wolf, S.A. Isard, Ann. Rev. Phytopathol. 45 (2007) 203–220.
2. M.M. Kennelly, D.M. Gadoury, W.F. Wilcox, P.A. Magarey, R.C. Seem, Phytopathol. 95 (2005) 1445–1452.
3. R.N. Strange, P.R. Scott, Ann. Rev. Phytopathol. 43 (2005) 83–116.
4. J.J. Burdon, P.H. Thrall, L. Ericson, Ann. Rev. Phytopathol. 44 (2006) 19–39.
5. T.E. Condeso, R.K. Meentemeyer, J. Ecol. 95 (2007) 364–375.
6. J. DeBokx, P. Piron, Potato Res. 20 (1977) 207–213.
7. L. Huber, T.J. Gillespie, Ann. Rev. Phytopathol. 30 (1992) 553–577.
8. B.A. McDonald, C. Linde, Ann. Rev. Phytopathol. 40 (2002) 349–379.
9. G.N. Agrios, Plant Pathology, Academic Press, San Diego, 2005.
10. J.E. Van Der Plank, Plant Diseases: Epidemics and Control, Academic Press, New York and London, 1963.
11. C.L. Campbell, L.V. Madden, Introduction to Plant Disease Epidemiology, Wiley, New York, 1990.
12. L.V. Madden, G. Hughes, F. van den Bosch, The Study of Plant Disease Epidemics, APS Press, St. Paul, MN, 2007.
13. R.J. Hijmans, G.A. Forbes, T.S. Walker, Plant Pathol. 49 (2000) 697–705.
14. M. Bergot, E. Cloppet, V. Pérarnaud, M. Déqué, B. Marçais, M.L. Desprez-Loustau, Glob. Change Biol. 10 (2004) 1539–1552.
15. D.R. Cooley, W.F. Wilcox, J. Kovach, S.G. Schloemann, Plant Dis. 80 (1996) 228–237.
16. M. McMullen, R. Jones, D. Gallenberg, Plant Dis. 81 (1997) 1340–1348.
17. J.C. Sutton, Can. J. Plant Pathol. 4 (1982) 195–209.
18. A.L. Andersen, Phytopathology 38 (1948) 595–611.
19. S.V. Beer, D.C. Opgenorth, Phytopathology 66 (1976) 317–322.
20. J.A. Kolmer, Ann. Rev. Phytopathol. 34 (1996) 435–455.
21. F. Lin, X.M. Chen, TAG 114 (2007) 1277–1287.
22. A. Ficke, D.M. Gadoury, R.C. Seem, Phytopathology 92 (2002) 671–675.
23. L.E. Hoffman, W.F. Wilcox, D.M. Gadoury, R.C. Seem, D.G. Riegel, Phytopathology 94 (2004) 641–650.
24. D.M. Gadoury, R.C. Seem, A. Stensvand, NY Fruit Quart. 2 (1995) 5–8.
25. K.A. Garrett, Climate Change and Plant Disease Risk, National Academies Press, Washington, DC, 2008, pp. 143–155.
26. K.A. Garrett, R.L. Bowden, Phytopathology 92 (2002) 1152–1159.
27. P.K. Anderson, A.A. Cunningham, N.G. Patel, F.J. Morales, P.R. Epstein, P. Daszak, Trends Ecol. Evol. 19 (2004) 535–544.
28. C.M. Brasier, BioScience 51 (2001) 123–133.
29. K.A. Garrett, S.P. Dendy, E.E. Frank, M.N. Rouse, S.E. Travers, Ann. Rev. Phytopathol. 44 (2006) 489–509.
30. S.E. Travers, M.D. Smith, J. Bai, S.H. Hulbert, J.E. Leach, P.S. Schnable, A.K. Knapp, G.A. Milliken, P.A. Fay, A. Saleh, K.A. Garrett, Frontiers Ecol. Environ. 5 (2007) 19–24.
31. B.A. Roy, S. Gusewell, J. Harte, Ecology 85 (2004) 2570–2581.
32. S. Chakraborty, J. Luck, G. Hollaway, A. Freeman, R. Norton, K.A. Garrett, K. Percy, A. Hopkins, C. Davis, D.F. Karnosky, CAB Rev.: Perspect. Agric. Vet. Sci. Nutr. Nat. Resour. 3 (2008). Article No. 054
33. S. Chakraborty, Australas. Plant Pathol. 34 (2005) 443–448.
34. M.P. Waldrop, M.K. Firestone, Microb. Ecol. 52 (2006) 716–724.

35. C.D. Harvell, C.E. Mitchell, J.R. Ward, S. Altizer, A.P. Dobson, R.S. Ostfeld, M.D. Samuel, Science 296 (2002) 2158–2162.
36. H. Scherm, Can. J. Plant Pathol. 26 (2004) 267–273.
37. M.J. Jeger, M. Pautasso, New Phytol. 177 (2008) 8–11.
38. H. Scherm, X.B. Yang, Phytopathalogy 85 (1995) 970–976.
39. A. Timmermann, J. Oberhuber, A. Bacher, M. Esch, M. Latif, E. Roeckner, Nature 398 (1999) 694–697.
40. H. Scherm, S.M. Coakley, Australas. Plant Pathol. 32 (2003) 157–165.
41. S.J. Bearchell, B.A. Fraaije, M.W. Shaw, B.D.L. Fitt, PNAS 102 (2005) 5438–5442.
42. M.W. Shaw, S.J. Bearchell, B.D.L. Fitt, B.A. Fraaije, New Phytol. 177 (2008) 229–238.
43. M.J. Zwankhuizen, J.C. Zadoks, Plant Pathol. 51 (2002) 413–423.
44. K.B. Baker, W.W. Kirk, J.M. Stein, J.A. Anderson, HortTechnology 15 (2005) 510–518.
45. A.O. Hannukkala, T. Kaukoranta, A. Lehtinen, A. Rahkonen, Plant Pathol. 56 (2007) 167–176.
46. G.D. Amman, in: W.J. Wattson (Ed.), The Role of the Mountain Pine Beetle in Lodgepole Pine Ecosystem: Impact on Succession, Springer-Verlag, New York, 1978.
47. S.W. Taylor, A.L. Carroll, in: T.L. Shore, J.E. Brooks, J.E. Stone (Eds.), Mountain Pine Beetle Symposium: Challenges and Solutions (October 30–31, 2003), Natural Resources Canada, Canadian Forest Service, Pacific Forestry Centre, Information Report BC-X-399, Victoria, BC, Kelowna, British Columbia, 2004, pp. 41–51.
48. K.E. Mock, B.J. Bentz, E.M. O'Neill, J.P. Chong, J. Orwin, M.E. Pfrender, Mol. Ecol. 16 (2007) 553–568.
49. W. Cranshaw, D. Leatherman, B. Jacobi, L. Mannex, Insects and Diseases of Woody Plants in the Central Rockies, Colorado State University Cooperative Extension, Fort Collins, CO, 2000.
50. A.L. Carroll, S.W. Taylor, J. Régnière, L. Safranyik, in: T.L. Shore, J.E. Brooks, J.E. Stone (Eds.), Mountain Pine Beetle Symposium: Challenges and Solutions (October 30–31, 2003), Natural Resources Canada, Canadian Forest Service, Pacific Forestry Centre, Information Report BC-X-399, Victoria, BC, Kelowna, British Columbia, 2004, pp. 223–232.
51. Ministry of Forests and Range – Province of British Columbia, Mountain Pine Beetle Action Plan 2006–2011, available from http://www.gov.bc.ca/for/.
52. I.A.S. Gibson, Ann. Rev. Phytopathol. 10 (1972) 51–72.
53. M.H. Ivory, Trans. Brit. Mycol. Soc. 50 (1967) 563–572.
54. D. Hocking, D.E. Etheridge, Ann. Appl. Biol. 59 (1967) 133–141.
55. A. Woods, K.D. Coates, A. Hamann, BioScience 55 (2005) 761–769.
56. K.A. Garrett, C.C. Mundt, Phytopathology 89 (1999) 984–990.
57. X. Chen, M. Moore, E.A. Milus, D.L. Long, R.F. Line, D. Marshall, L. Jackson, Plant Dis. 86 (2002) 39–46.
58. X.M. Chen, Can. J. Plant Pathol. 27 (2005) 314–337.
59. H. Tollenaar, B.R. Houston, Phytopathology 56 (1966) 787–790.
60. E.L. Sharp, Phytopathology 55 (1965) 198–203.
61. C. de Vallavieille-Pope, L. Huber, M. Leconte, H. Goyeau, Phytopathology 85 (1995) 409–415.
62. M.V. Wiese, Compendium of Wheat Diseases, APS Press, St. Paul, MN, 1987.
63. E.A. Milus, E. Seyran, R. McNew, Plant Dis. 90 (2006) 847–852.
64. S.G. Markell, E.A. Milus, Phytopathology 98 (2008) 632–639.

Index

A

Abiotic marine environment, 264–5
Absorption coefficients, 15
Acid–base regulation, 377–8
Adaption, 181, 219, 220, 273, 275, 276, 314, 316–17, 318, 319, 320–1, 383, 384, 426
Advanced very high resolution radiometer (AVHRR), 80
Aerosol(s), 23–4, 28, 32–3, 36–7, 72, 78, 80, 81, 88, 343, 381
  loading, 32–3, 37, 79–82
Alaskan salmon fisheries, 275
Altitudinal studies, 300
Amazon Forest-inventory network (RAINFOR), 302, 303
AMOC changing already?, 361–2
Annual thermal stress, 258
Annular modes (AM), 90, 146, 162, 178
Annular rings, 90
Antarctic, 6–8, 67, 115, 134, 160, 169, 178, 210, 356, 397, 398, 418
Antarctic Bottom water (AABW), 357
Ant-Arctic and Tundra lichens, 406
Anthropogenic, 14, 18, 28, 33, 38, 60, 71, 73, 79, 83, 98–9, 139, 155–6, 158–9, 165–6, 172–3, 201, 205–206, 208, 211, 234, 237, 238, 245, 246, 254, 256, 331, 372, 373–4, 381–2, 385–7, 393, 399
Anthropogenic forcing, 92–3, 153–4, 155, 169, 256, 331, 334
Anthropogenic influence, 36–7, 45, 73, 237, 245, 345, 351, 370, 398
Anthropogenic perturbations, 127, 384
Antropocene, 138–9
Arctic, 27, 83, 139, 169, 190, 199, 204, 206–207, 208, 209, 244, 245–6, 248, 286, 338, 344, 368, 402, 407, 418, 419
Arctic Oscillation (AO), 78, 178, 201, 244
Asteroids, 45, 68, 70, 71, 73
Astronomical modulation of climate, 135
Astronomical parameters, 104–12
Atlantic climate in the past, 358–9
Atlantic meridional overturning circulation (AMOC), 351, 353–7, 358, 359, 360, 361, 362, 363–4
Atlantic multidecadal oscillation (AMO), 238
Atmospheric circulation, 78–9, 84, 88–92, 119, 145–62, 166–7, 170, 349–50, 398
Atmospheric gases, 3–19
Atmospheric lifetime, 14, 15, 16, 17
Average temperature of the Earth, 5, 7

B

Baltic sea, 244, 276
Barnacle life cycle, 378
Baseline Surface Radiation Network (BSRN), 26, 28–9, 30
Benguela upwelling, 244
Bergman's rule, 201, 208–209
Bermuda Atlantic time-series (BATS), 371
Biodiversity, 206, 216, 235, 237–9, 246–7, 258, 263–5, 271, 272, 274–6, 284, 292, 299–300, 321, 358, 379, 380, 381, 383, 385
Biogeochemical cycling, 234, 247, 374
  and feedback, 381–3
Biogeographical changes, 238–40
Biogeography, 207, 240, 283–9
Bird ecology, 181–93
Black-body emission, 4, 5
Blackman's Principle of Limiting Factors, 308
Bleaching, 254–6, 258, 260, 266–7, 269, 272, 273, 274, 275, 375, 379, 381, 419, 420, 421
Body size, 199, 200, 201, 202, 209
Bottom of atmosphere (BOA), 22–4
Breeding:
  bird surveys, 182, 183
  grounds, 181, 185–8, 190, 192, 193
  period, 188, 189–90, 192
  ranges, 182–5, 188
  success, 190–2, 193, 243
Brewer–Dobson circulation, 80, 150
Brewer–Dobson stratospheric circulation, 79–80
Bryozoan density, 259
Bryozoan diversity, 259

C

Calcification, 237, 372, 375, 376–7, 380, 381, 382, 383, 384
Carbonate chemistry, 367–71, 375, 383, 386

Carbon dioxide (CO_2) concentration, 7, 22, 115, 117, 299, 308–309, 310, 316, 319, 320
Carbon isotope, 119–20, 128, 133–4
Cenozoic, 112, 119, 130, 133–4, 205
CF_2Cl_2, 9, 10, 11, 14
CH_4, 9, 10, 11, 12, 14, 15–16, 18
Changes in distribution, 282, 283–4, 300–301
Changes in phenology, 201, 247, 299–300
Chinese agriculture, 317
Chlorine monoxide (ClO) free radical, 8
Chlorofluorocarbons (CFCs), 3–4, 10
Circulation, driving mechanisms, 354–7
Cirrus hole, 58–60
Climate change models, 184, 311, 313, 314–15, 317–18
Climate models, 81, 83–4, 93, 129, 315, 359, 360–1, 363
Climatic oscillations, 104, 237–8, 245
Clouds, 8, 23–4, 32–3, 36, 38, 47–9, 54, 55, 58–9, 64–5, 67, 68, 70, 72, 80, 83, 148–9, 166, 204, 383, 407
  formation, 64–5, 68, 72
Clubroot disease, 310
CO_2, 4, 5, 6, 7, 8, 9, 10, 11, 12, 13, 14, 18, 22, 27, 79, 200, 234, 237, 319, 331, 350, 357–8, 359, 367–8, 370, 371, 372, 374, 376, 377–8, 379, 381, 382, 383, 384, 385–7, 415, 430–1
  levels, 9, 18, 331, 430–1
Coastal zone development, 291, 292
Coastline degradation, 409–21
Coasts at latitudinal extremes, 418–20
Cold water corals, 375, 380
Community composition, 292, 297, 301–302, 303
Competent pathogen, 426
Conducive environment, 426
Continuous plankton recorder (CPR), 238–9, 246
Convection-diffusion and Drift mechanisms, 60
Coral:
  density, 259
  diversity, 259
  ecosystems, 275, 279–80
  reef ecosystems, 253–60, 381
  species, 255, 256, 267, 269
Coriolis force, 147–9, 172–3, 352–3
Cosmic Rays (CR), 45–65
  cutoff rigidity variation, 60–3
  effects, 43–73
  intensity, 44, 45, 51–4
Crassulacean acid metabolisms (CAM) pathways, 308
Crop production, 307–21
Crossing Galactic arms, 71
Cumulative impacts, 272–4, 276

D

Daily temperature range (DTR), 37–8
Deep western boundary current (DWBC), 355, 361, 363–4
Distribution and abundance of mammals, 207–208
Dothistroma needle blight, 433–4

E

Early climate history, 131–2
Earth radiation Budget Experiment (ERBE), 48
Earth's albedo, 5, 23, 115–16
Earth's cloud coverage, 47–9
Earth's orbital characteristics, 44, 103–21
Eastern Caribbean coral reefs, 258
Eccentricity, 23, 103–104, 106, 108, 109, 110, 111, 113, 119–20, 135, 136
Echinoderm density, 259
El Chichon, 32, 78, 80
Electromagnetic radiation, 4–5
El Niño southern oscillation (ENSO), 83, 149, 162, 166, 170–2, 201, 237–8, 253–4, 268–9, 344, 350
Empirical orthogonal function (EOF), 168, 177, 178
Endemic lichens, 406
Energetic particle precipitation, 55–8
Eocene–Oligocene transition, 112
Eruptions, 32, 68, 73, 78, 83, 84, 88, 89, 90–2, 189
Europe, 27–8, 29, 62, 88, 134–5, 151, 169, 183–4, 185, 186–7, 189, 192, 220, 226, 238, 284–6, 289, 292, 310, 312, 314, 318–19, 334, 427
European continental shelf, 238–40
European station for time-series in the ocean at Canary Islands (ESTOC), 371
Evaporation rates, 33, 34–5, 37, 38, 131, 336
Extended Edited Cloud Report Archive (EECRA), 32
Extratropical, 82, 145, 146, 147, 149, 157, 158, 159–60, 161, 170
Extratropical Eddies, 155, 156–7, 158, 161
Extra-tropical patterns, 166–70
Extreme weather events, 288–9, 291, 311, 320, 432

F

Fauna of coral reefs, 258–60
Forbush decreases, 54–5

G

Galactic centre, 45, 65
Galactic cosmic rays, 44, 47–9, 58–60
Galactic Cosmic Ray solar cycles, 47–9
General circulation models (GCM), 90–2, 118–19, 128, 146, 162, 206, 353, 371
Genetic and Phenotypic changes, 216, 217, 218–19
Geological history of climate change, 127–39
Geomagnetic disturbances, 55–8
Geomagnetic field, 47, 56, 60–3
Glacial flow, 394
Global Brightening, 38
Global circulation model (GCM), 303, 319
Global Dimming, 22, 26–32, 33–8
Global Energy Balance Archive (GEBA), 26
Global radiation, 22, 24–5, 27, 31, 32–3, 36, 38, 432
Global and regional patterns of sea temperature, 343–4
Global surface temperature, 37–8, 68, 134
Global tropopause, 150, 155–6
Global warming, 3–18, 34–5, 36–7, 38, 58–60, 68–70, 92, 97, 129, 134, 150, 155, 156, 157, 158, 159, 172–3, 182, 188, 189, 192, 237, 254, 271, 289, 308, 342, 345, 357–8, 360–1, 363, 364, 368, 382, 395, 401, 402, 403, 405, 406, 407, 410, 411, 416, 418–20, 421
Global warming potential (GWP), 10, 11, 12, 13, 14, 17, 19
Great Barrier Reef, 254, 255, 258–9, 375, 419
Greenhouse Effect, 3–9, 12, 14, 47, 60, 67, 68, 73
Greenhouse gases, 3–4, 5–6, 9–15, 16, 17, 18, 27, 38, 72, 93, 127, 128, 129, 131, 153, 154, 159–60, 165–6, 178, 308, 343–4, 350, 354, 395
Greenhouse potential, 10
Greenhouse world, 119–20, 131
Greenland, 45, 46, 53, 67, 112, 136, 137, 329, 333, 344, 351, 358, 395–6, 398, 418
Greenland and Antarctic ice sheets, 136, 328, 329, 331, 391, 393, 394, 396, 399

H

Habitats, 184–5, 189, 201, 208, 209–10, 216–17, 218, 220, 221, 222, 223, 236–7, 246, 248, 258, 264, 265, 266, 267, 269, 271–2, 273, 274, 275, 276, 283, 291, 292, 321, 358, 379, 385, 401, 402–403, 405–407
Hadley cell (HC), 146, 148, 154, 162
Harmful algal bloom (HAB), 237
Hawaii ocean time-series (HOTS), 371, 372
Heat capacity of seawater, 349
Heat transport, 119, 129, 133, 134, 139, 354, 357–61, 363
Heatwave events, 289
High ground lichens, 406–407
High resolution climate models (HRCMs), 319
Holocene, 137, 138, 139, 256, 333, 334
How ice sheets work, 394–8
Hydrological Cycle, 36–7, 86–8, 118, 155, 171, 178, 328–9, 360

I

Ice ages, 104, 112–17, 119, 136, 193, 341, 344
Ice sheets, 80, 104, 109, 113, 115, 118, 132, 137, 138, 329, 330, 333, 338, 353, 360, 391–9
Increasing human utilisation of coastal zone, 413–14
Industrial Revolution, 6, 12, 13–14, 201
Infra-red radiation, 4, 5, 9
Inner radiation belt, 55–8
Insect pests, 225–6, 310, 432
Insolation, 83, 104, 105, 108, 109–12, 113, 114, 115–17, 118, 119, 136, 139, 316, 353
Intergovernmental Panel on Climate Change (IPCC), 153, 162, 270–1, 299–300, 325, 340, 393
International comprehensive ocean–atmosphere data set (ICOADS), 339–40
Interplanetary dust sources, 67–8
Intertidal biota, 284–8
Intertidal taxa, 284–6
Inter tropical convergence zone (ITCZ), 147–9, 162, 177
Invasive species, 224, 245–6
Ion generation, 54, 55

J

Jurassic, 133

K

Katmai eruption, 89
Krakatau eruption, 93

L

Lichens, 401–407
Life cycle events, 240–1, 247
Long-term climate trends, 129–31
Long-term datasets, 283–4

M

Magnetic storms, 54–5, 56
Maize, 307, 315, 316, 317–18

Mammal ecology, 197–211
Mammalian vermin, 310–11
Mammal morphology, 199, 208–209
Mammal thermoregulation, 197–9
Marine biodiversity, 263–76, 284, 380
Marine ectothermic species, 282
Marine and terrestrial responses, 236–7
Marsham family records, 299
Maunder minimum, 45, 46, 50–1, 54
Mediterranean, 88, 105, 117–18, 119, 161, 184, 186–7, 188, 240, 241, 244, 286, 291–2, 368, 374
Melting ice, 244, 328, 329
Meridional gradient, 115–16, 147
Meridional overturning circulation (MOC), 79, 345, 351, 353–7
Mesozoic, 106, 128, 133–4
Metabolic heat, 198
Metabolic rate, 198–9, 209, 243–4
Metabolism, 198, 199, 208–10, 235, 236, 308–309, 381
Migration, 181–2, 185–8, 192, 193, 286, 384
Migration routes, 186–7, 188
Milankovitch, 104, 109, 112, 113, 115, 117, 135
Milankovitch cycles, 135, 136–8, 139, 200, 341–2
Miocene, 115–16, 119, 120, 134
Monsoons, 117, 119, 166, 176–8
Mountain pine beetle, 217, 226, 433–4
Mt Pinatubo, 78–9, 80, 82, 84, 92, 343

N

National Oceanic and atmospheric administration (NOAA), 30, 83, 99, 167, 173, 355
Natural and Anthropogenic forcing, 345
Non-native species, 291–2, 381
Normalised difference vegetation index (NDVI), 299–300
North Atlantic deep water (NADW), 138, 355, 356–7, 358, 361
North Atlantic Oscillation (NAO), 88, 90, 99, 138, 166, 167–9, 170, 178, 186–7, 191, 201, 238, 241, 244, 245, 253–4, 268, 351
North Atlantic zooplankton, 267
Northern annular mode (NAM), 166
North sea, 185–6, 238, 240, 241, 243–4, 246, 267–8, 291, 374
Northward shifts, 182, 238–40, 285–6
Nutrient acquisition, 312

O

Obliquity, 103–104, 105, 106–109, 110, 111, 112, 113, 114, 115–18, 119, 120
Ocean:
  acidification, 120, 133–4, 237, 254, 260, 265, 291, 367–8
  carbon cycle, 381–2
  circulation, 129, 133, 349–50, 351–3, 358, 360–1, 362–3, 364
  currents, 133, 161, 236–7, 265, 271–2, 351–2, 353, 370
  nutrient cycle, 382–3
Ocean effects:
  global scale, 269–71
  local scale, 265–7
  regional scale, 267–9
Oceanic uptake of carbon, 357–8
Oceanographers' tools, 351–3
Odours, 312
Oligocene, 112, 119, 120, 134
Orbital induced climate change, 112
Overturning circulation, 79, 95–6, 155, 177–8, 345, 351, 353–6
Ozone (O_3), 3, 11, 14, 17, 24, 46, 70, 71–2, 79, 80, 83, 88, 89–90, 93, 150–1, 154, 159–60, 309, 360–1, 431

P

Pacific decadal oscillation (PDO), 158, 201, 245, 432–3
Pacific North American (PNA) pattern, 166, 170, 171–2, 203
Pacific Ocean, 118, 119–20, 245, 268
Paleocene, 119–20, 133–4, 200
Paleocene/Eocene Thermal maximum (PETM), 120
Pan evaporation, 22, 29, 34, 35, 36
Partial migration, 189
Pathogens, 222, 240, 272, 273, 299, 309–10, 311, 312, 320
Pelagic biodiversity, 245–6, 270
Pelagic ecosystems, 233–4, 235, 237–8, 246–7, 248
Pelagic phenology, 240–1
Pelagic productivity, 241–5
Penman equation, 34, 35
Perennial crops, 318
Phanerozoic glaciations, 131, 132–3
Phenology, 86, 200–201, 217, 219, 220, 240–1, 242, 247, 253, 275, 298, 299–300, 303, 311, 428
Phylogeography, 210

Phytoplankton, 72, 234, 235–7, 241–2, 243–8, 376, 382–3
Planck's Law, 22
Plankton abundance, 241–6
Planktonic ecosystems, 233–48
Plant:
  disease, 311, 425–3
  ecology, 297–303
  growth, 136–7, 200, 302–303
  pathogens, 425–3
  reproduction, 308–13, 315, 326–7, 429, 431
  responses to climate change, 298
Plasmodiophora brassicae, 310
Pleistocene, 113, 114, 115–16, 120, 136–7, 200, 210, 256
Pliocene, 105, 116, 129, 134, 139, 210, 256, 341, 344
Polar ozone and aerosol measurement (POAM), 80
Poleward expansion, 146, 149–50, 155–6
Positive forcing, 48
Potato late blight, 425–6
Precession, 103–104, 105, 106–109, 110, 111, 112, 113, 117, 118, 119, 135, 137, 138
Pressure gradient, 147
Primary greenhouse gases, 5–6, 9, 10
Proterozoic, 128, 129, 131–2
Pyranometer, 24

Q

Quasibiennial oscillation (QBO), 83, 85–6, 92, 173, 176
Quaternary icehouse, 134–5

R

Radiative efficiency, 11, 12, 13, 15, 19
Radiative forcing, 11, 12–13, 14, 18, 27, 29, 32, 33, 72, 79, 81, 82, 83, 86, 87, 94, 97, 98, 343–4
Rate determining process, 16
Recent impact of sea-level rise, 416–18
Relative sea-level rise, 412–13, 415, 417–18, 421
Reproduction, 181, 189–93, 201–202, 235, 287, 290–1, 292, 298, 308–13, 315, 426–7, 429, 431
Reproduction and recruitment, 290–1
Rice, 310, 316–17
Rising sea level, 291, 320, 325–34, 396, 410–411, 415
Role of ocean mechanisms, 337–41
Russia, 26, 33, 81, 187, 188, 313, 317

S

Saline water, 351, 354–5, 356
Salinity, 95, 96, 98–9, 118, 133, 138, 244, 271–2, 274, 285, 349–50, 356, 360, 368–9, 370, 376, 378, 416, 418
Scale of problems, 313–14
Sea-level:
  history, 333, 411
  and ice, 391–4
  rise, 93, 113–14, 254, 256, 266, 271, 273, 275, 325, 326, 327, 328, 329, 330–1, 333–4, 381, 391, 393, 395, 396, 397–8, 399, 410, 411–13, 415–18, 420, 421
  rise and coastal systems, 411–12
  rising, 325–32
Sea life, 233–48
Sea surface temperature (SST), 84, 87–8, 129, 149, 155, 158–9, 171, 172, 173, 174, 175, 176, 236, 237–8, 241–2, 246–7, 248, 254, 256, 258, 265, 270, 271, 286, 287, 291, 337–8, 339, 340, 341, 342, 343–4, 350, 351, 375, 384, 410–411, 416, 419, 420
Sea temperature, 29, 128, 244, 254, 256, 267, 283, 287–8, 341–3, 345
Sea temperature change, 337–45
Sexual selection, 192–3
SF_5CF_3, 10, 11, 14, 16, 18–19
Simulated climate change and plant disease, 430–1, 435
Soil erosion, 312, 320
Soil moisture, 36, 38, 223–4, 310, 318
Solar activity, 44, 45–7, 51–2, 53–4, 55–8, 63, 69–70, 71, 72
Solar flux energy, 5
Solar Irradiance, 25, 36–7, 45–6, 47, 49, 50, 51, 68, 93
Solar Radiation, 33–6, 37, 47, 70, 78, 300
Solar Surface radiation, 21–38
Sorghum, 315, 316, 317–18
South Africa, 62, 288, 314
Southern annular mode (SAM), 80, 88, 166
Southern Hemisphere (SH), 28, 88, 104, 112, 119, 132, 136, 137, 138, 148, 149, 153, 154, 158, 159, 160, 167, 170, 175, 287–8, 292, 299–300, 344, 352, 358
Space factors, 68–70, 71, 72, 73
Space weather, 43–73
Spatial changes, 218
Spatial climate variation, 206–209
Spatial mismatches, 216, 218, 220–2, 224–5
SST forcing, 158–9

Stefan–Boltzman equation, 22–3
Stefan's Law, 4–5
Stratosphere Mesosphere Explorer (SME), 80
Stratospheric Aerosol and Gas Experiment (SAGE), 80, 81
Stratospheric aerosol measurement (SAM), 80, 88, 166
Stratospheric warming, 84–92
Subtropical jets, 148, 149, 150, 158
Sunshine duration (SSD), 24–5, 31
Sun's irradiation flux energy, 44
Sunspots, 45, 49, 50, 51–3, 56, 57, 58, 60, 61, 69–70, 109
Supernova, 45, 70, 71–2
Surface radiation, 24–5, 26, 29, 37, 38
Susceptible host, 333, 426

T

Temporal changes, 217, 351, 358
Temporal mismatches, 217, 219–20, 222–4
Terrestrial water sources, 328–9
Thermal expansion, 328, 329–30, 393, 411, 420
Thermoregulation, 197–9
Thermotolerance, 290
Ti/Al index, 118–19
Top of the atmosphere (TOA), 23–4, 27, 32–3, 72, 78–9, 81, 82, 111, 353
Top–down mediated effects, 219, 222
Total ozone mapping Spectrometer (TOMS), 80, 150–1, 152
Total surface of clouds, 54
Tree ring data, 303
Tropical circulation, 82, 146, 147, 149–50, 155, 159–60, 161, 178
Tropical coral reef ecosystems, 254–8, 380
Tropical coral reefs, 253, 266–7
Tropical cyclones (TC), 166, 172–6, 416
Tropical Tropopause Heights, 155–6
Tropical widening, 146, 152, 153, 156, 157, 158, 159, 160
Troposphere, 5, 9–10, 11, 15, 16, 46, 70, 77–8, 79, 80, 84, 85–6, 88–9, 90–2, 99, 147, 148–9, 150, 156, 157, 160, 162
Tropospheric cooling, 84–6, 88, 89

U

Ultra-violet radiation, 16, 17, 24, 46, 70, 71–2
United States, 129, 161, 220, 226, 287, 381, 410–411, 435
USA, 29, 51, 54, 170, 191, 201–202, 203, 287, 291, 292, 301, 313, 315, 317, 318, 395, 412, 417, 419

V

Variable ocean, 350–1
Variable solar activity, 44
Variations of the Earth's orbital Characteristics, 44, 103–21
Vascular plants, 300–301, 406
Vegetation continuous description (VECODE), 118
Ventilation of the deep ocean, 354, 357–8
Vertical temperature structure, 157
Visible radiation, 4, 5
Volcanic activity, 77–99
Volcanic effect on atmospheric circulation, 88–92
Volcanic emissions, 77–8, 79
Volcanic eruptions, 32, 37, 77–8, 79–80, 82, 84–5, 88, 90, 91, 92, 93, 95–6, 98–9, 330–1, 343
Volcanic explosivity index (VEI), 77
Volcanic forcing, 79, 84, 85–6, 88, 90, 92, 97, 99
Volcanic impact on ocean, 83, 92–6

W

Walker circulation, 149, 155, 177
Warming, 36, 37, 38, 48, 65, 67, 68, 73, 79, 89, 90, 92, 93, 96, 98, 100, 101, 102, 103, 104, 106, 110, 115–16, 117, 118, 120, 129, 133, 134, 137, 139, 150–1, 155, 156–7, 158, 170, 178, 186, 222, 224, 226, 234, 235–6, 237–9, 240, 241, 243–4, 245, 246, 247, 253–4, 256, 264–7, 268, 269, 270–1, 273, 275, 284–5, 287, 289, 290, 291, 300, 316, 319, 337, 339, 340–1, 342, 343, 344, 345, 363, 381, 410, 419
Weather patterns, 165–78, 263–4, 268–9
Weed competition, 316
West Antarctic ice sheets (WAIS), 329, 397, 398
Wheat prices, 51–4, 72
Wheat stripe rust, 432, 434
Wien's Law, 4
Winners and losers, 193, 315–20, 430
World Radiation Data Center (WRDC), 26
World Radiation Monitoring Center (WRMC), 26

Z

Zooplankton communities, 244, 246–7

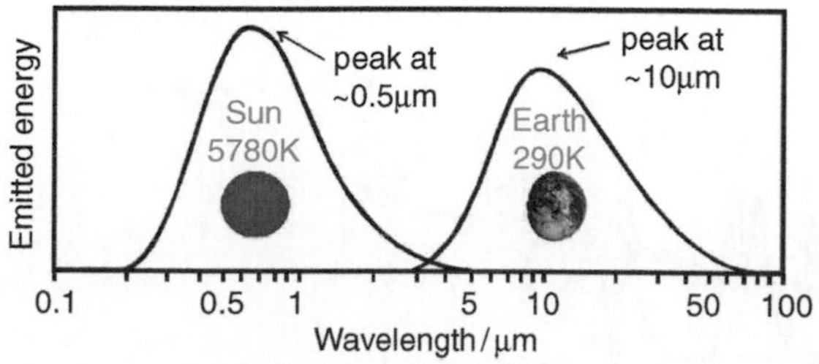

COLOR PLATE 1 Black-body emission curves from the sun ($T \sim 5780$ K) and the earth ($T \sim 290$ K), showing the operation of Wien's Law that $\lambda_{max} \propto (1/T)$. The two graphs are not to scale.

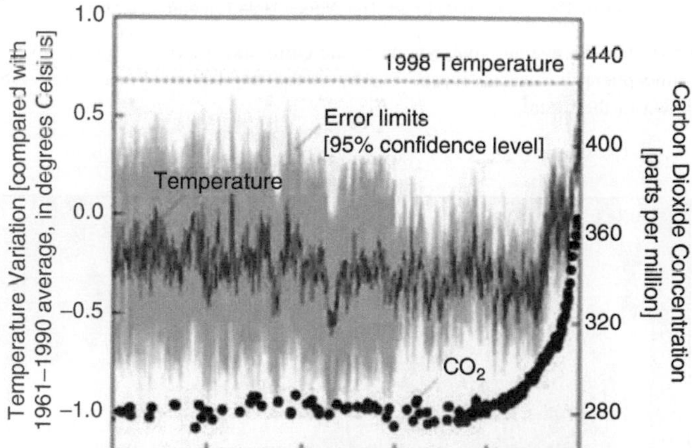

COLOR PLATE 2 The average temperature of the earth and the concentration level of CO_2 in the earth's atmosphere during the last 1000 a. (With permission from www.env.gov.bc.ca/air/climate/indicat/images/appendnhtemp.gif and www.env.gov.bc.ca/air/climate/indicat/images/appendCO2.gif)

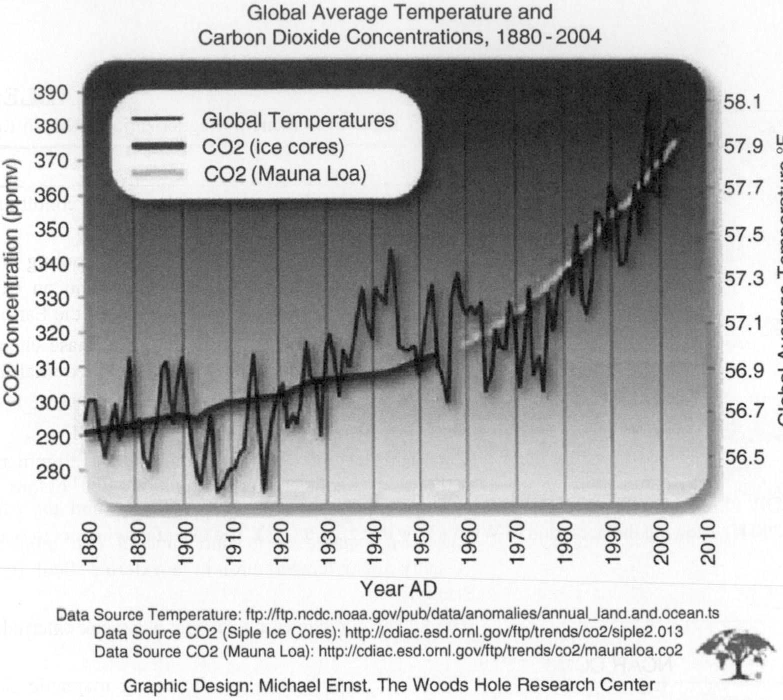

COLOR PLATE 3 The average temperature of the earth and the concentration level of CO_2 in the earth's atmosphere during the 'recent' history of the last 100 a. (With permission from the web sites shown in the figure.)

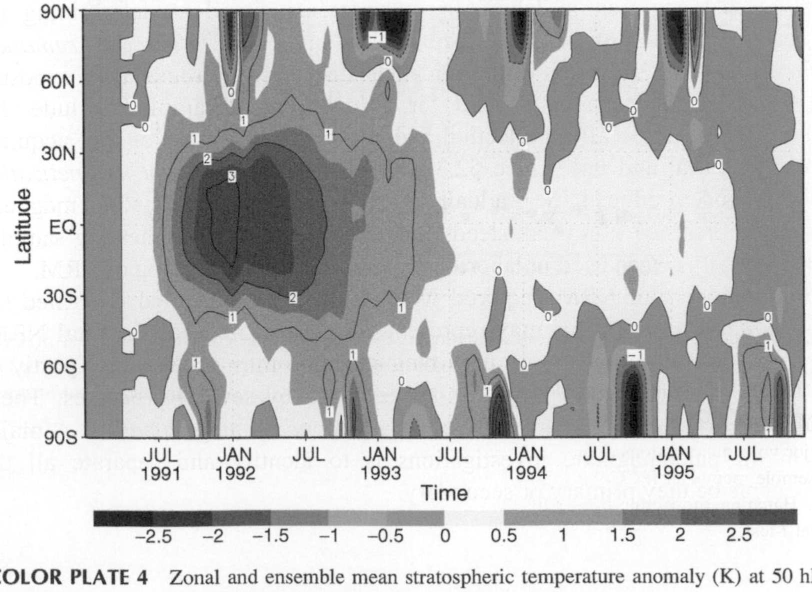

COLOR PLATE 4 Zonal and ensemble mean stratospheric temperature anomaly (K) at 50 hPa (at about 25 km) calculated with respect to control experiment.

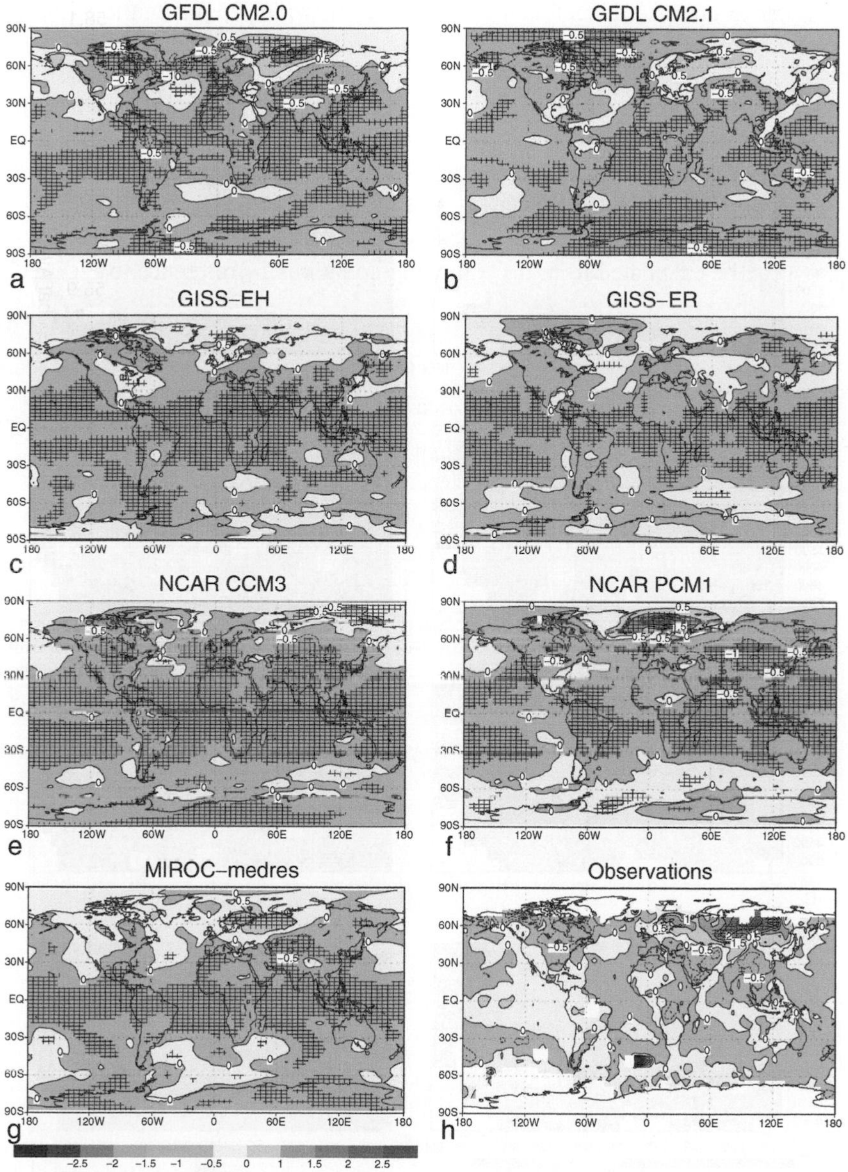

COLOR PLATE 5 Surface winter (DJF) air temperature anomalies (K) composited for nine major volcanic eruptions from 1883 until present and averaged for two seasons and all available ensemble members: IPCC AR4 model simulations (a–g); observations from HadCRUT2v dataset (h). Hatching shows the areas with at least 90% confidence level calculated using a two-tailed local *t*-test.

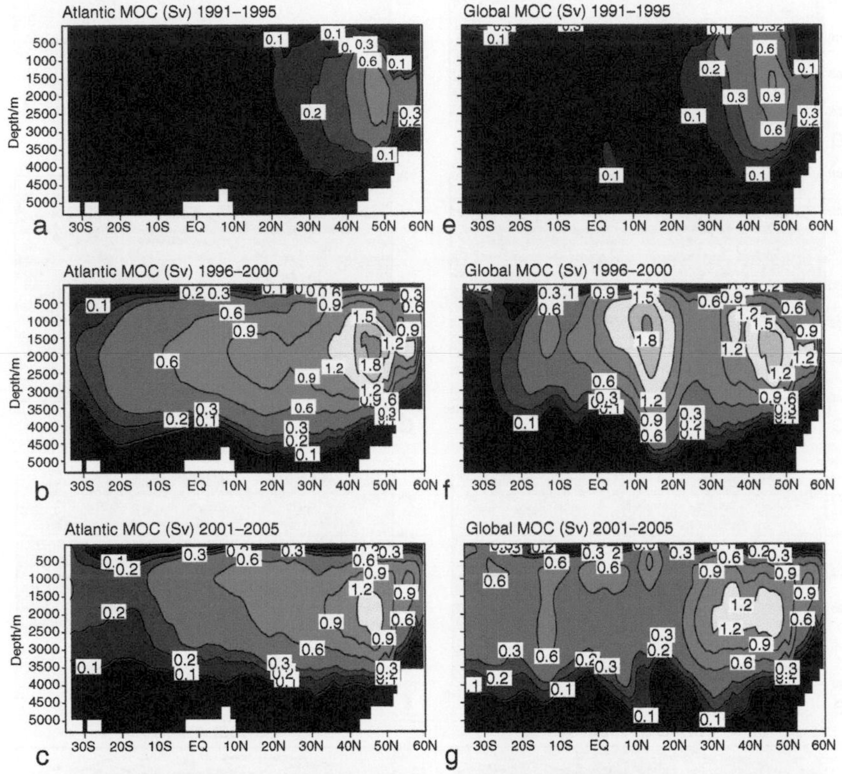

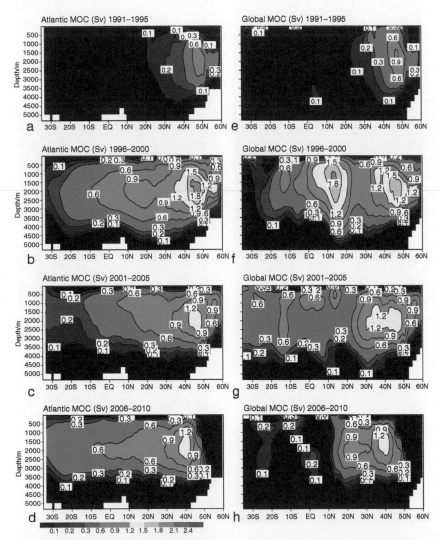

COLOR PLATE 6 The 5-year means MOC anomalies (Sv) from the Pinatubo ensemble averaged zonally over Atlantic basin (a–d) and over the globe (e–h).

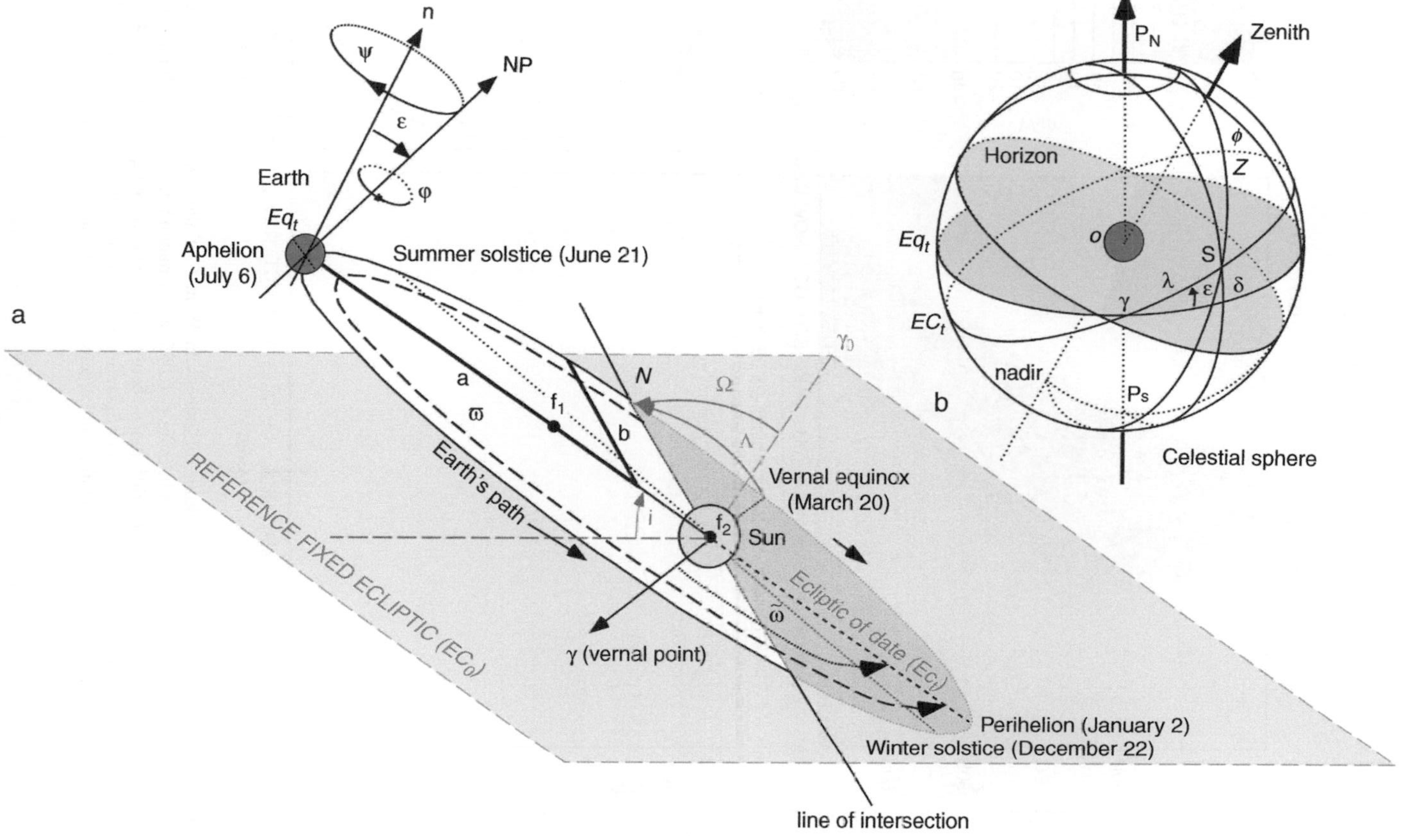

COLOR PLATE 7 Astronomical configuration of the Earth. (a) Elements of the Earth's orbital parameters (modified after [101]). (b) Position of a point (S) on the celestial sphere (modified after [38]). See text for explanations.

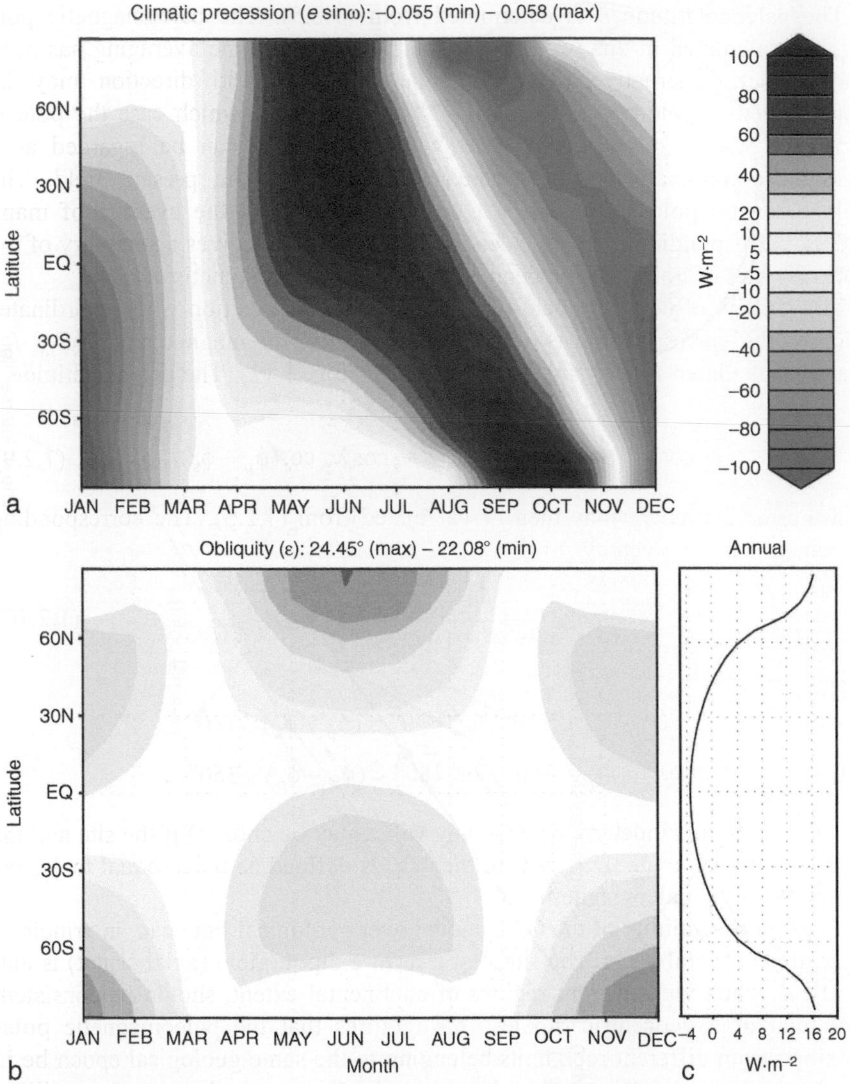

COLOR PLATE 8 Monthly incoming differences in $W \cdot m^{-2}$ at the top of the atmosphere. (a) Insolation differences between a minimum (−0.055) and a maximum (+0.058) climatic precession configuration. (b) Insolation differences between a maximum obliquity (Tilt = 24.45°) and a minimum obliquity (Tilt = 22.08°) configuration with zero eccentricity. (c) As in (b), but now plotted the annual incoming insolation differences. Solar constant = 1360 $W \cdot m^{-2}$.

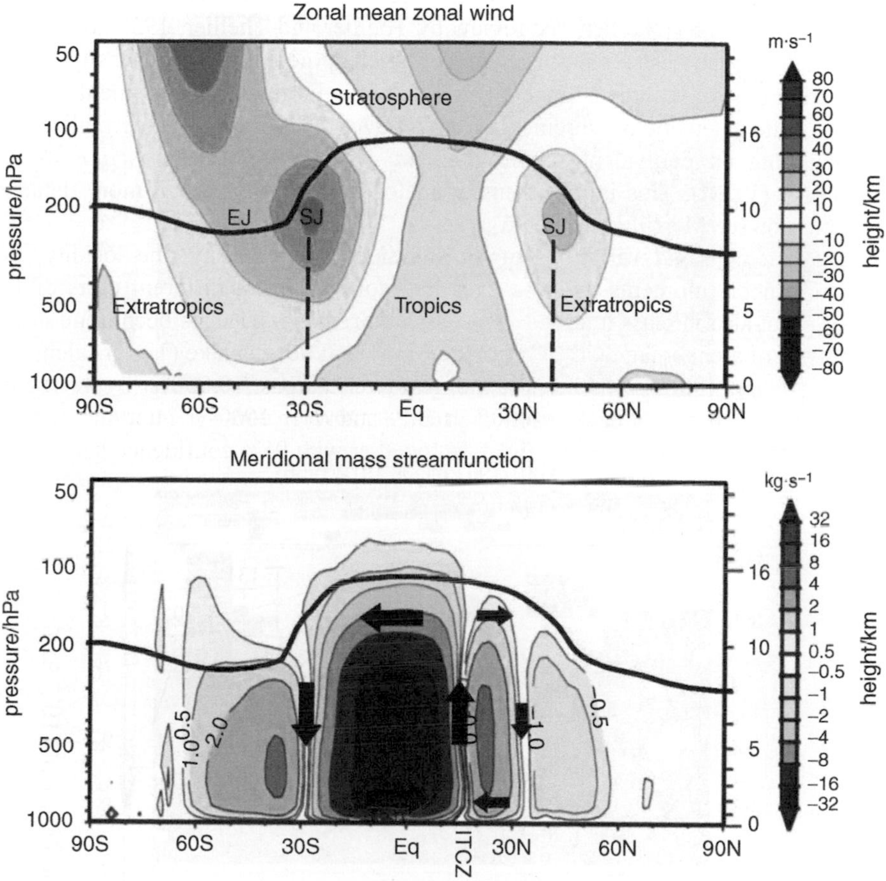

COLOR PLATE 9 Climatological mean circulation in the height–latitude plane during boreal summer (June, July, August) (JJA). Vertical axis is atmospheric pressure (in hPa) and height (in km) and horizontal axis is latitude (in degrees). The continuous black line denotes the thermally defined tropopause. (Top) Zonal mean zonal winds (in $m.s^{-1}$) derived from National Centers for Environmental Prediction/National Center for Atmospheric Research (NCEP/NCAR) reanalysis. The approximate position of the subtropical jet and the eddy driven jet is denoted by SJ and EJ, respectively. (Bottom) Mean meridional mass streamfunction (in $kg.s^{-1}$), with arrows indicating the direction and strength of the zonal mean overturning associated with the Hadley cell, with a strong winter cell in the SH and a weak summer cell in the NH.

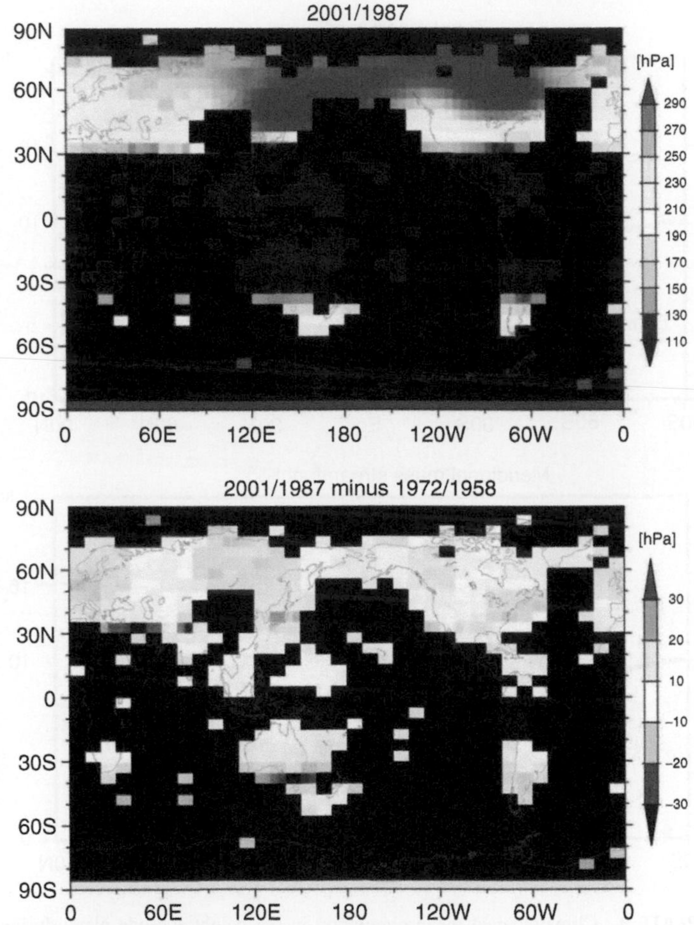

COLOR PLATE 10 Changes in tropopause pressure during boreal winter (December, January, February) derived from gridded radiosonde data HADRT V2.1 [93]. (Top) Absolute tropopause pressure (in hPa) averaged over 1987–2001. (Bottom) Differences in tropopause pressure (in hPa) between the late period 1987–2001 and an early period 1958–1972. Bluish grid points indicate that tropopause pressure is decreasing and tropopause heights are increasing. The bluish banded structures over southern Australia and southern Europe indicate a trend toward tropical tropopause conditions and thus a widening of the tropics. Adapted from [14].

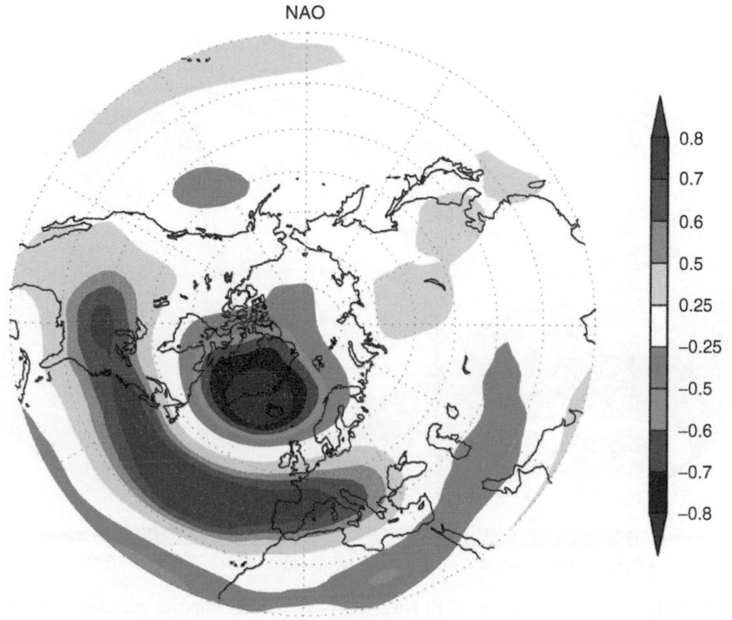

COLOR PLATE 11 Spatial pattern of the NAO as given by the temporal correlation between the Winter (DJFM) monthly standardised 50 kPa geopotential height anomalies at each point and the monthly teleconnection pattern time series from 1960 to 2000.

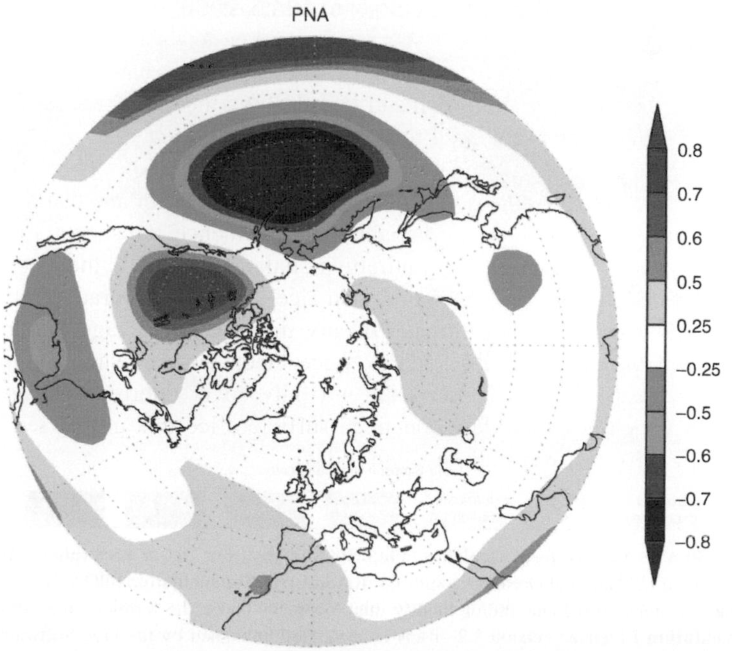

COLOR PLATE 12 Spatial pattern of PNA as given by the temporal correlation between the Winter (DJFM) monthly standardised 50 kPa geopotential height anomalies at each point and the monthly teleconnection pattern time series from 1960 to 2000.

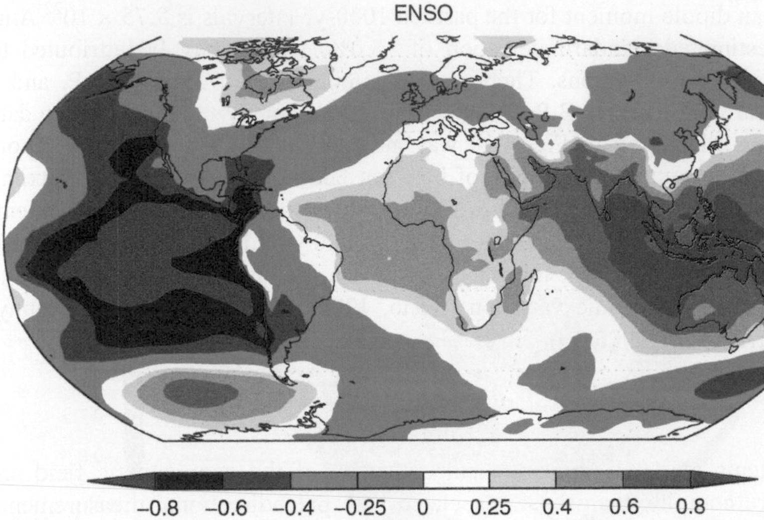

COLOR PLATE 13 Spatial pattern of El Niño as given by the temporal correlation between the annual (May–April) standardised SLP anomalies at each point and the monthly teleconnection pattern time series from 1960 to 2000.

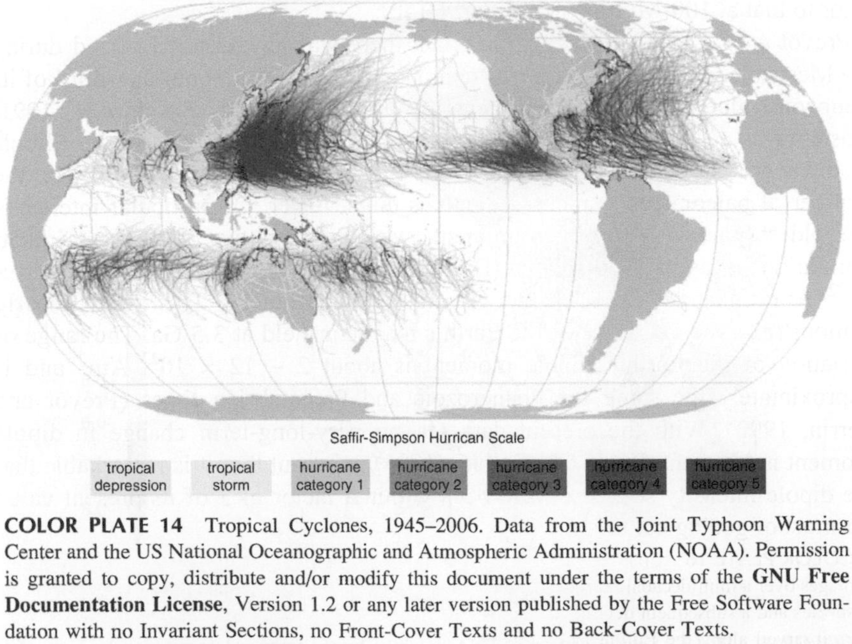

COLOR PLATE 14 Tropical Cyclones, 1945–2006. Data from the Joint Typhoon Warning Center and the US National Oceanographic and Atmospheric Administration (NOAA). Permission is granted to copy, distribute and/or modify this document under the terms of the **GNU Free Documentation License**, Version 1.2 or any later version published by the Free Software Foundation with no Invariant Sections, no Front-Cover Texts and no Back-Cover Texts.

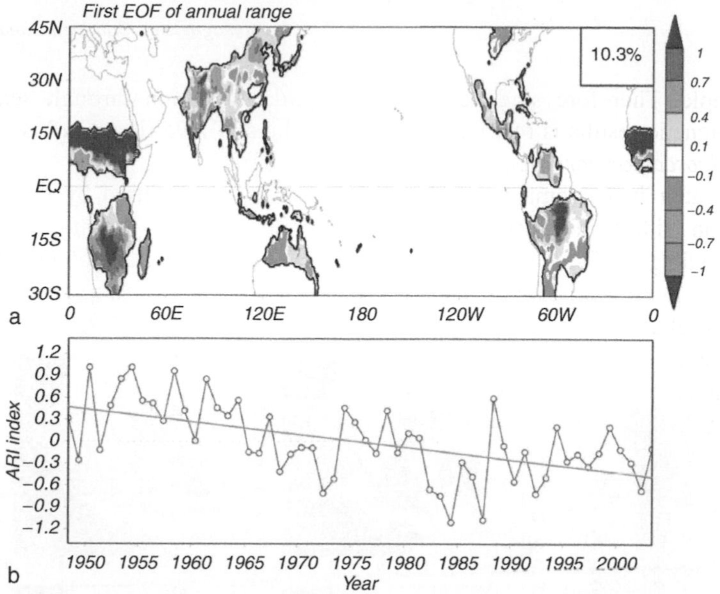

COLOR PLATE 15 (Reprinted from Fig. 3 of Ref. [45]). (a) The spatial pattern of the leading Empirical Orthogonal Function (EOF) mode of the normalised annual range precipitation anomalies over the global continental monsoon regions. The bold contour indicates the boundaries of the monsoon domain; (b) the corresponding principal component or annual range index (ARI) (© Cambridge Press; IPCC report, Chapter 3 and © AGU).

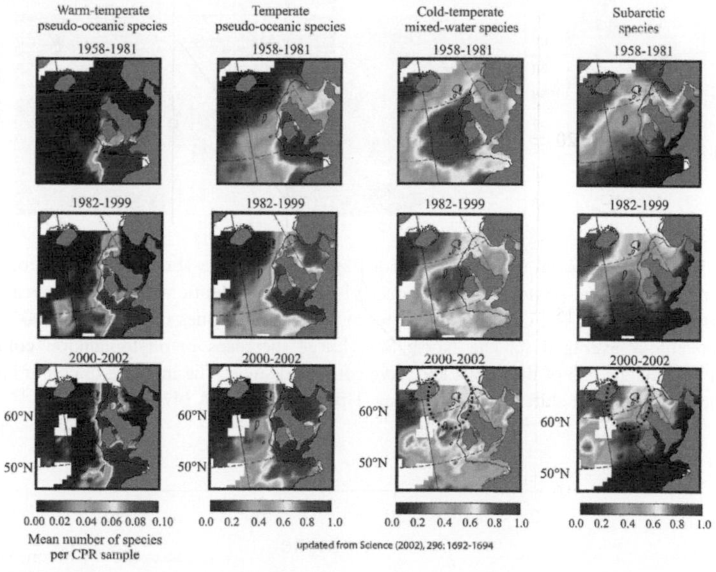

COLOR PLATE 16 Changes in the geographical distribution of four different plankton assemblages over a multidecadal period. There has been a rapid northerly movement of warm-temperate species and a subsequent decline in sub-arctic species over 40 years. Particularly rapid movement is observed along the European Continental Shelf. Data derived from the Continuous Plankton Recorder survey. Updated from Ref. [76].

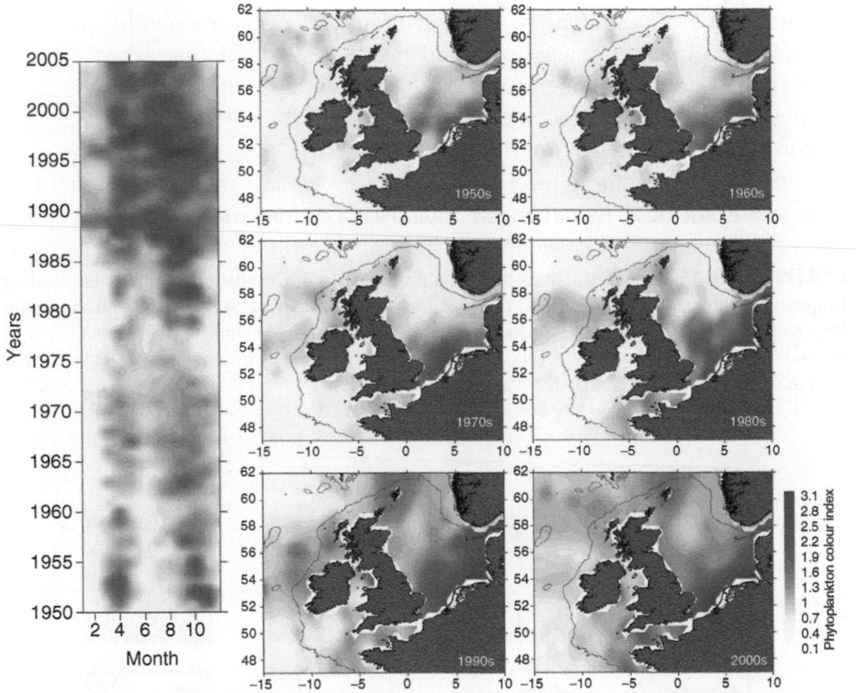

COLOR PLATE 17 Spatial-temporal maps of the changes in the abundance of phytoplankton colour (an index of total phytoplankton biomass) for the NE Atlantic averaged per decade from the 1950s to the present. The contour plot shows monthly mean values from 1950 to 2005 of phytoplankton colour averaged for the North Sea. Large increases in phytoplankton colour are observed towards the end of the 1980s and have continued since. The increase in colour has been associated with a regime shift in the North Sea. Updated from Ref. [47].

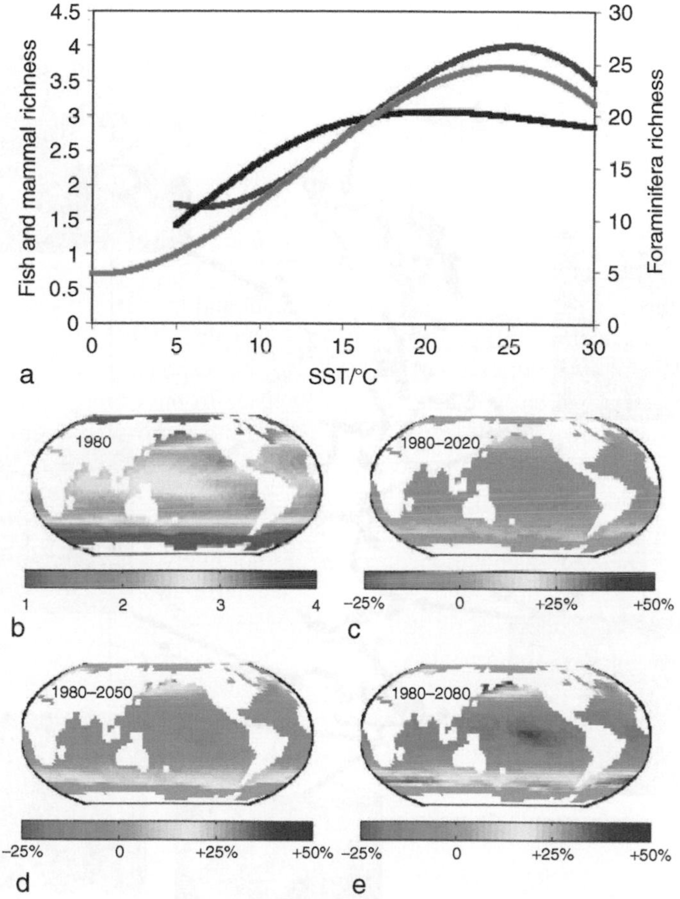

COLOR PLATE 18 Effects of sea surface temperature (SST) on marine pelagic biodiversity. (a) Empirical relationships between SST and the observed species richness of foraminiferan zooplankton (green, data from [62]), tuna and billfish (red, data from [17]) and genus richness of deep-water cetaceans (blue, data from [65]). Maps depict projected mean genus richness of deep water cetaceans in (b) 1980, and relative changes in richness projected to occur between (c) 1980 and 2020, (d) 1980 and 2050 and (e) 1980 and 2080 are shown. Changes are expressed as percents of the mean (over all ocean areas <65° latitude) diversity in 1980 minus one (as the minimum diversity is 1.0). Panels b–e are reprinted with permission from Ref. [65].

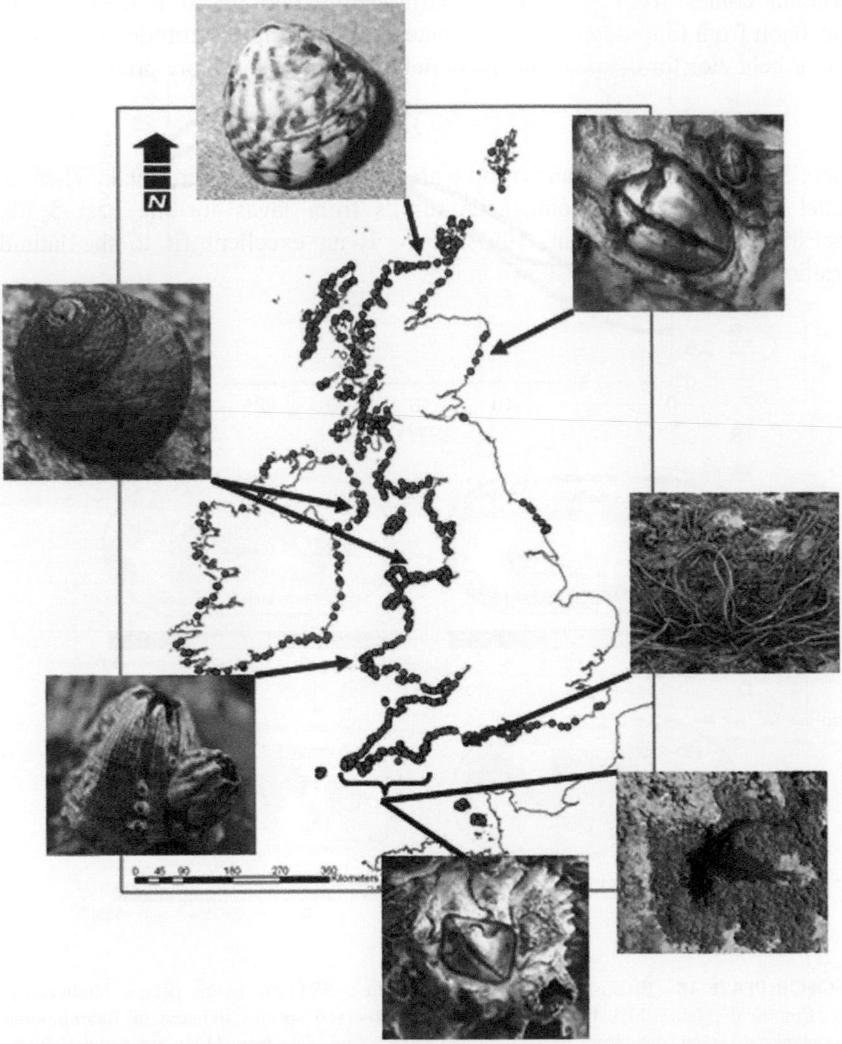

COLOR PLATE 19 Rocky shores in Britain where survey data has been collected in the 1950s and 2000s. Clockwise from bottom left, warm water species which have extended their northern limits; *Perforatus perforatus, Osilinus lineatus, Gibbula umbilicalis, Chthamalus montagui, Bifurcaria bifurcata.* Cold water species which have shown contractions in their southern range edge: *Alaria esculenta, Semibalanus balanoides.*

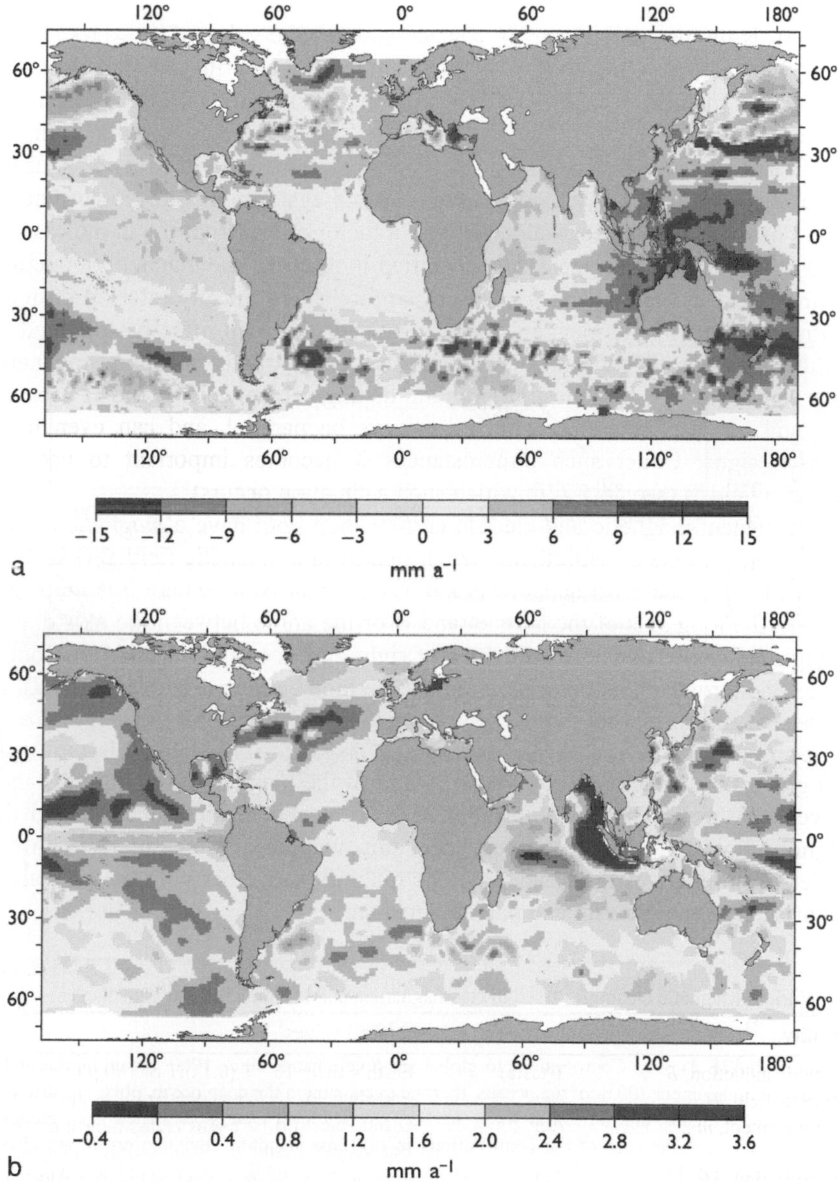

COLOR PLATE 20 Geographic patterns of sea-level change from (a) 1993 to 2003 (from Ref. [6], updated from Ref. [40]) and (b) 1955 to 2003 (from Ref. [6], updated from Ref. [41]).

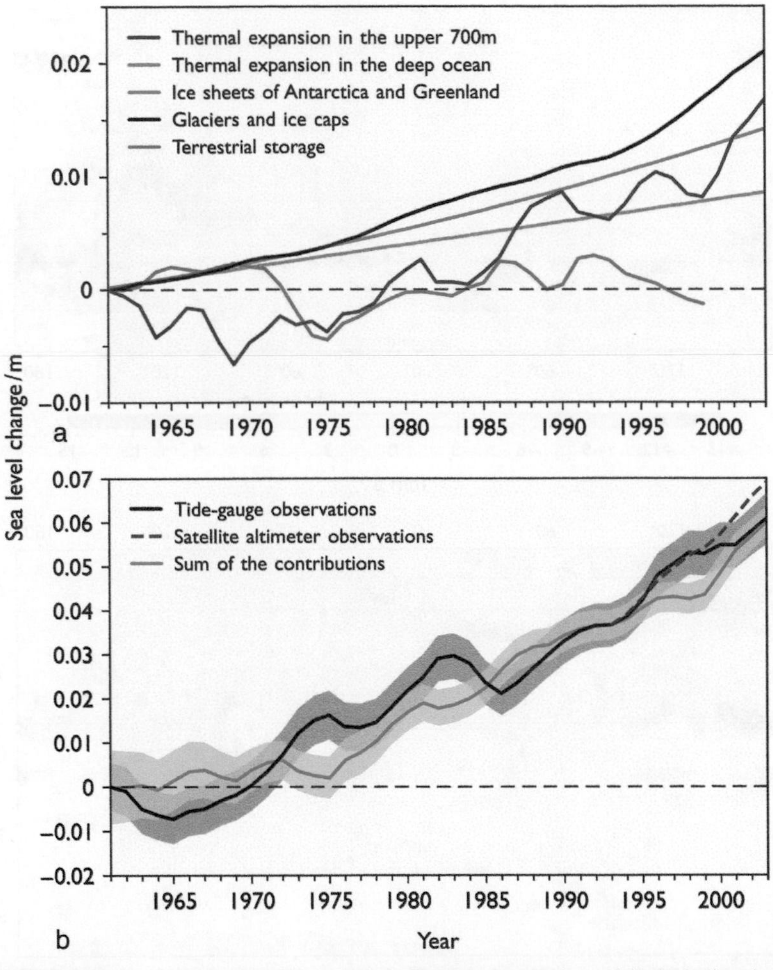

COLOR PLATE 21 (a) Contributions to global sea-level change since 1960, including thermal expansion in the upper 700 m of the oceans, thermal expansion in the deep ocean, polar ice sheets, glaciers and ice caps, and terrestrial water storage. (b) Sea-level change estimated from global measurements and the sum of the contributions in (a). One standard deviation errors are also shown. From Ref. [3].

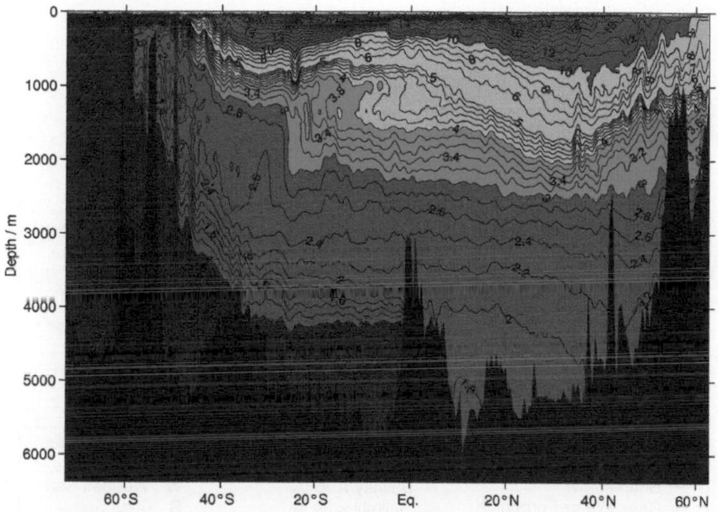

COLOR PLATE 22 Section of potential temperature along the meridional extent of the Atlantic. For temperatures less than 5 °C and greater than 5 °C, the black contours have a spacing of 0.2 and 1 °C, respectively. The red, indian red, salmon, cyan, light blue and dark blue areas denote temperatures above 16 °C, from 10 to 16 °C, from 4 to 10 °C, from 3 to 4 °C, from 1 to 3 °C and below 1 °C, respectively. Lowered temperature measurements acquired during three research expeditions – aboard *RV Ronald H. Brown* in 2003 (section A16N, PI: Bullister [PMEL]) and in 2005 (section A16S; PIs: Wanninkhof [NOAA]/Doney [WHOI]) and aboard *RV James Clark Ross* in 1995 (section A23; PIs: Heywood/King [NOCS]) – were joined together to compile this figure. Data source: Clivar and Carbon hydrographic data office (http://whpo.ucsd.edu/atlantic.htm). Adapted from a figure of Lynne D. Talley (http://sam.ucsd.edu/vertical_sections/Atlantic.html#a16a23).

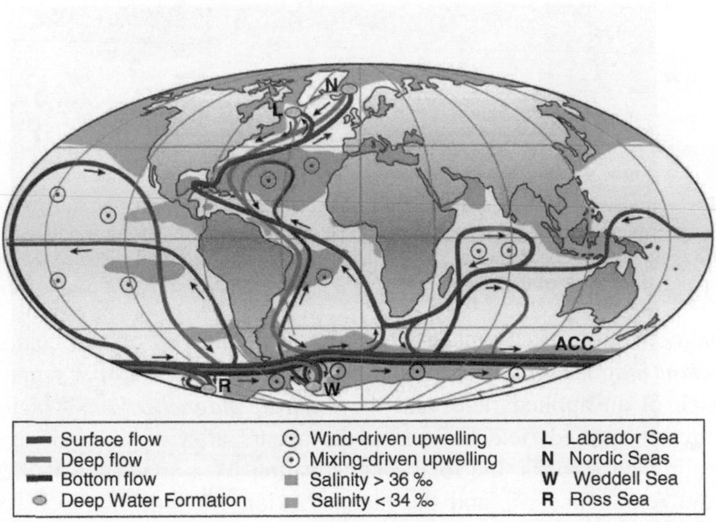

COLOR PLATE 23 Strongly simplified sketch of the global overturning circulation system. In the Atlantic, warm and saline waters flow northwards all the way from the Southern Ocean into the Labrador and Nordic Seas. By contrast, there is no deep water formation in the North Pacific and its surface waters are fresher. Deep waters formed in the Southern Ocean are denser and thus spread in deeper levels than those from the North Atlantic. Note the strongly localised deep water formation areas in comparison with the wide-spread zones of mixing-driven upwelling. Wind-driven upwelling occurs along the Antarctic Circumpolar Current (ACC). This figure has been published by Kuhlbrodt et al. [17].

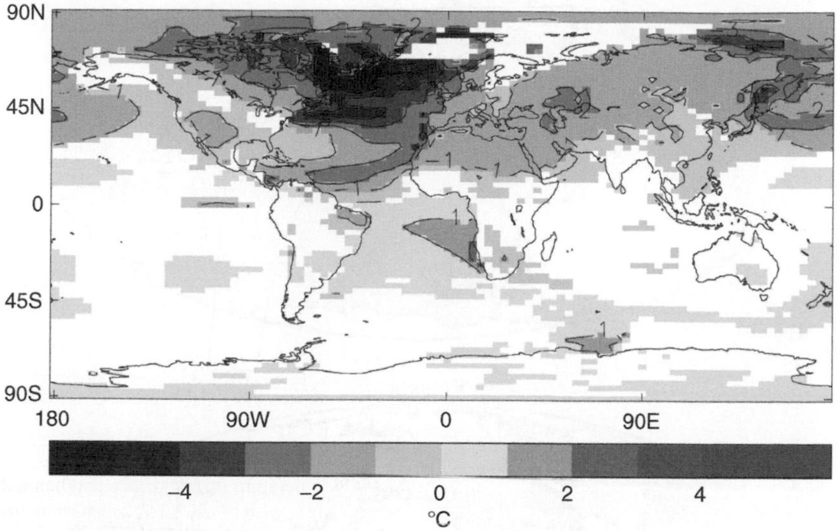

COLOR PLATE 24 Change in surface air temperature during the years 1920–1930 after the collapse of the AMOC in a water hosing experiment using the HadCM3 climate model. Areas where the anomaly is not significant have been masked. This figure has been published by Vellinga et al. [36].

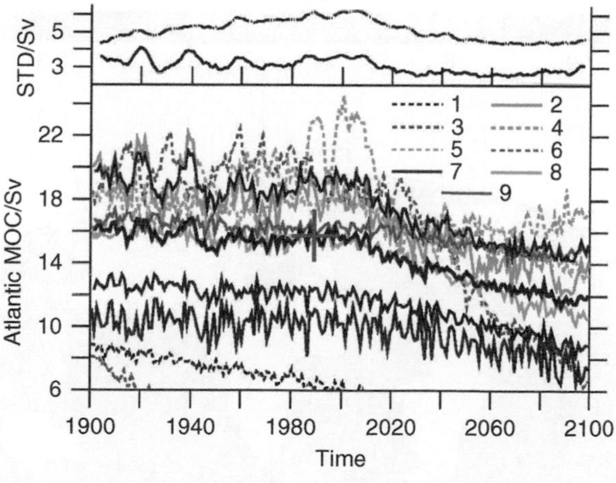

COLOR PLATE 25 Evolution of the AMOC as defined by the maximum overturning at 24°N for the period 1900–2100 in nine different climate models forced with the greenhouse gas emission scenario A1B. The AMOC evolutions of integrations with a skill score larger than one are shown as solid lines, those from models with a smaller skill score as dashed lines [39]. The weighted ensemble mean is shown by the thick black curve together with the weighted standard deviations (thin black lines). This figure was published by Schmittner et al. [39].

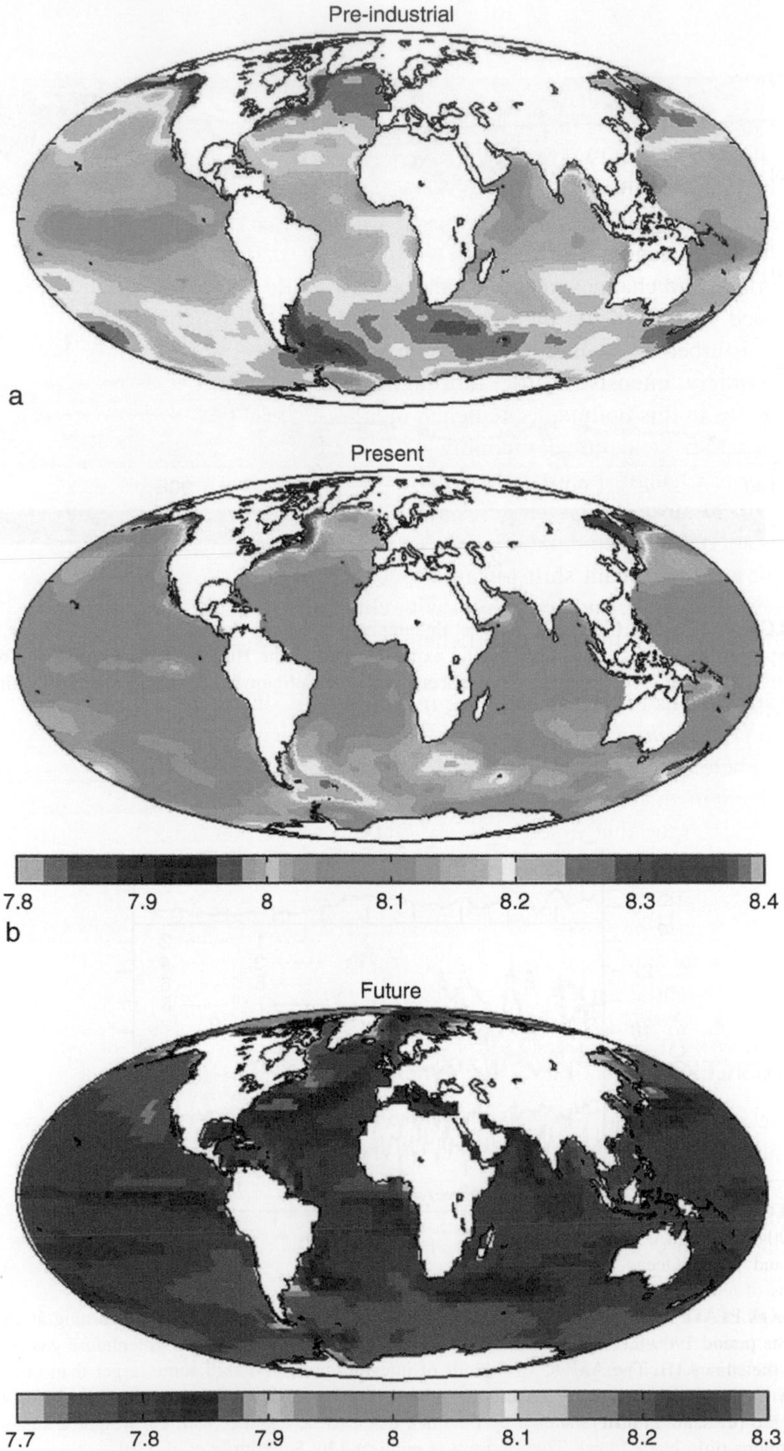

COLOR PLATE 26 (See legend on next page)

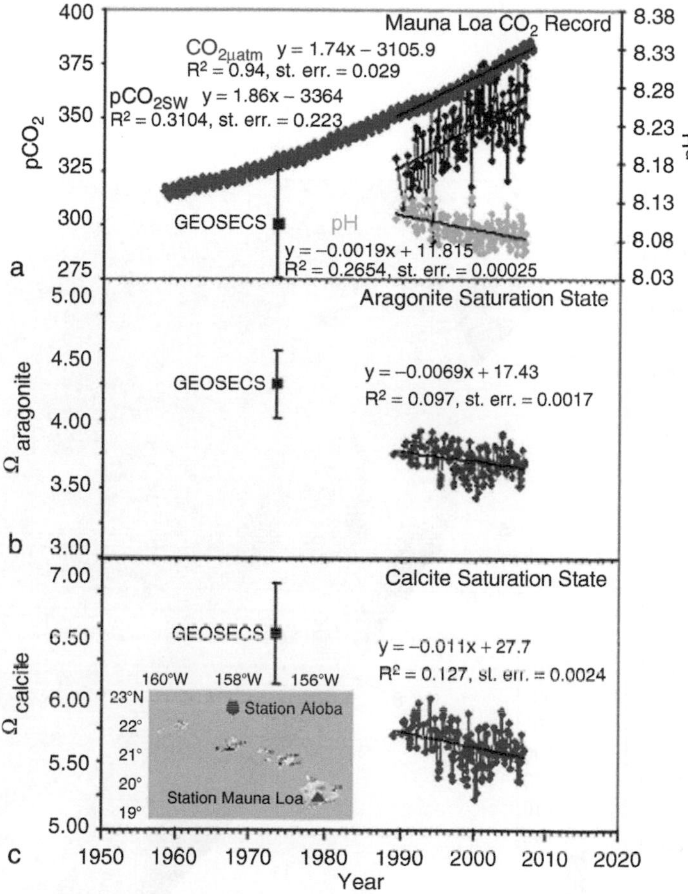

COLOR PLATE 27 (a) The Mauna Loa records of atmospheric CO_2 over the last 50 a with the pCO_{2SW} and surface ocean pH recorder during the last two decades from the Hawaii Ocean Time-Series (HOTS) and the resultant changes to (b) Agaronite saturation state and (c) calcite saturation state over the same period. From Doney et al. [17].

COLOR PLATE 26 (a) estimated pre-industrial (1700s) sea-surface pH and (b) present day (1990s) sea-surface pH, both mapped using data from the Global Ocean Data Analysis Project [5] and World Ocean Atlas climatologies; however, in the absence of estimated pre-industrial fields of temperature and salinity 1990s fields were used (although these contain a small signal from global warming). Note that GLODAP climatology is missing data in certain oceanic provinces (areas left white) including the Arctic Ocean, the Caribbean Sea, the Mediterranean Sea and the Malay Archipelago. (c) Predicted pH across the world's oceans for yr 2100 using the SOC model, which was part of the OCMIP-2 project [6] and used the IS92a CO_2 scenario. Note that the pH scale is different in (c). Courtesy of Andrew Yool (National Oceanography Centre, Southampton).

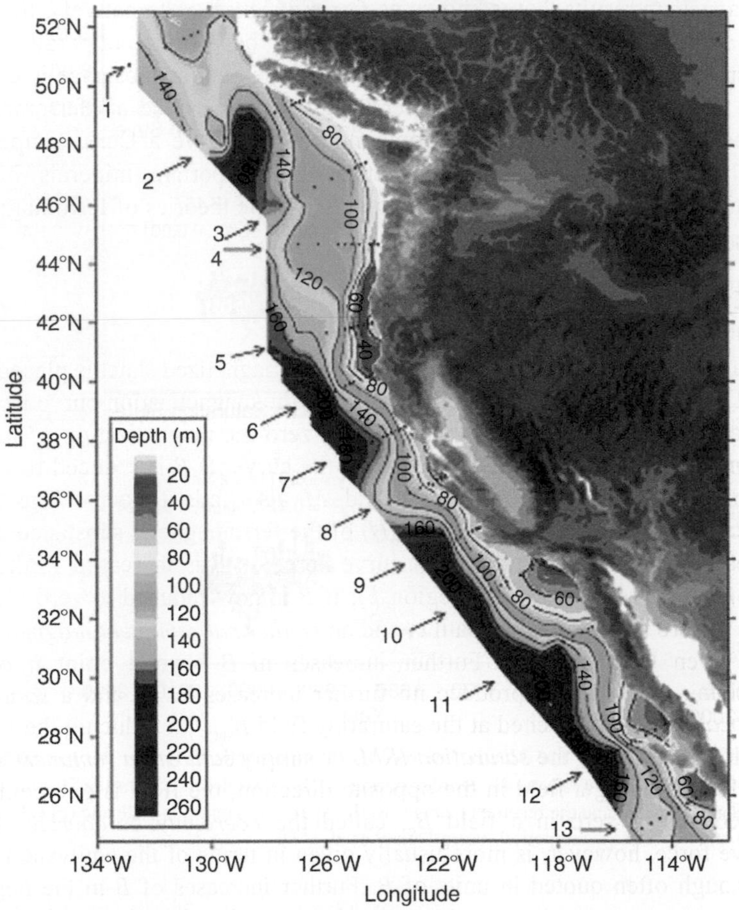

COLOR PLATE 28 Distribution of the depths of the undersaturated water (aragonite saturation < 1.0; pH < 7.75) on the continental shelf of western North America from Queen Charlotte Sound, Canada to San Gregorio Baja California Sur, Mexico. On transect line 5 the corrosive water reaches all the way to the surface in the inshore waters near the coast. The black dots represent station locations. From Feely et al. [13].

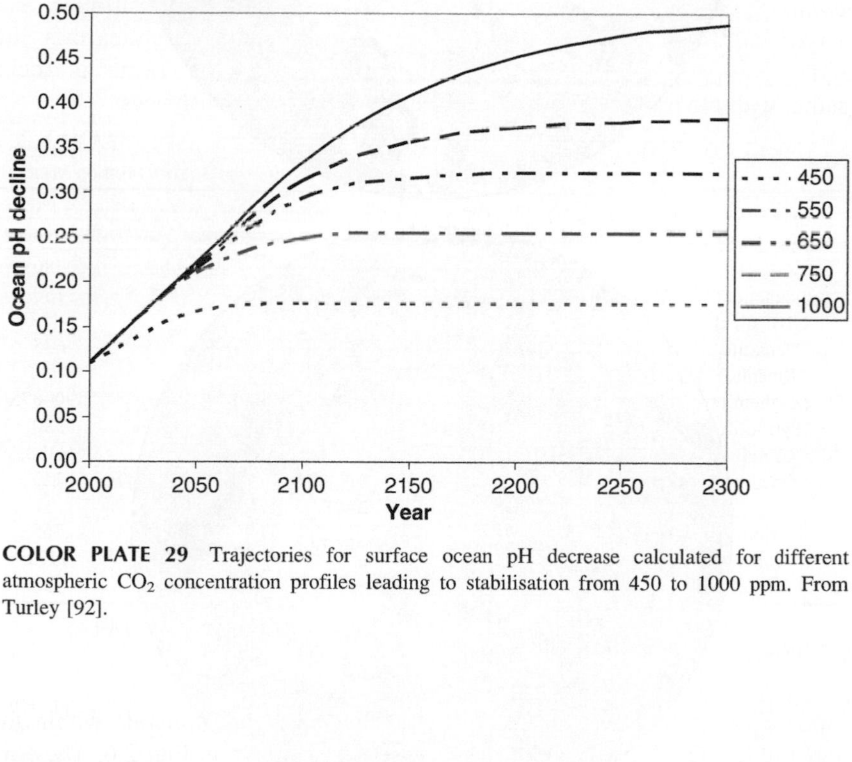

COLOR PLATE 29 Trajectories for surface ocean pH decrease calculated for different atmospheric CO_2 concentration profiles leading to stabilisation from 450 to 1000 ppm. From Turley [92].

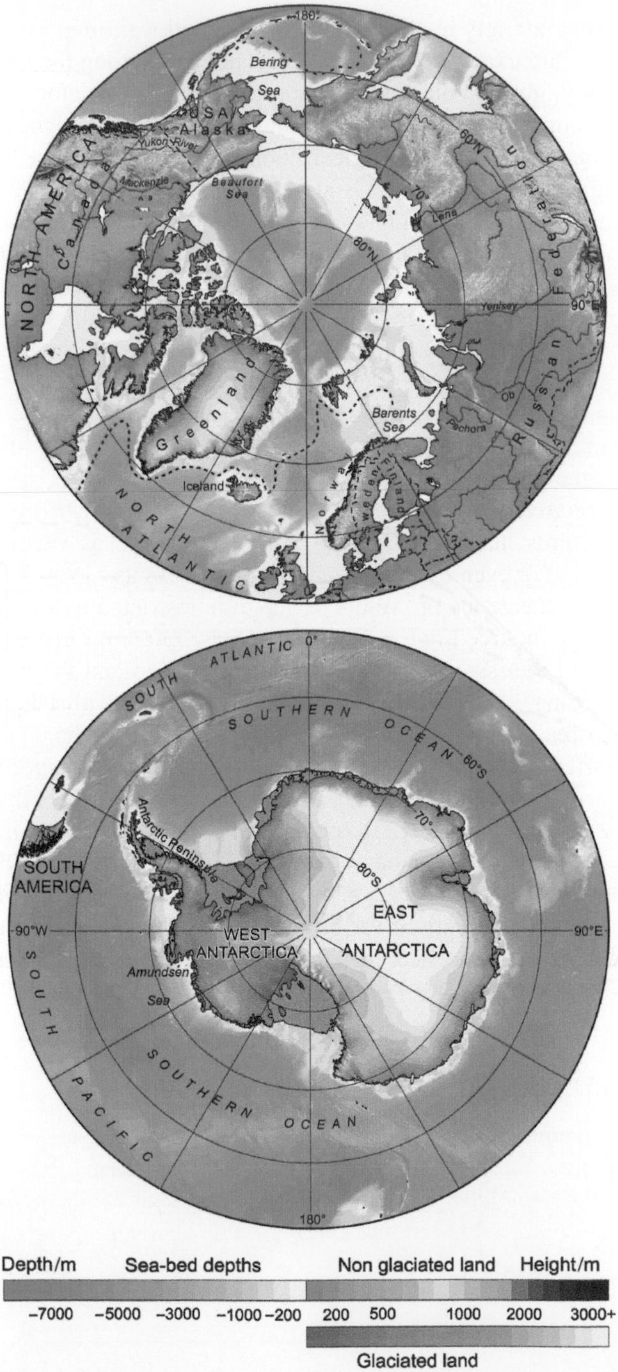

COLOR PLATE 30 Maps of the north and south polar regions showing ice sheets and place names used in the text. Modified from Ref. [19].

Printed and bound by CPI Group (UK) Ltd, Croydon, CR0 4YY

03/10/2024

01040419-0005